FUNDAMENTALS OF POLYMER PROCESSING

FUNDAMENTALS OF POLYMER PROCESSING

STANLEY MIDDLEMAN

*Professor of Chemical Engineering
and
Polymer Science and Engineering
University of Massachusetts, Amherst*

McGRAW-HILL BOOK COMPANY

*New York   St. Louis   San Francisco   Auckland   Bogotá   Düsseldorf
London   Madrid   Mexico   Montreal   New Delhi
Panama   Paris   São Paulo   Singapore   Sydney   Tokyo   Toronto*

This book was set in Times New Roman.
The editors were B. J. Clark and Douglas J. Marshall;
the cover was designed by Joan E. O'Connor;
the production supervisor was Leroy A. Young.
The drawings were done by Long Island Technical Illustrators.

FUNDAMENTALS OF POLYMER PROCESSING

8 9 0 BRBBRB 89

Library of Congress Cataloging in Publication Data

Middleman, Stanley.
    Fundamentals of polymer processing.

    Includes bibliographical references and index.
    1. Polymers and polymerization. I. Title.
TP1087.M5   668   76-48323
ISBN 0-07-041851-9

This book is dedicated to my students.

I learned much from my teachers,
more from my colleagues,
and most from my pupils.
Judah Ha Nasi
(Talmud)

Every writer making a secondary world $\cdots$ wishes to be a real maker, hopes that he is drawing on Reality, or that the peculiar quality of his secondary world (if not all the details) are derived from Reality, or are flowing into it. If he indeed achieves a quality that can fairly be described by the dictionary definition: "inner consistency of reality," it is difficult to conceive how this can be, if the work does not in some way partake of Reality.

J. R. R. Tolkien
(*On Fairy-Stories**)

*Published in the United States by Houghton Mifflin Company.

CONTENTS

PREFACE

The goal of this text is two-fold. It is written, first, in the hope that those individuals now (or soon to be) involved in the polymer processing industries will find it a useful vehicle for learning the fundamental principles by which one can analyze a wide variety of polymer flow processes. The second goal stems from my observation, as a teacher, that little opportunity exists, in most academic environments, for the student to take a coherent and meaningful course in *applied fluid dynamics* with an emphasis on flow processes dominated by viscous effects. I have tried to produce a text which will meet the first goal, and in doing so, provide motivation and opportunity for development of a useful course in this area.

The topics presented in the pages to follow provide a basis for one to two semesters of course material. I have found the content suitable to senior chemical engineering students looking for an elective course in this general area. Normally I have taught this material to first- and second-year graduate students in the Polymer Science and Engineering Department of the University of Massachusetts, Amherst. Since most of these students have a B.S. in chemistry, this often represents their first exposure to the peculiar mind of the engineer. Because of the lack of engineering background per se on the part of these students, I have included introductory material in mechanics, fluid dynamics, and heat and mass transfer, otherwise often found as a part of a B.S. chemical or mechanical engineering program. A brief survey of rheology is also included, although we normally provide a full one-semester course in this area. The teacher, then, has the option and the need to supplement this material if it does represent a first exposure for the students, or gloss over the derivations and emphasize the applications if the students have appropriate previous exposure to some topics. A large number of problems are included to provide practice, and a Bibliography is given which allows for extended study of specific areas of interest. This text is a *guide*; it does not suggest a static syllabus for a course.

In many ways I have been more concerned with the opportunity to expose students to the experience of modeling, rather than with the exposure to the specific topic of polymer processing. My own experience as a teacher tells me that most students are handicapped with the belief that the goal of their education is preparation to answer questions like "Find the derivative of x^2." Such questions, unlike those they will be paid to answer, have *a* correct answer. In teaching this material, and in writing this text, I have tried to emphasize the philosophy of modeling. My own personal approach is to attempt to convince my students that I am teaching a course in mythology, with a goal of developing the ability to create "useful" myths. Once the "grammar" and the other basic tools of writing myths (mathematical models) are learned, we devote most of our time to developing a sense of "plotting." With sufficient practice and exposure the student will learn how to develop mathematical models which can be expected to have some acceptable correspondence to reality.

This is not a comprehensive text in the field of polymer processing. Those readers who now work in the polymer manufacturing industries will surely find some of their favorite processes missing. This does not reflect the importance of a specific process—commercial importance was not the criterion for inclusion in this text. Rather, I have tried to treat processes which are important, but which also lend themselves to some degree of engineering analysis.

While this text grows out of some 15 years of experience in teaching, research and consulting, its final format and content reflect the strong input of several people who have given much of their thought to this project. My greatest debt is to Professor Morton Denn of the University of Delaware. I am fortunate to count him as a friend, and to be able to count on him for honest and constructive criticism. Professor Denn read the entire first draft of the manuscript and wrote a lengthy and detailed critique which provided the basis for improvement of the final draft. The failures that remain are my responsibility, and stem from stubbornness.

Dr. Donald Bigg of Battelle (Columbus Lab.) read the first draft of the manuscript and made suggestions to improve its content. Many of my graduate students have contributed to various sections of the text, but special mention must be made of Jehuda Greener and Mike Malone, who carried through the solutions of many of the models presented here, and who corrected many of my errors. Tom Mumley and Skip Rochefort provided some of the previously unpublished experimental results that appear in Chapter 8.

The final revision of the manuscript was carried out while I was a Visiting Scholar in the Chemical Engineering Department of Stanford University. I thank Professor Andreas Acrivos for making my stay there possible.

My final debt is to my wife Jo-Ann, and my daughters Melissa and Sharon, whose love sustained me through the insanity of authorship.

Stanley Middleman

FUNDAMENTALS OF POLYMER PROCESSING

POLYMER PROCESSING

Be sure to ask your teacher his reasons and sources.

Rashi

There is great diversity among those industries associated with polymeric materials. Some industries are principally concerned with the production of polymers from the raw (monomeric) materials. Others are concerned with the physical conversion of a polymeric material into a finished article. The diversity arises both from the wide range of properties of the multitude of commercially available polymers and from the great variety of physical processes that can transform a polymer to an article of commerce.

In the development of a process one needs a means of anticipating the behavior of the system, so that the relationships among the design and operating variables can be estimated. Only in this way can one hope to put together a process which can be made to operate successfully with a minimum of redesign. Mathematical modeling plays a central role in process development and design. This text is concerned with the methods of modeling a variety of important polymer processes.

Let us consider, in somewhat simplified form, the flowsheet of a hypothetical process for making a plastic film, as in Fig. 1-1. For this process we can list a number of steps, and for each step there are several questions which must be answered before elements of the system can be designed and integrated into a complete process. Let us look at some of these.

1

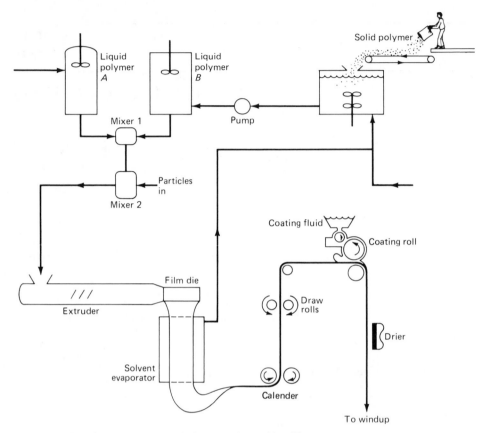

Figure 1-1 Flowsheet of a hypothetical process for making film.

- Dissolution: Solid polymer, in the form of chips, for example, is dissolved by contact with a solvent. What solvent is selected? Is the rate of solution dictated by the thermodynamic and other physical properties of the polymer-solvent pair, or can the rate be affected by such factors as chip size, intensity of stirring, and geometrical design of the stirrer and the tank?
- Pumping: Solution must be conveyed from one place to another. What is the relationship between the power requirements of the pump, the size of the piping, the flow rate, and the properties of the solution? Which properties of the solution are most relevant?
- Mixing of two liquids: How does one mix two liquids? How is the choice of mixing process dictated by the physical properties of the liquids? How does one define and measure mixedness? What properties of the liquids are most important in mixing-design considerations?
- Extrusion: An extruder is a pump which also can be used as a melter and a mixer. What are the basic principles of extruder design? When does one select an extruder over another type of pump?

- Flow through a die: A die is designed to produce a very specific set of conditions to be met by the emerging fluid, such as the film thickness, temperature, stress level, linear speed, surface gloss, etc. How is die design coupled with extruder design? In what detail must fluid properties be known in order to design a die?
- Calendering: This is a process of "squeezing" the film between two rolls for the purpose of making the film thinner and/or imparting surface characteristics to the film. How does the final film thickness depend on the calender design? What forces act on the film as it is calendered? Do these forces affect the flow of the film between the rolls?
- Drawing: This is a process of stretching the film by conveying the film between a set of rolls which have a greater linear speed than another set upstream. How do orientation and crystallinity change during the drawing process? Is significant heat generated during the stretching, and if so, how does one account for this in design?
- Coating: Another liquid may be coated on one side, or both, of the film. What factors determine the coating thickness? What choices are available for coating geometry?
- Drying: The coating may be a solution from which the solvent must be removed by application of heat. What factors affect the rate of heat transfer to a film moving through air? How long must the drying section be to produce film with a specified amount of residual solvent? What design factors, and physical properties, affect rate of evaporation?

Let us emphasize again that this is a hypothetical process, and the questions raised here are typical, but not comprehensive or inclusive of all the points that need to be raised in designing an integrated polymer process. It should be clear, however, that a *variety* of questions are raised. Many refer to fluid dynamics; some to heat and mass transfer. A knowledge of physical property data for polymers is essential.

Many of these questions refer to the so-called *transport phenomena*, the processes of momentum, heat, and mass transfer. Fluid dynamics is associated with momentum transport. It is with this topic of fluid dynamics that most of this book is concerned. No attempt will be made to answer *all* the questions raised above. However, an appreciation of the breadth and complexity of questions which must be answered in considering an *integrated* process provides a useful perspective for the study of specific elements of the process.

CHAPTER TWO

MODELING PHILOSOPHY

I am prepared for the worst, but hope for the best.

Disraeli

To analyze quantitatively the behavior of a fluid as it interacts with a processing system, it is necessary to formulate and solve the equations which describe the process. This is called *mathematical modeling*. The modeling process, in its most general form, begins when one writes the general *conservation* (of mass, momentum, and energy) equations that a system must obey, along with the *constitutive* equations which describe the properties of the material being processed. The conservation equations must be solved subject to certain constraints, the *boundary conditions*, which typically describe and proscribe such factors as the geometry of the system and forces or deformations imposed at or by the boundaries.

Generally, such mathematical formulations are too complex to admit simple analytical solutions, and often these problems require numerical (computer) solutions of such length and complexity that the effort is not justified. The term *modeling* will be used more specifically to refer to the modification of the mathematical formulation to yield a mathematical "model" that *is* tractable. In short, a simpler set of equations is used to describe the problem of interest.

How does one simplify a problem? By removing the offending complexity. Consider the coupled set of equations

$$\frac{1}{r}\frac{d}{dr}\left(r\mu\frac{du}{dr}\right) = -P_1 \tag{2-1}$$

$$\frac{1}{r}\frac{d}{dr}\left(r\frac{dT}{dr}\right) = -P_2\mu\left(\frac{du}{dr}\right)^2 \tag{2-2}$$

where $$\mu = e^{-T} \tag{2-3}$$

Boundary conditions are $u = T = 0$ on $r = 1$, and $du/dr = dT/dr = 0$ on $r = 0$. These equations arise in the analysis of viscous heating effects on capillary flow and will be discussed from a physical point of view in subsequent chapters. For the time being, let us look at the problem strictly from a mathematical point of view.

We may make the following statements about the problem:

- The single independent variable is r, so we are dealing with ordinary differential equations.
- The dependent variables (the "unknowns") are $u(r)$ and $T(r)$.
- Two parameters appear: P_1 and P_2.
- The equations are nonlinear: Eq. (2-1) has a product of $\mu(T)$ and a derivative of u, whereas Eq. (2-2) is quadratic in a derivative of u.
- The equations are coupled: if we use Eq. (2-3) to replace μ as a function of T in both Eqs. (2-1) and (2-2), we find *both* dependent variables u and T appearing in *both* differential equations.

Despite the coupled nonlinearities it is possible to obtain a formal analytical solution to this problem, as shown by Sukanek. However, the analytical solution requires use of a digital computer for the *evaluation* of the solution, and so is quite awkward to use if one wishes a rapid estimation of the effect of the parameters P_1 and P_2 on the solutions for u and T. Thus one might seek an approximate method for solution of these equations. Depending upon the method used, one will define a different model through each approximate solution.

Consider, for example, the approximation based on setting $P_2 = 0$. It is easily verified that the solutions are

$$T_0 = 0 \qquad u_0 = \frac{P_1}{4}(1 - r^2) \tag{2-4}$$

where we use a subscript 0 to denote a "zeroth-order" solution. It is obvious, immediately, that the effect of P_2 on T and u does not appear. There should be no surprise at this, since P_2 was set equal to zero and hence did not appear in the approximated form of Eq. (2-2). Only slightly less obvious is the fact that the effect of P_1 on T does not appear in this zeroth-order solution. In this case this occurs because the equations become uncoupled when $P_2 = 0$, allowing T_0 to be obtained directly from Eq. (2-2) alone, where P_1 does not appear.

Our goal here is not to discuss in any detail the solution of this particular problem. Rather, the point is to note that in the course of manipulating equations to simplify their solutions, we *can* go too far, with the result that we remove some of the features of interest from the problem. In the approximation cited (setting $P_2 = 0$) we have used a mathematical procedure known as "throwing out the baby with the bathwater." In order to remove the difficulty from the problem, we removed part of the reason for solving the problem at all.

The manner in which we develop approximate solutions often depends upon our goal. For example, if our primary interest is in an estimate of the dependence of $T(r)$ on P_1 and P_2, we might take the zeroth-order solution for $u(r)$, substitute that into the right-hand side of Eq. (2-2), and solve for $T(r)$. We have, then,

$$\frac{1}{r}\frac{d}{dr}\left(r\frac{dT}{dr}\right) = -P_2 e^{-T}\left(-\frac{P_1 r}{2}\right)^2 \tag{2-5}$$

This again uncouples the differential equations, but we are still left with a nonlinear equation for $T(r)$. We *could*, however, linearize the right-hand side by using T_0 in the exponential term, with the result that (recalling that $T_0 = 0$)

$$\frac{1}{r}\frac{d}{dr}\left(r\frac{dT}{dr}\right) = -\frac{P_2 P_1^2}{4}r^2 \tag{2-6}$$

whose solution is

$$T_1 = \frac{P_2 P_1^2}{64}(1 - r^4) \tag{2-7}$$

Now we have a solution which meets two objectives:

- It was very simply obtained.
- It shows the effect of P_1 and P_2 on T.

But is the solution an accurate representation of the original set of equations? This question must always be considered, but we must be careful not to focus on the question to the exclusion of an even more primitive question: Is the original set of equations an accurate representation of *reality?*

The former question will be taken up later when viscous heating problems are considered in Chap. 13. The latter, and more important, question is more difficult to answer because reliable experimental data are not available to a sufficient degree. This is not an uncommon problem. One is often in the position of developing models while lacking sufficient observational experience from which to draw some guidance. Still, one must often do *something,* and the *art* of modeling cannot be developed without the experience of trial and evaluation.

It is important, perhaps, to isolate here a basic statement about modeling: The ultimate test of a model is its correspondence with reality.

On many occasions we will commit mathematical indecencies in order to achieve some kind of tractable result. On some occasions the result will be so gross (in its correspondence with reality) that we will have to excuse ourselves and start over. But with some luck, some good sense, and some knowledge, we will sometimes achieve a useful result, with the final judgment being made with regard to the correspondence of the model with experience.

In summary, then, the goal of mathematical modeling is to produce an entity (in this case a set of equations and its solutions) which stands in place of the physical reality of interest, which lends itself to computation and analysis in some convenient form, and which bears some resemblance to the world of observational experience. In this light one might return to the quotation in the front of the book taken from a discussion by Professor Tolkien, one of the great creators of literary fantasy, which addresses itself to the problems of creating imaginary worlds as a background for his stories. If we replace the word *writer* by *engineer*, and substitute *mathematical model* for *secondary world*, then the creator of the hobbits of Middle-earth speaks to us who would create something equally fantastic: useful mathematical models of polymer processes.

BIBLIOGRAPHY

The viscous heating problem that leads to Eqs. (2-1) to (2-3) is discussed in detail in Chap. 13. An analytical solution is presented in

Sukanek, P. C.: Poiseuille Flow of a Power-Law Fluid with Viscous Heating, *Chem. Eng. Sci.*, **26**:1775 (1971).

An excellent discussion of the philosophy of modeling engineering systems is in

Petrie, C. J. S.: Mathematical Modelling and the Systems Approach in Plastics Processing: the Blown Film Process, *Polym. Eng. Sci.*, **15**:708 (1975).

This should be required reading.

THREE

CONTINUUM MECHANICS

There is no other source of knowledge but the intellectual manipulation of carefully verified observations.

Freud

Fluid dynamics is associated with the transfer of momentum from one region of a fluid to another. Momentum transfer involves events which are somehow embedded within the formalism of Newton's second law of motion. We have to develop that formalism now.

3-1 NEWTON'S SECOND LAW—POINT-MASS MECHANICS

$$\frac{d}{dt} m\mathbf{v} = \mathbf{F} \tag{3-1}$$

Newton's second law is embodied in the vector equation above. Such an equation has a dual character: There is first a certain *mathematical* content, or structure, which must be understood. Equation (3-1) is a *vector* equation, relating the vector $\mathbf{F}$ to the vector $d(m\mathbf{v})/dt$. Hence it is not one but three equations, of the form

$$\frac{d}{dt} mv_x = F_x \tag{3-1a}$$

$$\frac{d}{dt} mv_y = F_y \tag{3-1b}$$

$$\frac{d}{dt} mv_z = F_z \tag{3-1c}$$

The subscripts *x*, *y*, *z* indicate that we are actually applying Eq. (3-1) to the *components* of the vectors **v** and **F**. [We have arbitrarily assumed that a cartesian (*x*, *y*, *z*) coordinate system would be used.]

The other significant element of the mathematical character of Eq. (3-1) is that it is an *ordinary* differential equation (assuming that **v** and **F** are functions only of time).

In addition to its mathematical character, Eq. (3-1) embodies a *physical principle*, the *observation* that forces acting on a rigid body of mass *m* will cause momentum changes at a rate $d(m\mathbf{v})/dt$. For a *point mass*, i.e., for a body such that its dynamics are completely specified by the motion of the center of mass, the force **F** may be considered the sum or resultant of all forces acting on the body, and **v** is the velocity of the center of mass.

Note, incidentally, that we avoid writing the simpler, easier-to-memorize expression of Newton's second law, namely,

$$\mathbf{F} = m\mathbf{a} \tag{3-2}$$

Equation (3-2) is a valid simplification of Eq. (3-1) only if the mass of the body is constant. While this is often the case, the format of Eq. (3-2) still obscures the basic physical principle that Newton's second law relates forces to the time rate of change of *momentum*.

As an application of some of these ideas, consider a process for making nylon microspheres. An aqueous solution containing a diamine (e.g., hexamethylenediamine) is formed into droplets which are then contacted with diacid halide (e.g., sebacoyl chloride) in an organic phase. At the boundary between the two immiscible fluids an interfacial polycondensation occurs to produce a nylon membrane surrounding the aqueous drop. By controlling the time of contact (reaction time), it is possible to control the membrane thickness.

Let us develop a *model* of one aspect of the process based on the following design data:

- Aqueous drops of diameter 1000 μm = 0.1 cm are formed
- These drops contact the less dense organic phase by falling under gravity through a column of the organic phase contained in a cylinder of length 100 cm
- The aqueous drops enter the top surface of the organic column with a velocity of 100 cm/s

We want to find the *residence time*, which is simply the time required for a drop to fall the 100-cm length of the column.

Any model must be developed so as to be consistent with the known physics of the problem. Hence the first step of modeling is an elucidation of just what the physics of the process is expected to be. In this case, drops fall because the aqueous phase is denser than the surrounding organic phase. Thus we will need some quantitative information relevant to the motion of a drop through a fluid. Our intuition suggests that the continuous phase, because it is viscous, will retard the motion of a drop. Examination of the available information on the motion of a particle through a fluid suggests that, under certain conditions, the fluid exerts a retarding force given by

$$\mathbf{F} = -6\pi\mu a\mathbf{v} \tag{3-3}$$

where μ = viscosity of the continuous fluid

a = drop radius

$\mathbf{v}$ = relative velocity between drop and surrounding fluid

The minus sign arises since the force acts opposite the direction of motion.

One of the conditions for applicability of Eq. (3-3), which is known as Stokes' law, is that the drop be a *rigid* sphere. As part of the modeling process we assume that the formation of the nylon membrane creates a rigid surface on the drop. Whether this is so, or not, will determine in part the success of this model.

Two other forces acting on the drop are gravity, $\rho_d \frac{4}{3}\pi a^3 \mathbf{g}$, and buoyancy, $-\rho_c \frac{4}{3}\pi a^3 \mathbf{g}$, where the densities of the dispersed and continuous phases are ρ_d and ρ_c, respectively, and $\mathbf{g}$ is the acceleration of gravity.

We are tempted to treat the problem of motion in a fluid as a problem in point-mass mechanics. The resisting frictional force, for example, has been treated in such a way [through Eq. (3-3)] that, although it acts on the spherical interface between the two phases, it is expressed in terms of the velocity of the center of mass. Stokes' law, in other words, gives the *resultant* force of the surrounding fluid on the center of mass of the drop.

The presence of the surrounding fluid introduces another force that must be considered, and which arises from the fact that this problem is not strictly a point-mass mechanics problem. The motion of the particle induces motion in the surrounding fluid. If one attempts to accelerate the particle, it is necessary to overcome not only the inertia of the particle but also the inertia of part of the surrounding fluid. It can be shown that the effect of the inertia of the surrounding fluid can be exactly accounted for if the mass of the particle, in the momentum term on the left-hand side of Newton's second law, is increased by an amount equal to half the mass of the fluid displaced by the volume of the particle. This increment of mass is referred to as a *virtual mass*. By this device we again maintain the point-mass character of this particular problem.

Introduction of these ideas into Newton's second law gives

$$\frac{d}{dt}[\tfrac{4}{3}\pi a^3(\rho_d + \tfrac{1}{2}\rho_c)\mathbf{v}] = (\rho_d - \rho_c)\tfrac{4}{3}\pi a^3\mathbf{g} - 6\pi\mu a\mathbf{v} \tag{3-4}$$

or

$$\frac{dv}{dt} + \frac{9}{2}\frac{\mu}{\rho'a^2}v = \frac{\Delta\rho}{\rho'}g \tag{3-5}$$

where we have defined a hypothetical density as $\rho' = \rho_d + \tfrac{1}{2}\rho_c$, where we have written the density difference as $\Delta\rho = \rho_d - \rho_c$, and where we have written the vertical component of each term of the vector equation. This is a first-order linear ordinary differential equation, and an initial condition is required, namely,

$$v = v_0 \quad \text{at } t = 0$$

where v_0 is the initial velocity with which the drop enters the column. The solution of Eq. (3-5), satisfying this initial condition, is

$$v = \left(v_0 - \frac{2}{9}\frac{\Delta\rho\ ga^2}{\mu}\right)\exp\left(-\frac{9}{2}\frac{\mu}{\rho'a^2}\ t\right) + \frac{2}{9}\frac{\Delta\rho\ ga^2}{\mu} \tag{3-6}$$

To find the time required for the drop to fall a length L, we solve for t_L from

$$L = \int_0^{t_L} v(t)\ dt \tag{3-7}$$

Before carrying out this calculation, it is useful to examine the result for v [Eq. (3-6)] so as to understand the physical significance of each term.

The first thing to note is that the velocity is made up of two terms. The time-dependent term is related to the slowing down of the drop due to the retarding force **F**. We usually refer to such a term as the *transient* part of the solution, because the term, being exponentially decreasing, becomes successively smaller as time goes on. When the transient term is insignificant, the velocity is a constant,

$$v_t = \frac{2}{9}\frac{\Delta\rho\ ga^2}{\mu} \tag{3-8}$$

v_t is called the *terminal* velocity.

A question of practical importance is "How long a time is required for the drop to reach its terminal velocity?" To answer this, it is helpful to rewrite Eq. (3-6) in the form

$$\frac{v}{v_t} = 1 + \frac{v_0 - v_t}{v_t}\exp\left(-\frac{\Delta\rho\ g}{\rho'v_t}\right)t \tag{3-9}$$

It is clear that an infinite time is required to make the second term vanish *exactly*. It is more useful to find the time required to bring v to a value *nearly* equal to v_t. Let us arbitrarily choose $v/v_t = 1.05$. Then, if we can specify values for the parameters that appear in Eq. (3-9), we can answer the question. Let us take, in addition to values already given, the following:

$$g = 980\ \text{cm/s}^2 \qquad \rho_d = 1\ \text{g/cm}^3 \qquad \rho_c = 0.9$$

$$\Delta\rho = 0.1\ \text{g/cm}^3 \qquad \mu = 0.01\ \text{P} \qquad \rho' = 1.45$$

It follows that $v_t = 5.44$ cm/s, and the time to retard the drop to within 5 percent of its terminal velocity is only 0.47 s.

Returning to the earlier question, we can perform the integration indicated in Eq. (3-7) and find, for $L = 100$ cm, that the residence time is $t_L = 10.8$ s, by solving the integrated form of Eq. (3-7) for t_L:

$$t_L = \frac{L}{v_t} - \frac{(v_0 - v_t)\rho'}{\Delta\rho\ g}\left[1 - \exp\left(-\frac{\Delta\rho\ gt_L}{\rho'v_t}\right)\right] \tag{3-10}$$

We note immediately that the residence time is quite long in comparison to the transient time calculated just above, from which we conclude that most of the process occurs under conditions of steady velocity v_t. This fact suggests that we simplify the relationship between the residence time and other controllable parameters to the form

$$t_L = \frac{L}{v_t} = \frac{9\mu L}{2\,\Delta\rho\,ga^2} \tag{3-11}$$

which is quite a bit simpler to deal with than the transcendental algebraic equation [Eq. (3-10)] that results from the more exact application of Eq. (3-7). If we do so, however, we find that Eq. (3-11) gives a residence time of 18.4 s, which is nearly twice the value obtained from the exact solution of Eq. (3-10). Why have we suffered such a large error in calculating the residence time when the transient period of the dynamics takes up only a few percent of the total time? In this particular problem the error arises from the fact that the drop enters the column at a velocity ($v_0 = 100$ cm/s) nearly 20 times the terminal velocity. Even though viscous effects bring the drop to its terminal velocity in a very short time, the drop does travel a significant distance during the transient period. (In fact, the drop falls through nearly half the column before it reaches its terminal velocity.)

Thus we learn a lesson here with regard to modeling. A simplification with regard to one aspect of a problem does not necessarily produce corresponding simplifications in other related aspects of the problem. One of the great dangers in modeling *physical* problems is the tendency to focus on the superficial *mathematical* aspects of the model (e.g., the transient-time period is clearly very short compared to the total process time) and in doing so, to fail to think *physically* about the problem. In dealing with physical problems, one must never lose sight of the basic physical ideas. Mathematics is a tool which, once mastered, can be misused with great facility.

In a very limited sense, Eq. (3-10) provides a mathematical model for the interaction of some of the parameters of this process. We could use it to make estimates of the effect of changing various parameters on the residence time. With an additional model relating residence time to the membrane properties (e.g., thickness and mechanical strength), we would be in position to proceed with a rational first design of the process.

3-2 CONCEPT OF A CONTINUUM

Simple point-mass mechanics concerns itself with the behavior of rigid bodies. For the most part, one need only consider the motion of the center of mass of the body. Application of Newton's second law then leads to ordinary differential equations relating the position vector of a " point " to the single independent variable: time.

Now we have to introduce the concept of a continuous medium, or *continuum*. A continuum is a region of space through which properties such as temperature, pressure, density, and velocity may vary in a continuous manner. We suppose that we can indefinitely subdivide the region into smaller and smaller volumes and still retain the ability to define certain properties in that infinitesimal volume. Of course, this concept reaches its limitation when we approach a molecular scale, but the fact is that one does not usually need to go to such a fine scale in order to define the properties of interest in continuum mechanics.

Let us consider a few examples of continuous variables. Suppose we have a fluid confined to a container. For definiteness, suppose the fluid is water, containing a small amount of dissolved salt. What can we say about the "state" of the fluid?

One of the simplest descriptions is given if we know the mass of solution m, the volume of solution V, and the mass of salt m_s. In addition to these quantities, we can "derive" some additional information. For example, the average salt concentration is $\bar{c}_s = m_s/V$, and the average density of the solution is $\bar{\rho} = m/V$. But these are *average* quantities; they are "macroscopic" measures of the state of the fluid. They give no information regarding the *distribution* of salt throughout the fluid.

To characterize the distribution of properties we have to begin by considering some fixed point P in the space occupied by the fluid. Now let us imagine a small volume of fluid ΔV which encloses the point P. Within that differential volume there will be amounts of mass Δm and amounts of salt Δm_s. The average density of the material within ΔV is simply $\Delta m/\Delta V$.

Now suppose we consider

$$\lim_{\Delta V \to 0} \frac{\Delta m}{\Delta V}$$

By $\Delta V \to 0$ we must understand that ΔV becomes small compared to the volume V of the body but still remains larger than molecular volumes. Then we *define* the "local density" ρ as

$$\rho = \lim_{\Delta V \to 0} \frac{\Delta m}{\Delta V} \qquad \text{at the point } P$$

Similarly, the "local salt concentration" is

$$c_s = \lim_{\Delta V \to 0} \frac{\Delta m_s}{\Delta V} \qquad \text{at the point } P$$

Now let us think about these densities physically for a moment. We have not assumed that the fluid is homogeneous. Thus it is possible that c_s varies throughout the volume V. The same is true of ρ. Then, in general, $\rho = \rho(x, y, z)$ and $c_s = c_s(x, y, z)$, where (x, y, z) are coordinates of the point P. Furthermore, it is possible that these densities are dependent on time. For example, if we are cooling the fluid, and since density is a function of temperature, and temperature will be a function of time, then actually $\rho = \rho(x, y, z, t)$. In general, the quantities of interest in specifying the state of a continuous medium vary with time and also vary continuously throughout the volume. This leads to mathematical complications since generally we will have to deal with partial differential equations in order to describe the behavior of a continuum.

3-3 STRESS IN A CONTINUUM

Let us still think of our container of water and now consider *forces* acting within the fluid. The forces might arise from motion within the fluid which could be generated if we were stirring the solution. Generally, the forces arise from several sources. There might, for example, be "body forces" due to the action of gravity on the fluid. If flow is occurring, then the viscosity of the fluid will give rise to forces between adjacent regions of flow. And, in general, there may be a pressure distribution throughout the fluid which gives rise to forces.

Whatever the source, imagine a force acting on a small region of the fluid in the neighborhood of the point P. Let us consider a differential area ΔA containing the point P. Note that this area is really a *vector* quantity since it is characterized not only by a magnitude ΔA but also by an *orientation* described by the vector $\mathbf{n}$ normal to the area. Let the force acting on ΔA be $\Delta \mathbf{F}$. $\Delta \mathbf{F}$ is also a vector, of course, and in general $\Delta \mathbf{F}$ and $\mathbf{n}$ have different directions. Figure 3-1 shows the situation.

We could resolve the vector $\Delta \mathbf{F}$ into its three components in a coordinate system defined by $\mathbf{n}$ and a pair of axes in the plane of ΔA. Let an arbitrary pair of surface axes be $\mathbf{s}_1$ and $\mathbf{s}_2$. Then $\mathbf{s}_1, \mathbf{s}_2, \mathbf{n}$ define a cartesian coordinate system, and the components of $\Delta \mathbf{F}$ are called $(\Delta F_{1n}, \Delta F_{2n}, \Delta F_{nn})$.

The reason for the double subscripting is as follows. For any *given* force $\Delta \mathbf{F}$, there is an infinite number of areas ΔA, each characterized by a particular normal vector $\mathbf{n}$. Hence, when we resolve $\Delta \mathbf{F}$ into the $\mathbf{s}_1, \mathbf{s}_2, \mathbf{n}$ coordinate system, we must indicate (by the second subscript) which $\mathbf{n}$ vector, i.e., which area orientation, we have chosen, and by the first subscript we must indicate which axis $\mathbf{s}_1, \mathbf{s}_2$, or $\mathbf{n}$, of *that* coordinate system the force $\Delta \mathbf{F}$ has been resolved onto.

We will now *define* the stress *vector* at point P as

$$\mathbf{T} = \lim_{\Delta A \to 0} \frac{\Delta \mathbf{F}}{\Delta A}$$

If we consider

$$\lim_{\Delta A \to 0} \frac{\Delta F_{in}}{\Delta A} = T_{in}$$

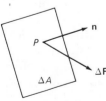

Figure 3-1 Definition sketch for consideration of a force $\Delta \mathbf{F}$ acting on a surface of area ΔA whose normal vector is $\mathbf{n}$.

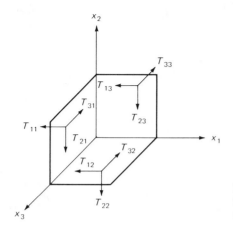

Figure 3-2 Definition sketch for resolution of the force **F** into the stress components acting on each of the three orthogonal surfaces.

where i refers to axes s_1, s_2, or n, then we have defined the *components* of the stress vector **T** *on the surface defined by* **n**. Note that components of the stress vector have two directions associated with them, one which characterizes the direction of Δ**F** and the other which gives the orientation of the surface ΔA, that is, **n**.

But **n** was arbitrary. It appears that we could define an infinite set of components of **T**. Does this mean that we cannot completely characterize force or stress in a continuum without specifying an infinite set of numbers?

The answer, of course, is no. We can show that the maximum information required is the components of **F** resolved onto *three* orthogonal surfaces. Let us consider, then, a set of cartesian axes x_1, x_2, x_3, or simply x_i in shorter notation, as shown in Fig. 3-2. The origin of the axes is point P, and the three orthogonal areas in the planes $x_1 x_2$, $x_2 x_3$, $x_3 x_1$ are understood to be differential areas, although our notation will no longer reflect this.

Let the force **F** have been resolved into stress components T_{11}, T_{21}, T_{31} in the face perpendicular to x_1; T_{12}, T_{22}, T_{32} in the face perpendicular to x_2; and T_{13}, T_{23}, T_{33} in the face perpendicular to x_3. We will prove the following: If we know the nine components of **T** defined above, we can obtain the components of **T** in *any* arbitrary surface.

Before proceeding we have to suffer one more annoyance: a convention whereby we can agree on the *sign* that is to be associated with a stress component. Our usual sign convention in mechanics is that velocities and forces whose action is in the positive direction of an axis are taken as positive. But by Newton's third law of action and reaction we have to recall that if the material on *one* side of a surface exerts a force on that surface, an equal and opposite force is exerted by material on the *other side* of the surface onto that surface. Our convention on signs will be the following:

The stress T_{ij}, due to action *by* material on the positive side of the surface *on* the material on the negative side, is positive if the line of action is along positive $\mathbf{x_i}$.

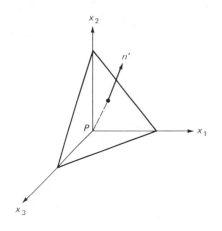

Figure 3-3 Definition sketch for consideration of the equilibrium of a tetrahedral-volume element.

Conversely, the stress exerted from the negative side of the surface on the material on the positive side is positive if the line of action is along negative $\mathbf{x}_i$.

If, in Fig. 3-2, the stresses are all those exerted *by forces from the negative sides* of the three planes, then the stresses whose directions are as indicated in the figure are positive stresses.

Our goal now is to show that if the nine components defined above are known, we can find the three components of stress in *any* arbitrary surface in the neighborhood of P. Consider a surface defined by a normal vector $\mathbf{n}'$ which cuts the coordinate axes $\mathbf{x}_i$, as shown in Fig. 3-3. The three orthogonal faces are acted upon by the stresses T_{ij} defined above. We are going to consider the equilibrium of the tetrahedral volume of material defined within the four surfaces shown in Fig. 3-3. Let us call the area of the slanted face $\Delta A'$. The area of the face normal to the x_1 axis is just the projection of $\Delta A'$ onto the 23 plane, or $a_{1n} \Delta A'$, where a_{1n} is the cosine of the angle between $\mathbf{x}_1$ and $\mathbf{n}'$.

Let us consider the components of force in the x_1 direction. On the face normal to $\mathbf{x}_1$ we will have a component $-T_{11}a_{1n} \Delta A'$. Note the minus sign, and review our convention. We are taking *forces* to be positive if their line of action is along positive $\mathbf{x}_i$. On the face normal to $\mathbf{x}_2$ we find $-T_{12} a_{2n} \Delta A'$, and on the third orthogonal face we find $-T_{13} a_{3n} \Delta A'$. Summing these three components, which all act in the $\mathbf{x}_1$ direction, we write

$$\Delta F_1 = -(T_{11}a_{1n} + T_{12}a_{2n} + T_{13}a_{3n}) \Delta A' \tag{3-12}$$

If we consider the contributions of the three orthogonal faces to forces in the other two directions, we find

$$\Delta F_2 = -(T_{21}a_{1n} + T_{22}a_{2n} + T_{23}a_{3n}) \Delta A' \tag{3-13}$$

and

$$\Delta F_3 = -(T_{31}a_{1n} + T_{32}a_{2n} + T_{33}a_{3n}) \Delta A' \tag{3-14}$$

These last three equations can be written in the following shorthand notation:

$$\Delta F_i = - \sum_{j=1}^{3} T_{ij} a_{jn} \ \Delta A' \tag{3-15}$$

This latter equation represents *three* equations for the components $\Delta F_1, \Delta F_2, \Delta F_3$.

Now let us think of a set of orthogonal coordinate axes x'_k, such that $x'_1 = n'$, and x'_2 and x'_3 are in the plane of the slanted face of the tetrahedron. We now want to resolve each force ΔF_i into its components in the x'_k coordinate system. This will again introduce the direction cosines a_{ik} between axes x_i and x'_k. It is easy to see that one finds

$$\Delta F'_k = \sum_{i=1}^{3} a_{ik} \ \Delta F_i = - \sum_i \sum_j a_{ik} a_{jn} T_{ij} \ \Delta A' \tag{3-16}$$

Now let the stress vector acting on the inclined plane (which is the same as the stress vector on the other three planes if ΔV is sufficiently small) be resolved into the x'_k coordinates to give T'_{kn}. Hence the *forces* in the x'_k directions, in terms of the stress T'_{kn}, are

$$\Delta F'_k = T'_{kn} \ \Delta A' \tag{3-17}$$

Then the sum of the force components on *all* four surfaces which enclose the volume ΔV, resolved into any of the three directions x'_k, may be written as

$$\left(T'_{kn} - \sum_i \sum_j a_{ik} a_{jn} T_{ij} \right) \Delta A'$$

According to Newton's second law, these terms must be balanced by any force of acceleration of the volume and by body forces such as gravity. In any event, such other forces are proportional to the *mass* of the tetrahedral-volume element or, equivalently, to its volume ΔV. Hence the equation of conservation of momentum takes the form

$$\left(T'_{kn} - \sum_i \sum_j a_{ik} a_{jn} T_{ij} \right) \Delta A' = O(\Delta V) \tag{3-18}$$

where $O(\Delta V)$ means "terms whose order of magnitude is proportional to ΔV."

But if Δx is a measure of the linear dimensions of the tetrahedron, we know that $\Delta A' = O((\Delta x)^2)$ while $\Delta V = O((\Delta x)^3)$. Hence if we take the limit as the tetrahedron shrinks to the point P, we find the volume terms vanishing faster than the area terms.

Newton's second law, then, leads us to conclude that, in the limit as the volume shrinks to the point P,

$$T'_{kn} = \sum_i \sum_j a_{ik} a_{jn} T_{ij} \tag{3-19}$$

But $\mathbf{n}'$ was an arbitrary vector, so Eq. (3-19) is valid for any orientation $\mathbf{n}'$, and so T'_{kn} are the components of the stress vector in *any* coordinate system $\mathbf{x}'_k$, in terms of the components in the $\mathbf{x}_i$ coordinates. In other words, if the nine components T_{ij} are known in *some* coordinate system, they are known in *all* coordinate systems. Thus the nine components T_{ij} are necessary and sufficient to completely define the state of stress at a point in a continuum.

The array, or matrix, of stress components may be written in the format

$$\mathbf{T} = \begin{pmatrix} T_{xx} & T_{xy} & T_{xz} \\ T_{yx} & T_{yy} & T_{yz} \\ T_{zx} & T_{zy} & T_{zz} \end{pmatrix} \tag{3-20}$$

$\mathbf{T}$ is called the stress *tensor*, and the T_{ij} are usually referred to as components of the stress tensor.

$\mathbf{T}$ is a *tensor* because, by *definition*, if any array of numbers transforms from one coordinate system to another according to Eq. (3-19), that array represents the components of a tensor.

Another simple argument based on principles of mechanics leads us to conclude that the stress tensor is symmetrical, which means simply that

$$T_{ij} = T_{ji} \tag{3-21}$$

Consider a small parallelepiped with sides dx_i, as in Fig. 3-4. The stress T_{12} gives rise to a moment about the axis $\mathbf{x}_3$ of magnitude $(T_{12}\ dx_1\ dx_3)\ dx_2$. The stress T_{21} gives rise to a moment of opposite sign of magnitude $(T_{21}\ dx_2\ dx_3)\ dx_1$. None of the stresses acting on any of the other four faces of the parallelepiped gives rise to moments about $\mathbf{x}_3$. The net moment due to stresses, then, is

$$M_3 = (T_{12} - T_{21})\ dx_1\ dx_2\ dx_3 = O((\Delta x)^3) \tag{3-22}$$

Body forces can also exert a moment about $\mathbf{x}_3$, but this will be of order $(\Delta x)^4$. The time rate of change of angular momentum will also be of order $(\Delta x)^4$. Hence, as $\Delta x \to 0$, equilibrium (i.e., conservation of angular momentum) requires that $M_3 \to 0$, or

$$T_{12} = T_{21} \tag{3-23}$$

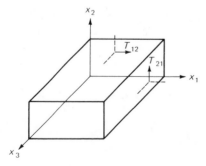

Figure 3-4 Definition sketch for proof of the symmetry of the stress tensor.

Similarly, by considering moments about x_1 and x_2, we can show $T_{23} = T_{32}$ and $T_{13} = T_{31}$.

The value of establishing symmetry is that we know that only six components (instead of nine) are needed to completely specify the state of stress at a point. Symmetry of the stress tensor must be considered an *assumption*, however. In principle, materials with an intrinsic *asymmetric structure*, such as liquid crystals, or dispersions of asymmetric particles, might be expected to show an asymmetric stress tensor. Despite these reservations one normally assumes the symmetry of **T**, and there is no strong experimental evidence to suggest that the failure of a model is due to a failure in this assumption.

3-4 THE DYNAMIC EQUATIONS (EQUATIONS OF MOTION)

We shall derive two basic conservation equations here. The first will express the principle of conservation of mass in a partial differential equation known as the continuity equation. The second equation expresses the principle of conservation of linear momentum in a continuous medium. It is an analog of Newton's second law which we used earlier in discussing point-mass mechanics. Both equations result from setting up an arbitrary, fixed "control volume" in the continuous medium and keeping track of the net flow of mass and momentum across the surfaces of the control volume.

Figure 3-5 shows the control volume, which we take as a parallelepiped aligned parallel to the axes of a cartesian coordinate system. The volume is to be thought of as differentially small, of magnitude $dV = dx_1\, dx_2\, dx_3$. In Fig. 3-5 the two sets of orthogonal faces are split for visual clarity. We may arbitrarily take one set of orthogonal faces to coincide with the coordinate planes.

The net rate of change of mass within the control volume is $(\partial/\partial t)\rho\, dx_1\, dx_2\, dx_3$. If there is to be a nonzero net rate of change of mass, it can

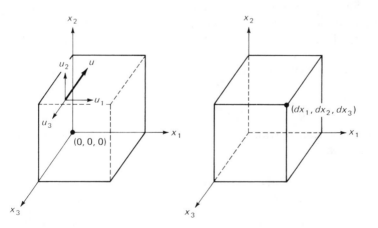

Figure 3-5 Volume element for derivation of continuity equation.

only arise from an unbalanced flow of material across the six faces of the parallelepiped. The volumetric flow rate across a surface is given by the product of the surface area and the velocity normal to that surface. Since we "define" a surface by its unit normal vector, the volumetric flow rate across a surface normal to the i direction will be $u_i \, dx_j \, dx_k$ ($i \neq j$ or k). The mass flow rate, then, will be $\rho u_i \, dx_j \, dx_k$ across that surface.

Let us consider a pair of parallel surfaces, such as the pair which are perpendicular to the x_1 axis. We arbitrarily take the x_1 coordinate to be $x_1 = 0$ for one surface and $x_1 = dx_1$ for the second. The mass flow rate across the first surface is

$$[\rho u_1 \, dx_2 \, dx_3]_{x_1=0}$$

and that across the second surface is

$$-[\rho u_1 \, dx_2 \, dx_3]_{x_1=dx_1}$$

The minus sign on the second term reflects the convention that if the velocity component u_1 is positive across the face at $x_1 = dx_1$, then the flow is directed *out* of the volume element and is considered a *negative* flow. The *net* flow across that pair of faces is just the difference

$$[\rho u_1 \, dx_2 \, dx_3]_{x_1=0} - [\rho u_1 \, dx_2 \, dx_3]_{x_1=dx_1}$$

Now, let us think of each bracketed term as some function of position x_1 and time t. We can then use Taylor's series to write the difference in a function evaluated at two points separated by a distance dx_1 and obtain

$$[F_1]_{x_1=0} - [F_1]_{x_1=dx_1} = -\frac{\partial}{\partial x_1}[F_1]_{x_1=0} \, dx_1$$

$$-\frac{1}{2}\frac{\partial^2}{\partial x_1^2}[F_1]_{x_1=0}(dx_1)^2 - \cdots \quad (3\text{-}24)$$

where we have simplified the writing not by filling in the brackets with the term $\rho u_1 \, dx_2 \, dx_3$ but by using instead F_1 to remind us that it is the x_1-directed flow term. Note that *partial* differentiation is indicated since each function $[F_1]$ might depend on time, as well as on x_1.

In a similar way we can write the mass flow rate across a surface perpendicular to the x_2 axis as

$$[F_2] = [\rho u_2 \, dx_1 \, dx_3]$$

and the *net* flow rate associated with a *pair* of surfaces parallel to each other, separated by a distance dx_2, becomes

$$[F_2]_{x_2=0} - [F_2]_{x_2=dx_2} = -\frac{\partial}{\partial x_2}[F_2]_{x_2=0} \, dx_2$$

$$-\frac{1}{2}\frac{\partial^2}{\partial x_2^2}[F_2]_{x_2=0}(dx_2)^2 - \cdots \quad (3\text{-}25)$$

Similarly we may write, for the final pair of surfaces that make up the volume element,

$$[F_3]_{x_3=0} - [F_3]_{x_3=dx_3} = -\frac{\partial}{\partial x_3}[F_3]_{x_3=0}\,dx_3$$

$$-\frac{1}{2}\frac{\partial^2}{\partial x_3^2}[F_3]_{x_3=0}(dx_3)^2 - \cdots \qquad (3\text{-}26)$$

The time rate of change of mass within the volume element must equal the sum of these three flow terms, with the result that

$$\frac{\partial}{\partial t}\rho\,dx_1\,dx_2\,dx_3 = \begin{aligned} &-\frac{\partial}{\partial x_1}(\rho u_1\,dx_2\,dx_3)\,dx_1 - O(dx)^4 \\ &-\frac{\partial}{\partial x_2}(\rho u_2\,dx_1\,dx_3)\,dx_2 - O(dx)^4 \\ &-\frac{\partial}{\partial x_3}(\rho u_3\,dx_1\,dx_2)\,dx_3 - O(dx)^4 \end{aligned} \qquad (3\text{-}27)$$

where $O(dx)^4$ means terms which are at least fourth-order products of the various dx_i's.

Since we regard the volume element to be fixed in space, the product $dV = dx_1\,dx_2\,dx_3$ is a constant, and if we divide both sides by the differential volume we find

$$\frac{\partial\rho}{\partial t} = -\left(\frac{\partial}{\partial x_1}\rho u_1 + \frac{\partial}{\partial x_2}\rho u_2 + \frac{\partial}{\partial x_3}\rho u_3\right) - O(dx) \qquad (3\text{-}28)$$

If the volume element is allowed to become very small ($dV \to 0$), the derivation above gives the mass balance in the infinitesimal neighborhood of the point that dV had surrounded. Terms of order (dx) vanish relative to the other terms in Eq. (3-28), and the final result is the continuity equation:

$$\frac{\partial\rho}{\partial t} = -\nabla\cdot\rho\mathbf{u} \qquad (3\text{-}29)$$

where the simpler vector format using the divergence operator has been introduced. In the special case of incompressible fluids (constant density), which we normally assume to be the case for liquids, the continuity equation becomes

$$\nabla\cdot\mathbf{u} = 0 \qquad \text{incompressible fluid} \qquad (3\text{-}30)$$

Thus the principle of conservation of mass is expressed in the form of Eq. (3-29) in general and in the form of Eq. (3-30) for the incompressible fluid.

Now let us consider a region of a continuum subject to stresses which may vary continuously throughout the small volume $dV = dx_1\,dx_2\,dx_3$, as shown in Fig. 3-6. We have again split the two sets of orthogonal faces for visual clarity.

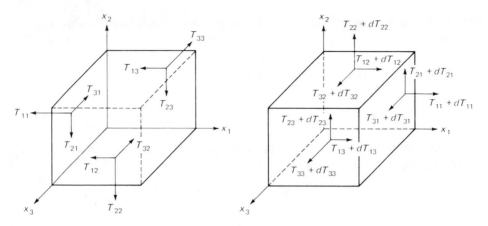

Figure 3-6 Volume element for derivation of equations of motion.

On the faces which lie in the coordinate planes we have the stresses T_{ij} as shown. The stresses on the second set of planes may differ from those on the first, since we allow T_{ij} to depend on the x_i coordinates. If the stresses vary continuously, we may write the stresses on one plane in terms of those on the plane parallel to it by using Taylor series, in the form

$$T_{ij}(dx_j) = T_{ij}(0) + \frac{\partial T_{ij}}{\partial x_j} dx_j + \frac{1}{2} \frac{\partial^2 T_{ij}}{\partial x_j^2} (dx_j)^2 + \cdots$$

$$= T_{ij}(0) + dT_{ij} \tag{3-31}$$

Now let us consider the net force acting in some direction, say, x_1, due to the system of stresses T_{ij}. We find, using Fig. 3-6 and the sign conventions,

$$[(T_{11} + dT_{11}) - T_{11}]\, dx_2\, dx_3 + [(T_{12} + dT_{12}) - T_{12}]\, dx_1\, dx_3$$
$$+ [(T_{13} + dT_{13}) - T_{13}]\, dx_1\, dx_2$$

to be the net x_1 component due to stresses. Using Eq. (3-31) we may write this as

$$S_1 \equiv \left(\frac{\partial T_{11}}{\partial x_1} dx_1\right) dx_2\, dx_3 + \left(\frac{\partial T_{12}}{\partial x_2} dx_2\right) dx_1\, dx_3$$
$$+ \left(\frac{\partial T_{13}}{\partial x_3} dx_3\right) dx_1\, dx_2 + \text{terms of } O(dx)^4 \tag{3-32}$$

In addition to surface stresses there may be a body force (usually due to gravity) proportional to the mass. If we take f_1 to be the component in the x_1 direction of the body force per unit mass, then the body force component is $\rho f_1\, dx_1\, dx_2\, dx_3$, where ρ is the density of the fluid in the volume element.

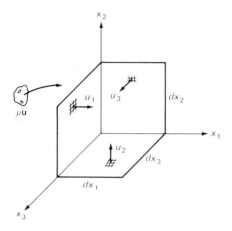

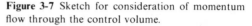

Figure 3-7 Sketch for consideration of momentum flow through the control volume.

If the element of volume being considered were a *rigid* body, Newton's second law would be completed by equating the stress term plus the body force to the acceleration term, which would have the form

$$\frac{\partial}{\partial t} \rho \, dx_1 \, dx_2 \, dx_3 u_1$$

But if we wish to apply our momentum principle to a continuous *fluid*, then clearly the volume element is not a rigid body, since we allow fluid to cross the boundaries of the volume element. It is this latter point, the fact that fluid crosses the boundaries of the control volume, which must now be considered.

If fluid crosses the surfaces of the volume element, then that flow carries momentum in and out of the volume element. Newton's second law really represents a balance relating forces acting on the volume to the time rate of change of momentum within the volume. In addition to the force terms, then, we must consider the net flow of momentum through the volume element. The momentum flow terms are handled in a similar manner to the mass flow terms in the derivation of the continuity equation. Figure 3-7 shows a definition sketch for the analysis.

Let **u** be the velocity vector in the neighborhood of the volume element. Any "parcel" of fluid crossing a surface of the control volume has a momentum (per unit volume) of $\rho\mathbf{u}$. Since we write the momentum principle in terms of *components* of the appropriate vectors (as we already have for surface forces and body forces), let us consider *only* the x_1 component of momentum flow. Then each parcel of fluid has a "density of x_1 momentum" equal to ρu_1. How do we calculate the *rate* at which momentum crosses each surface?

It is only necessary to multiply the density of momentum (which is what ρu_1 really is) by the velocity with which that parcel crosses each surface. The appropriate velocity is simply the component of **u** *normal* to each surface. Thus we find that the x_1 momentum flow into face $dx_2 \, dx_3$ is $(\rho u_1)u_1 \, dx_2 \, dx_3$, the x_1 momentum flow into face $dx_1 \, dx_3$ is $(\rho u_1)u_2 \, dx_1 \, dx_3$, and the x_1 momentum flow into face $dx_1\, dx_2$ is $(\rho u_1)u_3\, dx_1\, dx_2$. Note that in each case it is the same momentum density ρu_1, multiplying the velocity u_i normal to the face whose normal is $\mathbf{dx}_i$. The area factor in each case converts the velocity u_i to a volumetric flow rate normal to each face. Multiplying the momentum (in the x_1 direction) per unit volume ρu_1 gives the *rate* of flow of momentum across each face, which is what we are seeking.

The terms above represent momentum flows across the faces in the coordinate planes. Across the faces in the parallel planes at dx_1, dx_2, dx_3, we have, by Taylor's theorem again, $(\rho u_1)u_1\, dx_2\, dx_3 + d[(\rho u_1)u_1\, dx_2\, dx_3]$, etc. Consequently the *net* flow of x_1-directed momentum across all six surfaces of the control volume will reduce to

$$C_1 \equiv \frac{\partial}{\partial x_1}(\rho u_1 u_1)\, dx_2\, dx_3\, dx_1 + \frac{\partial}{\partial x_2}(\rho u_1 u_2)\, dx_1\, dx_3\, dx_2$$

$$+ \frac{\partial}{\partial x_3}(\rho u_1 u_3)\, dx_1\, dx_2\, dx_3 + \text{terms of order } (dx)^4 \qquad (3\text{-}33)$$

We call these the *convective terms*.

Now, and almost finally, the time rate of change of momentum within the volume element not only includes the convective *flow* terms above, which account for the net flow of momentum through the control surfaces, but must also include the time rate of change of the momentum *within* the volume, or $\partial/\partial t(\rho\, dx_1\, dx_2\, dx_3\, u_1)$. This idea is better understood if we notice that the convective terms vanish if the flow is *spatially homogeneous*, which means simply that all spatial derivatives $\partial/\partial x_1$, $\partial/\partial x_2$, $\partial/\partial x_3$ vanish. Yet there could still be unsteady-state effects causing momentum changes if $\mathbf{u} = \mathbf{u}(t)$.

Now we must put everything together. To save some paper, let us write the stress terms simply as S_1, from Eq. (3-32), and the convective terms as C_1 from Eq. (3-33). Then the momentum principle for the volume element becomes

$$\frac{\partial}{\partial t}(\rho u_1)\, dV + C_1 = S_1 + \rho f_1\, dV + 0(dx)^4 \qquad (3\text{-}34)$$

We have used $dV = dx_1\, dx_2\, dx_3$, and we must remember that both C_1 and S_1 are proportional to dV.

Now we simply divide both sides of Eq. (3-34) by dV and take the limit as $dV \to 0$. We assume that terms like $\partial(\rho u_1 u_2)/\partial x_2$ remain finite and well behaved in this limit. This is the essence of the continuum approximation: that spatial gradients do indeed make physical sense in a region of material that is macroscopically small but still molecularly large. If one goes through the limit exercise carefully (the proverbial and annoying "exercise for the reader"), one finds

$$\frac{\partial}{\partial t}(\rho u_1) + \frac{\partial}{\partial x_1}(\rho u_1 u_1) + \frac{\partial}{\partial x_2}(\rho u_1 u_2) + \frac{\partial}{\partial x_3}(\rho u_1 u_3)$$

$$= \frac{\partial T_{11}}{\partial x_1} + \frac{\partial T_{12}}{\partial x_2} + \frac{\partial T_{13}}{\partial x_3} + \rho f_1 \qquad (3\text{-}35)$$

All the above was carried out only for the x_1 component of momentum, forces, etc. But by an identical process one can derive two similar equations, and in fact all three can be written in the form

$$\frac{\partial}{\partial t}(\rho u_i) + \frac{\partial}{\partial x_j}(\rho u_i u_j) = \frac{\partial T_{ij}}{\partial x_j} + \rho f_i \qquad i = 1, 2, 3 \qquad (3\text{-}36)$$

where we agree to a *summation notation:* If a subscript is repeated, we sum over that subscript, i.e.,

$$\frac{\partial}{\partial x_j} T_{ij} = \sum_{j=1}^{3} \frac{\partial T_{ij}}{\partial x_j} = \frac{\partial T_{i1}}{\partial x_1} + \frac{\partial T_{i2}}{\partial x_2} + \frac{\partial T_{i3}}{\partial x_3} \qquad (3\text{-}37)$$

Equation (3-36) is the *dynamic equation* for a continuous medium. (It is really three equations.) It will form the foundation for subsequent analyses of flow processes. It will be convenient for such analyses to cast the dynamic equations into other *formats.* We emphasize the word *format* here because no new *physical* information comes of changing the form of the equations; it is simply a matter of convenience, and, in part, custom, to have the dynamic equations in other formats.

With the continuity equation we can rewrite the dynamic equations in the form

$$\rho\left(\frac{\partial u_i}{\partial t} + u_j \frac{\partial u_i}{\partial x_j}\right) = \frac{\partial T_{ij}}{\partial x_j} + \rho f_i \qquad (3\text{-}38)$$

The only physical statement required in going from Eq. (3-36) to Eq. (3-38) is that of conservation of mass. We have *not* assumed, for example, that the fluid is incompressible; that is, $\rho =$ constant. For the case of constant density, Eq. (3-38) remains unchanged. In this format the dynamic equations are known as *Cauchy's stress equations.*

Some comments on the terms that make up the left-hand side of Eq. (3-38) are in order. Physically, the term $\rho \, \partial u_i / \partial t$ represents the contribution to the momentum balance due to acceleration. It is analogous to the $d(mv)/dt$ term in point-mass or rigid-body mechanics. The term $\rho u_j \, \partial u_i / \partial x_j$ represents the contribution of convection to the momentum balance. We note that the convective terms are nonlinear in velocity; they involve a second-order product of velocity and velocity derivative. This nonlinearity presents a major problem in analysis of flows, since we cannot usually find analytical solutions of nonlinear equations and must resort to numerical methods if an exact solution is needed. We will find, however, that in many flows of particularly simple geometry the convective terms are exactly zero.

Because the left-hand side of the dynamic equations is proportional to mass (really to density ρ) times velocity (and velocity derivatives), these terms are some measure of the inertia associated with elements of fluid, and we refer to these as the *inertial terms* of the dynamic equations.

It is also possible to present an alternate interpretation of the inertial terms. Suppose we could isolate a particle of fluid, and move with it through the flow field, and observe changes in velocity while in this moving, or *lagrangian,* coordinate system. Then we would recognize changes due to two factors: Our velocity might change with time because there are real unsteady-state effects imposed on the flow, and in addition our velocity would *appear* to change with time because we would be moving spatially through regions of different velocity. But we could not separate these two effects immediately, without additional information.

Thus we would report a time derivative du/dt following the motion, i.e., in a lagrangian coordinate system. Later we might recognize that u was a function of t and position x. Just from the rules of calculus we could then say that if $\mathbf{u} = \mathbf{u}(t, \mathbf{x})$,

$$\frac{d\mathbf{u}}{dt} = \left(\frac{\partial \mathbf{u}}{\partial t}\right)_{\mathbf{x}} + \left(\frac{\partial \mathbf{u}}{\partial x_j}\right)_t \frac{dx_j}{dt} \qquad (Remember\colon \text{ Sum over } j) \qquad (3\text{-}39)$$

But if we move with the flow, then dx_j/dt is simply u_j, so

$$\frac{D\mathbf{u}}{Dt} = \frac{\partial \mathbf{u}}{\partial t} + u_j \frac{\partial \mathbf{u}}{\partial x_j} \qquad (3\text{-}40)$$

We change the notation to D/Dt because we have gone from the general *chain rule* of calculus in Eq. (3-39) to a special case, called the *derivative following the motion*, or *Stokes' derivative*. We note also that in Eq. (3-40) we have indicated that we are differentiating the *vector* $\mathbf{u}$. In fact, Eq. (3-40) is just a shorthand notation for three equations represented by

$$\frac{Du_i}{Dt} = \frac{\partial u_i}{\partial t} + u_j \frac{\partial u_i}{\partial x_j} \qquad \text{sum over } j \qquad (3\text{-}41)$$

which is now recognizable as the inertial terms of the dynamic equations.

Hence we often use the simpler *format*

$$\rho \frac{Du_i}{Dt} = \frac{\partial T_{ij}}{\partial x_j} + \rho f_i \qquad i = 1, 2, 3 \qquad (3\text{-}42)$$

to represent the three dynamic equations, or Cauchy's stress equations.

Another aspect of *format*, as opposed to *physical content*, has to do with our choice of coordinate system. The derivations given above have used a rectangular parallelepiped as a volume element, and cartesian coordinates as the basis for writing components of the vectors and tensors that describe velocity, force, and stress. We could derive the dynamic equations and the continuity equation in other coordinate systems as well. (See Prob. 3-3.) We will find that for most problem formulations there is a best choice of coordinate system, one which simplifies the format of the mathematical problem and in doing so renders the method of solution somewhat simpler.

In Table 3-1 we present the dynamic and continuity equations (for incompressible fluids) in the three most useful coordinate systems: cartesian, cylindrical, and spherical.

Table 3-1 Dynamic and continuity equations

Cartesian coordinates (x, y, z)

The dynamic equations:

x component $\quad \rho\left(\dfrac{\partial u_x}{\partial t} + u_x \dfrac{\partial u_x}{\partial x} + u_y \dfrac{\partial u_x}{\partial y} + u_z \dfrac{\partial u_x}{\partial z}\right) = -\dfrac{\partial p}{\partial x}$

$$+ \left(\dfrac{\partial \tau_{xx}}{\partial x} + \dfrac{\partial \tau_{yx}}{\partial y} + \dfrac{\partial \tau_{zx}}{\partial z}\right) + \rho g_x \quad (A)$$

y component $\quad \rho\left(\dfrac{\partial u_y}{\partial t} + u_x \dfrac{\partial u_y}{\partial x} + u_y \dfrac{\partial u_y}{\partial y} + u_z \dfrac{\partial u_y}{\partial z}\right) = -\dfrac{\partial p}{\partial y}$

$$+ \left(\dfrac{\partial \tau_{xy}}{\partial x} + \dfrac{\partial \tau_{yy}}{\partial y} + \dfrac{\partial \tau_{zy}}{\partial z}\right) + \rho g_y \quad (B)$$

z component $\quad \rho\left(\dfrac{\partial u_z}{\partial t} + u_x \dfrac{\partial u_z}{\partial x} + u_y \dfrac{\partial u_z}{\partial y} + u_z \dfrac{\partial u_z}{\partial z}\right) = -\dfrac{\partial p}{\partial z}$

$$+ \left(\dfrac{\partial \tau_{xz}}{\partial x} + \dfrac{\partial \tau_{yz}}{\partial y} + \dfrac{\partial \tau_{zz}}{\partial z}\right) + \rho g_z \quad (C)$$

The continuity equation:

$$\dfrac{\partial u_x}{\partial x} + \dfrac{\partial u_y}{\partial y} + \dfrac{\partial u_z}{\partial z} = 0$$

Cylindrical coordinates (r, θ, z)

The dynamic equations:

r component $\quad \rho\left(\dfrac{\partial u_r}{\partial t} + u_r \dfrac{\partial u_r}{\partial r} + \dfrac{u_\theta}{r}\dfrac{\partial u_r}{\partial \theta} - \dfrac{u_\theta^2}{r} + u_z \dfrac{\partial u_r}{\partial z}\right) = -\dfrac{\partial p}{\partial r}$

$$+ \left(\dfrac{1}{r}\dfrac{\partial}{\partial r}(r\tau_{rr}) + \dfrac{1}{r}\dfrac{\partial \tau_{r\theta}}{\partial \theta} - \dfrac{\tau_{\theta\theta}}{r} + \dfrac{\partial \tau_{rz}}{\partial z}\right) + \rho g_r \quad (A)$$

θ component $\quad \rho\left(\dfrac{\partial u_\theta}{\partial t} + u_r \dfrac{\partial u_\theta}{\partial r} + \dfrac{u_\theta}{r}\dfrac{\partial u_\theta}{\partial \theta} + \dfrac{u_r u_\theta}{r} + u_z \dfrac{\partial u_\theta}{\partial z}\right) = -\dfrac{1}{r}\dfrac{\partial p}{\partial \theta}$

$$+ \left(\dfrac{1}{r^2}\dfrac{\partial}{\partial r}(r^2\tau_{r\theta}) + \dfrac{1}{r}\dfrac{\partial \tau_{\theta\theta}}{\partial \theta} + \dfrac{\partial \tau_{\theta z}}{\partial z}\right) + \rho g_\theta \quad (B)$$

z component $\quad \rho\left(\dfrac{\partial u_z}{\partial t} + u_r \dfrac{\partial u_z}{\partial r} + \dfrac{u_\theta}{r}\dfrac{\partial u_z}{\partial \theta} + u_z \dfrac{\partial u_z}{\partial z}\right) = -\dfrac{\partial p}{\partial z}$

$$+ \left(\dfrac{1}{r}\dfrac{\partial}{\partial r}(r\tau_{rz}) + \dfrac{1}{r}\dfrac{\partial \tau_{\theta z}}{\partial \theta} + \dfrac{\partial \tau_{zz}}{\partial z}\right) + \rho g_z \quad (C)$$

The continuity equation:

$$\dfrac{1}{r}\dfrac{\partial}{\partial r}(ru_r) + \dfrac{1}{r}\dfrac{\partial u_\theta}{\partial \theta} + \dfrac{\partial u_z}{\partial z} = 0$$

Continued

Table 3-1—*Continued*

Spherical coordinates (r, θ, ϕ)

The dynamic equations:

r component
$$\rho\left(\frac{\partial u_r}{\partial t} + u_r \frac{\partial u_r}{\partial r} + \frac{u_\theta}{r}\frac{\partial u_r}{\partial \theta} + \frac{u_\phi}{r \sin\theta}\frac{\partial u_r}{\partial \phi} - \frac{u_\theta^2 + u_\phi^2}{r}\right)$$

$$= -\frac{\partial p}{\partial r} + \left(\frac{1}{r^2}\frac{\partial}{\partial r}(r^2 \tau_{rr}) + \frac{1}{r \sin\theta}\frac{\partial}{\partial \theta}(\tau_{r\theta}\sin\theta) + \frac{1}{r \sin\theta}\frac{\partial \tau_{r\phi}}{\partial \phi}\right.$$

$$\left. - \frac{\tau_{\theta\theta} + \tau_{\phi\phi}}{r}\right) + \rho g_r \quad (A)$$

θ component
$$\rho\left(\frac{\partial u_\theta}{\partial t} + u_r \frac{\partial u_\theta}{\partial r} + \frac{u_\theta}{r}\frac{\partial u_\theta}{\partial \theta} + \frac{u_\phi}{r \sin\theta}\frac{\partial u_\theta}{\partial \phi} + \frac{u_r u_\theta}{r} - \frac{u_\phi^2 \cot\theta}{r}\right)$$

$$= -\frac{1}{r}\frac{\partial p}{\partial \theta} + \left(\frac{1}{r^2}\frac{\partial}{\partial r}(r^2 \tau_{r\theta}) + \frac{1}{r \sin\theta}\frac{\partial}{\partial \theta}(\tau_{\theta\theta}\sin\theta)\right.$$

$$\left. + \frac{1}{r \sin\theta}\frac{\partial \tau_{\theta\phi}}{\partial \phi} + \frac{\tau_{r\theta}}{r} - \frac{\cot\theta}{r}\tau_{\phi\phi}\right) + \rho g_\theta \quad (B)$$

ϕ component
$$\rho\left(\frac{\partial u_\phi}{\partial t} + u_r \frac{\partial u_\phi}{\partial r} + \frac{u_\theta}{r}\frac{\partial u_\phi}{\partial \theta} + \frac{u_\phi}{r \sin\theta}\frac{\partial u_\phi}{\partial \phi} + \frac{u_\phi u_r}{r} + \frac{u_\theta u_\phi}{r}\cot\theta\right)$$

$$= -\frac{1}{r \sin\theta}\frac{\partial p}{\partial \phi} + \left(\frac{1}{r^2}\frac{\partial}{\partial r}(r^2 \tau_{r\phi}) + \frac{1}{r}\frac{\partial \tau_{\theta\phi}}{\partial \theta} + \frac{1}{r \sin\theta}\frac{\partial \tau_{\phi\phi}}{\partial \phi}\right.$$

$$\left. + \frac{\tau_{r\phi}}{r} + \frac{2\cot\theta}{r}\tau_{\theta\phi}\right) + \rho g_\phi \quad (C)$$

The continuity equation:

$$\frac{1}{r^2}\frac{\partial}{\partial r}(r^2 u_r) + \frac{1}{r \sin\theta}\frac{\partial}{\partial \theta}(u_\theta \sin\theta) + \frac{1}{r \sin\theta}\frac{\partial u_\phi}{\partial \phi} = 0$$

3-5 KINEMATICS

Kinematics refers to the analysis and description of motion. By contrast, *dynamics* is concerned with the relationship of motion to the *forces* causing, or accompanying, the motion. In this section we establish some purely kinematical ideas.

Let us consider a continuous medium in motion, characterized by a velocity vector **u** having components u_i in a cartesian coordinate system **x** centered at some point O. We will denote the components of velocity at the point O by u_{0i}. In the near neighborhood of O we expect that the velocity components can be expressed in terms of the kinematics *at* O by using Taylor's expansion, with the result that

$$u_i = u_{0i} + \left(\frac{\partial u_i}{\partial x_j}\right)_0 dx_j + O(dx_j)^2 \quad \text{sum over } j \quad (3\text{-}43)$$

Now we want to answer the following question: What kinematic quantities are necessary to tell us if there is *deformation* in the neighborhood of O? We call $\partial u_i / \partial x_j$ the velocity gradient of u_i in the x_j direction or, simply, the velocity gradient. For small dx_j we can neglect terms of order $(dx_j)^2$. Then the *relative motion* between material *at O* and material *near O* is

$$u_i - u_{Oi} = \left(\frac{\partial u_i}{\partial x_j}\right)_O dx_j \tag{3-44}$$

Now, for reasons that are clear only to professors, let us decompose the velocity gradient into the following arbitrary form:

$$\frac{\partial u_i}{\partial x_j} = \frac{1}{2}\left(\frac{\partial u_i}{\partial x_j} + \frac{\partial u_j}{\partial x_i}\right) + \frac{1}{2}\left(\frac{\partial u_i}{\partial x_j} - \frac{\partial u_j}{\partial x_i}\right) \tag{3-45}$$

If we define

$$\Delta_{ij} = \left(\frac{\partial u_i}{\partial x_j} + \frac{\partial u_j}{\partial x_i}\right) \tag{3-46}$$

and

$$\omega_{ij} = \left(\frac{\partial u_i}{\partial x_j} - \frac{\partial u_j}{\partial x_i}\right) \tag{3-47}$$

then we may write

$$u_i = u_{Oi} = \tfrac{1}{2}\,\Delta_{ij} + \tfrac{1}{2}\omega_{ij}\,dx_j \tag{3-48}$$

Thus we have decomposed the relative velocity into two parts. Why?

Consider first the motion associated with ω_{ij}, which we call $(du_i)_\omega$ and define by

$$(du_i)_\omega = \tfrac{1}{2}\omega_{ij}\,dx_j \tag{3-49}$$

Multiply both sides of this equation by dx_i to find

$$(du_i)_\omega\,dx_i = \tfrac{1}{2}\omega_{ij}\,dx_j\,dx_i \tag{3-50}$$

Keep in mind our summation convention; we must sum over repeated subscripts. Note that in Eq. (3-50) we find, on the right-hand side, a *pair* of repeated subscripts, so that there is a double summation implied there.

Since the subscript notation just serves as an index system for counting in the summation, the symbol used (format again) is irrelevant. We could change i and j to any other pair of symbols and still have the same meaning. In fact, we could change i to j and j to i, and still find that

$$\omega_{ij}\,dx_i\,dx_j = \omega_{ji}\,dx_j\,dx_i \tag{3-51}$$

But, from its definition in Eq. (3-47), we can see that

$$\omega_{ij} = -\omega_{ji} \tag{3-52}$$

Thus

$$\omega_{ij}\,dx_i\,dx_j = -\omega_{ij}\,dx_j\,dx_i \tag{3-53}$$

But this is possible only if both terms vanish, because, format aside, we are claiming some number equals its own negative. We conclude then that

$$\omega_{ij}\,dx_i\,dx_j = 0 \tag{3-54}$$

Referring to Eq. (3-49), we see that this implies

$$2(du_i)_\omega\,dx_i = 0 \tag{3-55}$$

But Eq. (3-55) is the scalar product of the vectors $(d\mathbf{u})_\omega$ and $d\mathbf{x}$:

$$2(d\mathbf{u})_\omega \cdot d\mathbf{x} = 0 \tag{3-56}$$

We are now within a few "buts" of our goal.

But the vanishing of a scalar product is just the consequence of the vectors being mutually perpendicular. If we recall that $d\mathbf{x}$ is just the position vector connecting the two points of interest, we conclude that the part of the motion

Table 3-2 Components of the rate-of-deformation tensor

Cartesian coordinates (x, y, z):

$$\Delta_{xx} = 2\frac{\partial u_x}{\partial x} \qquad \Delta_{xy} = \Delta_{yx} = \frac{\partial u_x}{\partial y} + \frac{\partial u_y}{\partial x}$$

$$\Delta_{yy} = 2\frac{\partial u_y}{\partial y} \qquad \Delta_{xz} = \Delta_{zx} = \frac{\partial u_x}{\partial z} + \frac{\partial u_z}{\partial x}$$

$$\Delta_{zz} = 2\frac{\partial u_z}{\partial z} \qquad \Delta_{yz} = \Delta_{zy} = \frac{\partial u_y}{\partial z} + \frac{\partial u_z}{\partial y}$$

Cylindrical coordinates (r, θ, z):

$$\Delta_{rr} = 2\frac{\partial u_r}{\partial r} \qquad \Delta_{r\theta} = \Delta_{\theta r} = r\frac{\partial}{\partial r}\left(\frac{u_\theta}{r}\right) + \frac{1}{r}\frac{\partial u_r}{\partial \theta}$$

$$\Delta_{\theta\theta} = 2\left(\frac{1}{r}\frac{\partial u_\theta}{\partial \theta} + \frac{u_r}{r}\right) \qquad \Delta_{\theta z} = \Delta_{z\theta} = \frac{\partial u_\theta}{\partial z} + \frac{1}{r}\frac{\partial u_z}{\partial \theta}$$

$$\Delta_{zz} = 2\frac{\partial u_z}{\partial z} \qquad \Delta_{zr} = \Delta_{rz} = \frac{\partial u_z}{\partial r} + \frac{\partial u_r}{\partial z}$$

Spherical coordinates (r, θ, ϕ):

$$\Delta_{rr} = 2\frac{\partial u_r}{\partial r} \qquad \Delta_{\theta r} = \Delta_{r\theta} = r\frac{\partial}{\partial r}\left(\frac{u_\theta}{r}\right) + \frac{1}{r}\frac{\partial u_r}{\partial \theta}$$

$$\Delta_{\theta\theta} = 2\left(\frac{1}{r}\frac{\partial u_\theta}{\partial \theta} + \frac{u_r}{r}\right) \qquad \Delta_{\phi\theta} = \Delta_{\theta\phi} = \frac{\sin\theta}{r}\frac{\partial}{\partial \theta}\left(\frac{u_\phi}{\sin\theta}\right) + \frac{1}{r\sin\theta}\frac{\partial u_\theta}{\partial \phi}$$

$$\Delta_{\phi\phi} = 2\left(\frac{1}{r\sin\theta}\frac{\partial u_\phi}{\partial \phi} + \frac{u_r}{r} + \frac{u_\theta\cot\theta}{r}\right) \qquad \Delta_{r\phi} = \Delta_{\phi r} = \frac{1}{r\sin\theta}\frac{\partial u_r}{\partial \phi} + r\frac{\partial}{\partial r}\left(\frac{u_\phi}{r}\right)$$

described by $(d\mathbf{u})_\omega$ is just motion perpendicular to the line connecting the points, which is simply *rigid rotation.*

Hence we have proven that the relative motion between two points in a flowing system, describable in terms of the velocity gradient components $\partial u_i/\partial x_j$, can be arbitrarily decomposed into a rigid *rotation* that depends on ω_{ij} and into a *deformation* whose components are given by Δ_{ij}.

We refer to Δ_{ij} as the components of the rate-of-deformation tensor. The importance of Δ_{ij} lies in the fact that we will assert, in Sec. 3-7, that the stress components τ_{ij} depend *only* on the rate of deformation components Δ_{ij} for a wide class of materials. In other words, it is Δ_{ij} which provides the essential connection between the dynamics and the kinematics of motion in a fluid.

Equation (3-46) gives the components Δ_{ij} of the rate-of-deformation tensor in cartesian coordinates. Table 3-2 gives the components in the three commonly used coordinate systems.

3-6 BOUNDARY CONDITIONS

The real flows of interest to us are of finite extent; mathematically we would speak of them as *bounded* flows. *Physically*, of course, they are bounded as well. Sometimes we have rigid boundaries imposed by the container or conduit, as in the case of the wall of a pipe. Sometimes we have *free* boundaries, as in the case of a jet of liquid issuing from the end of a capillary. On occasion we treat two-fluid problems, where a pair of immiscible liquids undergoes some deformation, and the interface between the pair of fluids is an unknown boundary whose position must be determined.

Certain physical principles dictate the conditions which must hold at the flow boundaries, and these boundary conditions must be specified before the dynamic equations can be solved. These conditions may be either kinematic or dynamic.

Kinematic Boundary Conditions: The No-slip Condition

Consider some solid moving relative to a fluid, and let the velocity of the solid be $\mathbf{U}_s$. It might be useful to think of the solid, just for the sake of concrete example, as the blade of an agitator immersed in a liquid.

The *no-slip* condition is a *model* which *asserts* that, at the solid-fluid boundary, there is no relative motion. This means that the fluid velocity vector $\mathbf{u}$ must satisfy the boundary condition

$$\mathbf{u} = \mathbf{U}_s \qquad \text{at the solid-fluid boundary} \qquad (3\text{-}57)$$

Let us look at an example that will occur later in the analysis of extrusion of molten polymers. The example also will serve to emphasize the *model* concept, particularly the idea that one must be aware that the *model* imposes conditions on the solution that are not necessarily inherent in the *real* system.

Figure 3-8 Two-dimensional rectangular channel with a moving upper surface.

Consider fluid enclosed in a two-dimensional rectangular channel, as shown in Fig. 3-8. The term *two-dimensional* refers to the assertion that no variations in the z direction occur, so that what happens in the xy plane is representative of the dynamics of the three-dimensional system. The upper surface moves at the velocity U in its own plane. The other three surfaces of the channel are stationary.

The no-slip condition implies the following:

$$
\begin{aligned}
u_x = 0 \qquad &\text{on} \begin{cases} x = 0 & 0 \le y < H \\ y = 0 & 0 \le x \le W \\ x = W & 0 \le y < H \end{cases} \\
u_y = 0 \\[1em]
u_x = U \qquad &\text{on } y = H \qquad 0 < x < W \\
u_y = 0
\end{aligned}
\tag{3-58}
$$

Note that we have left the velocity undefined at the upper corners, i.e., at $x, y = 0, H$ and W, H. As far as the *model* is concerned there is a discontinuity in velocity in the upper corners. As we approach the corner along a vertical boundary, the velocity is $u_x = 0$. As we approach the corner along the upper horizontal (and moving) boundary, the velocity is $u_x = U$.

Thus the model includes, because of the boundary conditions, a condition that is not physically realizable. This is what we might call a "mathematical artifact." It is important to recognize potential problems that may arise when simplifications motivated by the need to reduce the mathematical complexity are at variance with the physical reality of the system. In this specific case, we will find that no serious problem arises from the discontinuity at the corners.

Boundary conditions do not necessarily give values of the variables in terms of known quantities, as in the case treated above. Consider, as another example, the flow generated by the motion of a jet of one liquid through a second liquid in which it is immiscible. Assume the physical system is qualitatively as shown in Fig. 3-9.

Fluid 2 is contained within a cylindrical vessel. Fluid 1, immiscible in 2, and of higher density, is pumped from a circular inlet at A and flows down toward the outlet at B. It is assumed that only fluid 1 enters and leaves the system. Fluid 2 is entrained and caused to circulate by the motion of fluid 1. The geometry is assumed to be cylindrical, with the inlet and outlet both coaxial with the cylinder containing fluid 2.

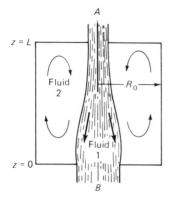

Figure 3-9 Sketch for analysis of a two-fluid problem.

Let us examine kinematic boundary conditions for this problem. For fluid 2, no-slip conditions imply

$$\mathbf{u}_2 = 0 \qquad \text{on} \begin{cases} r = R_0 & 0 \le z \le L \\ z = 0 & R_B \le r \le R_0 \\ z = L & R_A \le r \le R_0 \end{cases} \qquad (3\text{-}59)$$

Fluid 2 is also bounded by fluid 1, and the no-slip condition in this case states equality of velocity on either side of the interface:

$$\mathbf{u}_2 = \mathbf{u}_1 \qquad \text{on the interfacial boundary} \qquad (3\text{-}60)$$

Two points must be noted regarding this boundary condition. One is that the position of the interface is not known a priori; it is actually part of the solution being sought. The other point is that the velocity is not *specified*, it is simply *constrained*. In other words, while neither $\mathbf{u}_1$ nor $\mathbf{u}_2$ is known, they are constrained to be equal at their common boundary. In general, problems having boundary conditions like this are very difficult to solve.

Between the entrance and exit, fluid 1 contacts no solid boundary. However, there is a kinematic condition that can be imposed on fluid 1, arising from the assumed symmetry about the axis of the cylindrical coordinate system. Symmetry implies equality of the velocity profile on either side of the axis. The simplest mathematical formulation of this geometrical notion is the vanishing of the radial gradient of the axial velocity,

$$\left.\frac{\partial u_{1z}}{\partial r}\right|_{r=0} = 0 \qquad (3\text{-}61)$$

and the vanishing of the radial velocity component along the axis,

$$\left. u_{1r} \right|_{r=0} = 0 \qquad (3\text{-}62)$$

These are the only *kinematic* conditions that can be imposed in this example. Additional boundary conditions must be based on dynamic considerations.

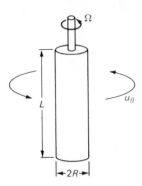

Figure 3-10 Cylinder rotating in a fluid.

Dynamic Boundary Conditions

Sometimes the *stresses* acting at flow boundaries are specified in the formulation of a problem. For example, consider a cylinder immersed in a fluid, as in Fig. 3-10. A torque $\mathcal{T}$ is applied, causing the cylinder to rotate about its axis at an angular velocity Ω.

The torque acting on the cylindrical surface of radius R imparts a reactive torque (in the sense of Newton's third law) to the surrounding fluid, thus setting the fluid in motion. The fluid, then, exerts a torque $-\mathcal{T}$ to the surface at R. But the torque is simply the shear force $T_{r\theta}|_R 2\pi RL$ times the "moment arm" R. (We ignore the forces acting on the circular faces normal to the cylinder axis, which would be small for a long cylinder, $L \gg R$). Thus a dynamic condition on the fluid, at $r = R$, takes the form

$$T_{r\theta}\bigg|_R = -\frac{\mathcal{T}}{2\pi R^2 L} \tag{3-63}$$

i.e., the shear stress is specified at the boundary.

A kinematic condition would be

$$u_\theta\bigg|_R = R\Omega \qquad (\Omega \text{ in rad/s}) \tag{3-64}$$

However, these two conditions are not independent. We cannot independently specify *both* the torque and the rotational speed. One will determine the other. In some problem formulations the kinematics are specified; in others the dynamics are taken as given.

Another type of dynamic boundary condition occurs at a free boundary separating two fluids. The condition follows from the physical statement that the stresses must be continuous across a boundary. Consider the flow of liquid entrained by the surface of a sheet being withdrawn from the surface of the liquid, as in Fig. 3-11. This type of problem occurs in the analysis of coating dynamics, and we will later carry out such an analysis. There is, of course, a simple kinematic

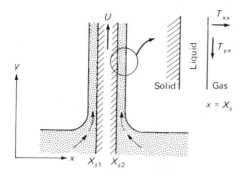

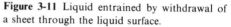

Figure 3-11 Liquid entrained by withdrawal of a sheet through the liquid surface.

condition to be satisfied in this problem, namely, no-slip at the interface between the liquid and the moving surface of the sheet:

$$u_x = 0 \quad \text{and} \quad u_y = U \quad \text{at } x = X_{s1}, X_{s2} \tag{3-65}$$

Let us consider a region of the sheet up above the dynamic meniscus formed near the surface. In that region the liquid coating is assumed to be approximately parallel to the sheet, i.e., of uniform thickness. We ask what kinds of stresses act on the outer coating surface, the interface between the liquid and the ambient gas.

The gas can exert two kinds of stresses on the boundary. A shear stress T_{yx} will be imposed since the gas has a finite (though small) viscosity and so resists the motion of the sheet and coating through it. Continuity of shear stress then takes the form

$$\text{Liquid } T_{yx}\bigg|_{X_s} = \text{gas } T_{yx}\bigg|_{X_s} \tag{3-66}$$

A common *model*, however, follows from the assertion that unless the speed U is quite high, a low-viscosity fluid such as a gas is not capable of exerting a significant shear stress. Hence a common form of shear-stress boundary condition is

$$T_{yx}\bigg|_{X_s} = 0 \quad \text{at the gas-liquid boundary} \tag{3-67}$$

Normal stresses do not necessarily vanish, since the external fluid is usually capable of exerting a pressure on the boundary. Thus another appropriate boundary condition for this problem would take the form

$$T_{xx}\bigg|_{X_s} = -p \tag{3-68}$$

The minus sign on pressure follows from our earlier convention on stress.

To review briefly, we have asserted that appropriate *dynamic* boundary conditions follow from the principle of continuity of stress across a boundary, whereas appropriate *kinematic* boundary conditions reflect the notion that velocity must be continuous at a boundary.

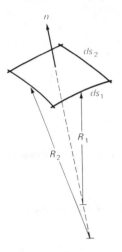

Figure 3-12 Differential area in the surface separating two immiscible fluids.

Interfacial Tension

When we talk of a "boundary" between two fluids we speak mathematically. We imply that it is possible to identify one fluid or the other on either side of a surface. If the fluids are miscible, the concept loses its utility, because the fluids will interpenetrate, by diffusion and convection, and no clearly defined boundary separates them.

If the fluids are immiscible, then a finite interfacial tension acts in a well-defined boundary of separation. We can show that this interfacial tension can give rise to a pressure *difference* across an interface, and so the dynamic condition on continuity of normal stress must reflect this fact.

We begin by relating the pressure difference due to interfacial tension to the geometry of the surface. Figure 3-12 shows a differential element of area within the boundary surface separating a pair of immiscible fluids. The area element is bounded by two pairs of parallel arcs of radii R_1 and R_2. If ds_1 is small enough, it can always be represented as the arc of some circle of radius R_1 and included angle $d\theta_1$.

Now, how does interfacial tension manifest itself? The interfacial tension σ can be thought of as a force per unit length acting across a line element in a

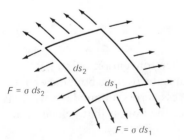

Figure 3-13 Surface tension forces acting along the edges of the differential area.

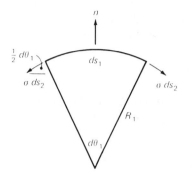

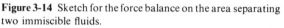

Figure 3-14 Sketch for the force balance on the area separating two immiscible fluids.

surface, the line of action of the force being normal to the line, and tangential to the surface.

If we isolate the differential area for the purpose of writing a force balance, we may consider it in terms of Fig. 3-13. Along each arc a distributed force acts uniformly, of magnitude $\sigma \, ds$. Let us now sum the components of these four forces in the direction of the normal **n** to the surface. The **n** component of those forces acting along the lines of length ds_2 is most easily calculated by using Fig. 3-14, which is a view normal to ds_1 and parallel to ds_2. Along each of the lines ds_2 is a force $\sigma \, ds_2$ tangent to the surface. The **n** component of each force is simply $(\sigma \, ds_2) \sin \left(\frac{1}{2} \, d\theta_1\right)$.

The angle $d\theta_1$ is related to the arc ds_1 through the radius of curvature of the arc, R_1:

$$ds_1 = R_1 \, d\theta_1 \tag{3-69}$$

For differential (i.e., small) angles, a good approximation is

$$\sin \left(\tfrac{1}{2} \, d\theta_1\right) = \tfrac{1}{2} \, d\theta_1 \tag{3-70}$$

Using these results, and summing the *two* forces along the pair of ds_2 lines, we find

$$F_2 = \frac{\sigma}{R_1} \, ds_1 \, ds_2 \tag{3-71}$$

By identical argument, the **n** component of forces acting along the pair of ds_1 lines is

$$F_1 = \frac{\sigma}{R_2} \, ds_1 \, ds_2 \tag{3-72}$$

Thus the net **n** component of force due to interfacial tension is

$$F_\sigma = \sigma \, ds_1 \, ds_2 \left(\frac{1}{R_1} + \frac{1}{R_2}\right) \tag{3-73}$$

If the fluids are *static*, then the only stresses on either side of the surface separating the two fluids are hydrostatic pressures, say, p_0 and p_i, where p_i is the pressure on the concave side ("inside") of the interface. These pressures also give rise to forces in the **n** direction of magnitudes $p_0 \, ds_1 \, ds_2$ and $-p_i \, ds_1 \, ds_2$. Hence, at equilibrium, the force balance on the differential surface results in

$$\left[p_0 - p_i + \sigma\left(\frac{1}{R_1} + \frac{1}{R_2} \right) \right] ds_1 \, ds_2 = 0 \tag{3-74}$$

In the limit as the differential area shrinks to a point, Eq. (3-74) gives the condition of equilibrium of normal stresses across an interface separating a pair of immiscible fluids:

$$p_i - p_0 = \sigma\left(\frac{1}{R_1} + \frac{1}{R_2} \right) \tag{3-75}$$

Physically, the equation tells us that interfacial tension causes an increased pressure on the "inside" of a surface, the magnitude depending on the radii of curvature of the surface. The geometrical boundary separating two immiscible fluids defines a region across which there is a discontinuity in fluid physical properties. Equation (3-75) shows that there is a corresponding discontinuity in normal stress across the boundary. Thus, to be precise, the dynamic boundary condition on normal stress is not one of continuity in the case of the boundary separating *immiscible fluids*, but rather one in which the normal stresses within each fluid on either side of the boundary differ at the boundary by a hydrostatic pressure given by Eq. (3-75).

Let us end this section with an application of some of these concepts, and motivate as well the need for information that will be presented in Sec. 3-7. We suppose that a spherical gas bubble of radius $R(t)$ is expanding within a large body of fluid. We seek a relationship between the pressure inside the bubble, P, and the rate of growth.

For our model of this process, let us assume that the gas has an insignificant viscosity, that the stresses in the region $0 \leq r < R(t)$ are strictly hydrostatic and uniform with respect to r in that region, and that gravitational effects are of no significance.

In the liquid outside the bubble, i.e., in the region $r > R(t)$, the velocity must satisfy the continuity equation, which we naturally choose to write in spherical coordinates for this problem:

$$\frac{1}{r^2} \frac{\partial}{\partial r} (r^2 u_r) + \frac{1}{r \sin \theta} \left[\frac{\partial}{\partial \theta} (u_\theta \sin \theta) + \frac{\partial u_\varphi}{\partial \varphi} \right] = 0 \tag{3-76}$$

If the flow is spherically symmetric, by which we mean that

$$\mathbf{u} = [u_r(r, t), 0, 0] \tag{3-77}$$

then the continuity equation gives the result

$$u_r = \frac{A}{r^2} \tag{3-78}$$

where A might be a function of time but is independent of the space variables.

The kinematic boundary condition at the bubble surface is continuity of the radial velocity:

$$u_r\bigg|_R = \frac{dR}{dt} = \dot{R} \qquad (3\text{-}79)$$

It follows then that

$$u_r = \frac{\dot{R}R^2}{r^2} \qquad (3\text{-}80)$$

Now let us examine another kinematic feature of the flow, the rate-of-deformation components, which in this case take the form (see Table 3-2)

$$\Delta_{rr} = -\frac{4\dot{R}R^2}{r^3}$$

$$\Delta_{\theta\theta} = \Delta_{\varphi\varphi} = \frac{2\dot{R}R^2}{r^3} \qquad (3\text{-}81)$$

All other $\Delta_{ij} = 0$

The existence of finite deformation rates in the liquid suggests that there will be stresses accompanying the deformation. Now we must relate the stresses associated with deformation to the deformation rates. We do so by introducing a model which is mathematically simple and consistent with physical experience.

We postulate the following: A particular stress component T_{ij} is related only to the corresponding rate-of-deformation component Δ_{ij}, and the relationship is linear. Thus we write

$$T_{ij} = -p\,\delta_{ij} + \mu\,\Delta_{ij} \qquad (3\text{-}82)$$

where δ_{ij} is defined so that $\delta_{ij} \equiv 1$ if $i = j$, and $\delta_{ij} \equiv 0$ if $i \neq j$. The term $-p\,\delta_{ij}$ reflects the observation that in the absence of deformation, when $\Delta \equiv \mathbf{0}$, it is still possible for normal stresses to exist. In a stationary fluid the normal stresses would correspond to a hydrostatic pressure, and our sign convention on stress leads us to write this as $-p$.

The coefficient μ is the viscosity of the fluid. At this stage we may regard Eq. (3-82) as a useful *postulate*, in that it will let us arrive at a result which can be subjected to experimental test. If the analysis does not lead to a realistic prediction of reality, we must then reexamine the postulates, the elements of the model, which led to the false prediction and attempt to improve the model.

Readers familiar with fluid dynamics will recognize Eq. (3-82) as a form of Newton's law of viscosity. Thus the previous few paragraphs could have been replaced by the statement: Let us assume, as part of this model, that the fluid outside the expanding bubble is newtonian.

We can now calculate T_{rr} in the outer fluid and find

$$T_{rr} = -p - \frac{4\mu\dot{R}R^2}{r^3} \qquad (3\text{-}83)$$

while the other two normal stresses are

$$T_{\theta\theta} = T_{\varphi\varphi} = -p + \frac{2\mu\dot{R}R^2}{r^3} \tag{3-84}$$

It would seem that the next step toward finding the pressure inside the bubble would be to impose the dynamic condition on continuity of *radial* stress, since the pressure P is, by definition, normal to surfaces, and so in this case directed radially. If the effect of surface tension is included, we find, at $r = R(t)$,

$$-P + \frac{2\sigma}{R} = T_{rr}\bigg|_R = -p_R - \frac{4\mu\dot{R}}{R} \tag{3-85}$$

[Note that if there were no deformation, this would give $2\sigma/R = P - p_R$, which is identical to Eq. (3-75).]

Recall now that our goal, as stated, is that we seek a relationship between pressure inside the bubble (P) and the rate of growth $(\dot{R}/R)$. Equation (3-85) does not quite achieve this, because we do not know the pressure $p(r)$ within the outer fluid, from which $p(R) = p_R$ would be known. How do we find $p(r)$?

We must now use the dynamic equations.

With the assumptions already made, we can show that the radial component of the dynamic equations takes the form (neglecting any effect of gravity)

$$\rho\left(\frac{\partial u_r}{\partial t} + u_r \frac{\partial u_r}{\partial r}\right) = \frac{1}{r^2}\frac{\partial}{\partial r}(r^2 T_{rr}) - \frac{T_{\theta\theta} + T_{\varphi\varphi}}{r} \tag{3-86}$$

If Eq. (3-80) for u_r and Eqs. (3-83) and (3-84) for the stresses are introduced, one can show that (see Prob. 3-8)

$$p_R = p_\infty + \rho(\ddot{R}R + \tfrac{3}{2}\dot{R}^2) \tag{3-87}$$

By p_∞ we mean the pressure far from the bubble. In a real physical system we might have a bubble in a relatively large body of fluid, and p_∞ might be taken as the ambient pressure acting on the (real finite) boundaries of the whole system.

Note that the contributions to pressure in Eq. (3-87) are independent of viscosity. If the derivation is studied (hence, Prob. 3-8) we find that these terms come out of the inertial terms of Eq. (3-86).

Our final result, then, is

$$P = \frac{2\sigma}{R} + \frac{4\mu\dot{R}}{R} + \rho(\ddot{R}R + \tfrac{3}{2}\dot{R}^2) + p_\infty \tag{3-88}$$

3-7 CONSTITUTIVE EQUATIONS

The dynamic equations embody the principle of conservation of momentum. They relate the velocity vector **u** to the stress tensor **T** and are valid for any fluid for which the continuum approximation is meaningful. There is no restriction as to whether the fluid is newtonian or nonnewtonian, viscoelastic or inelastic.

The continuity equation is strictly a *kinematic* constraint among the velocity gradients. No information about the stresses appears, and so there is no implication regarding the type of fluid for which it is valid other than with regard to the question of compressibility. In most processes of interest to us, we can regard the fluid as incompressible and use the continuity equation in the form given in the appropriate section of Table 3-1.

The boundary conditions are statements of *continuity* of the kinematic and dynamic variables, and again are not dependent on the type of fluid being considered.

Where, then, *does* the mechanical constitution of the fluid enter our analyses?

Let us first be convinced that we *do* need additional information. It seems reasonable to assert that a flow field is completely specified once the components of **u** and **T** are known as functions of the independent variables (position and time). **u** has three components, and the symmetric tensor **T** has six independent components, giving us *nine unknown functions* in all. But we have only *four* equations relating these functions: one continuity and three dynamic equations. The boundary conditions are not additional equations serving to relate the unknowns to the independent variables, since the boundary conditions are valid only at *specific* points (the boundaries) of the flow region; they are not valid *throughout* the flow field.

It would appear, then, that we are short five equations before we will even have formulated a problem which has a solution. These additional equations (there will in fact be six more) are called *constitutive equations*, and they relate components of **T** to the kinematics, usually through **Λ**, and thereby *define the type of fluid* for which a solution is being sought.

Introduction of Pressure

We begin by separating the stress tensor into two parts: a dynamic stress **τ**, related to the deformation of the fluid, and a normal stress of magnitude p. We write

$$\mathbf{T} = -p\,\boldsymbol{\delta} + \boldsymbol{\tau} \tag{3-89}$$

δ is called the *unit tensor*, because it is defined to have components of the form

$$\boldsymbol{\delta} = \begin{pmatrix} 1 & 0 & 0 \\ 0 & 1 & 0 \\ 0 & 0 & 1 \end{pmatrix} \tag{3-90}$$

Alternatively, we can write Eq. (3-89) in component form, as

$$T_{ij} = -p\,\delta_{ij} + \tau_{ij} \tag{3-91}$$

where $\delta_{ij} = 1$ if $i = j$, and $\delta_{ij} = 0$ if $i \neq j$.

Recall that the components of stress for which $i = j$ are the *normal* stresses, whereas the other components are acting in the plane of the surface on which they act and are called *shear* stresses. Thus, for shear stresses ($i \neq j$), we have $T_{ij} = \tau_{ij}$

by our definition [Eq. (3-91)]. The normal stresses T_{ij} $(i = j)$ include a contribution $- p$ which is identical in all three coordinate directions and which exists even if the dynamic stresses τ_{ij} vanish, i.e., under static conditions. The property of independence of direction is called *isotropy*.

An isotropic stress in a static fluid is what we normally call a *hydrostatic pressure*. We will still refer to p as "pressure" in Eq. (3-91), although it is not a priori obvious that it is the same pressure as defined in thermodynamics. (One reason for this uncertainty is that *deformation* implies nonequilibrium conditions, to which the principles and definitions of thermodynamics do not necessarily apply.)

Introduction of Eq. (3-89) trades the six unknown components of T for the six unknown components of τ. However, a *new* unknown is thereby introduced: the pressure p. Thus we have lost ground in attempting to balance unknown functions and equations; we are now short by *six* equations, and we shall find that these six equations will be the constitutive equations of the fluid.

Behavior of Fluids in Simple Flows

Before examining specific forms of constitutive equations, we should inquire into the types of responses these equations will have to accommodate in order to be considered useful.

Most flow processes are complex in some sense, usually in terms of the geometry of the boundaries or the motion imposed at, or by, the boundaries of the flow field. Because of such complexities, we usually *categorize* flow fields in terms of simple cases which provide a reference basis for the discussion and delineation of mechanical response. The two flows of greatest utility for this purpose are the *simple shear flow* and the *simple elongational flow*.

The simple shear flow (SSF) is defined in such a way that the rate-of-deformation tensor Δ has the form

$$\Delta = \dot{\gamma} \begin{pmatrix} 0 & 1 & 0 \\ 1 & 0 & 0 \\ 0 & 0 & 0 \end{pmatrix} \tag{3-92}$$

Note that this is a *kinematic* definition. The scalar $\dot{\gamma}$ is called the *shear rate*. This definition does not require that $\dot{\gamma}$ be a constant, although in some flows it may be. The simple shear flow can be achieved in several ways in the laboratory. For example, if fluid is confined to the annular space separating a pair of long concentric cylinders, and if one surface moves parallel to the other, then an SSF results so long as the flow remains laminar. This is true whether the cylindrical surface moves axially or rotationally. In the first case the velocity field must be $\mathbf{v} = [v_z(r), 0, 0]$, whereas in the latter case of rotational flow the velocity field must be given by $\mathbf{v} = [0, v_\theta(r), 0]$. In either case we have assumed angular symmetry (no θ dependence) and no z dependence of velocity. The latter assumption holds so long as no "end effects" occur associated with the presence of finite axial boundaries. If the cylinders are very long, by comparison to the distance across the annular gap

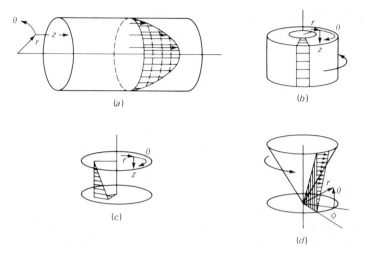

Figure 3-15 Configurations of several viscometric flows.

separating them, these end effects are found to be unimportant and to be confined to the regions near the ends of the cylinders.

Other such flows can be achieved, and because they form the basis of design of instruments for the measurement of viscosity, they are often referred to as *viscometric* flows. Figure 3-15 shows several viscometric flows.

It is possible to show that for a very general class of fluids (so-called *simple fluids*) the SSF is completely characterized by three material functions, and *if the kinematics and dynamics are at a steady state*, we may define these as Generalized viscosity coefficient:

$$\eta = \frac{T_{12}}{\dot{\gamma}} = \frac{\tau_{12}}{\dot{\gamma}} \tag{3-93}$$

primary normal stress coefficient:

$$\Psi_{12} = \frac{T_{11} - T_{22}}{\dot{\gamma}^2} \tag{3-94}$$

secondary normal stress coefficient:

$$\Psi_{23} = \frac{T_{22} - T_{33}}{\dot{\gamma}^2} \tag{3-95}$$

For steady-state SSF, these coefficients depend only on $\dot{\gamma}$.

The simple elongational flow (or simple extensional flow) (SEF) may be defined kinematically by a velocity field such that

$$\Delta = \dot{\varepsilon} \begin{pmatrix} 2 & 0 & 0 \\ 0 & -1 & 0 \\ 0 & 0 & -1 \end{pmatrix} \tag{3-96}$$

and the dynamics are completely characterized by a single material function, *defined in the steady-state* case as
Elongational (or extensional) viscosity,

$$\eta_e = \frac{T_{11} - T_{22}}{\dot{\epsilon}} \tag{3-97}$$

The scalar coefficient $\dot{\epsilon}$ is called the *principal extension rate.*

Elongational flows are more difficult to achieve in the laboratory, in part because end effects are more difficult to eliminate. In Sec. 3-6 the problem of flow surrounding an expanding gas bubble was considered. Review of that example should make it clear that this is an SEF. It is not a simple matter, however, to create such a flow under controlled conditions.

The pulling of a cylinder of material creates an elongational flow (see Prob. 3-22), but it should be clear that one cannot " pull " on a low-viscosity fluid, or "grip" it to hold one end in place. Nevertheless, some effort has gone into development of experimental techniques for creation of SEF, and bibliographical references are noted at the end of this chapter.

If one does achieve these flows in the laboratory, then one can *measure* the functions defined for them. Correspondingly, with a particular constitutive equation it is relatively easy to solve the dynamic equations for these flows and *calculate* the material functions that correspond to the constitutive equation. The comparison of *measured* and *calculated* material functions defined for simple flows is the usual means of evaluation of a constitutive equation.

Let us examine some typical experimental data to get an idea of how these material functions depend on the kinematics of the flows in which they are measured. Methods of obtaining such data are described in detail in references cited at the end of this chapter.

Figure 3-16a, b shows the viscosity as a function of shear rate for melts and solutions. Qualitatively the observations are the same: At sufficiently low shear rates most fluids exhibit a viscosity independent of shear rate; they are newtonian at low shear rate. As the shear rate increases, in most polymeric fluids, the viscosity falls off (often referred to as *shear thinning*), and on double-logarithmic coordinates one often observes nearly straight-line behavior at high shear rates. Since a straight line in logarithmic plotting means a *power relationship*, $\eta \approx \dot{\gamma}^p$, one often refers to this as *power law behavior.*

Figure 3-16c shows normal stress data. Such data are more difficult to obtain, by comparison to viscosity data, over a wide range of shear rates, and so a less complete picture emerges from examination of a set of data for a single fluid at a specific set of conditions. However, it is fairly well established that the normal stress coefficient behaves in a qualitative fashion similar to the viscosity function: constant Ψ at low shear rate, and nearly power law behavior at high shear rate.

Another feature of significance is the relative value of the primary coefficient Ψ_{12} to the secondary coefficient Ψ_{23}. Reliable data show Ψ_{23} to be opposite in sign, and usually quite a bit smaller in magnitude relative to Ψ_{12}.

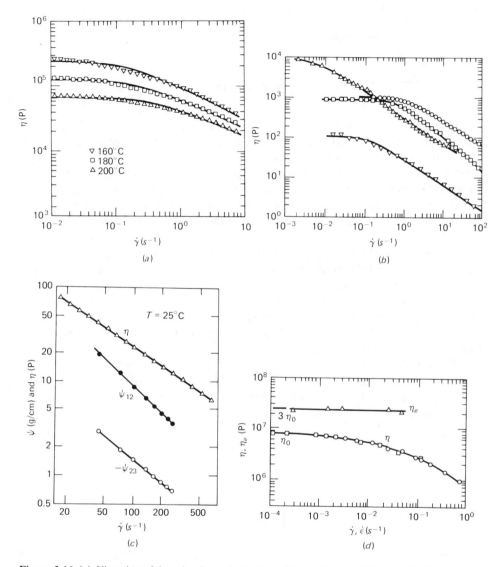

Figure 3-16 (*a*) Viscosity of low-density polyethylene. (*From Chen and Bogue*). (*b*) Viscosity of polymer solutions. (*Reproduced from Bird, Hassager, and Abdel-Khalik.*)

△ 2% polyisobutylene (*Huppler et al.*)

○ 5% polystyrene in Aroclor 1242 (*Ashare*)

▽ 0.75% polyacrylamide (Separan-30)

□ 7% aluminum soap (*Huppler et al.*)

(*c*) Viscosity and normal stress coefficients for 3% polyethylene oxide. (*Olabisi and Williams.*)
(*d*) Elongational viscosity of an isobutylene-isoprene rubber. (*Data of Stevenson reproduced from Bird, Hassager, and Abdel-Khalik.*)

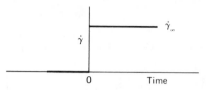

Figure 3-17 A step change in shear rate.

In Fig. 3-16*d* data for both shear and elongational viscosity are shown for a rubbery elastomer. The interesting feature here is the constancy of η_e as a function of $\dot{\epsilon}$, while the shear viscosity η shows shear-thinning behavior. Stevenson's experiment was one of uniaxial extension, and for a newtonian fluid one can show (see Prob. 3-29*a*) that $\eta_e = 3\mu$. The observation here is that η_e is equal to $3\eta_0$, where η_0 is the low-shear-rate asymptote of the $\eta(\dot{\gamma})$ curve. Depending on which function one examines, it would be possible to conclude incorrectly that this fluid is newtonian.

The functions shown in Fig. 3-16 are all defined for, and measured in, *steady-state* flows. It is of interest to examine the response of a fluid subjected to some simple *transient* flow field, too. We often find that the transient response is characteristic of a fluid and so provides another means of rheological characterization that is independent of the steady-state material functions. Furthermore, many real flow processes involve a changing flow field, and simple transient flows provide a means of modeling such processes.

The simplest transient flows are those defined by Eqs. (3-92) and (3-96), with the allowance that $\dot{\gamma}$ and $\dot{\epsilon}$ may be functions of time. The simplest function of time, what we might call the simplest "program" of transient deformation, is the *step function*, defined by

or

$$\frac{\dot{\gamma}(t)}{\dot{\gamma}_\infty} = \begin{cases} 0 & \text{for } t < 0 \\ \\ 1 & \text{for } t \geq 0 \end{cases} \qquad (3\text{-}98)$$
$$\frac{\dot{\epsilon}(t)}{\dot{\epsilon}_\infty}$$

and shown schematically in the sketch of Fig. 3-17. Of course, an *exact* step change is an idealization, but since that is just another name for *model*, we should not be held up by such a reservation.

For this idealized flow, we may define a set of material functions which characterize the behavior of a fluid subject to such a flow. In fact, we maintain the definitions of the four material functions of Eqs. (3-93) to (3-95) and (3-97), but we recognize that now these functions depend not only on the steady-state value of the deformation rate ($\dot{\gamma}_\infty$ or $\dot{\epsilon}_\infty$) but these material functions are now functions of the *time* of deformation as well. Thus we find it useful to use a slightly different notation and define the *stress-growth functions*:

$$\eta^+(t, \dot{\gamma}_\infty) = \frac{T_{12}(t, \dot{\gamma}_\infty)}{\dot{\gamma}_\infty} \qquad (3\text{-}99)$$

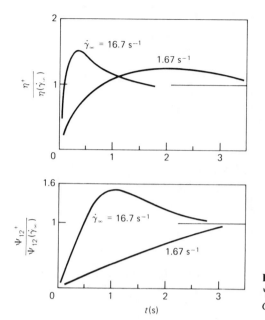

Figure 3-18 Stress-growth functions η^+ and Ψ_{12}^+ for 4% polystyrene solution. (*After Carreau.*)

$$\Psi_{12}^+(t, \dot{\gamma}_\infty) = \frac{T_{11}(t, \dot{\gamma}_\infty) - T_{22}(t, \dot{\gamma}_\infty)}{\dot{\gamma}_\infty^2} \tag{3-100}$$

$$\Psi_{23}^+ \text{ in a similar manner} \tag{3-101}$$

and

$$\eta_e^+(t, \dot{\epsilon}_\infty) = \frac{T_{11}(t, \dot{\epsilon}_\infty) - T_{22}(t, \dot{\epsilon}_\infty)}{\dot{\epsilon}_\infty} \tag{3-102}$$

with the understanding that Eq. (3-98) is part of this definition; i.e., these functions are defined only with respect to this particular flow.

Figure 3-18 shows stress-growth functions in shear. Qualitatively the most important feature is the possibility of "overshoot" of the steady-state stress. Quantitative features of importance are the magnitude of the overshoot, the time at which the maximum stress occurs, and the time at which stress equilibrium, steady state, occurs. Each of these features provides a means of characterization of the fluid. A general observation of stress-growth data suggests that the overshoot phenomenon occurs at high deformation rates, whereas a monotonic approach to the steady-state stress occurs at low deformation rates.

Although less experimental data are available, similar results are observed in transient elongational response, as shown in Fig. 3-19. The fluid is a melt of a low-density polyethylene. Of interest is the long time required for this fluid to achieve the steady-state stress level; at an extension rate of 10^{-3} s^{-1} the equilibrium stress is not reached for 1000 s. If such a fluid were being processed under conditions that the deformation rate changed in time intervals smaller than

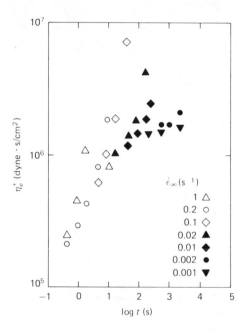

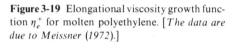

Figure 3-19 Elongational viscosity growth function η_e^+ for molten polyethylene. [*The data are due to Meissner (1972).*]

1000 s, the fluid might *never* be in a dynamic steady state, and steady-state rheological data, *the type of data normally available*, would be of no relevance to the design of the process. (That was significant; read it again!)

Too few sets of data are available for elongational flows to allow a conclusion that the observations of Fig. 3-19 are common to a broad class of polymeric fluids. Other material functions (e.g., all the "dynamic" functions of linear viscoelasticity: dynamic viscosity, loss modulus, storage modulus, etc.) may be defined for other types of simple kinematic situations. In some broad sense, *all* these material functions together help to characterize the behavior of a fluid subject to some deformation. From a practical point of view there are two questions that must be answered:

1. For a given process, what material functions are most relevant (i.e., essential) to the dynamic description of the process?
2. What types of constitutive equations characterize fluids whose behavior is similar to that illustrated in the preceding figures?

We turn first to consideration of the second question, recalling that in order to solve the dynamic equations we must have the T (or τ) dependence on the flow field, through Δ, for example. Having the material functions defined above, or any set, no matter how large, of material functions, still does not permit us to solve the dynamic equations that will describe a specific flow process. The material functions only characterize the fluid in *simple* flows. The constitutive equations characterize the fluid in *any* flow. (*Read this paragraph again.*)

Constitutive Equations for Viscous Fluids

Let us assert, postulate, claim, insist, or otherwise hope that there exists a class of fluids for which

$$\tau_{ij} = \eta \, \Delta_{ij} \tag{3-103}$$

where η is some scalar coefficient, but not necessarily a constant.

It is essential at the outset to understand the "philosophical status" of Eq. (3-103). It is a *definition* of a class of fluids. As a definition one cannot quarrel with it. We may *wonder* if such a fluid exists, but this uncertainty does not prevent us from *defining* the fluid.

The Bohr model of the atom is a similar philosophical construct. The model is not directly accessible to observation. But the *implications* of the model are consistent with experience, and so the model proves useful. Thus one may *define* a unicorn, and in the company of knowledgeable people should only expect to be asked "Yes, but is it a useful definition?" The question "Does the unicorn exist?" is not quite precise enough.

Indeed we know that fluids of the class defined by Eq. (3-103) do exist. In the simplest case, where the coefficient η is a *constant* μ, we have the newtonian fluid:

$$\tau_{ij} = \mu \, \Delta_{ij} \tag{3-104}$$

Let us examine some special consequences of newtonian behavior in a particular simple flow, plane Couette flow (PCF). Figure 3-20 shows the geometry of interest.

We consider a flow such that $\mathbf{u} = [u_x(t, y), 0, 0]$, subject to boundary conditions

$$u_x = \begin{cases} 0 & \text{on } y = 0 \\ U & \text{on } y = b \end{cases} \quad \text{for } t > 0$$

and an initial condition

$$u_x = 0 \quad \text{on } 0 \le y \le b \text{ at } t \le 0$$

When Eq. (3-104) is used to replace τ, and the continuity equation is used, the dynamic equations take the form

$$\rho \, \frac{\partial u_x}{\partial t} = -\frac{\partial p}{\partial x} + \mu \, \frac{\partial^2 u_x}{\partial y^2} \tag{3-105}$$

$$0 = -\frac{\partial p}{\partial y} \tag{3-106}$$

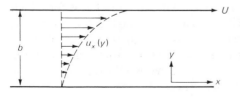

Figure 3-20 Sketch for analysis of transient flow in simple shear.

From Eq. (3-106) we find that p is not a function of y. If u_x is not a function of x, and if no pressure difference in the x direction is imposed on the system at the boundaries, then p is not a function of x either. Then we must find $u(t, y)$ as a solution to

$$\rho \frac{\partial u_x}{\partial t} = \mu \frac{\partial^2 u_x}{\partial y^2} \tag{3-107}$$

subject to the boundary and initial conditions stated above.

Equation (3-107) is easily solved by the method of separation of variables, or by use of the Laplace transformation, and the solution may be written in the form

$$u_x(t, y) = \frac{Uy}{b} - F(t, y) \tag{3-108}$$

The solution has been separated into two parts: a transient term F which decays to zero exponentially in time, and the steady-state term, which can be seen to be linear in y. In most cases of practical interest the transient term vanishes so rapidly that it need not be considered (see Probs. 3-14 and 3-15).

Let us examine, then, the implications of the steady-state solution

$$u_x = \frac{Uy}{b} \tag{3-109}$$

It follows easily that the only nonzero components of Δ are

$$\Delta_{xy} = \Delta_{yx} = \frac{U}{b} = \dot{\gamma} \tag{3-110}$$

We call $\dot{\gamma}$ the *shear rate*, and this definition is consistent with the more general definition of Eq. (3-92). In short, PCF is SSF.

Then, since the stresses τ_{ij} are proportional to the *corresponding* components of Δ_{ij} [Eq. (3-104)], the only nonzero components of τ are the shear stresses

$$\tau_{xy} = \tau_{yx} = \frac{\mu U}{b} \tag{3-111}$$

This is the form in which *Newton's law of viscosity* is usually stated in elementary treatments of fluid mechanics. The more rigorous definition is through Eq. (3-104), including the statement that μ is constant.

The most important implication of these results is that there are no normal stresses (apart from a uniform isotropic stress, pressure) developed in PCF of a newtonian fluid. The importance of this point is that there *are* fluids for which finite normal stresses are measured in PCF. Hence, Eq. (3-104) is too restrictive.

Let us go back to Eq. (3-103). η was referred to as a *scalar coefficient*. What we mean by this, specifically, is that η might be some function rather than, as in the case of the newtonian fluid, a constant. In this case we would refer to η as the *nonnewtonian viscosity*, or the *generalized viscosity*, coefficient.

Of what, then, do we expect η to be a function? Rather than try to justify some expectation, we will assume that η depends only on the kinematics of the flow, as defined through the components of $\boldsymbol{\Delta}$. But if η is a *scalar* function, how can it depend on a *tensor*?

Consider first a simpler, more familiar problem. The kinetic energy of a point mass depends on its velocity. Kinetic energy is not a vector; it has no direction associated with it. It has only a magnitude and so is a scalar function of velocity. But velocity *is* a vector. The resolution of this point lies in the fact that there is a scalar function of velocity, namely, the scalar product

$$u^2 = \mathbf{u} \cdot \mathbf{u} = u_x^2 + u_y^2 + u_z^2 \tag{3-112}$$

We have written $\mathbf{u} \cdot \mathbf{u}$ in cartesian coordinates, but we can show that the squared speed u^2 is independent of a coordinate system. A function of a vector that is independent of a coordinate system is called an *invariant*. Any arbitrary vector $\mathbf{v}$ has associated with it a *single* scalar function given by $\mathbf{v} \cdot \mathbf{v}$.

This fact should evoke the following question: Does a tensor have associated with it any invariant scalar functions? The answer is yes: three.

It is possible to show that there are three combinations of the components of $\boldsymbol{\Delta}$ which are scalar invariants. They are

$$I_\Delta = \Delta_{ii}$$

$$II_\Delta = \Delta_{ij}\,\Delta_{ij}$$

$$III_\Delta = \begin{vmatrix} \Delta_{11} & \Delta_{12} & \Delta_{13} \\ \Delta_{21} & \Delta_{22} & \Delta_{23} \\ \Delta_{31} & \Delta_{32} & \Delta_{33} \end{vmatrix} = \det \boldsymbol{\Delta} \tag{3-113}$$

In Eqs. (3-113) we use the summation convention on repeated subscripts. It is a simple matter to show that

$$I_\Delta = 0 \qquad \text{for an incompressible fluid} \tag{3-114}$$

Further, for the case of PCF, one can show that, even if η depends on the kinematics (see Prob. 3-18),

$$II_\Delta = 2\dot{\gamma}^2 \tag{3-115}$$

and
$$III_\Delta = 0 \tag{3-116}$$

At least for the PCF example we see that the assumption that η depends in some unspecified way on the kinematics can be replaced by the simpler relation

$$\eta = \eta(II_\Delta) = \eta(\dot{\gamma}) \tag{3-117}$$

We note that even for a nonnewtonian fluid of the type defined above, $\dot{\gamma}$ is a constant given by U/b (see Prob. 3-16).

If we again go back to the dynamic equations, we will find that now the shear stresses are given by

$$\tau_{xy} = \tau_{yx} = \eta\left(\frac{U}{b}\right)\frac{U}{b} \tag{3-118}$$

and the normal stresses are still zero. Thus the nonnewtonian fluid defined by Eq. (3-103) allows for a nonlinear relationship between shear stress and shear rate but still fails to predict the normal stresses commonly observed in shear flows of fluids which exhibit this nonlinearity in viscosity.

Because of the results developed above, we refer to Eq. (3-103) as a *purely viscous fluid*. In an SSF, as defined generally in Eq. (3-92) and as illustrated here specifically for the case of PCF, a viscous fluid *does* show a nonnewtonian viscosity since, by *definition*, η may be a *function* of shear rate.

Then, in the spirit of the philosophical statements made earlier, we would conclude that if normal stress effects are important, the purely viscous fluid is not a useful model for the description of shear flows.

Of course, not all flows of interest are simple shear flows. We have already considered the kinematics of a nonshear flow in Sec. 3-6—flow outside an expanding bubble. There we found that $\boldsymbol{\Delta}$ had *no shear* components, the only nonzero deformation rates being in the "normal" directions. Thus that flow was an example of SEF [Eq. (3-96)].

In such flows, of course, the purely viscous fluid does predict normal stresses, and indeed *only* normal stresses (see Probs. 3-21 and 3-22). It is essential to recognize, however, that these normal stresses are *viscous* in origin, since they clearly vanish in the absence of the viscosity function η.

Let us examine another means of introducing a nonlinearity into a constitutive equation. Consider a constitutive equation of the form

$$\boldsymbol{\tau} = \eta_0\,\boldsymbol{\Delta} + \beta\,\boldsymbol{\Delta}^2 \tag{3-119}$$

where η_0 and β are constants, and where we *define* $\boldsymbol{\Delta}^2$ to mean the tensor whose components, in cartesian coordinates, are

$$(\boldsymbol{\Delta}^2)_{ij} = \Delta_{ik}\,\Delta_{kj} \tag{3-120}$$

The effect of this kind of nonlinearity is most easily seen by considering the SSF defined by Eq. (3-92), for which it is easily shown that

$$\boldsymbol{\Delta}^2 = \dot{\gamma}^2\begin{pmatrix} 1 & 0 & 0 \\ 0 & 1 & 0 \\ 0 & 0 & 0 \end{pmatrix} \tag{3-121}$$

It follows, then, that the stress tensor has components given by

$$\boldsymbol{\tau} = \begin{pmatrix} \beta\dot{\gamma}^2 & \eta_0\dot{\gamma} & 0 \\ \eta_0\dot{\gamma} & \beta\dot{\gamma}^2 & 0 \\ 0 & 0 & 0 \end{pmatrix} \tag{3-122}$$

and we see that η_0 is still the viscosity, but now we find a finite normal stress coefficient of magnitude $\Psi_{23} = \beta$, while $\Psi_{12} = 0$.

By analogy to *algebraic* problems we might be tempted to continue this process and write τ in a power series in Δ of the form

$$\tau = \eta_0 \, \Delta + \beta \, \Delta^2 + \kappa \, \Delta^3 + \text{more powers of } \Delta \qquad (3\text{-}123)$$

However, it can be proved that any power of a tensor higher than second can be written in terms of lower powers with coefficients that depend on the invariants. For example, the cube of Δ may be written as

$$\Delta^3 = I_\Delta \, \Delta^2 - II_\Delta \, \Delta + III_\Delta \qquad (3\text{-}124)$$

Hence Eq. (3-119) is the *most general* form of algebraic constitutive equation, and it is clearly inadequate, since $\Psi_{12} = 0$ is not observed in most of the fluids of interest to us.

Algebraic constitutive equations are also clearly unable to describe *transient* material response. For example, a fluid defined by Eq. (3-119), subjected to a step change in shear rate, will exhibit a step change in stress and cannot show material functions of the kind observed in Fig. 3-18.

We must turn, then, to models which allow for more complex dynamic phenomena. These models are grouped under the term *viscoelastic fluids*.

Viscoelastic Fluids

Perhaps it is best to begin at the end, with an admission of failure. There is no constitutive equation which does a good job of predicting all the material functions that characterize the variety of simple flows that are useful in polymer process models. There are several constitutive equations which predict the behavior of *some* of the material functions with great accuracy, requiring only a small number of free constants to do so. Some of the "good" constitutive equations, however, are relatively complicated mathematically. If they are used with the dynamic equations, the resulting mathematical model is so complex that analytical solution is impossible, and numerical computation with a digital computer is so tedious as to not be worth the effort.

Thus one must learn the art of constitutive compromise. We must be able to examine a flow process and judge which elements of the kinematics are most essential to the behavior we wish to model. Then we must select a constitutive equation which is most appropriate to describe the dominant type of kinematics of that flow field. The resulting solution of the dynamic equations, for the velocities and stresses, must then be interpreted in light of the known weaknesses of the constitutive equation used. Finally, experience—the comparison of prediction and observation—must be the judge of success.

In contrast to the purely viscous models of the previous section, viscoelastic models must characterize two (usually) essential features of real viscoelastic fluids: the development of normal stresses in shearing flows, and transient phenomena of the type characterized not only by the stress growth functions η^+, Ψ^+, etc., but including classical stress relaxation phenomena as well.

Since we have used the word *classical*, let us begin with the classical viscoelastic model, the Maxwell model, expressible in the form

$$\tau_{ij} + \lambda \frac{\partial \tau_{ij}}{\partial t} = \eta_0 \, \Delta_{ij} \tag{3-125}$$

where λ and η_0 are constants.

This model is useless to our goals, except as a conceptual link to more adequate constitutive equations. It is useless because:

- It predicts newtonian behavior in shear
- It does not give normal stresses in shear
- Its transient behavior is monotonic in stress growth, contrary to experience at all but the lowest shear rates

Furthermore, it is not correct mathematically. It is possible to show that the ordinary time derivative $\partial/\partial t$ of a tensor τ is not a tensor!

The notion of time derivatives of tensors is a difficult one, and the reader who has not studied continuum mechanics may find the ensuing modifications of the Maxwell equation somewhat mystical. Several appropriate references are cited at the end of this chapter, but some relevant physical ideas can be introduced here before proceeding.

The *proof* that $\partial \tau/\partial t$ is not a tensor requires a firmer grounding in continuum mechanics and tensor analysis than is required for the rest of this text. It is not too difficult, however, to suggest the *nature* of the problem of time differentiation and the general method of its resolution. We can do this with a very simple example of a mechanics problem.

Imagine the turntable of a record player, and suppose two nails are driven partway into and perpendicular to the top surface. Let one nail coincide with the axis of rotation and the other be somewhere out toward the periphery of the disc. Now we stretch a rubber band so that it loops over and connects the two nails. The rubber band is in tension. We wish to describe the mechanical state of the rubber band.

Since the ends of the rubber band are fixed, and if we assume the rubber to be perfectly elastic so that no stress relaxation is allowed for, then it would follow that the tension in the rubber is independent of time. Now suppose that we wished to specify the tension in a coordinate system outside the material, fixed in the laboratory. Figure 3-21 shows the situation, and if the orientation of the rubber band is known, it is not very difficult to find the components of tension in the laboratory coordinate system. The components T_{xx} and T_{yy} are independent of time, of course.

Now suppose that the turntable rotates at a slow speed such that no centrifugal effects occur. Then the tension in the rubber band is unchanged. However, in the *fixed* coordinate system we now find that the components of tension are cyclic,

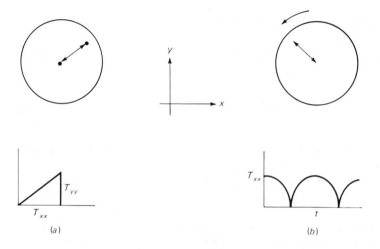

Figure 3-21 Sketch for consideration of stress in a rubber band attached to a turntable.

as suggested in Fig. 3-21b. For example, at the instant of time when the rubber band is aligned along the y axis, there is no x component T_{xx}. Thus we have a situation where $\partial T_{xx}/\partial t$ vanishes in one coordinate system but not in another.

The resolution of the problem is obvious in this simple case: In transforming from the coordinate system of the rubber band (which we might call a *material coordinate system*) to the fixed laboratory coordinate system, we need only account for the rotation of the rubber band with the turntable. In the more complex case of interest to us, where we wish to specify the state of stress in a continuously deforming continuous medium, we must also account in some way for the *deformation* of the material when we transform from a *material coordinate system* to the laboratory coordinate system. The difficulty lies in the fact that there is no *unique* mathematical way to do this.

In any physically meaningful mathematical statement certain basic principles must be observed. Some of these principles are so familiar that we rarely think of them. For example, we can only add terms if they are dimensionally consistent. Neither can we add terms which have a different mathematical " character," so we cannot add, for example, a vector and a scalar. Thus, if we wish to incorporate a time derivative of stress in a constitutive equation, we must find a way of writing a time derivative which preserves the physical idea of time rate of change while having the required mathematical character.

Two such time derivatives have been used, each of which has a somewhat different physical interpretation. One is called the *Oldroyd contravariant derivative*, whose cartesian components may be written as

$$\frac{\mathfrak{D}}{\mathfrak{D}t}\tau_{ij} = \frac{\partial \tau_{ij}}{\partial t} + u_k \frac{\partial \tau_{ij}}{\partial x_k} - \tau_{kj}\frac{\partial u_i}{\partial x_k} - \tau_{ik}\frac{\partial u_j}{\partial x_k} \tag{3-126}$$

It can be shown that the Oldroyd derivative gives the components, in fixed coordinates, of the time derivative as observed in a coordinate system which translates with and *deforms with the flow field.* Hence it is sometimes called the *codeformational derivative.*

Another time derivative that is a tensor is the Jaumann derivative, written in cartesian coordinates as

$$\frac{\mathscr{D}}{\mathscr{D}t} \tau_{ij} = \frac{\partial \tau_{ij}}{\partial t} + u_k \frac{\partial \tau_{ij}}{\partial x_k} - \tfrac{1}{2}\omega_{jm}\tau_{mi} - \tfrac{1}{2}\omega_{im}\tau_{mj} \tag{3-127}$$

The Jaumann derivative can be shown to give the components, in fixed coordinates, of the time derivative as observed in a coordinate system which translates with and *rotates* with the local rotation, as given by the vorticity tensor $\boldsymbol{\omega}$ defined earlier by Eq. (3-47). The Jaumann derivative is sometimes called the *corotational derivative.*

Thus it is possible to define a Maxwell model using the Oldroyd derivative, for example, which has the proper *mathematical* character to it, in the tensor format

$$\tau + \lambda \frac{\delta \tau}{\delta t} = \eta_0 \, \mathbf{\Delta} \tag{3-128}$$

The more difficult question is whether the equation has a realistic physical character to it.

For this fluid, then, let us find the stresses for the SSF, except that we will let $\dot{\gamma}$ be a step function of time, as in Fig. 3-17. It is not too difficult to show the following results (Prob. 3-35):

$$\tau_{12} = \eta_0 \dot{\gamma}_\infty (1 - e^{-t/\lambda}) \tag{3-129}$$

$$\tau_{11} = 2\eta_0 \lambda \dot{\gamma}_\infty^2 \left[1 - e^{-t/\lambda}\left(1 + \frac{t}{\lambda} \right) \right] \tag{3-130}$$

$$\tau_{22} = 0 \tag{3-131}$$

In terms of the appropriate material functions:

$$\eta^+ = \eta_0(1 - e^{-t/\lambda}) \tag{3-132}$$

$$\Psi_{12}^+ = 2\lambda\eta_0 \left[1 - e^{-t/\lambda}\left(1 + \frac{t}{\lambda} \right) \right] \tag{3-133}$$

$$\Psi_{23}^+ = 0 \tag{3-134}$$

Thus we find that the Maxwell model with the Oldroyd time derivative exhibits the following behavior:

η^+ increases monotonically (as is true of the classical Maxwell model)
Ψ_{12}^+ increases monotonically, reaching equilibrium somewhat later than η^+
$\Psi_{23}^+ = 0$ for all time (also true of the classical Maxwell model)

Now we have transient behavior, characterized by a relaxation time λ, but we still fail to predict overshoot phenomena. We *have* "achieved" a finite Ψ_{12}, so we have made some progress. The finite Ψ_{12} arises directly from the introduction of a mathematically appropriate time derivative.

As an example of a model-building exercise, consider a "superposition" of the two previous models, giving the constitutive equation in the form

$$\tau + \lambda \frac{\mathfrak{D}\tau}{\mathfrak{D}t} = \eta_0 \, \Delta + \beta \, \Delta^2 \tag{3-135}$$

For the transient SSF we find

$$\eta^+ = \eta_0\left(1 + \frac{\beta\lambda\dot\gamma^2}{\eta_0}\right)(1 - e^{-t/\lambda}) - \frac{\beta\lambda\dot\gamma^2}{\eta_0}\frac{t}{\lambda}e^{-t/\lambda} \tag{3-136}$$

$$\Psi_{12}^+ = \left[2\lambda\eta_0\left(1 + \frac{\beta\lambda\dot\gamma^2}{\eta_0}\right) - \beta\right](1 - e^{-t/\lambda})$$

$$- \left[2\lambda\eta_0\left(1 + \frac{\beta\lambda\dot\gamma^2}{\eta_0}\right)\frac{t}{\lambda} + \lambda^2\beta\dot\gamma^2\left(\frac{t}{\lambda}\right)^2\right]e^{-t/\lambda} \tag{3-137}$$

$$\Psi_{23}^+ = \beta(1 - e^{-t/\lambda}) \tag{3-138}$$

Now we find that both η^+ and Ψ_{12}^+ overshoot their equilibrium values, whereas Ψ_{23}^+ monotonically approaches a value equal to the parameter β. On the basis of experimental evidence cited earlier, we conclude that β is negative.

η^+ reaches its peak value at a time given by

$$\left(\frac{t}{\lambda}\right)_{\eta_{\max}^+} = -\frac{\eta_0}{\beta\lambda\dot\gamma^2} \tag{3-139}$$

Note that the time at which the maximum occurs decreases as the shear rate increases, which is qualitatively consistent with experience.

Ψ_{12}^+ reaches its peak value at a slightly later time given by

$$\left(\frac{t}{\lambda}\right)_{\Psi_{12,\,\max}^+} = -\frac{\eta_0}{\beta\lambda\dot\gamma^2}\left(1 + \sqrt{1 + \frac{\beta^2\dot\gamma^2}{\eta_0^2}}\right) \tag{3-140}$$

At equilibrium (i.e., steady state) we find the material functions are

$$\eta = \eta_0\left(1 + \frac{\beta\lambda\dot\gamma^2}{\eta_0}\right) \tag{3-141}$$

$$\Psi_{12} = 2\lambda\eta_0\left(1 + \frac{\beta\lambda\dot\gamma^2}{\eta_0}\right) - \beta \tag{3-142}$$

$$\Psi_{23} = \beta \tag{3-143}$$

Since β is apparently negative, both η and Ψ_{12} show shear-thinning behavior. If $\dot\gamma$ is allowed to be large, both η and Ψ_{12} become negative. This is clearly an artifact of the model, and it suggests that the results may only be reasonably valid for "small" shear rates.

The term *small*, in referring to the shear rate, is meaningless. It must be interpreted relative to some characteristic time for the fluid, such as λ, and the appropriate restriction would be to small values of $\lambda \dot{\gamma}$ (see Probs. 3-36 and 3-37).

Before we proceed, we should pause and recall our goal so that we do not lose sight of it in the midst of the surrounding algebra. We would like to establish a few constitutive equations that strike a compromise between complexity and utility. Equation (3-135) is a relatively simple and relatively realistic equation. We make that assertion based on the ease with which we can calculate some of the simple material functions, and based on the fairly good qualitative agreement between these predictions and our experience with polymeric fluids, at least insofar as *shear* behavior is concerned.

We can also examine the *elongational* response of a fluid defined by Eq. (3-135), and find (see Prob. 3-38), for a step change in elongation rate $\dot{\epsilon}$,

$$\frac{\eta_e^+}{3\eta_0} = \frac{1 + \beta\dot{\epsilon}/\eta_0 + 2\beta\dot{\epsilon}^2\lambda/\eta_0}{(1 - 2\lambda\dot{\epsilon})(1 + \lambda\dot{\epsilon})} + \frac{1 - \beta\dot{\epsilon}/\eta_0}{3(1 + \lambda\dot{\epsilon})} \exp\left[-\frac{(1 + \lambda\dot{\epsilon})t}{\lambda}\right]$$
$$-\frac{2}{3}\frac{1 + 2\beta\dot{\epsilon}/\eta_0}{1 - 2\lambda\dot{\epsilon}} \exp\left[-\frac{(1 - 2\lambda\dot{\epsilon})t}{\lambda}\right] \qquad (3\text{-}144)$$

The steady state is approached monotonically in this fluid, and at steady state the elongational viscosity function attains the value

$$\frac{\eta_e}{3\eta_0} = \frac{1 + \beta\dot{\epsilon}/\eta_0 + 2\beta\dot{\epsilon}^2\lambda/\eta_0}{(1 - 2\lambda\dot{\epsilon})(1 + \lambda\dot{\epsilon})} \qquad (3\text{-}145)$$

The most significant feature of this result is the appearance of a finite elongation rate, given by $\dot{\epsilon} = 1/2\lambda$, at which the elongational viscosity becomes unbounded. The question of whether this specific feature of the model we are evaluating is realistic is very difficult to answer since so few reliable data for η_e are available.

The data of Fig. 3-19, for a polymer melt, are not inconsistent with the *transient* [Eq. (3-144)] predictions but are too scanty to allow stronger confirmation. Data which are clearly inconsistent with Eq. (3-145) are shown in Fig. 3-22 for two different polymer solutions.

Since we are sometimes interested in processes where elongational flow is significant, by comparison to shear flow, we should be concerned about the use of a constitutive equation that predicts an infinite elongational viscosity, even if we recognize that the prediction could be an artifact of the model. The Maxwell model with the Jaumann (corotational) time derivative modifies this particular feature, and in the form

$$\tau + \lambda \frac{\mathscr{D}\tau}{\mathscr{D}t} = \eta_0 \,\boldsymbol{\Delta} \qquad (3\text{-}146)$$

predicts that $\eta_e = 3\eta_0$, which is in agreement with *some* data (e.g., Fig. 3-16*d*).

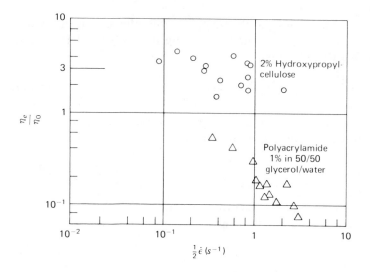

Figure 3-22 Elongational viscosity of two aqueous polymer solutions. (*Pearson.*)

Whereas the corresponding Oldroyd-Maxwell fluid [Eq. (3-128)] gives a *newtonian* shear viscosity [$\eta = \eta_0$ from Eq. (3-132), at steady state], the Jaumann-Maxwell fluid gives

$$\eta = \frac{\eta_0}{1 + (\lambda\dot{\gamma})^2} \tag{3-147}$$

and

$$\Psi_{12} = \frac{2\lambda\eta_0}{1 + (\lambda\dot{\gamma})^2} \tag{3-148}$$

While shear-thinning behavior is thus accommodated, neither prediction is realistic quantitatively. The dependence of viscosity on $\dot{\gamma}$ is not nearly so strong as Eq. (3-147) would suggest. In fact, according to Eq. (3-147), the shear stress function $\tau_{12}(\dot{\gamma})$ goes through a maximum at $\lambda\dot{\gamma} = 1$, and thereafter τ_{12} is a *decreasing* function of shear rate. Thus, while a *finite* elongational viscosity is predicted by a Maxwell model with a Jaumann derivative, the predictions on η and Ψ_{12} are so poor that one would not choose Eq. (3-146) in preference to Eq. (3-128). Furthermore, the shear and normal stresses are predicted to approach steady state (under a step change in shear rate) in an oscillatory manner, which is at variance with observations.

What, then, *does* one choose for a constitutive equation? At the present time the choice is made on the side of convenience with a modest nod toward realism. The problem, basically, is that if one uses a quantitatively realistic constitutive equation, and several are available, the resulting dynamic problem, consisting of the continuity equation, the dynamic equations, and the *set* of constitutive equations, forms a *system* of multiple, simultaneous, coupled, nonlinear partial differential equations, whose solution, except for nearly trivial flows, defies economical solution by computer.

The recommended compromise, at the time of writing this text, is the White-Metzner modification of the Maxwell model, utilizing the Oldroyd derivative in contravariant form. We write this as

$$\tau + \lambda(\text{II}_\Delta)\frac{\eth\tau}{\eth t} = \eta(\text{II}_\Delta)\,\Delta \tag{3-149}$$

We take both λ and η to be *functions* of the second invariant of the flow field, and the usual assumption is to relate λ to η through a "modulus" G, so that

$$\lambda = \frac{\eta}{G} \tag{3-150}$$

with G assumed to be independent of deformation rate.

This fluid gives the material functions in steady shear flow as

$$\eta = \eta(\text{II}_\Delta) \tag{3-151}$$

$$\Psi_{12} = \frac{2\eta^2}{G} \tag{3-152}$$

$$\Psi_{23} = 0 \tag{3-153}$$

with η an *unspecified* function of II_Δ. Thus we can select a function η from data and guarantee realistic shear-viscosity behavior. Furthermore, the prediction that $\Psi_{12} \approx \eta^2$ is found to be in good agreement with data for several materials. Since Ψ_{23} is observed to be small in many materials, the prediction of a zero value may not be a major drawback. Regardless of the functional choice of η, the elongational viscosity η_e will still become unbounded at $\dot{\epsilon} = 1/2\lambda$, so long as the Oldroyd derivative is used.

A common choice of a functional relationship for shear viscosity is the *power law*, which we may write in the form

$$\eta = K(\tfrac{1}{2}\text{II}_\Delta)^{(n-1)/2} \tag{3-154}$$

This is a two-parameter model which reflects the observation that at high shear rates the viscosity function, when plotted on double-logarithmic coordinates (as in Fig. 3-16a, b), is nearly a straight line. If the low-shear-rate region must be accommodated, a simple empirical three-parameter model of the form

$$\eta = \frac{\eta_0}{1 + (K/\eta_0)^{-1}(\tfrac{1}{2}\text{II}_\Delta)^{(1-n)/2}} \tag{3-155}$$

may be used. It should be apparent that at large values of II_Δ Eq. (3-155) yields the power law as an approximation.

We end this discussion of constitutive equations by offering some remarks relevant to the question "How do we decide between using a purely viscous model and a more complicated viscoelastic model?" In a loose sense the answer lies in evaluating the relative importance of elastic and viscous effects. If rheological data are available for the fluid of interest, then one can calculate the ratio of elastic stresses to shear stresses:

$$S_R = \frac{\tau_{11} - \tau_{22}}{2\tau_{12}} \tag{3-156}$$

In this particular format the ratio defines what is called the *recoverable shear* S_R. (The factor of 2 in the denominator is arbitrary but is usually included.) Of course, the rheological data that allow calculation of S_R would be taken in steady simple shear, so it is not clear that S_R would be a relevant parameter for any other type of flow.

As a rough rule of thumb, we may usually conclude that if S_R is much smaller than unity, then elastic stresses are unimportant relative to shear stresses, and a purely viscous constitutive equation may be adequate to describe the behavior of the fluid, *if the process involves shear rates comparable to those at which S_R is measured.*

It is interesting to examine S_R as predicted by viscoelastic models such as have been described above. We note first that we may write S_R as

$$S_R = \frac{\Psi_{12}\dot{\gamma}}{2\eta} \tag{3-157}$$

For an Oldroyd-Maxwell fluid we find [Eqs. (3-141) and (3-142), with $\beta = 0$]

$$S_R = \lambda\dot{\gamma} \tag{3-158}$$

For a Jaumann-Maxwell fluid [Eqs. (3-147) and (3-148)] we find the same result. Thus, another criterion that is useful in deciding whether viscoelasticity should be considered is in terms of the product of a relaxation time for the fluid and a characteristic deformation rate for the process. If $\lambda\dot{\gamma}$ is small we may expect elastic effects to be of minor importance.

One virtue of a criterion in terms of λ, rather than in terms of S_R, is that the normal stress data required to calculate S_R are often not available, especially at high shear rates. On the other hand, an estimate of a relaxation time is often available from dynamic mechanical (linear viscoelastic) studies of polymeric materials, which are more commonly carried out than are normal stress studies.

For *steady* flows, in both the eulerian and lagrangian sense, the product of relaxation time and characteristic deformation rate is often referred to as the Weissenberg number, Ws. We shall have occasion to refer to the Weissenberg number again. In the case of *unsteady* flows the growth and relaxation response of the fluid may be of major significance, and a characteristic parameter may be formed from the ratio of the relaxation time of the fluid to some appropriate time scale θ_D of the deformation. This is usually referred to as the Deborah number,

$$\text{De} = \frac{\lambda}{\theta_D} \tag{3-159}$$

If $\theta_D < \lambda$, the fluid will not have time to approach its steady response, and relaxation phenomena may dominate the process. Steady-state data may be of practically no relevance to the modeling of such a "rapid deformation," and a purely viscous model would not likely provide a reasonable constitutive equation.

PROBLEMS

3-1 Consider PCF generated between two infinite parallel plates, one of which moves relative to the other in its own plane. Let x_1 be the direction of motion, x_2 the coordinate normal to the plates, and x_3 the coordinate in the plane of the plates but normal to x_1.

Assume that $\mathbf{u} = (u_1, 0, 0)$ (i.e., the flow is unidirectional), and further assume that u_1 is not a function of x_3. Assume the system is at steady state. Assume constant density, and neglect body forces.

Using the physical principles of conservation of momentum and mass, show that the stresses T_{ij} are, at most, linear functions of the coordinates x_1 and x_2.

3-2 Derive Eq. (3-38) from Eq. (3-36) by using Eq. (3-29). Is incompressibility assumed?

3-3 (a) Derive the continuity equation in cylindrical polar coordinates, related to cartesian coordinates by

$$x = r \cos \theta \qquad r = (x^2 + y^2)^{1/2}$$

$$y = r \sin \theta \qquad \theta = \arctan \frac{y}{x}$$

$$z = z \qquad z = z$$

The appropriate control volume is shown in Fig. 3-23.

(b) Reduce the continuity equation derived in part a to its appropriate form in the case of constant density.

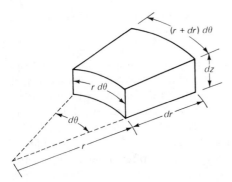

Figure 3-23 Volume element for cylindrical polar coordinates.

3-4 (a) For laminar flow along the axial direction within a circular tube of constant cross section, assume that $\mathbf{u} = (u_z, 0, 0)$, that is, $u_r = 0$ and $u_\theta = 0$. Further, assume axial symmetry, which means that $\mathbf{u}$ does not depend on θ. Show, using the continuity equation, that a consequence of these assumptions is that u_z is a function only of the radial coordinate r.

(b) Using the above results, write the dynamic equations for steady-state laminar flow in a circular tube, and prove that if T_{zz} is at most only a function of z, then T_{rz} is a linear function of r and is independent of z. You will need the assertion that T_{rz} is finite at $r = 0$.

3-5 (a) Consider the same flow as in Prob. 3-4. Let τ_R be the value of T_{rz} at the boundary between the fluid and the tube, i.e., at $r = R$. Let P_0 and P_L be the pressures at $z = 0$ and $z = L$, respectively. Making an axial force balance on the boundaries of the fluid, along the surface $r = R$ and on the surfaces $z = 0$, $z = L$, show that

$$\tau_R = \frac{R(P_0 - P_L)}{2L}$$

(b) Under what conditions is this result compatible with the result of Prob. 3-4b?

3-6 A spherically symmetric air bubble is growing within a large body of fluid. The bubble growth is characterized by the rate of growth of its radius $\dot{R}$. Assume that the flow in the fluid outside the bubble, i.e., in the region $R(t) \leq r < \infty$, is spherically symmetric. Then, taking $\mathbf{u} = (u_r, 0, 0)$, show that

$$u_r = \frac{\dot{R}R^2}{r^2}$$

follows strictly from kinematic considerations.

3-7 Consider the flow defined in Fig. 3-9 and discussed therein. Assuming no angular velocity, $u_{1\theta} = u_{2\theta} = 0$, and rotational symmetry, $\partial/\partial\theta = 0$, write the dynamic equations and the continuity equation for both fluids, along with the appropriate boundary conditions. Assume a steady-state flow.

3-8 Derive Eq. (3-87).

3-9 Give a physical interpretation of each term in Eq. (3-88).

3-10 Consider the relative importance of the terms in Eq. (3-88). Are there conditions under which inertial effects are much smaller than viscous effects? To answer this, *assume* that $\dot{R}/R_0 = k = $ constant, where R_0 is the bubble radius at time $t = 0$, and k is the "growth rate."

With this assumed model for the bubble kinematics, calculate the inertial terms and the viscous terms and compare them by examining their ratio. Show that a dimensionless combination of the parameters that characterize this problem provides a number whose size determines whether viscous or inertial forces dominate the dynamics. Restrict your "investigation" to times small enough that the bubble has not doubled in radius.

Make a similar comparison of the viscous terms to the surface tension terms.

3-11 For the case in Prob. 3-10 that $\dot{R}/R = k$, find the components of $\boldsymbol{\Delta}$ in spherical coordinates. If we define Δ_{rr} as the *strain rate*, find the relationship of the growth rate k to the strain rate.

3-12 An air bubble of initial size $R_0 = 0.1$ cm grows at a constant rate k such that $R/R_0 = 2$ at time $t = 0.1$ s. The liquid surrounding the bubble has a viscosity $\mu = 10$ P and a surface tension $\sigma = 50$ dynes/cm. Take the liquid density as 1 g/cm³. Take $p_\infty = 0$.

Find the pressure (in psi) inside the bubble as a function of time. Plot P versus t for $0 \leq t \leq 0.5$ s.

Does the pressure depend strongly on σ? (If σ were in error by 50 percent, what change would this cause in P?)

3-13 Work Prob. 3-12, but do so for the case where the liquid viscosity is only 1 P.

3-14 Find the function F in Eq. (3-108) by solving Eq. (3-107).

$$Answer: F = \frac{2}{\pi} \sum_{n=1}^{\infty} \frac{(-1)^{n+1}}{n} \sin \frac{n\pi y}{b} \exp\left(-\frac{n^2\pi^2\mu t}{\rho b^2}\right)$$

3-15 (a) Using the solution in Prob. 3-14 for F, show that a useful criterion of the time required to (nearly) achieve steady state in PCF is $t^* = t_\mu$, where $t_\mu = \rho b^2/\mu$ is a *viscous relaxation time*.

(b) How long a time is required to achieve steady state if $b = 1$ cm, $\rho = 1$ g/cm³, and $\mu = 10$ P?

(c) What would the viscous relaxation time be if $b = 0.1$ cm, $\rho = 1$ g/cm³, and $\mu = 1000$ P?

3-16 Show that the linear velocity profile in PCF at steady state is achieved for nonnewtonian fluids of the form of Eq. (3-103) so long as η is a function only of the components of $\boldsymbol{\Delta}$.

3-17 Consider a velocity field defined by $\mathbf{u} = x\mathbf{i} - y\mathbf{j}$, where $\mathbf{i}$ and $\mathbf{j}$ are unit vectors in the x and y directions.

(a) Write the components of Δ in cartesian coordinates.

(b) Write the three invariants of Δ.

(c) Transform **u** to plane polar coordinates.

(d) Write the components of Δ in plane polar coordinates.

(e) Using the components of Δ from part d, repeat part b and verify that the results are *invariant* to coordinate system.

3-18 Consider a flow for which $\mathbf{u} = [u_1(x_2), 0, 0]$ and show that Eqs. (3-114) to (3-116) hold. Assume incompressibility.

3-19 Consider a flow for which $\mathbf{u} = [u_1(x_2), 0, u_3(x_2)]$ and calculate the invariants of Δ. Assume incompressibility.

3-20 Calculate the invariants for the incompressible flow outside an expanding spherical bubble. Use Eqs. (3-77) and (3-78) for the velocity field.

3-21 For a fluid defined by the *power law* $\tau = \eta\,\Delta$, where $\eta = K\left|\frac{1}{2}\mathrm{II}_\Delta\right|^{(n-1)/2}$ and K and n are constants, find the pressure to expand a bubble at a constant rate $\dot{R}/R = k$. Neglect surface tension and inertial terms.

$$Answer: \quad Pp_\infty = \frac{4(2\sqrt{3})^{n-1}}{n} Kk^n$$

3-22 Consider uniaxial stretching of a cylinder, as shown in Fig. 3-24. Assume that $R(t)$ is independent of z.

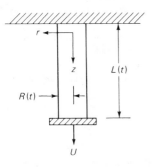

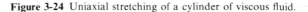

Figure 3-24 Uniaxial stretching of a cylinder of viscous fluid.

(a) Using only the condition of conservation of mass, show that the velocity field is "simple elongation," and

$$u_z = \frac{Uz}{L(t)} \qquad u_r = -\frac{Ur}{2L(t)}$$

(b) Give the components of Δ for this flow.

(c) Calculate the invariants for this flow.

(d) Neglecting surface tension and inertia, calculate the force F required to pull the cylinder for a newtonian fluid.

(e) If F is constant in part d, how does L change in time?

(f) How must L be programmed, i.e., give $L(t)$, such that the components of Δ are independent of time? Is this result true only for newtonian fluids?

(g) Rework (d) and (e) for a power law fluid, as defined in Prob. 3-21.

3-23 A "blow-molding" process requires the expansion of a "balloon" of molten polymer, as suggested in Fig. 3-25. As a model of this part of the process, consider the expansion of a spherical balloon of inside radius $R(t)$ and wall thickness W. Assume the fluid is newtonian, ignore inertial effects, and neglect surface tension. Assume also that $W \ll R$.

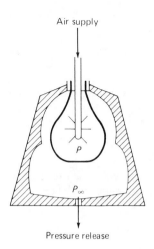

Air supply

P

P_∞

Pressure release

Figure 3-25 Blow molding: The balloon of molten polymer is expanded until it fills the mold cavity.

Show that the inflation pressure is given by

$$P = p_x + 12\mu \frac{\dot{R}}{R}\frac{W}{R}$$

and further, that

$$\frac{\dot{W}}{W} = -\frac{2\dot{R}}{R}$$

3-24 Rework Prob. 3-23, but assume that instead of a spherical balloon the appropriate geometrical model is that of a cylindrical tube which somehow remains a uniform cylinder during its expansion, without any change in length.

Answer: $P = p_x + 4\mu \dfrac{\dot{R}}{R}\dfrac{W}{R}$ $\dfrac{\dot{W}}{W} = \dfrac{-\dot{R}}{R}$

3-25 A spherical balloon of polymer is expanded under constant pressure of 115 psia against the atmosphere (15 psia). The fluid may be considered newtonian, with a viscosity of 10^6 P and a density of 1 g/cm^3. Take the initial values of the dimensions to be $R_0 = 2$ in, and $W_0 = \frac{1}{8}$ in.

Plot R versus t up to the time when $R = 6$ in. (*Hint:* Begin with the assumption that $\dot{R}/R$ is some constant, and show that this assumption is compatible with the dynamic and kinematic solutions given in Prob. 3-23.)

3-26 An estimate of the rate of deformation is desired for the processes described in Probs. 3-23 and 3-24. For each process calculate the *stretch-rate magnitude*, defined as

$$\dot{\epsilon} = \sqrt{\tfrac{1}{6}\mathrm{II}_\Delta}$$

Give the results in terms of the kinematic variables only, i.e., in terms of $\dot{R}/R$. Show that, for SEF [as defined by Eq. (3-96)], the definition of $\dot{\epsilon}$ above is consistent with the definition of $\dot{\epsilon}$ in Eq. (3-96).

3-27 What is the value, in units of s^{-1}, of $\dot{\epsilon}$ for the process of Prob. 3-25, as a function of time?

3-28 Rework Prob. 3-23 for a power law fluid as defined in Prob. 3-21.

3-29 For the following flows, calculate the elongational viscosity, defined as

$$\eta_e = \frac{T_{11} - T_{22}}{\frac{1}{2}\Delta_{11}}$$

using the newtonian constitutive equation $\tau = \mu \, \Delta$.

Take the kinematics to be steady state, that is, Δ independent of time.

(a) Uniaxial stretching of a cylinder (Prob. 3-22).

(b) Flow outside an expanding spherical bubble [use Eqs. (3-77) and (3-78)].

(c) $\mathbf{u} = 2\dot{\epsilon}(x_1, -\frac{1}{2}x_2, -\frac{1}{2}x_3)$.

(d) Expansion of a balloon (Prob. 3-23).

(e) Expansion of a tube of constant length (Prob. 3-24). This problem is not one of *simple elongation*. The kinematics are called *planar extension* when

$$\Delta = \dot{\epsilon} \begin{pmatrix} 2 & 0 & 0 \\ 0 & -2 & 0 \\ 0 & 0 & 0 \end{pmatrix}$$

3-30 For the data shown in Fig. 3-26, which were obtained by Stevenson in uniaxial stretching of a cylindrical sample of elastomer at constant elongation rate, calculate η_e at each stretch rate.

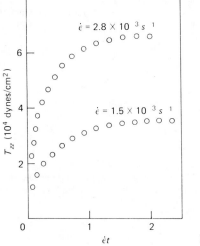

Figure 3-26 Data for Prob. 3-30.

3-31 Using the solution given in Prob. 3-14, calculate η^+ for PCF of a newtonian fluid.

3-32 The Maxwell model is dispensed with in the text following Eq. (3-125) with such brevity as to avoid any charge of cruelty. Yet it is beloved (if one judges on the basis of attention) by large groups of otherwise sensible polymer scientists. Comment on the difference in goals of the polymer physical chemist and the polymer process (chemical) engineer, and justify the concern (or lack thereof) of each with regard to the Maxwell model.

3-33 Find Δ^2 [Eq. (3-120)] for SSF and for SEF.

3-34 Equations (3-126) and (3-127) are valid as definitions of time derivatives of *any* second-order tensor (in cartesian components). Replace τ by Δ in those definitions, and find the components of each time derivative of Δ for two cases: SSF and SEF. Compare the results with Δ^2 from Prob. 3-33.

3-35 Establish Eqs. (3-129) to (3-131).

3-36 It is sometimes useful to rewrite algebraically complicated equations in terms of dimensionless variables and parameters. Rewrite Eqs. (3-136) to (3-138) in terms of the following:

A dimensionless time: $\quad T = \dfrac{t}{\lambda}$

A dimensionless shear rate: $\Gamma = \lambda\dot{\gamma}$

A dimensionless parameter: $B = \dfrac{\beta}{\lambda\eta_0}$

Now, prepare a set of graphs of

$$\frac{\eta^+}{\eta_0} \text{ versus } T \quad \text{ and } \quad \frac{\Psi_{12}^+}{2\lambda\eta_0} \text{ versus } T$$

for $\Gamma = 0.1, 1, 2$ and $B = -0.1, -0.5$.

3-37 Using the same variables as in Prob. 3-36, plot the steady-state values of η/η_0 and $\Psi_{12}/2\lambda\eta_0$ as functions of Γ, with B as a parameter. Use $B = -0.1$ and -0.5, and carry Γ out to the point that $\eta/\eta_0 = 0.1$.

3-38 Using Eq. (3-135), calculate η_e^+, and plot $\eta_e/3\eta_0$ versus $E = \lambda\dot\epsilon$, with $B = \beta/\lambda\eta_0$ as a parameter. Plot $\eta_e^+/3\eta_0$ versus T for $E = 0.1$ and 0.4.

3-39 Derive Eqs. (3-147) and (3-148), and find, in addition, the function Ψ_{23} for the Jaumann-Maxwell fluid [Eq. (3-146)].

3-40 Find η_e for the fluid defined by Eq. (3-149), with η given by Eq. (3-155) and λ by Eq. (3-150). Is $d\eta_e/d\dot\epsilon$ positive for all choices of material constants?

3-41 Repeat Prob. 3-40, but use the Jaumann derivative in Eq. (3-149). Give a qualitative sketch of a double-logarithmic plot of η_e versus $\dot\epsilon$.

3-42 Find the transient shear functions η^+, Ψ_{12}^+, and Ψ_{23}^+ for the Jaumann-Maxwell fluid [Eq. (3-146)].

BIBLIOGRAPHY

3-4 The dynamic equations (equations of motion)

A good general reference covering much of the material in this chapter as well as in several others is

Bird, R. B., W. E. Stewart, and E. N. Lightfoot: "Transport Phenomena," John Wiley & Sons, Inc., New York, 1960.

3-7 Constitutive equations

A more comprehensive treatment of this and related topics is

Middleman, S.: "The Flow of High Polymers," Interscience Publishers, a division of John Wiley & Sons, Inc., New York, 1968.

On the topic of elongational flows, recommended reading includes

Dealy, J. M.: Extensional Flow of Non-Newtonian Fluids—A Review, *Polym. Eng. Sci.*, **11:**433(1971).

as well as some of the experimentally oriented papers noted below.
The data shown in Fig. 3-16 are from

Chen, I-J., and D. C. Bogue: Time-Dependent Stress in Polymer Melts and Review of Viscoelastic Theory, *Trans. Soc. Rheol.*, **16:**59(1972).
Stevenson, J. F.: Elongational Flow of Polymer Melts, *AIChE J.*, **18:**540(1972).
Huppler, J. D., E. Ashare, and L. A. Holmes: Rheological Properties of Three Solutions. Part I. Non-Newtonian Viscosity, Normal Stresses, and Complex Viscosity, *Trans. Soc. Rheol.*, **11:**159 (1967); Part II. Relaxation and Growth of Shear and Normal Stresses, ibid, **11:**181(1967).
Ashare, E.: "Rheological Properties of Monodisperse Polystyrene Solutions," Ph.D. thesis, University of Wisconsin, Madison, 1968.
Olabisi, O., and M. C. Williams: Secondary and Primary Normal Stresses, Hole Error, and Reservoir Edge Effects in Cone-and-Plate Flow of Polymer Solutions, *Trans. Soc. Rheol.*, **16:**727(1972).

An excellent survey article which deals principally with viscoelastic constitutive equations utilizing the corotational (Jaumann) time derivative, and which compares theory with data, is

Bird, R. B., O. Hassager, and S. I. Abdel-Khalik: Co-Rotational Rheological Models and the Goddard Expansion, *AIChE. J.*, **20**:1041(1974).

The data of Fig. 3-18 may be found in

Carreau, P. J.: Rheological Equations from Molecular Network Theories, *Trans. Soc. Rheol.*, **16**:99 (1972).

The data of Fig. 3-19 are discussed and compared with theory in the Bird, Hassager, and Abdel-Khalik paper noted above. The data are from

Meissner, J.: Deformationsverhalten der Kunststoffe im flüssigen und im festen Zustand, *Kunststoffe*, **61**:576, 688(1971).

A related paper in English is

————: Modifications of the Weissenberg Rheogoniometer for Measurement of Transient Rheological Properties of Molten Polyethylene under Shear. Comparison with Tensile Data. *J. Appl. Polym. Sci.*, **16**:2877(1972).

The data of Fig. 3-22 are from

Pearson, G. H.: "Elongational Flow of Polymer Solutions," Ph.D. thesis, Univ. of Massachusetts, Amherst, 1975.

FOUR

DIMENSIONAL ANALYSIS IN DESIGN AND INTERPRETATION OF EXPERIMENTS

A circle may be small, yet it may be mathematically as beautiful and perfect as a large one.

Disraeli

In many polymer processes the geometry of the boundaries is so complex that it is not possible to obtain analytical solutions to the dynamic equations, even for the newtonian fluid. Numerical solutions, using the digital computer, may likewise be impractical in terms of time and expense. (We note that the reservation on "time," in this case, refers principally to programming time. A complex problem, of a type which has some novelty to it relative to the experience of those who take responsibility for its solution, may easily require several person-months for the development of a working programming scheme.) In many cases, especially when the process of interest becomes "pathological" while in production, some kind of analysis is required within a relatively short time span. Often a well-designed experiment can provide suitable answers with regard to questions of design and operation.

Especially in the case of process equipment which is in active commercial production, it is often impossible to do experiments with the actual system of interest. Likewise, cost and time considerations may make it impossible to purchase or build an identical system. One often turns, then, to experiments with scale models. In this section we examine rational methods of designing and interpreting experiments carried out on a physical scale different from that of the real system of interest.

4-1 THE PRINCIPLE OF DYNAMIC SIMILARITY

Suppose we have a fluid characterized by a viscoelastic model of the form

$$\tau + \lambda \frac{\delta \tau}{\delta t} = \eta_0 \, \Delta \tag{4-1}$$

where we take λ and η_0 to be constants. We may write the set of dynamic equations in the vector notation [Eq. (3-42) modified slightly by introducing τ for **T** and using, as the body force, the gravitational acceleration **g**].

$$\rho \frac{D\mathbf{u}}{Dt} = -\nabla p + \nabla \cdot \tau + \rho \mathbf{g} \tag{4-2}$$

and the continuity equation is, assuming incompressibility,

$$\nabla \cdot \mathbf{u} = 0 \tag{4-3}$$

Our first goal is to rewrite this set of equations in terms of dimensionless variables. This is done by first selecting a set of *characteristic parameters* of the specific part of the system that is being modeled. Usually the system is characterized by geometric and kinematic parameters. Some length scale L is usually selected as a geometric parameter. Typical choices might be the diameter of a tank where a fluid is being mixed, or the separation between the "lips" of a die through which fluid is being pumped. For a kinematic parameter one often selects a velocity U that characterizes the system, such as the linear velocity at the tip of an impeller stirring a fluid, or the average velocity through a die. We shall see that so long as L and U are well defined, the choice is not critical.

Let us assume, then, that some choice of L and U has been made. We define the following dimensionless variables:

All space variables are divided by L. If cartesian coordinates are used, for example, then the new variables become

$$x' = \frac{x}{L}$$

$$y' = \frac{y}{L} \quad \text{or} \quad \mathbf{x'} = \frac{\mathbf{x}}{L} \text{ in vector format}$$

$$z' = \frac{z}{L}$$

All components of velocity are divided by U, so that a dimensionless velocity becomes

$$\mathbf{u'} = \frac{\mathbf{u}}{U}$$

A dimensionless time is defined as

$$t' = \frac{Ut}{L}$$

All components of stress are made dimensionless by dividing by $\eta_0 \, U/L$, while pressure is normalized with ρU^2, with the result that

$$p' = \frac{p}{\rho U^2}$$

$$\tau' = \frac{\tau L}{\eta_0 \, U}$$

Then a dimensionless rate-of-deformation tensor $\mathbf{\Delta}'$ takes the form

$$\mathbf{\Delta}' = \frac{L}{U} \mathbf{\Delta}$$

Vector differential operators like ∇, which involve terms of the form $\partial/\partial x$, etc., become dimensionless by

$$\nabla' = L\nabla$$

If the steps outlined above are carried out through all the terms of Eqs. (4-1) to (4-3), a new set of *dimensionless* equations may be written as

$$\tau' + \text{Ws} \, \frac{\partial \tau'}{\partial t'} = \mathbf{\Delta}' \tag{4-4}$$

$$\frac{D\mathbf{u}'}{Dt'} = -\nabla'p' + \frac{1}{\text{Re}} \nabla' \cdot \tau' + \frac{1}{\text{Fr}} \frac{\mathbf{g}}{g} \tag{4-5}$$

$$\nabla' \cdot \mathbf{u}' = 0 \tag{4-6}$$

The following dimensionless groups (other than the dependent and independent variables) now appear:

- A Weissenberg number, $\text{Ws} = U\lambda/L$, which in a loose sense is a ratio of elastic stresses to viscous stresses in the flow.
- A Reynolds number, $\text{Re} = \rho UL/\eta_0$, which may be seen (again, loosely) to be a ratio of inertial stresses (convection of momentum) ρU^2 to viscous stresses $\eta_0 \, U/L$.
- A Froude number, $\text{Fr} = U^2/gL$, which gives a measure of the relative importance of inertial effects to gravitational effects. In the Froude number, the magnitude of $\mathbf{g}$ appears, and $\mathbf{g}/g$ is then a unit vector in Eq. (4-5).

No dynamic problem is completely defined solely through Eqs. (4-1) to (4-6). The boundary and initial conditions must also be considered, and when they are made dimensionless additional groups may appear. For example, if surface tension appears in a boundary condition, a Weber number $We = \rho U^2 L/\sigma$ will appear in the dimensionless boundary conditions. If a flow involves a pair of fluids, then two sets of equations of the form of Eqs. (4-1) to (4-6) are required. Usually one uses the same U and L for both sets of equations and finds a pair of Reynolds numbers and a pair of Weissenberg numbers. Depending on the boundary conditions at the interface between the pair of fluids, one will usually find a ratio of the two viscosities, and possibly a ratio of the relaxation times, appearing in the final formulation of the problem. In addition, purely geometric dimensionless parameters will occur through the boundary conditions, such as the ratio of length to width of a die, or the ratio of length to diameter of a pipe, or the ratio of impeller diameter to tank diameter in a mixer.

Now we make a deceptively simple, but very important, statement. The dependent variables of the set of Eqs. (4-1) to (4-6) depend *only* on those independent variables, and those parameters, that appear in the equations *and the boundary conditions*. Hence we can write

$$\tau' = \tau'(\mathbf{x}', t'; \text{ Re, Fr, Ws, geometrical parameters, and any other dimensionless}$$
groups that enter through the boundary conditions of the specific
problem) $\qquad$ (4-7)

and likewise for $\mathbf{u}'$ and p'.

The significance of this idea is the following. Suppose we subject two fluids to "similar processes." The word *similar* is to be taken in its usual meaning but also specifically to imply *geometric similarity* of the boundaries of the system, such as the shape of the mixing element in the two systems or the shape of the die through which the fluid is moving. The *scale*, i.e., the *size*, may be different, but the shape must be the same. The latter is guaranteed if dimensionless geometric parameters are the same in both systems.

Suppose we choose corresponding characteristic parameters U and L for each system. By *corresponding* we do not mean having the same magnitude but rather that, apart from the fact that the scale of the two systems is different, the definition of U and L is the same in the two systems.

Suppose, further, that the two fluids are describable by the same form of constitutive equation. Again, we allow for the possibility that η_0 and λ may be different in *value* in each fluid, but the *form* of applicable constitutive equation must be the same.

Each system is then completely defined by a set of equations of the form of Eqs. (4-1) to (4-6), and each solution satisfies Eq. (4-7). If *all* dimensionless parameters are identical in the two systems—not just the geometric parameters but also the dynamic parameters such as Re and Ws—then we say that the two systems satisfy *dynamic similarity*.

Our final statement, which follows directly from Eq. (4-7), is that the dependent *dimensionless* variables are identical in dynamically similar systems.

Thus, while a solution of the appropriate equations may not be possible, we are assured that if an experiment is performed on some system, that experiment gives information about dynamically similar systems. This principle of dynamic similarity can then be used in the design and interpretation of experiments with scale models of process equipment.

4-2 SOME APPLICATIONS

Pressure Drop for Flow through a Film Die

Preparations are being made to produce film by extruding a molten polymer through a film die 10 ft in width. The die has not yet been delivered by the manufacturer, but to design a control system for the process it is necessary to have a good estimate of the expected relationship between pressure drop and flow rate through the die. The internal geometry of the die is quite complex, and it is not likely that a mathematical model can be developed in a short period of time which can estimate pressures with better precision than a factor of 2. The decision is made to perform a series of experiments with a scale model of the die. We wish to examine some problems associated with design and interpretation of suitable experiments.

The following physical property data are available:

- The fluid may be characterized by Eq. (4-1), with $\eta_0 = 80$ P, $\lambda = 0.01$ s, and $\rho = 1$ g/cm^3.
- The model die is one-tenth scale of the full-size system.
- Expected flow rates in the full-scale system are such that the linear speed of extruded melt at the die exit is in the range of 10 to 100 cm/s.

Boundary conditions on this flow would specify no slip on the interior surfaces of the die. Only *geometric* dimensionless groups (shape factors) would appear in such boundary conditions.

It is usually the case that gravitational forces play no role in the dynamics of this kind of extrusion process. Hence no Froude number similarity need be considered.

Since the model die is "to scale," this implies geometric similarity. To ensure *dynamic* similarity, then, it is necessary to perform experiments in the model that match the Reynolds and Weissenberg numbers characteristic of the full-scale flow.

Let us choose L to be the width of the die, 300 cm in full scale and 30 cm in model scale. For the characteristic velocity U we will use the linear speed of the extruded melt.

We can make the following calculations of the expected range of Reynolds and Weissenberg numbers at full scale:

U	L	ρ	η_0	λ	Ws	Re
10	300	1	80	0.01	0.00033	37.5
100	300	1	80	0.01	0.0033	375

Suppose we inquire into how we design a one-tenth scale experiment to have the same range of Ws and Re. We require, at the lowest flow rate,

$$\text{Re} = \frac{UL\rho}{\eta_0} = 37.5 \tag{4-8}$$

$$\text{Ws} = \frac{U\lambda}{L} = 0.00033 \tag{4-9}$$

We have already fixed L at 30 cm. Hence we must pick values of U, λ, and η_0/ρ to satisfy Eqs. (4-8) and (4-9). (Since ρ for most liquids of interest has such a small range of variability, its exact value is not a very important factor, and we simply incorporate ρ into the viscosity. η_0/ρ is called the *kinematic viscosity*.) Since we have two equations which constrain these three variables, there is no unique "solution," no unique set of experimental conditions which will give dynamic similarity.

The situation is somewhat more complicated than this latter comment would suggest, however. If we divide one constraint by the other, we find

$$\frac{\text{Re}}{\text{Ws}} = \frac{L^2\rho}{\lambda\eta_0} = 1.125 \times 10^5 \tag{4-10}$$

Since L is specified to be 30 cm, this gives

$$\lambda\frac{\eta_0}{\rho} = 8 \times 10^{-3} \tag{4-11}$$

The significance of this result is the fact that λ and η_0/ρ are not independent; they are constrained to satisfy Eq. (4-11). Since both λ and η_0/ρ are rheological properties of a fluid, it is not at all obvious that one *can* find any fluid with properties that satisfy Eq. (4-11). In particular, it is obvious that the actual fluid of interest cannot be used, since for that fluid

$$\lambda\frac{\eta_0}{\rho} = 0.01\left(\tfrac{80}{1}\right) = 8 \times 10^{-1}$$

Let us suppose that three fluids are prepared, and a series of experiments (pressure drop versus flow rate) are carried out in the scale model. The (hypothetical) results are tabulated in Table 4-1.

The pressure drop versus flow rate data have been normalized so that we can plot $\Delta P/\rho U^2$ versus Re, with Ws as a parameter. Figure 4-1 shows the results.

At the extremes of the full-scale process (Re, Ws = 37.5, 3.3×10^{-4} and Re, Ws = 375, 33×10^{-4}) we can locate "operating points" with reasonable confidence, since we have closely matched the Re, Ws range in the model experiment. We have arbitrarily chosen an operating point at an Re of 150 (corresponding to $U = 40$) for which Ws is 13.3×10^{-4}. The closest Ws in the model experiment was 16.5×10^{-4}. We have located our best estimate of the expected $\Delta P/\rho U^2$ by assuming that the value would be above the "data" for

Table 4-1 Results of experiments in scale model of die*

	U	$10^4 \, \Delta P$	$10^4 \, Ws$	Re	$10^4 \, \Delta P/\rho U^2$
$\eta = 1$†	1	1	0	30	1
	2	2.1	0	60	0.52
$\lambda = 0$†	5	4.8	0	150	0.19
	10	9	0	300	0.09
$\eta = 1$	1	0.5	3.3	30	0.5
	2	0.96	6.6	60	0.24
$\lambda = 0.01$	5	2.4	16.5	150	0.096
	10	4.0	33	300	0.04
$\eta = 0.5$	1	0.3	3.3	60	0.3
	2	0.6	6.6	120	0.15
$\lambda = 0.01$	5	1.5	16.5	300	0.06
	10	3.0	33	600	0.03

* All units in cgs. $L = 30$, $\rho = 1$ in all experiments.
† Newtonian.

$Ws = 16.5 \times 10^{-4}$ but below the extrapolation that would pass through $Ws = 6.6 \times 10^{-4}$.

From these results, then, we estimate that the pressure drops would be as shown below:

Full-scale system—predicted results

U	Re	$10^4 \, Ws$	$10^4 \, \Delta P/\rho U^2$	$10^4 \, \Delta P$
10	37.5	3.3	0.42	42
40	150	13.3	0.12	192
100	375	33	0.036	360

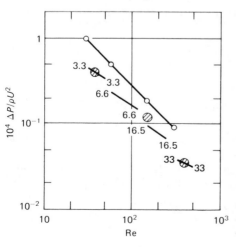

Figure 4-1 Pressure drop versus flow rate data from Table 4-1, plotted in dimensionless form. Numbers within the figure represent values of $10^4 \, Ws$; the decimal point represents the particular datum point. Large shaded circles represent the extremes and one intermediate point of the full-scale operation.

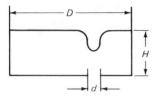

Figure 4-2 Geometry of a tank of fluid draining through the outlet pipe.

Draining a Large Tank

After a batch reaction a polymerization tank is to be drained. As material is pumped out through the bottom pipe a vortex may form at the surface. When the level H is low, the vortex may be sucked into the pipe, entraining air. This must be avoided (see Fig. 4-2).

We wish to find the minimum liquid level H^* before air entrainment as a function of the volumetric drawoff rate Q by performing a model study. The expected full-scale operation is characterized by

$$Q = 50 \text{ to } 100 \text{ gal/min}$$

$$\rho = 1 \text{ g/ml}$$

$$\eta = 100 \text{ P, and the fluid may be considered newtonian}$$

$$D = 10 \text{ ft and } d = 1 \text{ ft}$$

Design a model experiment using a 1-P oil as the model fluid. Take $\rho = 1$ g/ml for the oil.

In this system the vortex shape may depend on gravity, so we should expect to have to deal with both the Reynolds number and the Froude number. While there is a free surface in this problem, we know from experience that surface tension effects will not be important in a large-scale system. (The radius of curvature would be too large to give rise to any significant pressure effect.)

If dynamic similarity is attained between the model and the tank, then we expect that

$$\left(\frac{H^*}{D}\right)_{\text{model}} = \left(\frac{H^*}{D}\right)_{\text{tank}} \tag{4-12}$$

when

$$\text{Re}_{\text{model}} = \text{Re}_{\text{tank}} \tag{4-13}$$

and

$$\text{Fr}_{\text{model}} = \text{Fr}_{\text{tank}} \tag{4-14}$$

We will use D as the characteristic length scale in this system. For a linear velocity U we could use the mean velocity in the drawoff pipe,

$$U' = \frac{4Q}{\pi d^2} \tag{4-15}$$

It will be simpler, however, to define U as

$$U = \frac{Q}{D^2} \qquad (4\text{-}16)$$

The difference between these two definitions is clearly that

$$U' = \frac{Q}{D^2} \frac{D^2}{d^2} \frac{4}{\pi} = kU \qquad (4\text{-}17)$$

The constant k is just $4D^2/\pi d^2$, which is just some pure number dependent on the shape of the tank. If we have geometric similarity then k will be the same in both the tank and the model. Hence we can arbitrarily use Eq. (4-16) for the definition of U. Furthermore, we do not have to use a consistent set of units (i.e., we can use gallons, feet, and poise) so long as we use the same units to describe both the model and the tank, because this will again only introduce *scale factors*, this time as conversion factors.

Hence we will define Re and Fr as

$$\text{Re} = \frac{Q\rho}{D\eta_0} \qquad (4\text{-}18)$$

$$\text{Fr} = \frac{Q^2}{D^5} \qquad (4\text{-}19)$$

(Note that since g is the same in both systems, we will drop it from the definition of Fr.)

At 100 gal/min, we find

$$\text{Re}_{\text{tank}} = \frac{100(1)}{10(100)} = 0.10 \qquad (4\text{-}20)$$

(Note that because of inconsistent units the *magnitude* of Re is meaningless.)

$$\text{Fr}_{\text{tank}} = \frac{(100)^2}{(10)^5} = 0.10 \qquad (4\text{-}21)$$

We are apparently constrained to use a model fluid for which $\eta_0 = 1$ P. For convenience let us suppose that we build a scale model with $D = 1$ ft. To achieve an Re of 0.10, then, we need to use a Q of [see Eq. (4-18)]

$$Q = \frac{D\eta_0 \text{ Re}}{\rho} = \frac{1(1)(0.10)}{1} = 0.1 \text{ gal/min}$$

But then

$$\text{Fr}_{\text{model}} = \frac{Q^2}{D^5} = \frac{(0.1)^2}{(1)^5} = 0.01$$

which is not the same as the Fr_{tank}. What happened?

We cannot arbitrarily fix *both* the size of the model and the viscosity of the fluid. Since there are two constraints to be satisfied [Eqs. (4-20) and (4-21)], we must leave two variables unspecified initially. Thus, from the equations, D and Q must satisfy

$$\frac{Q\rho}{D\eta_0} = 0.1 \tag{4-22}$$

$$\frac{Q^2}{D^5} = 0.1 \tag{4-23}$$

If η_0/ρ is specified to be 1, then there is a unique solution of this pair of equations, namely,

$$D = 10^{-1/3} \text{ ft} \qquad Q = 10^{-4/3} \text{ gal/min}$$

Thus if we are constrained to use a fluid for which $\eta_0/\rho = 1$, then we must use a D of 0.465 ft and a Q of 0.047 gal/min to simulate the tank conditions at 100 gal/min.

At the lower flow rate of 50 gal/min we have

$$\frac{Q\rho}{D\eta_0} = 0.05 \tag{4-24}$$

$$\frac{Q^2}{D^5} = 0.025 \tag{4-25}$$

For $\eta_0/\rho = 1$ we find $D = 0.465$ ft, as before, and $Q = 0.024$ gal/min to simulate the tank conditions at 50 gal/min. We can make similar calculations at intermediate flow rates.

Thus the experiments should be performed in a model of $D = 0.465$ ft at drawoff rates in the range $0.024 \le Q \le 0.047$ gal/min. The relationship between the observed H^*_{model} and the expected H^*_{tank} is then, from Eq. (4-12),

$$H^*_{\text{tank}} = \frac{D_{\text{tank}}}{D_{\text{model}}} H^*_{\text{model}} \tag{4-26}$$

or

$$H^*_{\text{tank}} = \frac{10}{0.465} H^*_{\text{model}} = 21.5 H^*_{\text{model}}$$

A potential problem, that one might anticipate from physical intuition but that cannot be inferred from the mathematical analysis, is in the probability that H^*_{model} may be quite small and hence subject to significant error in measurement. For example, if H^*_{model} is 0.5 ± 0.125 in, the predicted H^*_{tank} is approximately 11 ± 3 in. In a 10-ft-diameter tank an uncertainty of 3 in in depth involves a significant mass of fluid, and the degree of uncertainty may be unacceptable. Such a problem will arise any time the scale factor between the model and the real system is very large.

Friction Factors

For flow through dies and conduits the principal information required is often the pressure drop as a function of flow rate. From Eq. (4-7), it follows that such information may be cast in the form

$$\frac{\Delta P}{\rho U^2} = F(\text{Re, Fr, geometrical parameters, ...})$$

Depending upon the form of constitutive equation, additional groups such as Ws may appear. The function F is not given through the principle of dynamic similarity; we only know that $\Delta P/\rho U^2$ depends on several dimensionless groups.

In some relatively simple cases it may be possible to obtain further information beyond what can be extracted from the similarity principle, even though the dynamic equations remain too complex to yield to analytical solution. We introduce a method called *inspectional analysis* and illustrate it in a relatively simple case.

Let us consider steady-state flow through a conduit of uniform cross section. For simplicity assume the fluid is newtonian, of viscosity μ. The cross section is arbitrary in shape, but it is assumed that the cross section does not vary in shape or scale with respect to the z axis (see Fig. 4-3).

It will be most convenient to use cylindrical coordinates, and it follows that the axial component of the dynamic equations, for a newtonian fluid in steady flow, and assuming no angular component of velocity u_θ, takes the form

$$\rho\left(u_r \frac{\partial u_z}{\partial r} + u_z \frac{\partial u_z}{\partial z}\right) = -\frac{\partial p}{\partial z} + \mu\left[\frac{1}{r}\frac{\partial}{\partial r}\left(r \frac{\partial u_z}{\partial r}\right) + \frac{\partial^2 u_z}{\partial z^2} + \frac{1}{r^2}\frac{\partial^2 u_z}{\partial \theta^2}\right] \quad (4\text{-}27)$$

If the conduit is sufficiently long the velocity field becomes fully developed, meaning that u_z is not a function of z for large z. From the continuity equation, since

$$\frac{\partial u_z}{\partial z} + \frac{1}{r}\frac{\partial}{\partial r}(ru_r) = 0 \quad (4\text{-}28)$$

it is clear that when the flow becomes fully developed the radial velocity u_r must vanish.

Equation (4-27) will be simplified first by eliminating u_r through the continuity equation, which is rearranged to the form

$$-u_r = \frac{1}{r}\int_0^r r' \frac{\partial u_z}{\partial z}\,dr' \quad (4\text{-}29)$$

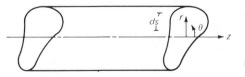

Figure 4-3 Section of a conduit of uniform cross section.

The second simplification is an approximation. In the viscous terms on the right-hand side of Eq. (4-27), $\partial^2 u_z / \partial z^2$ will vanish when the flow is fully developed, and is expected to be the smaller of the two viscous terms under all conditions. As a consequence, Eq. (4-27) is rewritten in the form

$$\rho\left(-\frac{\partial u_z}{\partial r}\frac{1}{r}\int_0^r r'\frac{\partial u_z}{\partial z}\,dr' + u_z\frac{\partial u_z}{\partial z}\right) = -\frac{\partial p}{\partial z} + \mu\left[\frac{1}{r}\frac{\partial}{\partial r}\left(r\frac{\partial u_z}{\partial r}\right) + \frac{1}{r^2}\frac{\partial^2 u_z}{\partial \theta^2}\right] \tag{4-30}$$

Since there are two unknowns u_z and p, a second equation is required. Again, an approximation will be made that is exact for fully developed flow. If the radial component of the dynamic equations is examined and u_r is assumed to be zero, it follows that

$$\frac{\partial p}{\partial r} = 0 \tag{4-31}$$

Subject to the validity of these approximations, u_z and p are found as solutions of Eqs. (4-30) and (4-31). Since the equations are too difficult to solve analytically, we seek some information through the method of inspectional analysis.

The method begins by normalizing the equations by defining dimensionless variables

$$u^* = \frac{u_z}{U} \qquad r^* = \frac{r}{D}$$

$$z^* = \frac{z}{D} \qquad p^* = \frac{p - p_0}{\rho U^2}$$

In the above, we take D as some characteristic linear dimension of the cross section, p_0 is the pressure (assumed known) at the conduit entrance $z = 0$, and U is the average velocity defined by

$$U = \int_0^{2\pi} \int_0^{R(\theta)} u_z r\,dr\,\frac{d\theta}{A_c} \tag{4-32}$$

where A_c is the cross-sectional area normal to z.

The normalized equations have the form

$$-\left(\frac{\partial u^*}{\partial r^*}\frac{1}{r^*}\int_0^{r^*} r'^*\frac{\partial u^*}{\partial z^*}\,dr'^* + u^*\frac{\partial u^*}{\partial z^*}\right)$$

$$= -\frac{\partial p^*}{\partial z^*} + \frac{1}{\text{Re}}\left[\frac{1}{r^*}\frac{\partial}{\partial r^*}\left(r^*\frac{\partial u^*}{\partial r^*}\right) + \frac{1}{r^{*2}}\frac{\partial^2 u^*}{\partial \theta^2}\right] \tag{4-33}$$

$$\frac{\partial p^*}{\partial r^*} = 0 \tag{4-34}$$

where $\text{Re} = UD\rho/\mu$ is a Reynolds number.

The boundary conditions are

$u^* = 0$ on $r^* = R^*(\theta)$, the conduit perimeter

$p^* = 0$ at $z^* = 0$, the conduit inlet

$u^* = 1$ at $z^* = 0$ (flat velocity profile at the inlet)

If one simply did dimensional analysis it would be possible to write

$$u^* = u^*(r^*, \theta, z^*, \mathrm{Re}) \tag{4-35}$$

and little else of value would follow.

The method of inspectional analysis seeks to remove all parameters from the equations. It is not always possible to do so, but in this case if a new axial variable is defined as

$$z^{**} = \frac{z^*}{\mathrm{Re}} \tag{4-36}$$

the equations become identical to Eqs. (4-33) and (4-34), except that z^* is replaced by z^{**} and Re does not appear in the equations or boundary conditions. Now one can write

$$u^* = u^*(r^*, \theta, z^{**}) \tag{4-37}$$

$$p^* = p^*(z^{**}) \tag{4-38}$$

and these will be sufficient to establish more useful results than follow from the simpler dimensional analysis.

Let us make a force balance across the conduit over an axial length L. The pressure drop ΔP over a length L must be balanced by the action of the shear stress at the conduit boundary:

$$\Delta P A_c = \int_0^L \oint_s \tau_{zn} \, ds \, dz \tag{4-39}$$

The shear stress τ_{zn} is the stress whose action is in the axial direction z, acting on an element of conduit surface whose normal is **n**. ds is the arc length along the perimeter s.

For a newtonian fluid we can write

$$\tau_{zn} = \mu \, \Delta_{zn} \tag{4-40}$$

If the shape of the perimeter is given in the form $r = R(\theta)$, then it would be possible, in theory, to relate Δ_{zn} to $\Delta_{z\theta}$ in cylindrical coordinates. It is unnecessary to do so in this development.

Now we introduce dimensionless variables as before, and in addition we normalize s with D and note that A_c would be proportional to D^2. After rearranging some of the algebra, Eq. (4-39) would be found to have the form

$$\frac{\Delta P}{\rho U^2} = \frac{1}{\mathrm{Re}} \int_0^{L/D} \oint_{s^*} \Delta_{zn}^* \, ds^* \, dz^* \tag{4-41}$$

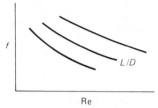

Figure 4-4 Qualitative sketch suggesting f (Re) with L/D as a parameter.

If we had stopped at Eq. (4-35) we could now state that

$$\frac{\Delta P}{\rho U^2} = F(\text{Re}, L/D) \tag{4-42}$$

While $\Delta P/\rho U^2$ is a suitable dimensionless pressure drop, it is customary when dealing with problems of steady flow through pipes and conduits of uniform cross section to define a *friction factor* as (see Prob. 4-8)

$$f = \frac{1}{4} \frac{D}{L} \frac{\Delta P}{\frac{1}{2}\rho U^2} \tag{4-43}$$

We still can say no more than $f = f$ (Re, L/D) at this stage, and the functional behavior, of course, is not given through dimensional analysis. Still, one could conclude that if a series of experiments were carried out in geometrically similar systems with newtonian fluids, a compact way to present the data would be through a family of curves as suggested in Fig. 4-4. This figure simply reflects graphically the statement that $f = f$ (Re, L/D). The shape of the curves cannot be inferred from that statement, however.

Now let us rearrange Eq. (4-41) in the following way: Replace dz^* by $dz^{**} = dz^*/\text{Re}$. Note that the upper limit of integration must then change from L/D to $(L/D)/\text{Re}$. Multiply both sides by Re $(D/2L)$. The result is

$$f \, \text{Re} = \frac{1}{2} \frac{\text{Re}}{L} \frac{D}{L} \int_0^{L/\text{Re }D} \oint_{s^*} \Delta_{zn}^{**} \, ds^* \, dz^{**} \tag{4-44}$$

where Δ_{zn}^{**} is the rate of deformation obtained by using Eq. (4-37), so that

$$\Delta_{zn}^{**} = \Delta_{zn}^{**}(\theta, z^{**}) = \Delta_{zn}^{**}(s^*, z^{**}) \tag{4-45}$$

For the special case of fully developed flow, for which Δ_{zn}^{**} does not depend on z^{**}, it follows, from Eq. (4-44), that

$$f \, \text{Re} = K \tag{4-46}$$

where

$$K = \frac{1}{2} \oint \Delta_{zn}^{**} \, ds^* = \text{constant} \tag{4-47}$$

K would depend only on the shape of the cross section.

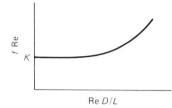

Figure 4-5 Sketch of f Re as inferred from inspectional analysis.

More generally, Eq. (4-44) may be put in the form

$$f \, \mathrm{Re} = \frac{\mathrm{Re}\, D}{L} F\left(\frac{\mathrm{Re}\, D}{L}\right) \tag{4-48}$$

The function F depends on the manner in which Δ^{**} depends on s^* and z^{**}, but since both integrals are definite integrals the function depends only on the parameter Re D/L.

Equation (4-48) is a much stronger result, because it tells us that we can take data and present them as a graph of the form of Fig. 4-5. A *single* curve suffices, and we now know that the curve is asymptotic to some constant at low values of Re D/L. (How do we know the asymptote is at *low* values?)

The advantage of proving that Fig. 4-5 is valid, in comparison to Fig. 4-4, is that fewer experiments are required to establish the functionality of the single line in Fig. 4-5. Thus experimental work may be guided by inspectional analysis and made more efficient.

PROBLEMS

4-1 Insofar as dimensional analysis is concerned, will the use of the Oldroyd time derivative alter any of the conclusions drawn by using the Jaumann derivative?

4-2 If, instead of Eq. (4-1), the fluid of interest is described by Eq. (3-119), what additional dimensionless parameters may appear in the formulation of the equations that describe a flow process?

4-3 Show that if τ is made dimensionless with ρU^2, Eqs. (4-4) and (4-5) have a different format. Would this alter any subsequent conclusions? Suppose p had been made dimensionless with $\eta_0 U/L$, as was τ. What would Eqs. (4-4) and (4-5) look like?

4-4 Carry out a dimensional analysis of the dynamics of a power law fluid by replacing Eq. (4-1) with

$$\tau = K(\tfrac{1}{2}\mathrm{II}_\Delta)^{(n-1)/2}\, \Delta$$

Give the dimensionless forms of the constitutive equation, dynamic equations, and continuity equation, and then simply list the dimensionless groups that appear in those equations.

4-5 Many "fluid" foods are described by a constitutive equation of the form

$$\tau = \tau_Y + K(\tfrac{1}{2}\mathrm{II}_\Delta)^{(n-1)/2}\, \Delta \qquad \text{if } \tau > \tau_Y$$

where τ_Y is a *yield-stress tensor*. The restriction to $\tau > \tau_Y$ is meant to imply that unless the stresses exceed the yield stress, no deformation occurs. This is a form of plastic behavior, then; the "fluid" is an elastic solid until it "yields" to become a fluid.

What dimensionless groups characterize the dynamics of such a material?

4-6 In a scale model of a sheeting die the following data are obtained:

ΔP, dynes/cm^2	Q, ml/s
10^5	2
3×10^5	5.5
10^6	15

The fluid can be considered newtonian, with $\eta_0 = 10^6$ P and $\rho = 1$ g/ml. The die exit is rectangular, of width $W = 10$ cm and with a lip separation of $H = 1$ cm.

(a) Use this information to calculate pressure drops to be expected in a sheeting die for which $W = 0.1$ cm, $H = 0.01$ cm, using the same fluid. The volume flow rate will be in the range 0.02 to 0.05 ml/s.

(b) If a less viscous fluid is used in the scale model, say, $\eta_0 = 10^4$ P, what range of values of Q would you study?

(c) For part b, calculate the expected range of ΔP. Does this result suggest any difficulties in performing the experiments?

4-7 Using Eqs. (4-18) and (4-19), show that the diameter of the scale model is related to the viscosity of the model fluid by

$$D = k'\eta_0^{2/3}$$

If we have a model for which $D = 1$ ft, what viscosity must be used?

4-8 Consider steady, fully developed flow in a long circular pipe. Make a force balance and show that f, as defined in Eq. (4-43), is

$$f = \frac{\tau_R}{\frac{1}{2}\rho U^2}$$

where τ_R is the shear stress at the pipe wall.

FIVE

SIMPLE MODEL FLOWS

There are many fine ideals which are not realizable, and yet we do not refrain from teaching them.

Smolenskin

We often use idealized models to form a basis for describing more complex systems. In flow processes the idealization may be with respect to

- Type of fluid
- Geometry
- Boundary conditions
- Thermodynamics (usually in the assumption of isothermal flow)

One set of models that we use a great deal in polymer-flow-process design and analysis involves most of these factors. In this section we set out the solutions for several flow processes and offer a few brief examples of applications. It will be seen that these simple models, or combinations and modifications of them, recur in the analyses of a variety of polymer flow processes to be described subsequently.

We divide the flows into two classes. In the first, usually called Poiseuille flow, the velocity field is generated by applying an external pressure to the fluid, and the boundaries of the system are rigid and stationary. The classical example is flow through a circular pipe across whose ends a pressure difference ΔP is maintained. The second class of flows is that in which there is *no* pressure gradient imposed on the system, but the boundaries may move in such a way as to induce a flow field;

85

the action of viscous effects is to allow the moving boundary to *drag* fluid along with it. Hence these flows are often called *drag flows*, but we will usually refer to them as *Couette flows*.

If sufficient simplifications are made, it is possible to obtain an analytical solution of the dynamic equations to give the velocity vector **u**. From the velocity field additional information follows easily. One usually wants the volumetric flow rate Q, and in the case of pressure-driven flows the relationship between Q and ΔP is desired. The relationship is often most conveniently cast into dimensionless form as a friction factor–Reynolds number relationship.

Because we will often be dealing with nonnewtonian fluids it is important to have a measure of the shear rate. Usually the shear rate varies throughout the flow field, and we simply calculate a "nominal" shear rate, often the maximum shear rate in the system, which gives a useful estimate for characterizing the state of deformation of the fluid.

5-1 POISEUILLE FLOW (PRESSURE FLOW)

Long Circular Pipe

Assume isothermal, laminar, fully developed flow down the axis of a long pipe of circular cross section. The flow has angular symmetry, so that $u_\theta = 0$ and $\partial/\partial\theta = 0$. For fully developed flow the radial velocity vanishes, $u_r = 0$, and the axial velocity u_z is independent of axial position z. Let the pipe diameter be $D = 2R$ and the length be L. For a newtonian fluid, defined by

$$\tau_{ij} = \mu \, \Delta_{ij} \tag{5-1}$$

the dynamic equations, subject to the assumptions above, take the form

$$0 = -\frac{\partial p}{\partial z} + \frac{\mu}{r}\frac{\partial}{\partial r}\left(r\frac{\partial u_z}{\partial r}\right) \tag{5-2}$$

$$0 = -\frac{\partial p}{\partial r} \tag{5-3}$$

From Eq. (5-3) it follows that p is not a function of radial coordinate; at most,

$$p = p(z) \tag{5-4}$$

Since u_z is a function only of r, when Eq. (5-2) is written in the form

$$\frac{dp}{dz} = \frac{\mu}{r}\frac{d}{dr}\left(r\frac{du_z}{dr}\right) \tag{5-5}$$

it is seen that the left-hand side is not a function of r, while the right-hand side is not a function of z. The only way this can be true is if both sides are constant:

$$\frac{dp}{dz} = c = \frac{\mu}{r} \frac{d}{dr}\left(r \frac{du_z}{dr}\right) \tag{5-6}$$

It follows that

$$p = cz + a \tag{5-7}$$

The boundary conditions on p will be taken in the form $p = 0$ at $z = L$ and $p = \Delta P$ at $z = 0$.

It follows that

$$c = -\frac{\Delta P}{L} \tag{5-8}$$

When the second-order differential equation for u_z is solved [the right-hand part of Eq. (5-6)], one finds that (assuming no slip at $r = R$)

$$u_z = \frac{\Delta P R^2}{4\mu L}\left[1 - \left(\frac{r}{R}\right)^2\right] \tag{5-9}$$

Once the velocity profile is obtained, several measures of interest may be calculated.

The volumetric flow rate Q is

$$Q = \int_0^R u_z(r)2\pi r \; dr = \frac{\pi \; \Delta P R^4}{8\mu L} \tag{5-10}$$

(This is known as the *Hagen-Poiseuille law*.)

The friction factor [Eq. (4-43)] f is

$$f = \frac{16}{\text{Re}} = \frac{D \; \Delta P}{2L\rho U^2} \tag{5-11}$$

where the Reynolds number for this flow is defined as

$$\text{Re} = \frac{UD\rho}{\mu} \tag{5-12}$$

In this definition U is the average velocity given by

$$U = \frac{4Q}{\pi D^2} \tag{5-13}$$

The shear rate is given by

$$\dot{\gamma} = \frac{\Delta P r}{2\mu L} \tag{5-14}$$

There is no shear along the center line ($r = 0$), and the maximum shear rate, often called the *nominal shear rate*, is given by

$$\dot{\gamma}_R = \frac{R\,\Delta P}{2\mu L} = \frac{8U}{D} \tag{5-15}$$

For a nonnewtonian fluid the power law provides a good model, from which it can be shown that

$$u_z(r) = \frac{nR}{1+n}\left(\frac{R\,\Delta P}{2KL}\right)^{1/n}\left[1 - \left(\frac{r}{R}\right)^{1+1/n}\right] \tag{5-16}$$

The volumetric flow rate is

$$Q = \frac{n\pi R^3}{1+3n}\left(\frac{R\,\Delta P}{2KL}\right)^{1/n} \tag{5-17}$$

and the nominal shear rate is

$$\dot{\gamma}_R = \frac{2(1+3n)}{n}\frac{U}{D} \tag{5-18}$$

Flow between Parallel Plates

The geometry is as shown in Fig. 5-1. Otherwise the conditions of flow are as in the case of pipe flow. Let the velocity vector be $\mathbf{u} = [u_x(y), 0, 0]$. For the newtonian fluid the dynamic equations give

$$0 = -\frac{\partial p}{\partial y} \tag{5-19}$$

$$0 = -\frac{\partial p}{\partial x} + \mu\frac{\partial^2 u_x}{\partial y^2} \tag{5-20}$$

One easily finds the velocity profile to be

$$u_x = \frac{B^2\,\Delta P}{8\mu L}\left[1 - \left(\frac{2y}{B}\right)^2\right] \tag{5-21}$$

It follows that the volumetric flow rate (per unit width W in the z direction) is

$$\frac{Q}{W} = \frac{B^3\,\Delta P}{12\mu L} \tag{5-22}$$

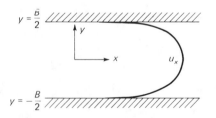

Figure 5-1 Geometry for flow between infinite parallel plates.

The friction factor is

$$f = \frac{24}{\text{Re}} \tag{5-23}$$

where $\text{Re} = UB\rho/\mu$ and the friction factor is defined as

$$f = \frac{B \, \Delta P}{L} \frac{1}{\frac{1}{2}\rho U^2} \tag{5-24}$$

The average velocity is given by $U = Q/BW$.

The nominal shear rate is

$$\dot{\gamma}_B = \frac{6U}{B} \tag{5-25}$$

For the power law fluid the corresponding results are found to be

$$u_x = \frac{nB}{2(1+n)} \left(\frac{B \, \Delta P}{2KL}\right)^{1/n} \left(1 - \left|\frac{2y}{B}\right|^{1+1/n}\right) \tag{5-26}$$

$$\frac{Q}{W} = \frac{nB^2}{2(1+2n)} \left(\frac{B \, \Delta P}{2KL}\right)^{1/n} \tag{5-27}$$

Axial Annular Flow

In this case the fluid is confined between concentric cylinders of length L and radii R_i and R_0. We assume the cylinders are stationary, and the velocity vector is

$$\mathbf{u} = [0, 0, u_z(r)] \tag{5-28}$$

so that $u_r = u_\theta = 0$. The dynamic equations are identical to those for the case of the circular pipe treated in Eqs. (5-2) and (5-3), and Eqs. (5-4) through (5-8) are valid for this case also. The solution to Eq. (5-6), subject to the boundary conditions $u_z = 0$ on $r = R_i$ and R_0, is

$$u_z = \frac{\Delta P R_0^2}{4\mu L} \left[1 - \left(\frac{r}{R_0}\right)^2 + \frac{1 - \kappa^2}{\ln(1/\kappa)} \ln \frac{r}{R_0}\right] \tag{5-29}$$

where $\kappa = R_i/R_0 < 1$.

The volumetric flow rate is

$$Q = \frac{\pi \, \Delta P R_0^4}{8\mu L} \left[1 - \kappa^4 - \frac{(1 - \kappa^2)^2}{\ln(1/\kappa)}\right] \tag{5-30}$$

The shear rate at the outer cylinder can be used as a nominal shear rate and is found to be

$$\dot{\gamma}_{R_0} = \frac{4U}{R_0 - R_i} G(\kappa) \tag{5-31}$$

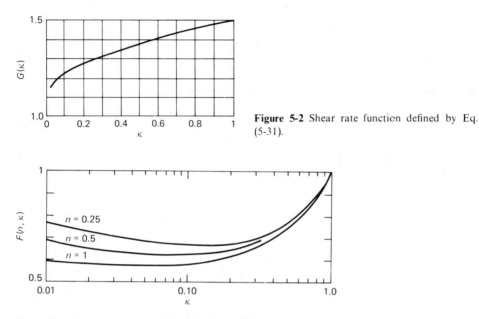

Figure 5-2 Shear rate function defined by Eq. (5-31).

Figure 5-3 Flow rate function defined by Eq. (5-32).

The function $G(\kappa)$ is plotted in Fig. 5-2. We note that the case $\kappa = 1$ gives the result for flow between parallel plates.

For the power law fluid one can, of course, obtain the velocity profile, but the method is somewhat tedious algebraically. We will not really need that result, although it may be useful to present the volumetric flow rate, which is

$$Q = \frac{n\pi R_0}{1 + 2n}(R_0 - R_i)^{2 + 1/n}\left(\frac{\Delta P}{2KL}\right)^{1/n}F(n, \kappa) \tag{5-32}$$

where $F(n, \kappa)$ is given in Fig. 5-3.

Flow in a Rectangular Duct

Figure 5-4 shows the geometry. We consider steady, laminar, fully developed flow down the z axis of a duct whose cross section in the xy plane is rectangular. From the dynamic equations we find that the axial pressure gradient is constant:

$$-\frac{\partial p}{\partial z} = \frac{\Delta P}{L} \tag{5-33}$$

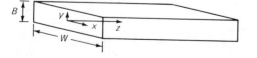

Figure 5-4 Geometry for flow within a rectanular duct.

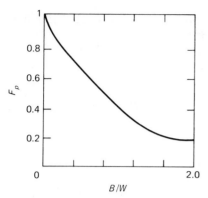

Figure 5-5 Shape factor for newtonian flow in a rectangular duct.

and the velocity profile $u_z(x, y)$ is the solution to

$$\frac{\partial^2 u_z}{\partial x^2} + \frac{\partial^2 u_z}{\partial y^2} = \frac{\Delta P}{\mu L} \tag{5-34}$$

subject to boundary conditions $u_z = 0$ on all four walls.

The solution is given by an infinite (Fourier) series, and the most useful result is the volumetric flow rate, found to be

$$Q = \frac{W B^3 \Delta P}{12 \mu L} F_p \tag{5-35}$$

F_p is a *shape factor*, given by an infinite series as a function of the *aspect ratio* W/B. The shape factor is plotted in Fig. 5-5.

For the power law fluid one must solve a nonlinear partial differential equation (the dynamic equation) to find u_z. No analytical solution is possible, and

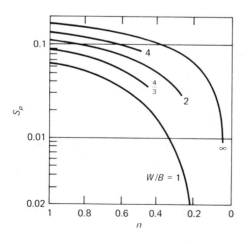

Figure 5-6 Shape factor for power law flow in a rectangular duct.

numerical methods must be used. The solution can then be numerically integrated to give Q, and the result may be cast in the form

$$Q = WB^2 \left(\frac{B \, \Delta P}{2KL} \right)^{1/n} S_p \qquad (5\text{-}36)$$

where S_p is a shape factor which depends on n and W/B. Figure 5-6 gives the function S_p.

Flow in Odd-shaped Ducts

In Chap. 4 it was shown that regardless of the shape of the cross section, so long as the shape does not vary down the axis of the duct, it is possible to express the pressure drop–flow rate relationship in the general form, *for laminar flow only*,

$$f \, \text{Re} = \text{constant} \qquad (5\text{-}37)$$

It was stated that arbitrary choices may be made for length scales, velocity scales, etc., the only effect of the choice being in the magnitude of the constant. We emphasize that the "constant" may be a function of geometric shape factors. Here we show that if some simple dynamic ideas are introduced into the calculation of the friction factor, there is a naturally occurring length scale, the hydraulic radius, which may be used in the definition of f and Re.

We assume fully developed flow down the axis of a duct of length L, whose cross section is some arbitrary shape of area A and perimeter Π. The axial component of shear stress along the perimeter will be denoted simply as τ. A force balance gives

$$A \, \Delta P = L \oint \tau \, d\Pi \qquad (5\text{-}38)$$

It is understood that τ may be a function of position along the perimeter, and the integration is along the closed contour of length Π.

Now let us define the average shear stress $\bar{\tau}$ as

$$\bar{\tau} = \frac{1}{\Pi} \oint \tau \, d\Pi \qquad (5\text{-}39)$$

and hence rewrite the force balance as

$$A \, \Delta P = \bar{\tau} L \Pi \qquad (5\text{-}40)$$

(Keep in mind that A is the cross-sectional area and $L\Pi$ is the area along which the shear stress acts.)

If we decide to always define the friction factor as

$$f = \frac{\bar{\tau}}{\frac{1}{2}\rho U^2} \qquad (5\text{-}41)$$

(and this is, in fact, the custom which is usually followed), then it follows that

$$f = \frac{\Delta P r_h}{L} \frac{1}{\frac{1}{2}\rho U^2} \tag{5-42}$$

where r_h is called the *hydraulic radius*, and is defined by

$$r_h = \frac{A}{\Pi} \tag{5-43}$$

We note that for a circular pipe r_h is given by $D/4$, and Eq. (5-42) gives the usual definition of the friction factor, the *Fanning* friction factor. One occasionally finds a friction factor defined without the factor of 4 which relates r_h and D, and in using an equation or graph which presents f as a function of Reynolds number, it is essential to check the definition of f.

By custom, the Reynolds number for flow down a duct is defined as

$$\mathrm{Re} = \frac{4r_h U \rho}{\mu} \tag{5-44}$$

so that in the case of the circular pipe the usual definition, with D the characteristic length, is obtained.

With these definitions for f and Re, one can prepare simple graphical presentations of the dependence of the f Re product on shape factors for a variety of cross-sectional shapes. In some cases analytical solutions of the dynamic equations are possible; in others it is necessary to use numerical methods or variational methods.

Figures 5-7 and 5-8 give f Re for ducts of rectangular and annular shape.

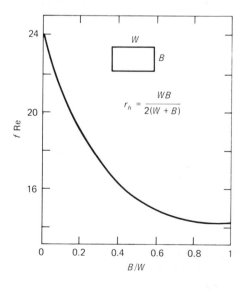

Figure 5-7 f Re for newtonian flow in a rectangular duct.

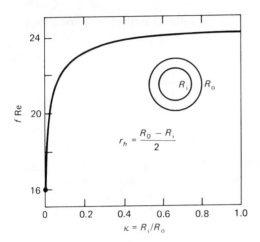

Figure 5-8 f Re for newtonian flow in an annular duct.

These particular figures contain the same information given earlier in Eq. (5-30) for an annular duct and in Eq. (5-35) (with Fig. 5-5 for F_p) for a rectangular duct.

An interesting feature of the graphical display, and to some extent its advantage over the analytical results, is that it is possible to see (literally) when flow in these complex-shaped ducts can be approximately represented by simpler models. In the case of the rectangular duct, for example, we see that when the width of the duct is great compared to the depth, the value of f Re is 24, which is the exact result for flow between infinite parallel planes [Eq. (5-23)]. Of more practical value, we see that when $B/W < 0.1$ the error in using the infinite-plate model for relating flow rate to pressure drop is less than 15 percent.

A similar examination of Fig. 5-8 shows that f Re approaches the infinite-parallel-plate result as κ approaches unity. Some thought reveals this to be ex-

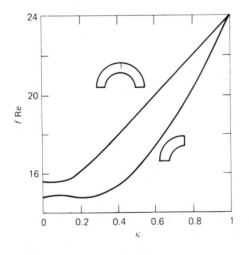

Figure 5-9 f Re for newtonian flow in segments of an annular duct.

pected. Of significance is the indication that the parallel plate model is a good approximation for $\kappa > 0.5$.

Figure 5-9 shows f Re for ducts which are sectors of a circle.

Short Tubes: Flow Not Fully Developed

In Chap. 4 it was shown that for laminar flow under conditions that the flow field was not fully developed, the f Re product was not constant but some function of Re D/L. In a real system, of course, the flow is not fully developed because the fluid enters the duct from some other region, and the velocity field must rearrange itself from its configuration in the upstream region. The axial distance over which this occurs is known as the *entrance region.*

Theoretical analysis of the entry problem is quite difficult because the dynamic equations involve both u_z and u_r as functions of r and z. The θ equation may be ignored, but one is left with a set of three partial differential equations for the unknowns u_z, u_r, and p. The two dynamic equations are nonlinear, and no analytical solution is possible. However, numerical solutions can be carried out, and several features of flow in "short" tubes can be examined.

Perhaps the first question is "What constitutes a 'short' tube?" For a circular cross section the axial length L_e required for the center-line velocity to adjust to within 99 percent of its fully developed value is found theoretically to be

$$\frac{L_e}{D} = 0.59 + 0.056 \text{ Re} \tag{5-45}$$

for the case that the upstream velocity profile (i.e., at the pipe entrance) is flat. The corresponding result for flow between parallel plates is

$$\frac{L_e}{B} = 0.63 + 0.044 \text{ Re} \tag{5-46}$$

The interesting point here is that no matter how small the Reynolds number, a finite length of the order of the diameter or width of the duct is required to produce the fully developed flow profile.

If the flow is not fully developed the f Re product is increased, and Fig. 5-10 shows its dependence on Re D/L.

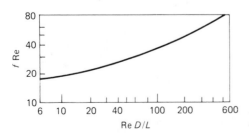

Figure 5-10 f Re for newtonian flow in the entry region of a circular pipe.

5-2 COUETTE FLOW (DRAG FLOW)

We turn now to consideration of flows where the velocity field is generated by the motion of the boundaries. No pressure gradient is imposed *on* the system, although in some cases the flow field may itself *establish* a pressure gradient. As in Sec. 5-1, flows where the geometry is particularly simple will be examined to provide models that will be useful in later considerations of polymer processes.

Plane Couette Flow

In this flow, generated between infinite parallel planes one of which moves relative to the other in its own plane, the dynamic equations for laminar isothermal flow reduce to

$$\frac{d}{dy}\tau_{xy} = 0 \tag{5-47}$$

Figure 5-11 shows a definition sketch for the analysis.

The solution of this equation is a constant shear stress. On the assumption that, at steady state, the shear stress is a function only of the shear rate of deformation, it follows then that the shear rate must also be constant. If it is assumed that the flow field is $\mathbf{u} = [u_x(y), 0, 0]$, it follows then that

$$\frac{du_x}{dy} = \text{constant} \tag{5-48}$$

The solution of this equation, satisfying the boundary conditions $u_x = 0$ at $y = 0$ and $u_x = U$ at $y = B$, is

$$u_x = \frac{Uy}{B} \tag{5-49}$$

The volumetric flow rate, per unit width W in the z direction, may be written in the form

$$\frac{Q}{UBW} = \frac{1}{2} \tag{5-50}$$

Equation (5-50) is valid for nonnewtonian as well as newtonian fluids.

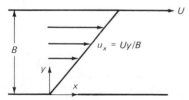

Figure 5-11 Geometry for plane Couette flow.

Circular Annular Couette Flow

In this flow the fluid is confined in the annular space between concentric cylinders of radii R_i and R_0. We assume that the inner cylinder rotates with linear velocity $R_i \Omega$, thereby generating an isothermal laminar flow for which $\mathbf{u} = [0, u_\theta(r), 0]$. We assume angular symmetry and no flow in the z direction.

For any fluid the dynamic equations take the form

$$\frac{-\rho u_\theta^2}{r} = -\frac{\partial p}{\partial r} \tag{5-51}$$

$$\frac{1}{r^2}\frac{\partial}{\partial r} r^2 \tau_{r\theta} = 0 \tag{5-52}$$

Thus we know that

$$p = \int \rho \frac{u_\theta^2}{r} \, dr \tag{5-53}$$

and

$$\tau_{r\theta} = \frac{a}{r^2} \tag{5-54}$$

where a is some constant.

Let us illustrate the solution for a power law fluid. Then

$$\tau_{r\theta} = K\left[-r\frac{d}{dr}\left(\frac{u_\theta}{r}\right)\right]^n = \frac{a}{r^2} \tag{5-55}$$

Solving for the derivative one finds

$$-\frac{d}{dr}\left(\frac{u_\theta}{r}\right) = \frac{1}{r}\left(\frac{a}{Kr^2}\right)^{1/n} = br^{-2/n-1} \tag{5-56}$$

or

$$\frac{u_\theta}{r} = \frac{br^{-2/n}}{2/n} + c \tag{5-57}$$

Appropriate boundary conditions are $u_\theta = R_i \Omega$ at $r = R_i$ and $u_\theta = 0$ at $r = R_0$ from which b and c may be calculated.

The final result for u_θ may be written in the form

$$\frac{u_\theta}{R_i\Omega} = \frac{r}{R_i}\frac{1-(R_0/r)^{2/n}}{1-\kappa^{-2/n}} \tag{5-58}$$

where, as before, $\kappa = R_i/R_0$. For the newtonian fluid one need only set $n = 1$ in Eq. (5-58).

As a nominal shear rate we may use the value at the inner (rotating) cylinder and find that (see Prob. 5-16)

$$\dot{\gamma}_{R_i} = \left[r\frac{d}{dr}-\left(\frac{u_\theta}{r}\right)\right]_{r=R_i} = \frac{(2/n)\Omega}{1-\kappa^{2/n}} \tag{5-59}$$

The case where the inner cylinder rotates is useful in several applications. The more general case where both cylinders rotate at different speeds is given in the problems at the end of this chapter.

Axial Annular Couette Flow

In this flow, as shown in Fig. 5-12, the fluid is dragged down the axis in the annular space between coaxial cylinders, where the inner cylinder is moving at velocity U in the z direction. The dynamic equations for this case reduce to

$$\frac{d}{dr} r\tau_{rz} = 0 \tag{5-60}$$

from which it follows that

$$\tau_{rz} = \frac{a}{r} \tag{5-61}$$

For a power law fluid

$$\tau_{rz} = K\left(-\frac{du_z}{dr}\right)^n \tag{5-62}$$

and it follows that the velocity profile, subject to boundary conditions $u_z = U$ at $r = R_i$ and $u_z = 0$ at $r = R_0$, is given by

$$\frac{u_z}{U} = \frac{1}{\kappa^q - 1} \left[\left(\frac{r}{R_0}\right)^q - 1\right] \tag{5-63}$$

where $q = 1 - 1/n$. The volumetric flow rate can be found and written in the form

$$\frac{Q}{2\pi R_0(R_0 - R_i)U} = \frac{1}{q+2} \frac{1 - \kappa^{q+2}}{(1-\kappa)(\kappa^q - 1)} - \frac{1+\kappa}{2(\kappa^q - 1)} \tag{5-64}$$

The reason for writing this in the form given above is that it is now similar to the solution for the case of PCF [Eq. (5-50)], since $R_0 - R_i$ is analogous to B, and the " unit width " W is similar to $2\pi R_0$. It can be shown that the right-hand side of Eq. (5-64) approaches the value of $\frac{1}{2}$ in the limit as $\kappa \to 1$, as should be expected (see Prob. 5-22).

The newtonian model cannot be obtained from the power law solution in this case upon setting $n = 1$ ($q = 0$). The newtonian results are

$$\frac{u_z}{U} = \frac{\ln (r/R_0)}{\ln \kappa} \tag{5-65}$$

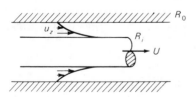

Figure 5-12 Geometry for axial annular Couette flow.

and

$$\frac{Q}{2\pi R_0 (R_0 - R_i) U} = -\frac{2\kappa^2 \ln \kappa - \kappa^2 + 1}{4(1 - \kappa) \ln \kappa} \tag{5-66}$$

The nominal shear rate (arbitrarily taken at the inner surface $r = R_i$) is given by

$$-\frac{du_z}{dr}\bigg|_{R_i} = \dot{\gamma}_{R_i} = -\frac{U}{R_0 - R_i} \frac{q(1 - \kappa)\kappa^{q-1}}{(\kappa^q - 1)} \tag{5-67}$$

For the newtonian case

$$\dot{\gamma}_{R_i} = -\frac{U}{R_0 - R_i} \frac{1 - \kappa}{\kappa \ln \kappa} \tag{5-68}$$

Again, the format of Eqs. (5-67) and (5-68), involving the factor $U/(R_0 - R_i)$, is dictated by the desire to have a format similar to that of the plane case, for which $\dot{\gamma} = U/B$ follows directly from Eq. (5-49). Since calculations are so much simpler for the plane geometry, it is useful to have an idea of the errors involved in using a plane approximation for the flow rate or for the nominal shear rate. We can illustrate this point in the case of the nominal shear rate.

For PCF the shear rate is uniform across the flow region and is given by U/B. For the newtonian case of axial annular flow, the shear rate varies across the annulus. Figure 5-13 shows the behavior predicted by differentiating Eq. (5-65).

The significant point is that the shear rate is highest at the moving surface, lowest at the fixed surface, and significantly different from the uniform value of the case $\kappa = 1$, for κ values less than $\kappa = 0.8$. However, if one only needs a rough estimate of the magnitude of the shear rate in the annular system, it is clear that even for κ values as small as $\kappa = 0.3$ the simple calculation of $U/(R_0 - R_i)$ is a reasonable order of magnitude estimate of the actual shear rate.

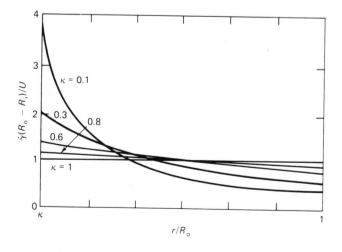

Figure 5-13 Radial variation of shear rate across an annular duct for newtonian Couette flow.

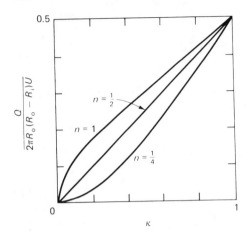

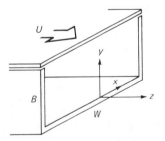

Figure 5-14 Dimensionless volumetric flow rate for axial Couette flow of a power law fluid in an annular duct.

By similar procedures one can examine the corresponding situation for the power law fluid.

One may also examine the volumetric flow rate, as given in Eqs. (5-64) or (5-66), to determine how good an estimate the PCF model provides. Figure 5-14 shows the dimensionless volumetric flow rate $Q/[2\pi R_0(R_0 - R_i)U]$ as a function of κ. Again it is clear that the plane model provides an order of magnitude estimate for the annular case but is accurate only for κ near unity. The more strongly nonnewtonian the fluid is, the more the plane model deviates from the accurate solution for the annular flow.

Flow down a Rectangular Channel

This is similar to the case of PCF except that there is a finite width W bounded by stationary walls which retard the flow induced by the motion of the upper plane (see Fig. 5-15). As in the pressure flow case one must solve a *partial* differential equation, and it is most convenient to put the solution into a format that allows comparison with Eq. (5-50). Thus we solve

$$0 = \frac{\partial^2 u_z}{\partial x^2} + \frac{\partial^2 u_z}{\partial y^2} \tag{5-69}$$

Figure 5-15 Geometry for Couette flow in a rectangular duct.

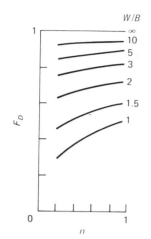

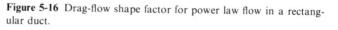

Figure 5-16 Drag-flow shape factor for power law flow in a rectangular duct.

subject to $u = 0$ on $y = 0$, $x = \pm\frac{1}{2}W$, and $u_z = U$ on $y = B$. The solution is found in the form of an infinite (Fourier) series, and the volumetric flow rate follows from

$$Q = \int_0^B \int_{-W/2}^{W/2} u_z(x, y) \, dx \, dy \qquad (5\text{-}70)$$

We write the solution as

$$Q = \tfrac{1}{2}UBWF_D \qquad (5\text{-}71)$$

where F_D is a drag-flow shape factor that depends only on the aspect ratio of the duct. Figure 5-16 gives F_D, and includes as well the results for the power law case.

5-3 APPLICATIONS TO ANALYSIS OF WIRE COATING

A wire-coating operation is designed in the following manner. Molten polymer is fed to a circular die, along whose axis a wire is continuously drawn. Figure 5-17 shows the geometry of interest. We seek the relationship between the coating thickness δ and the other parameters that characterize the problem.

The motion of the wire entrains the polymer and draws it through the die. This is a case of axial annular drag flow. We assume steady-state isothermal flow

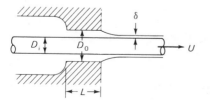

Figure 5-17 Geometry of a wire-coating die.

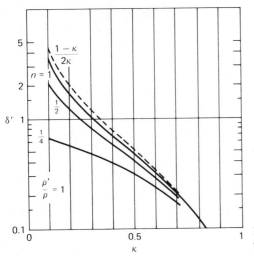

Figure 5-18 Dimensionless coating thickness in wire coating.

of a power law fluid. Some distance downstream of the die the polymer is transported as a rigid coating and so has a velocity identical to the wire, U. The mass flow rate of coating is simply

$$m = \rho\pi[(R_i + \delta)^2 - R_i^2]U \qquad (5\text{-}72)$$

where ρ is the downstream coating density. By continuity, m must be identical to the mass flow rate associated with the drag flow through the die:

$$m = \rho'Q \qquad (5\text{-}73)$$

where ρ' is the fluid density at the conditions within the die. For Q we may use the result given in Eq. (5-64). When m is eliminated between Eqs. (5-72) and (5-73), one obtains a quadratic equation for δ in the form

$$\delta'^2 + 2\delta' - \frac{\rho'}{\rho}\frac{2}{\kappa}\left(\frac{1}{\kappa} - 1\right)H(\kappa, q) = 0 \qquad (5\text{-}74)$$

where $\delta' = \delta/R_i$, and $H(\kappa, q)$ is the right-hand side of Eq. (5-64). [Figure 5-14 gives $H(\kappa, q)$.] Solving for δ', one finds

$$\delta' = \frac{\delta}{R_i} = \left[1 + \frac{\rho'}{\rho}\frac{2}{\kappa}\left(\frac{1}{\kappa} - 1\right)H(\kappa, q)\right]^{1/2} - 1 \qquad (5\text{-}75)$$

Figure 5-18 shows δ' as a function of these parameters.

One important conclusion that follows from these results is that the maximum coating thickness possible under the conditions of this model is

$$\delta_{max} = \tfrac{1}{2}(R_0 - R_i) \qquad (5\text{-}76)$$

which is achieved only in the limit of $\kappa \to 1$ (Prob. 5-28).

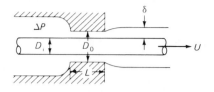

Figure 5-19 Wire coating with an imposed pressure.

Another interesting feature, which follows from Eq. (5-75), is the independence of δ from the velocity U. Indeed, it would appear that the only way in which δ may be varied is through changes in the geometry of the die for a given fluid.

But suppose one designed a die and then found that the observed coating thickness was not exactly at the desired level. Alternatively, suppose that a change in the end-use conditions dictated a change in coating thickness. The system as described by Eq. (5-75) is too inflexible to allow control over δ without a die change.

There is, however, a means of introducing the desired feature of control over the coating thickness in the same die. This is achieved by imposing a pressure on the fluid upstream of the die. We illustrate some features of this problem by using a newtonian model, which will simplify the algebra and make it possible to investigate several aspects of the solution. The power law case is most conveniently done numerically.

The flow may be defined as shown in Fig. 5-19. A pressure drop ΔP is imposed on the fluid, and the net flow is the sum of that induced by the drag of the wire and that imposed by the pressure itself. For a newtonian fluid these two contributions are independent and additive. Hence we may begin by adding the Q for the pressure flow [Eq. (5-30)] to the drag flow contribution [Eq. (5-66)], with the result written as

$$Q = \frac{\pi R_0^2 U}{2} \left(\frac{1 - \kappa^2}{\ln (1/\kappa)} - 2\kappa^2 \right) + \frac{\pi \, \Delta P R_0^4}{8\mu L} \left[1 - \kappa^4 - \frac{(1 - \kappa^2)^2}{\ln (1/\kappa)} \right] \quad (5\text{-}77)$$

If Eq. (5-72) for the mass flow rate is equated to $\rho'Q$, as in the pure drag flow case above, one may solve for the (dimensionless) coating thickness and find

$$\delta' = \frac{\delta}{R_i} = (1 + f_d + f_p)^{1/2} - 1 \quad (5\text{-}78)$$

where

$$f_d = \frac{\rho'}{\rho} \left(\frac{1 - \kappa^2}{2\kappa^2 \ln (1/\kappa)} - 1 \right) \qquad f_p = \frac{\rho'}{\rho} \Phi \frac{1}{8\kappa^2} \left[1 - \kappa^4 - \frac{(1 - \kappa^2)^2}{\ln (1/\kappa)} \right]$$

and

$$\Phi = \frac{\Delta P R_0^2}{\mu U L}$$

Figure 5-20 shows δ' as a function of κ, with Φ as a parameter. For a given die

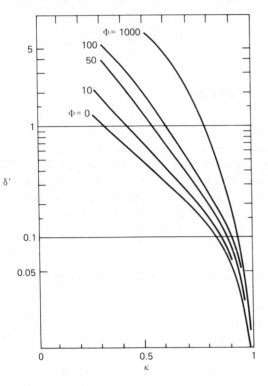

Figure 5-20 Dimensionless coating thickness in wire coating with an imposed pressure: newtonian fluid.

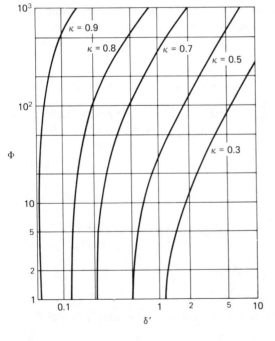

Figure 5-21 Pressure required to produce a given coating thickness: newtonian fluid.

(κ) and at a given pressure drop (Φ) one may find δ' from the graph. Alternatively one may rearrange Eq. (5-78) to solve for the pressure term as

$$\Phi = \frac{8\kappa^2}{\left[1 - \kappa^4 - \dfrac{(1-\kappa^2)^2}{\ln (1/\kappa)} \right]} \frac{\rho}{\rho'} (\delta'^2 + 2\delta' - f_d) \tag{5-79}$$

Figure 5-21 shows Φ as a function of δ' with κ as a parameter; this format is more useful if one specifies the coating thickness (δ') and seeks the pressure required to achieve this coating.

One feature of design significance that can be seen on inspection of Fig. 5-21 relates to the sensitivity of the coating thickness to pressure and velocity fluctuations. The nearly vertical character of the $\Phi(\delta')$ curves implies a small slope for $d\delta'/d\Phi$, that is, a relative insensitivity of coating thickness to the parameter Φ, which (from its definition above) is seen to involve three parameters that may vary (perhaps unintentionally) in a process: ΔP, μ, and U. Thus the system is fairly insensitive to inadvertent process fluctuations, but by the same token, if one wishes to change δ' for a fixed die, it is necessary to make a large (order of magnitude) change in Φ.

It would appear then that imposing a pressure onto the drag flow gives the desired capability of making small changes in the coating thickness (a sort of "fine tuning"), but if large changes are desired one must either change the die or have the capability of varying operating conditions (particularly U and ΔP) over a wide range.

Let us finish this aspect of the wire-coating example with a numerical calculation. Suppose we have a die whose dimensions are $R_0 = 0.1$ cm, $R_i = 0.07$ cm, $L = 1$ cm. The fluid viscosity at die conditions is 10 P. We wish to coat the wire so that $\delta = 0.021$ cm, and the production rate is such that the wire speed is 100 ft/s.

We may calculate δ' and find $\delta' = \delta/R_i = 0.021/0.07 = 0.30$. For this die we have $\kappa = 0.7$.

Quick inspection of Fig. 5-20 shows that a finite value of Φ is required, somewhere between 10 and 50. From Fig. 5-21 we find that

$$\Phi = 30 = \frac{\Delta P R_0^2}{\mu U L}$$

Solving for ΔP (remember to convert U to centimeters per second) one can find $\Delta P = 90 \times 10^6$ dynes/cm$^2 \doteq 1300$ psi. Suppose it is necessary to have a coating uniformity of better than ± 1 percent. Fluctuations in the wire speed U will lead to coating nonuniformities through a variation in Φ. What is the maximum tolerable variation in U under these operating conditions?

We can begin by deriving a useful general result. We go back to Eq. (5-78) and find

$$\frac{d\delta'}{df_p} = \frac{\frac{1}{2}}{\delta' + 1} \tag{5-80}$$

We need $d\delta'/d\Phi$, which follows from

$$\frac{d\delta'}{d\Phi} = \frac{d\delta'}{df_p}\frac{df_p}{d\Phi} = \frac{\frac{1}{2}}{\delta' + 1}\frac{f_p}{\Phi} \tag{5-81}$$

The relative (or fractional) change in δ' with respect to some fractional change in Φ (which corresponds to a change in U) is

$$\frac{d\ln\delta'}{d\ln\Phi} = \frac{\Phi}{\delta'}\frac{d\delta'}{d\Phi} = \frac{1}{2}f_p\frac{1}{\delta'(1 + \delta')} \tag{5-82}$$

In this example the operating conditions are such that $f_p = 0.23$ and $\delta' = 0.3$. Thus we find that

$$\frac{d\ln\delta'}{d\ln\Phi} = 0.29$$

or the fractional change in δ', due to a change in Φ, is only 29 percent of the fractional change in Φ. Hence, if δ' must be held to ± 1 percent, we can tolerate speed variations in the wire travel of as much as 3.4 percent.

Now suppose a process change is ordered, requiring an increased output of coated wire, which will be achieved by increasing U to 300 ft/s. If δ is to be unchanged, and the same die is used, what ΔP is required? What effect does this have on sensitivity of δ to U?

Since δ' and κ are unchanged we still require $\Phi = 30$. Since U is increased by a factor of 3, the pressure will have to be increased by the same factor. Thus a pressure increase, to 4000 psi, is required. Since no change in δ' or f_p occurs it is seen, from Eq. (5-82), that there is no change in sensitivity of δ' to U.

Some Failures of This Model

In dealing with models such as this one, there is always the danger of getting so involved in the details of the model that one loses sight of the fact that the model represents an idealization—an approximation—of reality. One particular feature that should be mentioned here is the role that surface tension may play in this coating problem. In the limit of $\kappa \to 0$ one is dealing with the problem of *withdrawal coating*, which will be considered later. In that problem a sheet or wire (or fiber) is withdrawn (usually vertically) from a free surface of the coating bath. Here surface tension plays a strong role in establishing the dynamic meniscus from which the wire is withdrawn.

The case of combined pressure and drag flow was solved above only for the newtonian fluid. The power law case is much more complicated and has to be worked out numerically. We note here, and discuss the point in detail later, that one cannot superimpose power law solutions [Eqs. (5-64) and (5-32)] for drag flow and pressure flow, as was done with the newtonian fluid. A different type of combined flow (helical flow) is considered in Sec. 5-5, and in that case a useful approximate analytical solution is possible.

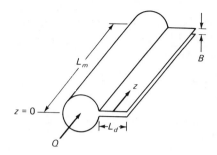

Figure 5-22 End-fed sheeting die.

5-4 ANALYSIS OF UNIFORMITY OF FLOW FROM AN END-FED SHEETING DIE

This example considers the behavior of an end-fed sheeting die, shown schematically in Fig. 5-22. Fluid is supplied to one end of a manifold which may be considered to be a circular pipe of radius R and length L_m, closed at one end $(z = L_m)$. A slit runs along the manifold, providing the inlet to a sheeting die of length L_d. The separation between the parallel die surfaces is B.

Fluid enters the manifold and moves toward the closed end under the action of an axial pressure gradient. There is a "leakage" flow into the slit. Because of the axial gradient in the manifold the pressure drop across the slit is not uniform. As a result the flow rate through the slit may be a function of z, with the resultant production of nonuniformities in the flow leaving the sheeting die. We wish to examine the magnitude of such nonuniform flow and how the nonuniformity depends upon geometric and rheological parameters.

The model begins with a mass balance on an element of volume shown in Fig. 5-23. The difference between the volumetric flow rates at any pair of surfaces normal to the manifold axis, separated by an axial distance dz, is

$$Q_z - Q_{z+dz} = Q_x \tag{5-83}$$

where Q_x is the volumetric flow rate down the die axis x, across a width dz.

For small dz, we may write this in the form

$$-\frac{\partial Q_z}{\partial z} dz = Q_x \tag{5-84}$$

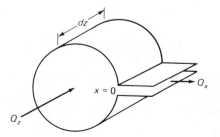

Figure 5-23 Mass balance on a volume element of the end-fed die.

Now we assume that, apart from the material balance above, Q_z and Q_x are independent, and each is given by the simple model for pipe flow (for Q_z) or for flow between parallel planes (for Q_x). Thus we have, for Q_z,

$$Q_z = \frac{n\pi R^3}{1 + 3n} \left[\frac{R(-dP/dz)}{2K} \right]^{1/n} \tag{5-85}$$

where we have used Eq. (5-17), assuming power law flow, and replaced $\Delta P/L$ with the local pressure gradient, of magnitude $-dP/dz$.

For Q_x we write, using Eq. (5-27),

$$Q_x = dz \frac{nB^2}{2(1 + 2n)} \left[\frac{B(-dP_x/dx)}{2K} \right]^{1/n} \tag{5-86}$$

where the width W in this case is dz, and where $\Delta P/L$ is replaced by the gradient in the x direction of the pressure P_x.

Now it is necessary to connect the pressures P and P_x. This is done by neglecting any entrance effects associated with the flow from the manifold into the die, from which it follows that

$$P_x(z) = P(z) \qquad \text{at } x = 0 \tag{5-87}$$

Consistent with this assumption we also take

$$\frac{-dP_x}{dx} = \frac{P(z)}{L_d} \tag{5-88}$$

Putting these ideas together we have

$$\frac{-dQ_z}{dz} = \frac{nB^2}{2(1 + 2n)} \left[\frac{BP(z)}{2KL_d} \right]^{1/n}$$

$$= \frac{\pi R^3}{1 + 3n} \left(\frac{R}{2K} \right)^{1/n} \left(-\frac{dP}{dz} \right)^{1/n - 1} \frac{d^2P}{dz^2}$$

or

$$\frac{d^2P}{dz'^2} - \beta^2 \frac{P^{1/n}}{\left(-\frac{dP}{dz'} \right)^{1/n - 1}} = 0 \tag{5-89}$$

where

$$\beta^2 = \frac{n(1 + 3n)}{2\pi(1 + 2n)} \frac{B^2 L_m^{1/n + 1}}{R^{3 + 1/n}} \left(\frac{B}{L_d} \right)^{1/n} \tag{5-90}$$

and $z' = z/L_m$.

Equation (5-89) is a nonlinear ordinary differential equation whose solution is subject to the boundary conditions $P = P_0$ at $z' = 0$ and $dP/dz' = 0$ at $z' = 1$. The second boundary condition follows from the assertion that the pipe is closed at $z = L_m$. Hence there is no flow, and no pressure gradient, at that point. An interesting point is that β, and so P, is independent of K. Thus the level of viscosity is not important, but the power law index is.

The solution of Eq. (5-89) must be obtained numerically. It is useful to examine the newtonian limit, which can be treated analytically, and to use that solution for consideration of factors that affect uniformity. For $n = 1$ we find

$$\frac{P}{P_0} = e^{-\beta z'} + e^{-\beta} \frac{\sinh \beta z'}{\cosh \beta} \tag{5-91}$$

A suitable measure of uniformity would be the ratio of the minimum and maximum flow rates across the width of the die. Since, for the newtonian fluid, Q_x is proportional to $P(z')$, we only need to examine

$$E = \frac{P(L_m)}{P(0)} = (\cosh \beta)^{-1} \tag{5-92}$$

We see that E is near unity for small β, so an appropriate design strategy is to select dimensions such that

$$\beta^2 = \frac{2B^3 L_m^2}{3\pi R^4 L_d} \ll 1 \tag{5-93}$$

5-5 COMBINED FLOW (HELICAL FLOW) OF A POWER LAW FLUID

In subsequent chapters we will examine some processes which involve axial annular pressure flow in combination with circular annular drag flow. Figure 5-24 shows the geometry of interest. For the newtonian fluid the two components of velocity u_θ and u_z are independent and may be written down by inspection of results already presented in earlier sections of this chapter. For a nonnewtonian fluid the two components are not independent, and the analysis of the flow becomes mathematically complex.

Physically the situation is straightforward. Suppose we think of the axial flow as the primary flow. The solution for the axial flow alone is easily found, and Eq. (5-32) gives the $Q(\Delta P)$ relationship. Now suppose that the outer cylinder is set in rotation *while* the axial pressure gradient is maintained. With a newtonian fluid there is no interaction between the two components of flow. This can be seen in the independence of the θ and z components of the equations of motion for a newtonian fluid subject to this flow. For a nonnewtonian fluid, however, for which

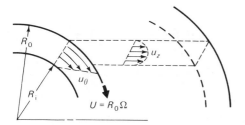

Figure 5-24 Geometry for helical flow in an annulus.

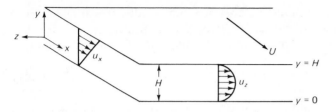

Figure 5-25 Plane approximation for the helical flow problem.

the viscosity is a function of the second invariant of Δ, say, $\eta = \eta(II_\Delta)$, the rotation of the outer cylinder increases the magnitude of II_Δ and thereby alters the viscosity of the fluid. Since most of the polymeric fluids of interest are shear thinning $(d\eta/d\ II_\Delta < 0)$, we can expect *increased* axial flow as a consequence of the angular drag flow induced by the rotating cylinder.

We introduce here a very simple model of the dynamics of this system, but one which should prove to be quite useful in some subsequent applications. The first simplification is geometric. We consider only ratios $R_i/R_0 = \kappa$ close enough to unity that the plane approximation may be used. The new geometry is shown in Fig. 5-25.

As a constitutive equation we choose the power law, written in the form [note Eq. (3-154)]

$$\tau = K(\tfrac{1}{2}II_\Delta)^{(n-1)/2}\Delta \tag{5-94}$$

where, for this flow,

$$\tfrac{1}{2}II_\Delta = \left(\frac{du_x}{dy}\right)^2 + \left(\frac{du_z}{dy}\right)^2 \tag{5-95}$$

The dynamic equations reduce to

$$0 = -\frac{\Delta P}{KL} + \frac{d}{dy}\left[(\tfrac{1}{2}II_\Delta)^{(n-1)/2}\frac{du_z}{dy}\right] \tag{5-96}$$

$$0 = \frac{d}{dy}\left[(\tfrac{1}{2}II_\Delta)^{(n-1)/2}\frac{du_x}{dy}\right] \tag{5-97}$$

and the boundary conditions are

$$u_x = U \qquad u_z = 0 \qquad \text{on } y = H$$

$$u_x = u_z = 0 \qquad\qquad \text{on } y = 0$$

The following dimensionless variables simplify the problem:

$$\varphi = \frac{u}{U} \qquad \zeta = \frac{y}{H} \qquad II'_\Delta = \left(\frac{U}{H}\right)^{-2}II_\Delta$$

$$\alpha = \frac{\Delta P\ H^{n+1}}{KLU^n}$$

The dimensionless equations now become

$$0 = -\alpha + \frac{d}{d\zeta}\left[(\tfrac{1}{2}\mathrm{II}'_\Delta)^{(n-1)/2}\frac{d\varphi_z}{d\zeta}\right] \tag{5-98}$$

$$0 = \frac{d}{d\zeta}\left[(\tfrac{1}{2}\mathrm{II}'_\Delta)^{(n-1)/2}\frac{d\varphi_x}{d\zeta}\right] \tag{5-99}$$

$$\varphi_x = 1 \qquad \varphi_z = 0 \qquad \text{on } \zeta = 1$$

$$\varphi_x = \varphi_z = 0 \qquad \text{on } \zeta = 0$$

Because of symmetry of φ_z about $\zeta = \tfrac{1}{2}$, and to avoid fractional roots of negative numbers, we consider the solution for φ_z in the region $\tfrac{1}{2} < \zeta < 1$, where $d\varphi_z/d\zeta > 0$. (Note that the coordinate system of Fig. 5-25 shows flow in the $-z$ direction.)

Equation (5-99) may be integrated once to give

$$(\tfrac{1}{2}\mathrm{II}'_\Delta)^{(n-1)/2}\frac{d\varphi_x}{d\zeta} = c_1 \tag{5-100}$$

whereas integration of Eq. (5-98) gives

$$(\tfrac{1}{2}\mathrm{II}'_\Delta)^{(n-1)/2}\frac{d\varphi_z}{d\zeta} = \alpha(\zeta - \tfrac{1}{2}) \tag{5-101}$$

[The symmetry condition at $\zeta = \tfrac{1}{2}$ has been used to evaluate the integration constant in Eq. (5-101).]

It follows that

$$\frac{d\varphi_x}{d\zeta} = \frac{c_1}{\alpha(\zeta - \tfrac{1}{2})}\frac{d\varphi_z}{d\zeta} \tag{5-102}$$

which may be used to uncouple the two velocity gradients, with the result that

$$\frac{d\varphi_z}{d\zeta} = \alpha(\zeta - \tfrac{1}{2})[\alpha^2(\zeta - \tfrac{1}{2})^2 + c_1^2]^{(1-n)/2n} \tag{5-103}$$

and

$$\frac{d\varphi_x}{d\zeta} = c_1[\alpha^2(\zeta - \tfrac{1}{2})^2 + c_1^2]^{(1-n)/2n} \tag{5-104}$$

By integration, using the boundary condition that $\varphi_z = \varphi_x = 0$ at $\zeta = 0$, we find

$$\varphi_z = \alpha\int_0^\zeta (\zeta - \tfrac{1}{2})[\alpha^2(\zeta - \tfrac{1}{2})^2 + c_1^2]^{(1-n)/2n}\, d\zeta \tag{5-105}$$

and

$$\varphi_x = c_1\int_0^\zeta [\alpha^2(\zeta - \tfrac{1}{2})^2 + c_1^2]^{(1-n)/2n}\, d\zeta \tag{5-106}$$

The constant c_1 may be evaluated from $\varphi_x = 1$ at $\zeta = 1$, which leads to

$$1 = c_1 \int_0^1 [\alpha^2(\zeta - \tfrac{1}{2})^2 + c_1^2]^{(1-n)/2n} \, d\zeta \qquad (5\text{-}107)$$

For a fixed choice of n and α, c_1 may be found by numerical integration of Eq. (5-107). Then numerical integration of Eqs. (5-105) and (5-106) gives φ_z and φ_x.

Of particular interest will be the axial volumetric flow rate per unit width in the $-z$ direction:

$$\frac{Q}{W} = -2UH \int_0^{1/2} \varphi_z \, d\zeta \qquad (5\text{-}108)$$

and the shear stress at the moving surface:

$$\tau_{xy} \bigg|_{y=H} = K(\tfrac{1}{2}\mathrm{II}_\Delta)^{(n-1)/2} \frac{du_x}{dy} \bigg|_{y=H} = K\left(\frac{U}{H}\right)^n c_1 \qquad (5\text{-}109)$$

[Use has been made of Eq. (5-100).] Figure 5-26 shows the results of these computations, which are consistent with our expectations.

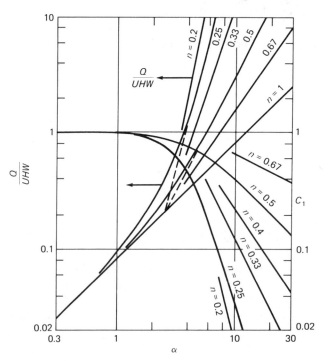

Figure 5-26 Dimensionless flow rate and shear stress functions for helical power law flow: plane approximation.

This problem is of interest from a secondary point of view, that of the philosophy of mathematical modeling. While Eqs. (5-103) and (5-104) cannot be integrated analytically, there are two *asymptotic* cases ($\alpha \to 0$ and $\alpha \to \infty$) which can be handled exactly. For small α ($\alpha \ll c_1$) Eqs. (5-103) and (5-104) simplify to

$$\frac{d\varphi_z}{d\zeta} = c_1^{(1-n)/n}\alpha(\zeta - \tfrac{1}{2}) \tag{5-110}$$

and

$$\frac{d\varphi_x}{d\zeta} = c_1^{1/n} \tag{5-111}$$

The solution to Eq. (5-111), satisfying the boundary conditions on φ_x, is

$$\varphi_x = \zeta \tag{5-112}$$

and c_1 is found to be $c_1 = 1$ for any n.

The solution to Eq. (5-110) is then

$$\varphi_z = \frac{\alpha}{2}(\zeta^2 - \zeta) \tag{5-113}$$

From Eq. (5-108) we find

$$\frac{Q}{UHW} = \frac{\alpha}{12} \qquad \text{for } \alpha \to 0 \tag{5-114}$$
$$c_1 = 1$$

In a similar manner, for the case $\alpha \gg c_1$, we can find (see Prob. 5-32)

$$\frac{Q}{UHW} = \frac{n(\alpha/2)^{1/n}}{2(1 + 2n)} \qquad \text{for } \alpha \to \infty \tag{5-115}$$
$$c_1 = \frac{1}{n}\left(\frac{2}{\alpha}\right)^{(1/n)-1}$$

Figure 5-26 shows these asymptotic limits. It should be apparent, from the way the asymptotic lines intersect, that there is actually a fairly narrow region, where α is neither large nor small (say, $1 < \alpha < 3$), where it is necessary to obtain the numerical solution to the general equations [Eqs. (5-103) and (5-104)]. The identification of asymptotic cases often can save considerable time in planning numerical solutions to complex problems.

Example 5-1 A melt is being pumped through a tubing die which has a rotating inner mandrel. The stationary outer cylindrical surface of the die has a 3-in diameter. The mandrel is a cylinder of 2.90-in diameter. The "land" of the die (the axial length of the annulus) is 3 in.

A pressure drop of 250 psi is imposed across the die land. Find the output under the conditions

$$n = \tfrac{1}{2} \qquad K = 1 \text{ lb} \cdot \text{s}^{1\cdot 2}/\text{in}^2 \qquad \Omega = 10 \text{ rpm}$$

What would the output be with no rotation?

In this problem $R_0/R_i = 1.5/1.45 = 1.03$. Thus $\kappa = 0.97$ and we expect the plane approximation to be accurate.

We find $U = (2\pi\Omega R_0)/60 = 1.57$ in/s (note we converted Ω to radians per second). We calculate α and find

$$\alpha = \frac{\Delta P(R_0 - R_i)^{n+1}}{KLU^n} = 0.74$$

From Fig. 5-26 we find $Q/UHW = 0.062$ or (noting $W = 2\pi R_0$)

$$Q = 0.062(1.57)(0.05)(2\pi)(1.5) = 0.046 \text{ in}^3/\text{s}$$

If there were no rotation ($\alpha = \infty$) then we have axial annular Poiseuille flow. For large α we see from Eq. (5-115) that the asymptotic solution gives

$$\frac{Q}{UHW} = \frac{\alpha^2}{32} = \frac{\Delta P^2 (R_0 - R_i)^3}{32K^2L^2U}$$

or $Q = 0.0128 \text{ in}^3/\text{s}$.

Hence, the output is increased by a factor of 3.6 through rotation of the mandrel.

Example 5-2 For Example 5-1 calculate the increased power expenditure due to rotation of the mandrel. The power associated with the axial (pressure) flow is just the product of pressure drop and flow rate:

$$Q \, \Delta P = \begin{cases} 3.2 \text{ in} \cdot \text{lb/s} = 4.8 \times 10^{-4} \text{ hp} & \text{(no rotation)} \\ 11.5 \text{ in} \cdot \text{lb/s} = 17.5 \times 10^{-4} \text{ hp} & \text{(at 10 rpm)} \end{cases}$$

In the case of rotation it is also necessary to expend power to drive the mandrel. This power requirement is just the product of the rotational speed and the *force* in the direction of rotation (which is a shearing force), and so we find [noting Eq. (5-109)]

$$\tau_{xy}\Big|_{y=R_i} (2\pi R_i L)U = K\left(\frac{U}{R_0 - R_i}\right)^n c_1 (2\pi R_i LU)$$

For $\alpha = 0.74$ we find $c_1 = 1$ from Fig. 5-26. Hence the mandrel power is found to be 240 in · lb/s or 0.036 hp. The increase in power requirement is a factor of 79, which is hardly an efficient way to obtain an increase in flow rate of 3.6 times. We note, for example, that in the absence of rotation, for $n = \frac{1}{2}$, Q is proportional to ΔP^2. Thus we could increase Q by a factor of 3.6 by increasing ΔP by a factor of 1.9. This would require an increased power expenditure of a factor of 6.8. We shall see subsequently that there might be other reasons, than increasing output, for creating this type of flow field. Problem 5-33 presents a situation where rotation is used to reduce the pressure drop required to achieve a specified flow rate.

5-6 COMBINED FLOW (PARALLEL FLOW) OF A POWER LAW FLUID

In Sec. 5-3 a wire-coating situation was modeled by combining the pressure and drag flow solutions for a newtonian fluid. It was pointed out that this simple combination or superposition of flows is not valid for nonnewtonian fluids. The reason for this is clear if we examine the dynamic equations for this flow. For a power law fluid in laminar steady flow between infinite parallel surfaces, the velocity must satisfy

$$0 = -\frac{\Delta P}{KL} + \frac{d}{dy}\left\{ \left[\left(\frac{du_z}{dy}\right)^2 \right]^{(n-1)/2} \frac{du_z}{dy} \right\} \tag{5-116}$$

subject to boundary conditions

$$u_z = \begin{cases} 0 & \text{on } y = 0 \\ U & \text{on } y = B \end{cases}$$

If pressure flow *alone* occurred the velocity would satisfy

$$0 = -\frac{\Delta P}{KL} + \frac{d}{dy}\left\{ \left[\left(\frac{du_P}{dy}\right)^2 \right]^{(n-1)/2} \frac{du_P}{dy} \right\} \tag{5-117}$$

$$u_P = \begin{cases} 0 & \text{on } y = 0 \\ 0 & \text{on } y = B \end{cases}$$

If drag flow *alone* occurred the velocity would satisfy

$$0 = \frac{d}{dy}\left\{ \left[\left(\frac{du_D}{dy}\right)^2 \right]^{(n-1)/2} \frac{du_D}{dy} \right\} \tag{5-118}$$

$$u_D = \begin{cases} 0 & \text{on } y = 0 \\ U & \text{on } y = B \end{cases}$$

If u_z in Eq. (5-116) is to be simply the sum of u_P and u_D, then it would have to be true that the function $u_P + u_D$ would satisfy Eq. (5-116). But if $u_P + u_D$ is substituted for u_z in Eq. (5-116) and if Eqs. (5-117) and (5-118) are accounted for, it can be seen that the right-hand side of Eq. (5-116) does not vanish, and so $u_P + u_D$ is *not* a solution.

Having stated what $u_P + u_D$ is *not*, we now wish to investigate what it *is*. In particular, we wish to know how good an approximation $u_P + u_D$ might be to the solution of Eq. (5-117), even though we admit it is not exact.

At first glance it would appear that one could simply integrate Eq. (5-116) and produce an analytical solution. In part this is true, but a complicating feature enters the problem which, ultimately, leads to the necessity for some numerical computation in the problem. We can see this qualitatively by considering the shape of the expected velocity profiles.

Figure 5-27 shows several possible velocity profiles. To the left- and right-hand extremes, respectively, are the pure drag and pure pressure velocity profiles.

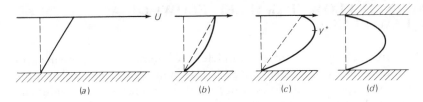

Figure 5-27 Velocity profiles in parallel flow.

Profile b would occur if a relatively small pressure were imposed on the system in the direction of flow. Profile c would occur if a larger pressure drop were imposed. The relevant point is that under some conditions the profile has a maximum *between* the surfaces. Hence the velocity gradient du_z/dy changes sign at some point $y*$. Since a fractional root occurs in Eq. (5-116), it is necessary to know a priori the sign of the terms in the equation. The complication arises because the point $y*$ is not known a priori; it is a function of the pressure drop (in a dimensionless parameter) and a function of n.

Solution of Eq. (5-116) has been carried out, and two references are cited at the end of this chapter. Here we only present the solution in a format suitable for examining the error that would arise if superposition of the solutions to Eqs. (5-117) and (5-118) were used.

Instead of examining the velocity profiles themselves let us work in terms of the volume flow rate Q. Assuming superposition is valid, we write

$$Q = Q_D + Q_P = \tfrac{1}{2}UBW + \frac{nWB^2}{2(1+2n)}\left(\frac{B\,\Delta P}{2KL}\right)^{1/n} \tag{5-119}$$

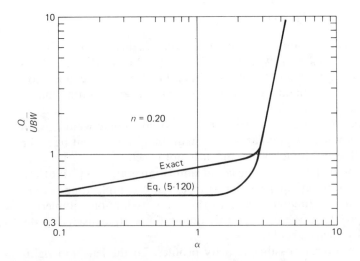

Figure 5-28 Pressure drop–flow rate relationship for parallel flow of a power law fluid.

which is just the sum of Eqs. (5-27) and (5-50). It is convenient to nondimensionalize Eq. (5-119) to the form

$$\frac{Q}{UBW} = \frac{1}{2} + \frac{n}{(1 + 2n)2^{1 + 1/n}} \alpha^{1/n} \tag{5-120}$$

where

$$\alpha = \frac{\Delta P B^2}{K(U/B)^{n-1}UL} \tag{5-121}$$

Figure 5-28 shows Eq. (5-120) for the case $n = 0.2$ and compares it to the exact solution of Eq. (5-116). As expected, the approximate solution is accurate at the asymptotes corresponding to $\alpha \to 0$ and $\alpha \to \infty$. Significant errors occur in the intermediate range of α.

PROBLEMS

5-1 Solve Eq. (5-6) to obtain Eq. (5-9). Indicate clearly the *two* boundary conditions on u_z as a function of r.

5-2 The Hagen-Poiseuille law [Eq. (5-10)] shows a very strong dependence of volumetric flow rate on the tube radius. If the measured value of radius were in error by 10 percent, what would the expected error in Q be?

5-3 Beginning with the dynamic equations, derive Eq. (5-16).

5-4 Show that the dependence of volumetric flow rate on tube radius is stronger for a power law fluid than for a newtonian fluid in the usual case $n < 1$. If the measured value of radius were in error by 10 percent, what would the expected error in Q be for $n = 0.2$?

5-5 Two circular tubes of identical length are connected to the same liquid reservoir in which a power law fluid is held. The tubes differ in radii by a factor of 2. When a pressure drop ΔP is imposed, the volumetric flow rates from the two tubes differ by a factor of 40. What is the value of n? How different are the nominal shear rates in this case?

5-6 Derive Eq. (5-26). Why is it necessary to use the absolute value signs on $2\gamma/B$ in this equation?

5-7 Derive Eq. (5-29).

5-8 Derive Eq. (5-31) and show that

$$G(\kappa) = \frac{1 - \dfrac{1 - \kappa^2}{2 \ln (1/\kappa)}}{\dfrac{1 + \kappa^2}{1 - \kappa} - \dfrac{1 + \kappa}{\ln (1/\kappa)}}$$

5-9 A power law fluid has a value of $n = \frac{1}{2}$. Give the cgs units of K. In those units, K has a value of 10^4. This fluid is to be pumped through a rectangular pipe of length 10 cm for which $B = 0.1$ cm and $W = 0.4$ cm. Give the expected pressure drop in pounds per square inch. Calculate the nominal shear rate, using Eq. (5-25). The volumetric flow rate is 1 ml/s.

5-10 Derive an expression for the nominal shear rate for power law flow between parallel plates, beginning with Eq. (5-26). What error was made using Eq. (5-25) in Prob. 5-9? What is the significance of the error?

5-11 Some references give the friction factor for laminar pipe flow as $f = 64$ Re. Assuming Re is still defined as $UD\rho/\mu$, how is their f defined?

5-12 A viscous newtonian fluid is being pumped through a circular pipe which has a solid separator

Figure 5-29 Cross section of straight pipe with divider down the axis.

running down the axis, as shown in Fig. 5-29. Thus the flow is really that corresponding to a pair of semicircular pipes. Use Fig. 5-9 to calculate a $\tilde{K}$ factor, defined by

$$\tilde{K} = \frac{\Delta P(\text{divided pipe})}{\Delta P(\text{open pipe})} \qquad \text{at constant } Q \text{ and } D$$

5-13 Using Eq. (5-17), show that the friction factor can be written in the form $f = 16/\text{Re}_n$, where a modified Reynolds number is defined thereby. Show that the Reynolds number may be considered to be of the form

$$\text{Re}_n = \frac{UD\rho}{\eta}$$

where $\eta = \tau_R/\dot{\gamma}_R$ and $\dot{\gamma}_R$ is given in Eq. (5-18).

5-14 Take the form of $G(\kappa)$, given in Prob. 5-8, and show that for very small κ a good approximation is

$$G(\kappa) \approx \frac{(-\ln \kappa) - \frac{1}{2}}{(-\ln \kappa) - 1}$$

How small must κ be for the value of the nominal shear rate to be within 10 percent of the value for an empty pipe?

5-15 Show that a wire along the z axis of a circular pipe exerts a major effect on the flow rate out of apparent proportion to its size. Begin by showing that the reduction in flow rate, relative to the case $\kappa = 0$ (an open pipe), is

$$\tilde{K} = 1 + \frac{1}{\ln \kappa} \qquad \text{for } \kappa \ll 1$$

Find how small κ must be so that $\tilde{K} = 0.99$. Do the case of pressure flow, where the wire is stationary.

5-16 Derive Eq. (5-59). What error would be made if, instead of this result, one used

$$\dot{\gamma} = \frac{R_i \Omega}{R_0 - R_i} = \frac{\Omega}{1/\kappa - 1}$$

to estimate the shear rate? Answer for the case that $\kappa = 0.5$, and $n = 1$ and $n = \frac{1}{2}$.

5-17 Show that the shear rate given in Prob. 5-16 is obtained as the limit of Eq. (5-59) when $\kappa \to 1$.

5-18 Give the general solution for circular annular Couette flow of a power law fluid when both cylinders are rotating at frequencies Ω_0 and Ω_i.

5-19 In Eq. (5-62) is it necessary to be concerned about the sign of du_z/dr? Why? Why was the negative sign inserted in that specific case?

5-20 Derive Eqs. (5-63) and (5-65).

5-21 What happens to Eq. (5-64) when $n = \frac{1}{3}$? Derive the correct result for that case.

5-22 Show that the right-hand side of Eq. (5-64) approaches $\frac{1}{2}$ as $\kappa \to 1$. Do the same for Eq. (5-66).

5-23 Consider flow of a newtonian fluid in a rectangular duct of aspect ratio W/B. How large must the aspect ratio be so that the volumetric flow rate is within 10 percent of the value predicted by the assumption of $W/B = \infty$ (infinite plates)? Answer for both drag flow and pressure flow. Are the answers much different if the fluid is power law with $n = \frac{1}{2}$?

5-24 Using Eq. (5-68), find the tension (defined as force per cross-sectional area) required to draw a wire through a coating die under the following conditions:

$$R_0 = 0.1 \text{ cm} \qquad R_i = 0.07 \text{ cm} \qquad L = 1 \text{ cm}$$

$$\mu = 10 \text{ P} \qquad U = 100 \text{ ft/s}$$

Is this tension comparable to the yield stress of steel? Nylon?

5-25 Suppose a wire-coating die were designed as shown in Fig. 5-30. Let the dimensions be

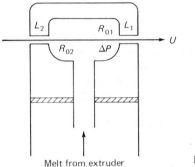

Figure 5-30 Wire-coating die.

$R_i = 0.02$ cm, $R_{01} = 0.04$ cm, $R_{02} = 0.022$ cm, $L_1 = 0.3$ cm, $L_2 = 1$ cm. Let U be 100 ft/s, and take the fluid to be newtonian, with $\mu = 100$ P.

If we want $\delta = 0.02$ cm, what must ΔP be (in pounds per square inch)?

Under these conditions, will there be a significant leakage flow out the rear of the die, where the wire enters?

5-26 For the problem as stated in Prob. 5-25, give the sensitivity of the coating thickness to variations in U and ΔP. If U can be held to ± 2 percent but ΔP might vary by ± 20 percent, what might be the expected percentage variability in the coating thickness δ?

5-27 Referring to the wire-coating analysis, we can see that in the limit of $\kappa \to 1$, $\delta = \frac{1}{2}(R_0 - R_i)$ at $\Delta P = 0$. For the geometry given in Prob. 5-25:
  (a) What would δ be at $\Delta P = 0$?
  (b) What ΔP would give $\delta = \frac{1}{2}(R_0 - R_i)$?

5-28 Prove the assertion of Eq. (5-76).

5-29 It is desired to achieve a coating thickness of $\delta = 0.02$ cm onto a wire of radius $R_i = 0.2$ cm by using an annular die of outer radius $R_0 = 0.23$ cm and length $L = 0.5$ cm. The coating may be considered newtonian with a viscosity of 200 P. Find the pressure ΔP at a wire speed $U = 50$ ft/s.

5-30 A PVC coating resin is nonnewtonian, as the data in Fig. 5-31 indicate. It is coated onto a wire of

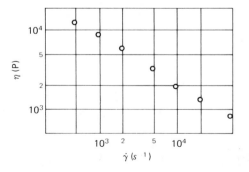

Figure 5-31 Data for Prob. 5-30.

diameter 0.04 in moving at 1000 ft/min through a die of diameter 0.06 in. Find the coating thickness in the absence of an applied pressure. Calculate the nominal shear rate, and compare it to the range of shear rate over which the rheological data were obtained. Work the problem two ways, and compare the answers, by assuming:

(a) The fluid is power law

(b) The fluid is newtonian, with a viscosity given by the nonnewtonian viscosity observed at the nominal shear rate.

5-31 Calculate the uniformity index E of Eq. (5-92) for an end-fed die for which $R = 0.5$ in, $L_m = 40$ in, $B = 0.01$ in, and $L_d = 0.5$ in.

5-32 Derive Eqs. (5-114) and (5-115).

5-33 In a tubular extrusion system it is necessary to achieve an output of 3 lb/h from the tubing die described in Example 5-1. The polymer melt, at the extrusion conditions, has power law parameters $n = \frac{1}{2}$, $K = 10$ lb·s$^{1/2}$/in^2. From mechanical design constraints it is necessary that the imposed pressure drop be less than 2500 psi. Find the mandrel speed necessary to reduce the pressure drop to this level at the specified flow rate. What horsepower is expended in rotating the mandrel?

5-34 A wire is coated in an annular die operating under the conditions of Prob. 5-29, but the coating fluid obeys the power law, with $K = 1$ lbf·s$^{0.2}$/in^2 and $n = 0.2$. If we want $\delta = 0.02$ cm, what must ΔP be (in pounds per square inch)? Use Fig. 5-28, and compare the exact and approximate solutions.

5-35 Rework Prob. 5-34 but use the newtonian analysis with μ evaluated at the nominal shear rate. Compare the pressure with that from the more exact model.

5-36 Modify Eq. (5-120) for the case where the pressure gradient *opposes* the drag flow, and plot Q/UBW versus α. (This analysis comes up in the theory of extrusion. Figure 6-11 gives the exact solutions for this case.) Compare the simple solution to the exact solution for $n = 0.2$ and $n = 0.5$.

5-37 The equation below is derived by Parnaby and Worth† for steady isothermal flow through a die whose boundaries are cones with a common apex, as shown in Fig. 5-32. Give the derivation in detail.

$$\Delta P = \left(\frac{Q}{\Omega \pi}\right)^n \frac{2K}{3n \tan \beta \, R_0^{3n}} (1 - Y^{3n})$$

$$\Omega(s, \kappa) = \int_\kappa^1 |\lambda^2 - \xi^2|^{s+1} \xi^{-s} \, d\xi \qquad (s = 1/n)$$

† *Proc. Inst. Mech. Eng.*, **188**: 357 (1974).

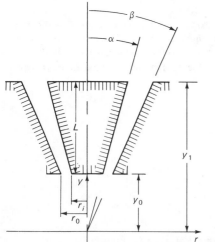

Figure 5-32 Conical die.

λ is found from

$$\int_{\kappa}^{\lambda} \left(\frac{\lambda^2}{\xi} - \xi \right)^s d\xi = \int_{\lambda}^{1} \left(\xi - \frac{\lambda^2}{\xi} \right)^s d\xi$$

For $\kappa > 0.6$ good approximations to λ and Ω are found from

$$\lambda = 0.5(1 + \kappa)$$

$$\Omega = \frac{(1 + \kappa)(1 - \kappa)^{s+2}}{2(s + 2)}$$

5-38 The following data have been obtained on a low-density polyethylene at 190°C flowing through a die whose boundaries are cones (the geometry is given in Fig. 5-32).

Q, cm^3/s	ΔP, N/m^2
0.84	2.5×10^6
1.5	3.4×10^6
2.0	4.0×10^6

$\alpha = 0.086$ rad
$\beta = 0.117$ rad
$r_o = 0.6$ cm
$r_i = 0.44$ cm
$L = 5.1$ cm

The melt is power law, with $n = \frac{1}{3}$ and $K = 162{,}000$ N·s$^{1/3}$/m^2.

(a) Do the cones have a common apex?

(b) Use the theory given in Prob. 5-37 to predict $Q(\Delta P)$ and compare to the data.

(c) Use a mean annular gap and the theory for annular flow and repeat part b. Define clearly what kind of mean you use (arithmetic, geometric, hydraulic, etc.).

5-39 *Melt index* (MI) is often used as a measure of the viscosity of olefins such as polyethylene. It is measured as the number of grams of polymer extruded in 10 min from a cylindrical die of diameter 0.0825 in and length 0.315 in under a pressure of 43.25 psi at 190°C.

(a) Is MI directly related to viscosity, or might it depend on flow properties related to unsteady viscous and/or elastic flow?

(b) Assume that MI *is* related to viscosity, free of any experimental artifact. Plot viscosity, in poise, versus MI, for $0.1 < MI < 10$.

(c) Could two fluids with the same MI have different $\eta(\dot{\gamma})$ curves?

(d) What is the magnitude of the shear rate in the MI experiment?

(e) The MI of a particular low-density polyethylene is given as $MI = 1.4$. The zero shear viscosity was measured in a cone-and-plate instrument at 190°C and found to be $\eta_0 = 1.4 \times 10^5$ P. Is this consistent with part b?

BIBLIOGRAPHY

5-1 Poiseuille flow (pressure flow)

The f Re curves in Figs. 5-7 to 5-10 are adapted from those given in Chap. 6 of

Kays, W. M.: "Convective Heat and Mass Transfer," McGraw-Hill Book Company, New York, 1966.

A numerical study of entrance lengths in newtonian fluids is given in

Atkinson, B., M. P. Brocklebank, C. C. H. Card, and J. M. Smith: Low Reynolds Number Developing Flows, *AIChE J.*, **15**: 548 (1969).

and Eqs. (5-45) and (5-46) are taken from there.

5-4 End-fed sheeting die

The power law case is treated in

McKelvey, J. M., and K. Ito: Uniformity of Flow from Sheeting Dies, *Polym. Eng. Sci.*, **11**: 258 (1971).

5-5 Helical flow of a power law fluid

Solutions for a viscoelastic fluid model are given in

Tanner, R. I.: Helical Flow of Elastico-Viscous Liquids, *Rheol. Acta*, **3**: 21 (1963).

5-6 Combined flow (parallel flow) of a power law fluid

The case of a pressure gradient opposing the drag flow arises in the theory of extrusion and is discussed in

Kroesser, F. W., and S. Middleman: The Calculation of Screw Characteristics for the Extrusion of Non-Newtonian Melts, *Polym. Eng. Sci.*, **5**: 1 (1965).

The general case is given in

Flumerfelt, R. W., M. W. Pierick, S. L. Cooper, and R. B. Bird: Generalized Plane Couette Flow of a Non-Newtonian Fluid, *Ind. Eng. Chem. Fundam.*, **8**: 354 (1969).

CHAPTER SIX

EXTRUSION

Failures are made only by those who fail to dare, not by those who dare to fail.

Binstock

An extruder is a pump. It is a versatile machine, capable of performing other operations in concert with its pumping function. Figure 6-1 shows details of a typical screw extruder. If the extruder is fed with solid polymer chips or beads, a melting operation is normally achieved within a few diameters downstream of the feed inlet. This operation is often referred to as *plasticating*, and such an extruder is a *plasticating extruder*. If the feed is a fluid, usually a molten polymer, the extruder is called a *melt extruder*. If dissimilar polymers, or polymer plus another fluid, or polymer plus pigment or filler is fed to the extruder, the machine serves the additional function of a mixer.

In any of these operations the goal of the extruder is to produce a homogeneous molten material at a flow rate, pressure, and temperature suitable for the next operation in the process line. The next stage is usually the formation of a solid polymeric article.

The formation process may be continuous, such as when the polymeric fluid is forced through a wire-coating die. The process may also be discontinuous when the fluid is intermittently forced into a mold, as in the production of bottles, combs, or automobile bumpers.

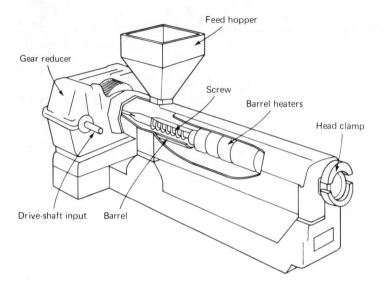

Figure 6-1 Cutaway view of a typical single-screw extruder.

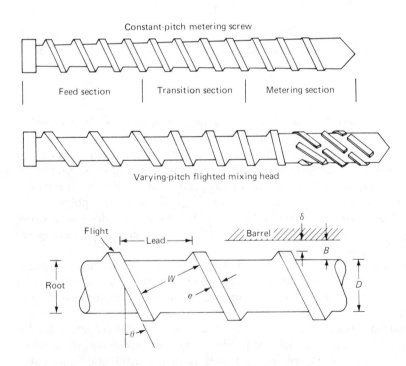

Figure 6-2 Extruder screw geometry.

Part of the versatility of the extruder lies in the design of screws which may be specialized with respect to the function to be served, such as mixing, metering, or removal of volatile solvent. The design of the screw may also vary with the type of plastic resin to be handled. Figure 6-2 shows some typical screw geometries and details which define screw parameters.

In this chapter the analysis of the fluid dynamics of extrusion will be considered. The melting process is not treated here.

6-1 NEWTONIAN ISOTHERMAL ANALYSIS

Under practical circumstances extrusion usually involves a molten polymer flowing under nonisothermal conditions. In developing models of melt extrusion it is useful to begin with the simplest case, namely, isothermal newtonian extrusion. A geometric simplification makes it possible to treat this case analytically, and the resulting "simplified theory" then serves as a basis for comparison of subsequent modifications to the theory.

The geometric simplification begins upon "unwinding" the helical screw channel, as shown in Fig. 6-3. The relative motion of the screw and the barrel becomes equivalent to the steady motion of a plane at an angle θ to the helical axis z. Thus a drag flow is generated with components in the x and z directions.

The dynamic equations in the x and z directions will be written with the assumption that the inertial terms are unimportant, a reasonable assumption in highly viscous fluids. For the steady state we write

$$0 = -\frac{\partial p}{\partial x} + \mu\left(\frac{\partial^2 u_x}{\partial x^2} + \frac{\partial^2 u_x}{\partial y^2}\right) \tag{6-1}$$

$$0 = -\frac{\partial p}{\partial z} + \mu\left(\frac{\partial^2 u_z}{\partial x^2} + \frac{\partial^2 u_z}{\partial y^2}\right) \tag{6-2}$$

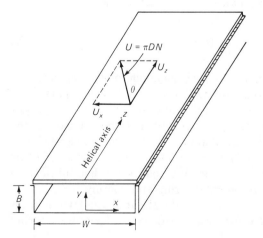

Figure 6-3 Geometry of the "unwound" helical screw channel. The upper surface, representing the barrel, moves at an angle θ to the helical axis and causes flow down the channel (U_z) as well as transverse to it (U_x).

We assume that the screw geometry is uniform in the z direction and that u_z is not a function of z. The pressure gradient $\partial p/\partial z$ is replaced by the constant value

$$\frac{\partial p}{\partial z} = \frac{\Delta P}{Z} \tag{6-3}$$

where ΔP is the pressure *rise* down the extruder, and Z is the *helical* length.

Equation (6-2) is identical to that treated earlier for the flow in a rectangular duct. The boundary conditions include $u_z = U \cos \theta$ at $y = B$. Thus we have a case of combined pressure and drag flow in the z direction, and since the fluid is newtonian we may simply superimpose the solutions given in Chap. 5 [Eqs. (5-35) and (5-71)] and write

$$Q = \tfrac{1}{2} U_z WBF_D - \frac{WB^3}{12\mu} \frac{\Delta P}{Z} F_p \tag{6-4}$$

where $U_z = U \cos \theta$. Note that the pressure flow is a " backflow " *opposed* to the drag flow. In most extruders the screw has a large aspect ratio (W/B), as a consequence of which F_D and F_p are nearly unity.

Because the barrel velocity has a component $U_x = U \sin \theta$, there is a *transverse* flow $u_x(x, y)$. For a large aspect ratio we would expect u_x to show only a weak dependence on x, except in the neighborhood of the walls (the screw flights). As an approximation, then, we replace Eq. (6-1) with

$$0 = - \frac{\partial p}{\partial x} + \mu \frac{\partial^2 u_x}{\partial y^2} \tag{6-5}$$

subject to conditions

$$u_x = \begin{cases} 0 & \text{on } y = 0 \\ -U_x & \text{on } y = B \end{cases}$$

The solution is found upon integrating twice with respect to y and is

$$u_x = - U_x \frac{y}{B} - \frac{1}{2\mu} y(B - y) \frac{\partial p}{\partial x} \tag{6-6}$$

In this integration the pressure gradient $\partial p/\partial x$ has been taken to be independent of the y coordinate. This is equivalent to assuming that the velocity u_x is parallel to the surfaces at $y = 0$ and B. Again, of course, this cannot be true in the neighborhood of the screw flights, but it might be a good approximation over most of the channel if the aspect ratio is large. This type of approximation, where parallel flow is assumed, is usually referred to as a *lubrication approximation* because of its utility in the modeling of lubrication systems. We shall see other lubrication approximations in subsequent chapters on calendering and coating.

Even though this solution ignores the presence of walls at $x = \pm\tfrac{1}{2} W$, the walls exert one effect that must be accounted for, even at values of x far removed from the walls: There can be no *net* flow of material in the x direction. Thus a constraint on u_x is

$$\int_0^B u_x \, dy = 0 \tag{6-7}$$

This constraint requires that the pressure gradient take the form

$$\frac{\partial p}{\partial x} = -\frac{6\mu U_x}{B^2} \tag{6-8}$$

with which the velocity may be written as

$$u_x = U_x \frac{y}{B}\left(2 - \frac{3y}{B}\right) \tag{6-9}$$

Equation (6-9) is an approximation that is valid, at best, over the middle portion of the screw away from the walls. The x component of velocity makes no contribution to the extruder output Q, and so the approximation is of no relevance to that aspect of the model. The velocity components *are* necessary, however, in the calculation of the power requirement for the extruder, which we next consider.

In general, if a force $\mathbf{F}$ is required to move an object at a velocity $\mathbf{U}$, the force does work at a rate

$$\dot{W} = \mathbf{F} \cdot \mathbf{U} \tag{6-10}$$

and $\dot{W}$ is the power expended by that process. In the case of extrusion, the motion of the barrel relative to the screw requires a power input

$$\dot{W} = \int_0^Z \int_0^W \left(\tau_{xy}\bigg|_B U_x + \tau_{zy}\bigg|_B U_z\right) dx \, dz \tag{6-11}$$

The shear stresses are given by

$$\tau_{zy} = \mu\left(\frac{\partial u_z}{\partial y} + \underline{\frac{\partial u_y}{\partial z}}\right) \tag{6-12}$$

$$\tau_{xy} = \mu\left(\frac{\partial u_x}{\partial y} + \underline{\frac{\partial u_y}{\partial x}}\right) \tag{6-13}$$

In Eq. (6-12) the underlined term vanishes by the assumption of uniform flow down the z axis. In Eq. (6-13) the underlined term is neglected as being small, which is true except possibly near the channel boundaries at $x = \pm\frac{1}{2}W$. Equation (6-9) is useful in allowing an approximate calculation of that part of the power associated with u_x, the transverse flow.

In Eq. (6-12) we need $u_z(y)$. Here it is most useful to again approximate the flow as if W/B were very large. In that case the velocity u_z is easily found to be of the same form as u_x, namely,

$$u_z = U_z \frac{y}{B} - \frac{1}{2\mu} y(B - y)\frac{\Delta P}{Z} \tag{6-14}$$

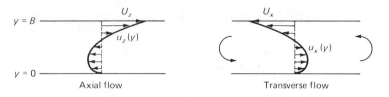

Figure 6-4 Velocity profiles in axial and transverse components. The axial flow can have a region of negative velocity near the root of the screw, as shown, if the back pressure is high enough. The transverse flow is a closed circulation and so always has both positive and negative components.

Figure 6-4 shows the velocity components u_z and u_x.

It follows that the power, according to this simplified theory, is given by

$$\dot{W} = \left(\frac{4\mu U_x^2}{B} + \frac{\mu U_z^2}{B} + \frac{U_z B}{2} \frac{\Delta P}{Z} \right) ZW \tag{6-15}$$

Thus Eqs. (6-4) and (6-15) relate the basic *performance variables* that characterize extruder performance: output Q, the pressure buildup ΔP, and the power $\dot{W}$. Since these are three variables, a third equation is required to provide a unique solution.

If the extruder simply discharged the fluid at its downstream end from an unrestricted outlet, there would be no mechanism by which a pressure gradient would develop. This would be the case of *free discharge*: $\Delta P = 0$. Normally there is a die of some kind at the downstream end of the extruder, and the flow restriction imposed by the die is what actually establishes the pressure rise down the extruder axis. Thus the additional relationship among the variables is the pressure drop across the die associated with the extruder output. It is assumed that all the output passes through the die.

For newtonian fluids most dies would behave in such a way that

$$Q = \frac{k}{\mu} \Delta P \tag{6-16}$$

where k is related to the die geometry. Equation (6-16) is called the *die characteristic*.

It is most convenient to rewrite Eq. (6-4) in the form

$$Q = AN - C \frac{\Delta P}{\mu} \tag{6-17}$$

where the screw speed N (rpm) has been introduced through the relationship

$$N = \frac{U}{\pi D} \tag{6-18}$$

Equation (6-17) is called the *screw characteristic*. The parameters A and C are strictly geometric and may be found on comparison of Eqs. (6-17) and (6-4).

One may easily solve for Q and ΔP from Eqs. (6-16) and (6-17) and find

$$Q = \frac{Ak}{C+k}\,N \tag{6-19}$$

and

$$\Delta P = \frac{\mu A}{C+k}\,N \tag{6-20}$$

(The interesting result, unexpected at first glance but easily rationalized, is the independence of output from the viscosity of the fluid.)

Finally, the power may be written in the form

$$\dot{W} = E\mu N^2 Z + AN\,\Delta P \tag{6-21}$$

where

$$E = \frac{\pi^3 D^3 \sin \theta}{B}(1 + 3 \sin^2 \theta) \tag{6-22}$$

In this last expression it has been assumed that the screw is single-flighted, which means that *one* channel is wrapped helically down the extruder axis. In that case W and D are dependent, and

$$W = \pi D \sin \theta \tag{6-23}$$

if the thickness of the screw flights is ignored in comparison to W.

Two extreme conditions provide useful measures of extruder performance. If there is no die resistance ($k \to \infty$), we have the case of *open discharge*, at which the maximum output of the extruder is achieved. From Eq. (6-19) it is clear that

$$Q_{\max} = AN \tag{6-24}$$

At the other extreme, if the die resistance is very large ($k \to 0$), the extruder develops its maximum pressure but produces no output. We see that in this case

$$\Delta P_{\max} = \frac{\mu A}{C}\,N \tag{6-25}$$

Variable Channel Depth

Most screws do not have uniform geometry down the helical axis. Most commonly the depth B varies in some way according to the purpose of the extruder and the type of material it handles. We illustrate here a simple case, where B is a linear function of z.

If the channel depth is a function of z, it follows that u_z (and u_x) is a function of z, and Eq. (6-2) is not exactly correct. However, if the z dependence of B is not too great, we might expect that Eq. (6-2) and its solution would be a good approximation, with the modifications that B should be left as a variable $B(z)$ and the pressure gradient $\partial p/\partial z$ should be retained as a function of z.

Thus the simplest approximation would be to go back to Eq. (6-14) and write

$$u_z = U_z \frac{y}{B(z)} - \frac{1}{2\mu} y[B(z) - y] \frac{\partial p}{\partial z} \tag{6-26}$$

The volumetric flow rate across any plane normal to the helical axis z would still be constant, and

$$Q = \tfrac{1}{2} U_z W B(z) - \frac{WB^3(z)}{12\mu} \frac{\partial p}{\partial z} \tag{6-27}$$

Solving for the pressure gradient, one finds

$$\frac{\partial p}{\partial z} = \frac{6\mu U_z}{B^2(z)} - \frac{12\mu Q}{W} \frac{1}{B^3(z)} \tag{6-28}$$

Now, as a specific example, let us assume that $B(z)$ is a linear function:

$$B(z) = B_0 + \frac{B_z - B_0}{Z} z \tag{6-29}$$

Then Eq. (6-28) may be integrated over the helical axis, from $z = 0$ to $z = Z$, to give (after some algebraic rearrangement)

$$Q = \tfrac{1}{2} U_z W \frac{B_0 B_z}{\frac{1}{2}(B_0 + B_z)} - \frac{W}{12\mu} \frac{(B_0 B_z)}{\frac{1}{2}(B_0 + B_z)} \frac{\Delta P}{Z} \tag{6-30}$$

This still has the form of Eq. (6-17), and if the appropriate coefficients are redefined then Eqs. (6-19) and (6-20) still hold.

Comparison to Experiment

A $\tfrac{3}{4}$-in Brabender Extruder is available, having the following dimensions (all linear dimensions in inches)

$$D = 0.75 \qquad Z = 49.3 \qquad B_0 = 0.150 \Big|$$
$$W = 0.596 \qquad \theta = 17.7° \qquad B_z = 0.066 \Big| \text{linear variation}$$

A newtonian fluid of viscosity $\mu = 20$ P and density $\rho = 1.4$ g/ml is extruded through a cylindrical die of diameter $d = \frac{1}{32}$ in and length $L = \frac{3}{32}$ in.

In order to predict Q and ΔP we must first calculate

$$A = \tfrac{1}{2}\pi D W \langle B_1 \rangle \cos \theta \qquad C = \frac{W \langle B^3 \rangle}{12Z} \qquad k = \frac{\pi d^4}{128L}$$

where $\langle B_1 \rangle = B_0 B_z / [\tfrac{1}{2}(B_0 + B_z)]$ and $\langle B^3 \rangle = (B_0 B_z)^2 / [\tfrac{1}{2}(B_0 + B_z)]$.

The results are $A = 6.1 \times 10^{-2}$ in^3, $C = 9.15 \times 10^{-7}$ in^3, and $k = 2.38 \times 10^{-7}$ in^3, from which we find

$$Q \text{ g/min} = 0.289N \text{ rpm}$$

$$\Delta P \text{ psi} = 0.264N \text{ rpm}$$

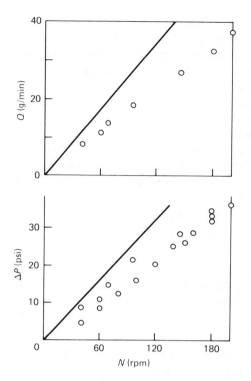

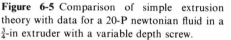

Figure 6-5 Comparison of simple extrusion theory with data for a 20-P newtonian fluid in a $\frac{3}{4}$-in extruder with a variable depth screw.

Figure 6-5 compares these predictions to experimental data. The predicted linearity of both Q and ΔP as functions of N is observed in the data. However, both sets of data fall below the predictions by almost exactly the same factor, 1.51. The source of this error is not evident, but it is likely to be due to several factors. Inaccuracy in the channel depth B is amplified in calculation of the parameter C, which varies with B^3. It is not known how accurate the manufacturer's specifications on B are.

6-2 NEWTONIAN ADIABATIC ANALYSIS

In a highly viscous fluid one often observes a significant temperature rise due to the dissipative action of the shearing forces acting on the fluid. The extent of the temperature rise can be controlled to some degree by using the surfaces which confine the flow as a medium for heat transfer. Commercial extruders normally allow for heat transfer at the barrel surface, and some screws are designed with a facility for heat exchange as well. Thus the ultimate temperature rise in a fluid undergoing extrusion depends on the mode of thermal operation of the system, as well as on the fluid dynamics itself.

The question of how one attempts to control the temperature of a fluid undergoing extrusion depends in part on the effect of the flow field on the temperature. Temperature rises could be so high that one might have to *cool* the barrel of the extruder under some operating conditions. The basis for beginning such a thermal analysis is the calculation of the *adiabatic temperature rise*, which assumes no heat transfer across the fluid boundaries. This is the maximum temperature rise in the absence of external heat exchange.

The major effect that must be accounted for in the nonisothermal analysis is the strong dependence of viscosity on temperature. As a consequence the assumption that the flow field is independent of the z coordinate is not strictly correct. As in the case of variable $B(z)$, we will again assume that a good approximation is that the velocity u_z satisfies, at any value of z,

$$u_z = U_z \frac{y}{B} - \frac{1}{2\mu} y(B - y) \frac{\partial p}{\partial z} \tag{6-31}$$

with the understanding that $\mu = \mu(z)$.

If this expression is integrated across the channel cross section to give Q, we find, upon introducing the geometric parameters A and C, that

$$Q = AN - CZ \frac{1}{\mu} \frac{\partial p}{\partial z} \tag{6-32}$$

If this expression is integrated with respect to z, we find

$$\Delta P = \frac{AN - Q}{CZ} \int_0^z \mu(z) \, dz \tag{6-33}$$

Now it is necessary to introduce the temperature rise into the analysis. We begin with the assumption that an input of power $d\dot{W}$ gives rise to an increase in the thermal energy of the fluid, such that

$$\frac{d\dot{W}}{dz} = \rho C_p Q \frac{dT}{dz} \tag{6-34}$$

[Equation (6-34) neglects the relatively small amount of work being done on the fluid at the rate $Q \, dP$.]

If we modify Eq. (6-21) slightly for this case, we can also write the power input as

$$\frac{d\dot{W}}{dz} = EN^2 \mu \tag{6-35}$$

where, again, the small contribution of a *flow-work* term $AN \, dP/dz$ is dropped for simplicity. From these last two equations we may write

$$\mu \, dz = \frac{\rho C_p Q}{EN^2} \, dT \tag{6-36}$$

This makes it possible to write the pressure drop as

$$\Delta P = \frac{AN - Q}{CZ} \frac{\rho C_p Q}{EN^2} \int_{T_0}^{T_z} dT$$

or

$$Q = AN - \frac{ECZN^2}{\rho C_p} \frac{\Delta P}{Q \, \Delta T} \qquad (6\text{-}37)$$

Equation (6-37) is an *adiabatic screw characteristic*. It is really a quadratic equation for $Q(\Delta P, \Delta T)$.

Since the viscosity enters the analysis at least implicitly, we must introduce a model for $\mu(T)$. The simplest realistic model is of the form

$$\mu = ae^{-bT} \qquad (6\text{-}38)$$

This lets us write an equation for $T(z)$ from Eq. (6-36), with the result that

$$\frac{\rho C_p Q}{EN^2} dT = ae^{-bT} dz \qquad (6\text{-}39)$$

After we multiply both sides by e^{bT}, this equation may be integrated to give

$$e^{b \, \Delta T} = \frac{b\mu_0 EN^2 Z}{\rho C_p Q} + 1 \qquad (6\text{-}40)$$

where $\mu_0 = ae^{-bT_0}$ is the viscosity at the entrance to the extruder ($z = 0$) where $T = T_0$. It will be useful to define a variable $\chi(\Delta T)$ as

$$\chi = e^{b \, \Delta T} \qquad (6\text{-}41)$$

and to consider χ (instead of ΔT) as the primary temperature-rise variable. Equation (6-40) is most conveniently rearranged to the form explicit in Q:

$$Q = \frac{\mu_0 bEN^2 Z}{\rho C_p(\chi - 1)} \qquad (6\text{-}42)$$

If the die sees fluid of viscosity $\mu(T_z) = \mu_z$, then the die characteristic may be written as

$$Q = \frac{k \, \Delta P}{\mu_z} = \frac{k\chi \, \Delta P}{\mu_0} \qquad (6\text{-}43)$$

In the screw characteristic [Eq. (6-37)], if we replace ΔP using the die characteristic and if ΔT is replaced by $\chi(\Delta T)$, we find

$$Q = AN - \frac{ECZN^2 b\mu_0}{\rho C_p k} \frac{1}{\chi \ln \chi} \qquad (6\text{-}44)$$

If Q is eliminated by using Eq. (6-42), the result is an equation for χ:

$$\chi \ln \chi \left(N_2 - \frac{1}{\chi - 1} \right) = N_1 \qquad (6\text{-}45)$$

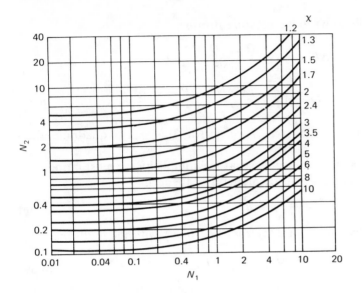

Figure 6-6 χ as a function of N_1 and N_2.

where $N_1 = C/k$, and $N_2 = A\rho C_p/NZ\mu_0 bE$. For any value of N_1 and N_2 a solution for χ may be obtained. The most convenient format for presentation of solutions is that of Fig. 6-6.

Once χ is obtained one may calculate ΔT from Eq. (6-41). Q follows from Eq. (6-42), which may be written more concisely as

$$Q = \frac{AN}{N_2(\chi - 1)} \tag{6-46}$$

and $\dot{W}$ may be calculated from the integrated form of Eq. (6-34),

$$\dot{W} = \rho C_p Q \, \Delta T \tag{6-47}$$

Example 6-1 In Sec. 6-1 an example was worked involving extrusion of a fairly low-viscosity fluid. It was found that the simple isothermal theory was in error to some extent. Could that error be associated with viscous heating?

To answer the question let us calculate the expected adiabatic temperature rise for that case.

We need the following additional information:

$$b = 0.025 \text{ K}^{-1} \qquad \rho C_p = 300 \text{ in} \cdot \text{lbf/in}^3 \cdot \text{K}$$
$$E \text{ [from Eq. (6-22)]} = 55.5 \text{ in}^2 \qquad \text{(using } \bar{B}_1 \text{ for } B\text{)}$$

We find $N_1 = 3.8$, and $N_2 > 10^3$ at $N = 200$ rpm. This N_2 value is so large that Fig. 6-6 cannot be used. However, it is possible to show that for very large N_2 (compared to $1 + N_1$)

$$\chi - 1 = \frac{1 + N_1}{N_2} \tag{6-48}$$

The resulting χ is so close to unity that we conclude that there is no perceptible temperature rise for this example.

Example 6-2 A single-flighted screw extruder has the following dimensions:

$$D = 2 \text{ in} \qquad B = 0.2 \text{ in}$$

$$L = 15 \text{ in} \qquad \theta = 29°$$

It pumps a molten newtonian fluid of viscosity $\mu = 0.2 \text{ lbf·s/in}^2$, evaluated at the inlet temperature. Take the fluid specific gravity as 0.95 and the heat capacity as $\rho C_p = 300 \text{ in·lbf/in}^3 \cdot \text{K}$. The temperature dependence of viscosity is exponential, with $b = 0.025 \text{ K}^{-1}$.

For screw speeds up to 120 rpm, plot the adiabatic temperature rise (in degrees Celsius), the output (in pounds per hour), and the horsepower as functions of N. Assume open discharge.

We begin by calculating the parameters A and E.

For $\theta = 29°$, $\sin \theta = 0.49$ and $\cos \theta = 0.88$. For free discharge, $k = \infty$.

From the definitions presented earlier,

$$A = \tfrac{1}{2}\pi DWB \cos \theta = \tfrac{1}{2}\pi^2 D^2 B \sin \theta \cos \theta = 1.6 \text{ in}^3$$

$$E = \frac{\pi^3 D^3 \sin \theta (1 + 3 \sin^2 \theta)}{B} = 1025 \text{ in}^2$$

The helical length is $Z = L/\sin \theta = 31$ in. Since $k = \infty$, $N_1 = 0$.

$$N_2 = \frac{A \rho C_p}{N Z \mu_0 bE} = \frac{2.9}{N}$$

with N in revolutions per *second*. For $N_1 = 0$ we have $\chi = 1 + 1/N_2$ by inspection of Eq. (6-45). Once χ is found we may calculate the desired quantities from

$$\Delta T = \frac{1}{b} \ln \chi = 40 \ln (1 + 0.35N)$$

$$Q = \frac{\mu_0 bEZ}{\rho C_p} \frac{N^2}{\chi - 1} = \frac{A}{N_2} \frac{N}{\chi - 1}$$

$$= \frac{0.56N^2}{\chi - 1} \text{ in}^3/\text{s} = \frac{70N^2}{\chi - 1} \text{ lb/h}$$

$$\dot{W} = \rho C_p Q \, \Delta T = 300Q \, \Delta T \text{ in·lbf} = \frac{300}{12(550)} Q \, \Delta T \text{ hp}$$

Figure 6-7 shows these results.

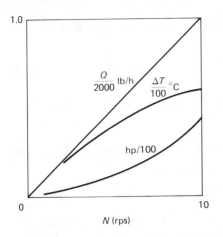

Figure 6-7 Results calculated for Example 6-2.

Example 6-3 A tubing die is designed as shown in Fig. 6-8. At $N = 120$ rpm find the output, pounds per hour, and the temperature of the extrudate. The feed temperature is 275°F. Assume that the extruder and the fluid are those of Example 6-2. The tubing die has dimensions

$$R_i = 0.250 \text{ in} \qquad R_0 = 0.300 \text{ in} \qquad L = 3 \text{ in}$$

Assume that the die is insulated and there is no temperature change from the end of the screw to the die outlet. Assume the screw operates adiabatically.

First we must find the constant k in the die equation $Q = k \, \Delta P / \mu$. Since the die is annular we use Eq. (5-30), from which we find

$$k = \frac{\pi R_0^4}{8L} \left[1 - \kappa^4 - \frac{(1 - \kappa^2)^2}{\ln(1/\kappa)} \right] = 6.1 \times 10^{-6} \text{ in}^3$$

The parameters A and E were calculated in Example 6-2. The constant C is found to be

$$C = \frac{WB^3}{12Z} = \frac{\pi DB^3 \sin \theta}{12Z} = 0.66 \times 10^{-4} \text{ in}^3$$

Then $N_1 = C/k = 10.9$.

We found in Example 6-2 that $N_2 = 2.9/N$ (with N in revolutions per second). At 120 rpm = 2 rps, $N_2 = 1.45$.

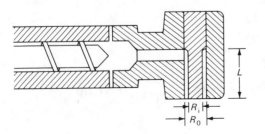

Figure 6-8 Geometry of a tubing die for Example 6-3.

From Fig. 6-6 we find $\chi = 4.7$. This gives

$$\Delta T = 40 \ln 4.7 = 62°C$$

$$Q = \frac{70}{3.7} N^2 = 76 \text{ lb/h}$$

The temperature of the extrudate is $275 + 1.8(62) = 387°F$.

We may also find the pressure drop across the die using the die characteristic in the form

$$\Delta P = \frac{\mu_0 e^{-b \Delta T} Q}{k}$$

with the result that $\Delta P = 4241$ psi.

6-3 OPTIMAL DESIGN

Even in the simplest model of extrusion, the isothermal case with uniform screw geometry, the output depends on a large number of variables. Equation (6-19) may be written as

$$Q = \frac{\pi^2 D^2 NB \sin \theta \cos \theta}{2 + \pi DB^3 \sin^2 \theta / 6kL} \tag{6-49}$$

It is not difficult to see that the dependence of Q on B shows a maximum; there is some channel depth B^0 which maximizes the output, all other parameters being held constant. Similarly, there is a helix angle $\theta°$ which maximizes Q.

We may talk about *optimal design* here in a limited sense. Let us suppose that D and L are fixed, as well as the die characteristic parameter k. Then B^0 may be found as the solution to $(\partial Q/\partial B)_\theta = 0$, from which it is found that B^0 depends on θ and is

$$B^0 = \left(\frac{6kL}{\pi D \sin^2 \theta} \right)^{1/3} \tag{6-50}$$

In a similar manner we can consider $(\partial Q/\partial \theta)_B = 0$ and find $\theta°$ from

$$\sin^2 \theta° = \left(2 + \frac{\pi DB^3}{12kL} \right)^{-1} \tag{6-51}$$

A *pair* of values of B and θ that maximizes Q must satisfy both Eqs. (6-50) and (6-51), and one finds that

$$\theta* = 30°$$

$$B* = \left(\frac{24kL}{\pi D} \right)^{1/3} \tag{6-52}$$

The output at this maximum is

$$Q^* = \frac{\sqrt{3}}{12} \pi^2 D^2 B^* N \tag{6-53}$$

Thus we find that a 30° helix angle optimizes the output of the extruder, for isothermal newtonian flow.

But *output* is not the only *performance variable* of interest. The economics of extrusion also involves the *power* required to pump the fluid. In the isothermal case the power input creates two effects: the axial flow which produces the output and the transverse flow which simply circulates and mixes the fluid but does not create any output.

That part of the power requirement associated with the output is given by the term $Q\,\Delta P$. We may then consider the ratio

$$\epsilon = \frac{Q\,\Delta P}{\dot{W}} \tag{6-54}$$

as a measure of the *efficiency* of the extruder. If ϵ is near unity then nearly all the power produces output.

If it is assumed that the optimum B^0 is used, then it can be shown that ϵ depends on θ according to

$$\epsilon = \frac{\cos^2 \theta}{3(1 + \sin^2 \theta)} \tag{6-55}$$

The maximum value of ϵ is only $\frac{1}{3}$ and occurs when $\theta = 0$. A zero helix angle is a trivial case, giving no output, as Eq. (6-49) shows. The efficiency ϵ falls off monotonically as θ increases. At 30°, for example, ϵ is two-thirds its maximum value.

Since there are different criteria of optimality, it is not clear how one chooses the "best" helix angle. Under some conditions, for example, one might wish to enhance the homogeneity of the melt by allowing for more transverse mixing. The usual practices result in helix angles that are typically in the range of 10 to 25°.

Under some circumstances one may wish to maximize the pressure developed by the extruder or, at least, to know what that maximum pressure may be. If conditions are isothermal the problem is trivial, since the pressure increases with N monotonically according to Eq. (6-20).

For adiabatic flow, however, the pressure may go through a maximum as screw speed increases, if the heating effect is sufficient to reduce the viscosity by a significant amount. It is possible to determine the operating conditions at which the maximum pressure occurs.

We begin with the adiabatic screw characteristic, Eq. (6-37), and divide through both sides by AN. The ratio $g \equiv Q/AN$ normalizes Q to its maximum value, which would be achieved at open discharge. The adiabatic screw characteristic takes the form

$$g = 1 - \frac{ECZb}{\rho C_p A^2} \frac{\Delta P}{g \ln \chi} \tag{6-56}$$

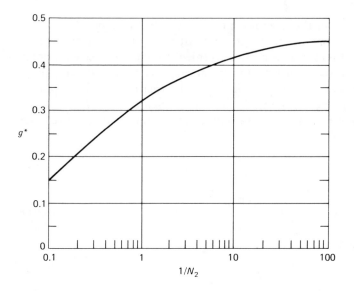

Figure 6-9 Solution of Eq. (6-59).

We eliminate χ through the adiabatic energy balance, Eq. (6-40), which may be written in the form

$$\chi - 1 = \frac{1}{gN_2} \tag{6-57}$$

where N_2 was defined through Eq. (6-45). Equation (6-56) may be written in the form

$$\frac{\Delta P}{a_2} = g(1 - g) \ln\left(1 + \frac{1}{gN_2}\right) \tag{6-58}$$

where $a_2 = \rho C_p A^2 / ECZb$. On differentiating ΔP with respect to g to find the maximum pressure, it may be seen that the value of g at which the maximum pressure occurs must satisfy

$$\ln\left(1 + \frac{1}{g^*N_2}\right) = \frac{1 - g^*}{1 - 2g^*} \frac{1}{1 + g^*N_2} \tag{6-59}$$

Figure 6-9 gives $g^*(N_2)$. The maximum pressure then follows upon substituting g^* for g in Eq. (6-58).

6-4 NONNEWTONIAN ISOTHERMAL ANALYSIS

For nonnewtonian fluids, even with geometric simplifications and the assumption of isothermal flow, it is not possible to find simple analytical expressions for output and pressure, such as are given in Eqs. (6-19) and (6-20). Graphical procedures provide the most convenient means of treating such problems.

In the isothermal case for the newtonian fluid, the *output variables* Q and ΔP are the simultaneous solutions of the die characteristic and the screw characteristic equations (6-16) and (6-17). It is most convenient to rewrite these equations in dimensionless form and then illustrate the graphical procedure. Following that, the nonnewtonian case will be presented in a parallel treatment of a graphical solution method. Then the nonisothermal case may be treated.

We begin by defining a dimensionless output and a dimensionless pressure drop:

$$\Pi_Q = \frac{Q}{U_z BW} \tag{6-60}$$

$$\Pi_P = \frac{\Delta P \, B^2}{\mu U_z Z} \tag{6-61}$$

For convenience we recall here that

$$U_z = \pi DN \cos \theta \tag{6-62}$$

$$W = \pi D \sin \theta \tag{6-63}$$

$$Z = \frac{L}{\sin \theta} \tag{6-64}$$

The screw characteristic [Eq. (6-17)] is easily seen to have the form

$$\Pi_Q = \tfrac{1}{2} - \tfrac{1}{12}\Pi_P \tag{6-65}$$

The die characteristic then takes the form

$$\Pi_Q = \frac{\Pi_P}{12N_1} \tag{6-66}$$

where N_1 was defined earlier [below Eq. (6-45)]. Then the operating conditions, at the selected value of N, are the coordinates of the point of intersection of Eqs. (6-65) and (6-66), as sketched in Fig. 6-10.

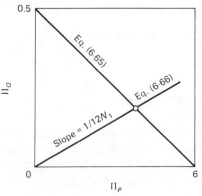

6 **Figure 6-10** Plot of dimensionless screw and die characteristics for isothermal newtonian extrusion.

For the nonnewtonian fluid both the screw and die characteristics are nonlinear. The die characteristic comes from the appropriate model for pressure flow. If we assume that the power law is a suitable model, then we may write the die characteristic in the form

$$Q = \frac{k\,\Delta P}{\bar{\mu}_d} \tag{6-67}$$

where $\bar{\mu}_d$ is the "apparent" viscosity of the fluid at the conditions in the die. In general we can calculate $\bar{\mu}_d$ if we have an estimate for the nominal shear rate in the die, using

$$\bar{\mu}_d = K\dot{\gamma}_d^{n-1} \tag{6-68}$$

For a given die geometry one can estimate $\dot{\gamma}_d$ at a given flow rate. For example, if the die is a circular tube, then from Eqs. (5-17) and (5-18) we can show that

$$k = \frac{n\pi R^4}{2L(1 + 3n)} \tag{6-69}$$

and

$$\bar{\mu}_d = K\left[\frac{(1 + 3n)Q}{n\pi R^3}\right]^{n-1} = K\left(\frac{QR}{2kL}\right)^{n-1} \tag{6-70}$$

Since Q may be unknown a priori, we must estimate Q to find $\bar{\mu}_d$. Trial and error may be necessary in some cases. While Eq. (6-67) has a linear *format*, we note that since $\bar{\mu}_d$ depends upon Q, the die characteristic will be nonlinear.

Now we turn to calculation of the screw characteristic. For a power law fluid under combined pressure and drag flow in a rectangular channel, the dynamic equations take the form (neglecting inertial terms)

$$0 = -\frac{\partial p}{\partial x} + K\frac{\partial}{\partial y}\left[(\tfrac{1}{2}\mathrm{II}_\Delta)^{(n-1)/2}\frac{\partial u_x}{\partial y}\right] \tag{6-71}$$

$$0 = -\frac{\partial p}{\partial z} + K\frac{\partial}{\partial y}\left[(\tfrac{1}{2}\mathrm{II}_\Delta)^{(n-1)/2}\frac{\partial u_z}{\partial y}\right] \tag{6-72}$$

where

$$\tfrac{1}{2}\mathrm{II}_\Delta = \left(\frac{\partial u_x}{\partial y}\right)^2 + \left(\frac{\partial u_z}{\partial y}\right)^2 \tag{6-73}$$

In writing these equations we have assumed that there *is* a transverse flow u_x but that the aspect ratio is large enough that u_x and u_z are independent of x and are functions only of y. We note that since the viscosity depends on II_Δ, which depends on both u_x and u_z, Eqs. (6-71) and (6-72) are *coupled*, i.e., both variables appear in each. Analytical solution is impossible in this case, and numerical methods must be used.

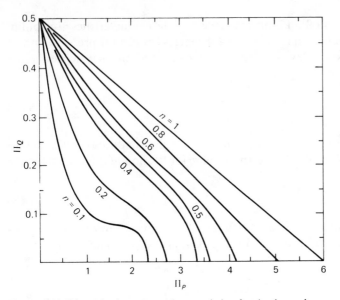

Figure 6-11 Dimensionless screw characteristics for isothermal power law extrusion. Transverse flow neglected.

It is possible to uncouple the equations with the approximation that the transverse flow is unimportant, in which case we must solve

$$0 = -\frac{\partial p}{\partial z} + K\frac{\partial}{\partial y}\left\{\left[\left(\frac{\partial u_z}{\partial y}\right)^2\right]^{(n-1)/2}\frac{\partial u_z}{\partial y}\right\} \tag{6-74}$$

In this one-dimensional model, if we assume that the screw channel is uniform down the helical axis z, we have a nonlinear ordinary differential equation to solve, which we may write as

$$0 = -\frac{\Delta P}{Z} + K\frac{d}{dy}\left(\left|\frac{du_z}{dy}\right|^{n-1}\frac{du_z}{dy}\right) \tag{6-75}$$

The absolute value sign replaces the operation of squaring and taking the $(n-1)/2$ root. (The point here is that one cannot take the fractional root of a negative number without getting imaginary numbers, and the velocity gradient du_z/dy will, in general, change sign in the region $0 \le y \le B$.)

The solution of Eq. (6-75) has been carried out, and the most important result is the presentation of the screw characteristics given in Fig. 6-11. Π_Q and Π_P are defined as in the newtonian case, except that in Π_P one uses $\bar{\mu}_s$ at the nominal shear rate in the screw, given by

$$\dot{\gamma}_s = \frac{U_z}{B} \tag{6-76}$$

so that

$$\bar{\mu}_s = K\left(\frac{U_z}{B}\right)^{n-1} \tag{6-77}$$

The graphical method of solution then requires plotting Π_Q versus Π_P for both the screw and the die, with the intersection giving the operating point. The die characteristic [Eq. (6-67)] may be converted to Π form as in Eq. (6-66) for the newtonian case. However, we must note that now we have *two* viscosities, $\bar{\mu}_s$ at the screw shear rate and $\bar{\mu}_d$ at the die shear rate. If we agree to define Π_P based on the screw viscosity $\bar{\mu}_s$, then the die characteristic takes the form

$$\Pi_Q = \frac{1}{12N_1} \Pi_P \frac{\bar{\mu}_s}{\bar{\mu}_d} \tag{6-78}$$

Example 6-4 An extruder of dimensions $D = 2$ in, $B = 0.2$ in, $L = 40$ in, and $\theta = 29°$ is used to produce plastic rod from a die of diameter $d = \frac{1}{16}$ in and length $L_d = \frac{1}{4}$ in. Assume the melt is at uniform temperature, and at that temperature the melt obeys the power law, with $n = \frac{1}{2}$ and $K = 1$ lbf·s$^{1/2}$/in^2. These power law parameters were obtained for shear rates in the range 10^3 to 10^4 s^{-1}.

Find the output and the die pressure at 30 rpm.

In this case it is simplest to calculate the die characteristic directly as $Q(\Delta P)$ and then convert the screw characteristic to a $Q(\Delta P)$ curve.

For the die: $\quad Q = \dfrac{n\pi R^3}{3n + 1}\left(\dfrac{R\,\Delta P}{2KL}\right)^{1/n} \quad$ [Eq. (5-17)]

$$Q \text{ in}^3/\text{s} = 7.5 \times 10^{-8}\,\Delta P^2 \text{ psi}$$

For the screw: $\quad Q = WBU_z\Pi_Q = \pi^2 D^2 BN \sin\theta \cos\theta\,\Pi_Q$

$$Q \text{ in}^3/\text{s} = 1.7\Pi_Q$$

$$\Delta P = \frac{Z\bar{\mu}_s U_z}{B^2}\Pi_P = \frac{Z\bar{\mu}_s \pi DN \cos\theta}{B^2}\Pi_P$$

$$\bar{\mu}_s = K\left(\frac{\pi DN \cos\theta}{B}\right)^{n-1} = 0.27 \text{ lbf·s/in}^2$$

$$\Delta P \text{ psi} = 1730\Pi_P$$

From Fig. 6-11, for $n = \frac{1}{2}$, we can pick off pairs of Π_P, Π_Q values for plotting the screw characteristic. Representative values are given as follows:

Π_P	Π_Q	ΔP, psi	Q, in^3 s
1	0.32	1730	0.53
2	0.23	3460	0.38
3	0.12	5190	0.20

From Fig. 6-12 one finds the intersection to be $Q = 0.48$ in^3/s $= 62$ lb/h, and $\Delta P = 2700$ psi.

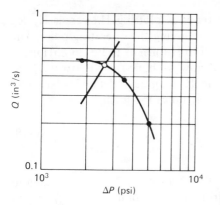

Q (in³/s)

0.1
10³   10⁴
           Δ*P* (psi)   **Figure 6-12** Graphical solution of Example 6-4.

One should always check the shear rates in the die and extruder against the shear rates over which the rheological data were obtained to ensure that they are in the same range of values. For the die,

$$\dot{\gamma}_d = \frac{1+3n}{n}\frac{Q}{\pi R^3} = 2.3 \times 10^4 \text{ s}^{-1}$$

For the screw,

$$\dot{\gamma}_s = \frac{U_z}{B} = \frac{\pi D N \cos\theta}{B} = 14 \text{ s}^{-1}$$

This latter calculation suggests a potential problem, since the rheological data begin at shear rates 2 orders of magnitude above the screw shear rate. The effect of this feature is illustrated in Prob. 6-9.

6-5 NONNEWTONIAN ADIABATIC ANALYSIS

When nonisothermal effects are introduced to the nonnewtonian analysis, the problem becomes quite complex. It is not possible to work with an analytical form for the screw characteristic, and so the analysis carried out in Sec. 6-2 cannot be paralleled for the power law fluid.

It is possible to solve the nonnewtonian adiabatic case numerically, but the introduction of additional (thermal) parameters makes it impossible to present the results *concisely*, as in the newtonian case, with a few equations and a single graph. It is possible, however, to present a concise approximate method by modifying the simpler analyses presented in Secs. 6-2 and 6-4.

We begin by noting that Eqs. (6-43) and (6-44), for the newtonian adiabatic case, may be put in dimensionless form, with the results

Screw:
$$\Pi_Q = \tfrac{1}{2} - \tfrac{1}{12}\Pi_P \frac{\chi - 1}{\ln \chi} \tag{6-79}$$

Die: $$\Pi_Q = \frac{1}{12N_1}\,\chi\Pi_P \tag{6-80}$$

From the definition of χ [Eq. (6-41)] we can see that

$$\chi = \frac{\mu_0}{\mu_z} \tag{6-81}$$

that is, χ is the ratio of the viscosities at the temperatures of the extruder inlet and the die, respectively. This indicates that the die characteristic, when plotted as Q versus ΔP, should involve the viscosity at the *die* conditions. This is logical and in fact is exactly consistent with Eq. (6-43).

The interpretation of the screw characteristic [Eq. (6-79)] is a little more complicated in the adiabatic case. The factor involving χ, which multiplies Π_P, can be expressed as

$$\frac{\chi - 1}{\ln \chi} = \frac{1 - \mu_0/\mu_z}{\ln\left(\mu_z/\mu_0\right)} = \frac{\mu_{LM}}{\mu_z} \tag{6-82}$$

where μ_{LM} is the *log mean viscosity* defined by

$$\mu_{LM} = \frac{\mu_z - \mu_0}{\ln\left(\mu_z/\mu_0\right)} \tag{6-83}$$

It is not difficult to show that μ_{LM} approaches the *arithmetic* average of μ_z and μ_0 when μ_z and μ_0 are not too different. Otherwise, μ_{LM} lies between μ_0 and μ_z. The pressure term, then, may be written in the form

$$\frac{\Pi_P(\chi - 1)}{\ln \chi} = \frac{\Delta P B^2/U_z Z}{\mu_0 \mu_z/\mu_{LM}} = \frac{\Delta P B^2}{\langle\mu\rangle U_z Z} \tag{6-84}$$

where we define an *adiabatic average* viscosity as

$$\langle\mu\rangle = \frac{\mu_0 \mu_z}{\mu_{LM}} \tag{6-85}$$

If Eq. (6-84) is compared to the definition of Π_P [Eq. (6-61)], we see that Eq. (6-79) may be written as

$$\Pi_Q = \tfrac{1}{2} - \tfrac{1}{12}\langle\Pi_P\rangle \tag{6-86}$$

where the brackets $\langle\ \rangle$ on Π_P remind us that we use $\langle\mu\rangle$ in Eq. (6-61).

Then we may use the same screw characteristic as in the isothermal case but with a modification of the viscosity according to Eq. (6-85). Thus the *newtonian adiabatic* case could be solved through a graphical procedure as follows:

1. Calculate N_1 and N_2, and from Fig. 6-6 find ΔT.
2. Plot Q versus ΔP from the die characteristic, using μ_z at the die temperature $T_z = T_0 + \Delta T$.

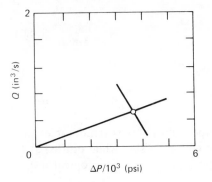

Figure 6-13 Graphical solution of Example 6-5.

3. Plot Q versus ΔP from the screw characteristic, using $\langle\mu\rangle$ as defined in Eq. (6-85).
4. The intersection gives the operating point $(Q, \Delta P)$.

Example 6-5 Rework Example 6-3 by using the graphical procedure just described. The necessary parameters were calculated in Example 6-3.

For $N = 2$ rps, we found $N_1 = 10.9$, $N_2 = 1.45$. From Fig. 6-6 we found $\chi = 4.7$, or $\Delta T = 62°$.

The die characteristic obeys

$$Q = \frac{k}{\mu_z}\Delta P = 1.4 \times 10^{-4}\ \Delta P \text{ psi}$$

which is plotted in Fig. 6-13.

The screw characteristic [Eq. (6-79)] is most easily plotted by calculating $\Pi_Q = \frac{1}{2}$ or $Q = AN$ at $\Delta P = 0$ and

$$\Pi_P = 6\frac{\ln \chi}{\chi^{-1}}$$

or
$$\Delta P = 6\frac{\ln \chi}{\chi^{-1}}\frac{U_z Z\mu_0}{B^2} \quad \text{at } Q = 0$$

The end points of the screw characteristic are found to be $(\Delta P, Q) = $ (4280 psi, 0) and (0, 3.2 in^3/s). This line is plotted in Fig. 6-13, and the resulting intersection is found to be

$$\Delta P = 3600 \text{ psi}$$

$$Q = 0.5 \text{ in}^3/\text{s} = 65 \text{ lb/h}$$

in good agreement with the results given in Example 6-3.

The extension of this method to the nonnewtonian case will be made in the following way. We assume a power law model, with K given as a function of temperature according to

$$\frac{K}{K_0} = e^{-b(T-T_0)} \tag{6-87}$$

while n is assumed independent of T, at a fixed shear rate. These assumptions are in reasonable agreement with experience.

A further assumption is that Fig. 6-6 holds for power law fluids if μ_0, which appears in N_2, is calculated at the nominal screw shear rate and at the inlet temperature, so that

$$\bar{\mu}_0 = K_0 \left(\frac{U_z}{B} \right)^{n-1} \tag{6-88}$$

Then the procedure is as follows:

1. Calculate N_1 and N_2, and from Fig. 6-6 find ΔT.
2. Plot Q versus ΔP from the die characteristic, using $\bar{\mu}_z$ at the die temperature $T_z = T_0 + \Delta T$ and at the die shear rate, so that

$$\bar{\mu}_z = K_0 e^{-b\,\Delta T} \dot{\gamma}_d^{n-1} = \frac{K_0}{\chi} \dot{\gamma}_d^{n-1} \tag{6-89}$$

3. Determine the appropriate value of n, and plot Q versus ΔP from the corresponding power law screw characteristic, using $\langle \mu \rangle$ evaluated at the screw shear rate, with an *adiabatic average* $\langle K \rangle$ defined so that

$$\langle \mu \rangle = \langle K \rangle \left(\frac{U_z}{B} \right)^{n-1}$$

$$= \frac{K_0 K_z}{K_{\mathrm{LM}}} \left(\frac{U_z}{B} \right)^{n-1}$$

$$= K_0 \frac{\ln \chi}{\chi - 1} \left(\frac{U_z}{B} \right)^{n-1} \tag{6-90}$$

4. The intersection gives the operating point $(Q, \Delta P)$.

Example 6-6 Dow Styron 666 (a general-purpose polystyrene) is to be extruded using the extruder of Example 6-2. Find the flow rate and temperature rise *at open discharge* for a screw speed of 12 rpm. The melt enters the extruder at 170°C.

Take ρC_p as 300 in · lbf/in³ · K. Viscosity data are given in Fig. 6-14.

As before, $A = 1.6$ in³, $E = 1025$ in², and $Z = 31$ in. Since $k = \infty$ at open discharge, $N_1 = 0$.

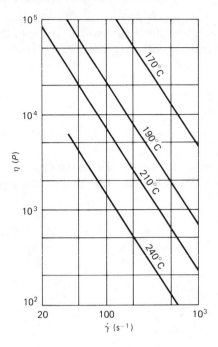

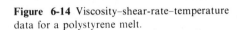

Figure 6-14 Viscosity–shear-rate–temperature data for a polystyrene melt.

From the rheological data presented we can find

$$n = \tfrac{1}{3}$$

$$K_0 = 4 \times 10^5 \text{ dyne} \cdot \text{s}^{1/3}/\text{cm}^2 = 5.8 \text{ lbf} \cdot \text{s}^{1/3}/\text{in}^2$$

$$b = 0.066 \,^{\circ}\text{C}^{-1}$$

We now carry out the steps outlined above.

$$N_2 = \frac{A\rho C_p}{NZ\bar{\mu}_0 bE}$$

We must calculate $\bar{\mu}_0$.

$$\bar{\mu}_0 = K_0 \left(\frac{U_z}{B}\right)^{n-1} = 5.8 \left(\frac{\pi DN \cos\theta}{B}\right)^{-2\cdot 3}$$

$$= 1.86 \text{ lbf} \cdot \text{s}/\text{in}^2$$

$$N_2 = 0.62$$

For $N_1 = 0$,

$$\chi = 1 + \frac{1}{N_2} = 2.6$$

$$\ln\chi = 0.96 \qquad \Delta T = \frac{1}{b}\ln\chi = 14\,^{\circ}\text{C}$$

This completes step 1.

For the case of open discharge $\Delta P = 0$, and so $\Pi_P = 0$.

From Fig. 6-11, for $n = \frac{1}{3}$, we find $\Pi_Q = \frac{1}{2}$, or (taking a density of 60 lb/ft^3)

$$Q = AN = 40 \text{ lb/h}$$

At *open discharge*, Q is unaffected by fluid properties and is independent of temperature rise or the degree of nonnewtonian character of the fluid. This is a consequence of the *one-dimensional* drag flow assumptions.

Example 6-7 Repeat Example 6-3 for the fluid described in Example 6-6. Give ΔP and ΔT at an output of 20 lb/h, and specify the screw speed N required to produce this output.

In this formulation of the problem the screw speed is unknown. Trial and error is required to find the value of N which produces the specified output. We found in Example 6-6 that an output of 40 lb/h is achieved at $N = 12$ rpm. The die resistance will reduce the output from this level. We will proceed by calculating the output at several values of N and then interpolate from these results to the specified output.

Let $N = 12$ rpm $= 0.2$ rps. As in Example 6-6, $N_2 = 0.62$. Since N_1 is a purely geometric parameter, we may use the value found in Example 6-2, $N_1 = 10.9$. From Fig. 6-6 we find $\chi = 9.5$. Then $\ln \chi = 2.25$ and $\Delta T = 34°$C. This completes step 1.

The die characteristic is

$$Q = \frac{k}{\mu_z} \Delta P = \frac{k\chi}{K_0} \dot{\gamma}_d^{n-1} \Delta P$$

Since we have the equivalent result for the power law fluid in Eq. (5-32), it is easier to use that directly in the form

$$Q = \frac{n\pi R_0}{1 + 2n} (R_0 - R_i)^{2 + 1/n} \frac{F(n, \kappa)}{[2(K_0/\chi)L]^{1/n}} \Delta P^{1/n}$$

From Fig. 5-3 we find $F(\frac{1}{3}, 0.833) = 0.93$. The die characteristic becomes

$$Q = 1.1 \times 10^{-9} \Delta P^3$$

(*Note:* Q is in cubic inches per second and ΔP in pounds per square inch.) This is plotted on Fig. 6-15.

From Fig. 6-11, for $n = \frac{1}{3}$, some representative values of Π_Q and Π_P are

Π_Q	0.4	0.3	0.2	0.1
Π_P	0.40	0.75	1.5	2.5

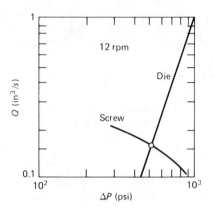

Figure 6-15 Graphical solution of Example 6-7.

To convert this to a screw characteristic in terms of ΔP and Q it is necessary to calculate $\langle \mu \rangle$.

$$\langle \mu \rangle = K_0 \frac{\ln \chi}{\chi - 1} \left(\frac{U_z}{B} \right)^{n-1}$$

$$= 5.8 \left(\frac{2.25}{8.5} \right) (5.53)^{-2/3} = 0.49 \text{ lbf·s/in}^2$$

$$\Delta P = \frac{\langle \mu \rangle U_z Z \Pi_P}{B^2} = 420 \Pi_P$$

$$Q = U_z BW \Pi_Q = 1.1(0.2)(3.05) \Pi_Q = 0.67 \Pi_Q$$

With the Π_P, Π_Q values given above we can plot the screw characteristic. Figure 6-15 shows the screw and die characteristics for $N = 12$ rpm. The intersection of the curves gives $Q = 0.16$ in^3/s, or 20 lb/h (at a density of 60 lb/ft^3). The choice of 12 rpm for this calculation was convenient since it allowed use of several numbers from Example 6-6. Normally it would be necessary to repeat the procedure and find Q at several values of N, and then, by interpolation, pick the N that produces the specified output. In this case it is accidental that the first choice of N gives the right output. We find, then, that at $N = 12$ rpm we have a pressure drop of 520 psi and a temperature rise (adiabatic) of 34°C.

6-6 EVALUATION OF SOME OF THE PREVIOUS KINEMATIC AND GEOMETRIC SIMPLIFICATIONS

In the previous sections we have examined extrusion models based on simplifications of the kinematics of the flow field as well as on idealizations of the geometry of the extruder. Here we examine the relaxation of some of these idealizations and thereby attempt to evaluate their importance. Some of the geometric simplifications can be relaxed, and more realistic models presented, with only a minor increment of effort. The justification for having made the simplifications at all, in such cases, was somewhat weak but lay in the desire to use the simplest possible geometry for evaluation of the more important rheological and thermal features of the extruder model.

The principal kinematic assumption used in earlier sections, the neglect of the transverse flow in all but the newtonian isothermal case, can be relaxed only at great computational expense, and it is important to evaluate the role of the transverse flow under realistic operating conditions. We begin, then, with a consideration of transverse flow. We will consider subsequently several of the geometric simplifications which can be relaxed, and assess their importance.

Transverse Flow

For isothermal power law extrusion the appropriate equations were given earlier: Eqs. (6-71) to (6-73). Kinematic boundary conditions specify no motion at the stationary surfaces. At the moving surface the appropriate boundary conditions are

$$u_z = \pi DN \cos \theta \qquad (6\text{-}91)$$
$$u_x = -\pi DN \sin \theta \qquad \text{at } y = B \qquad (6\text{-}92)$$

It is not too difficult to see that when the equations of motion and these boundary conditions are nondimensionalized the screw characteristics have the functional form

$$\Pi_Q = \Pi_Q(\Pi_P, n, \theta) \qquad (6\text{-}93)$$

where Π_Q and Π_P have been previously defined. We see immediately that θ enters the screw characteristic as an independent variable when transverse flow is included.

Numerical solutions of this problem have been carried out, and Fig. 6-16 shows screw characteristics for $\theta = 15°$. When transverse flow is considered it is common to redefine the Π's as

$$\Pi'_Q = \Pi_Q \cos \theta \qquad (6\text{-}94)$$

$$\Pi'_P = \Pi_P(\cos \theta)^n \qquad (6\text{-}95)$$

which corresponds to using $U = \pi DN$ as the characteristic velocity, instead of $U_z = \pi DN \cos \theta$, in nondimensionalizing the equations.

For the newtonian fluid, for which the transverse flow has no effect on the screw characteristic, we observe the expected result at open discharge:

$$\Pi'_Q = \tfrac{1}{2} \cos \theta \qquad (6\text{-}96)$$

For $n < 1$ we find that the screw characteristics are qualitatively similar to those for one-dimensional flow, given earlier in Fig. 6-11. A quantitative comparison is given in Fig. 6-17 (in terms of Π rather than Π') and it is seen that the effect of

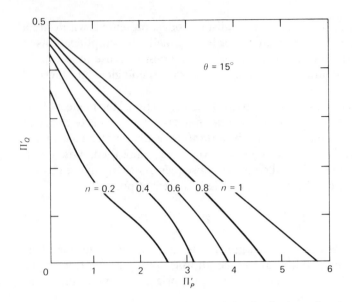

Figure 6-16 Dimensionless screw characteristics for isothermal power law extrusion. Transverse flow included. $\theta = 15°$.

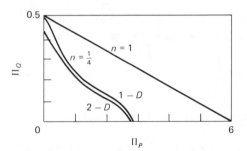

Figure 6-17 Comparison of one- and two-dimensional isothermal screw characteristics for a power law fluid. $\theta = 17.7°$.

transverse flow is to reduce the output. The extent of reduction is small, however, and not of sufficient magnitude to justify the additional numerical work required for its calculation. If we were concerned only with extruder *output*, we would not be motivated to include transverse flow in the model.

With regard to the power requirement we can return to Eq. (6-15), which includes the contribution of transverse flow. If we define a dimensionless power parameter as

$$\Pi_{\dot{w}} = \frac{\dot{W}B}{WZU_z^2\mu} \tag{6-97}$$

we may rearrange Eq. (6-15) (after using the screw characteristic to eliminate ΔP) and find

$$\Pi_{\dot{w}} = 4 - 6\Pi_Q + \underline{4\tan^2\theta} \tag{6-98}$$

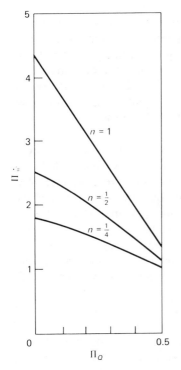

Figure 6-18 Dimensionless power parameter as a function of flow rate, including the effect of transverse flow for $\theta = 17.7°$.

The underlined term gives the contribution of the transverse flow to the power requirement. It is clear that the *relative* importance of transverse flow depends upon Π_Q, but unless θ is quite small the neglect of the transverse flow contribution gives a significant error.

Numerical solutions which include the transverse flow have been carried out for typical values of n and θ, and it is found that the power is underestimated by a lesser amount than in the case $n = 1$ if the one-dimensional power-law velocity field is used instead of the two-dimensional (transverse flow included) solution. In either case, whether transverse flow is included or not, the power law solution must be carried out numerically.

Fenner has solved the dynamic equations, including the transverse flow, and presents several examples of results in his comprehensive monograph on modeling of melt extrusion. Figure 6-18 shows $\Pi_{\dot{w}}$ for the newtonian and two power law cases, for $\theta = 17.7°$. $\Pi_{\dot{w}}$ is defined, for the power law fluid, by

$$\Pi_{\dot{w}} = \frac{\dot{W} B^n}{WZKU_z^{1+n}} \tag{6-99}$$

This definition of $\Pi_{\dot{w}}$ corresponds to Eq. (6-97), with an *apparent* viscosity evaluated at the nominal shear rate in the screw channel.

From Fig. 6-18 it is clear that if one attempted to use the newtonian power formula [Eq. (6-98)] with an apparent viscosity based on the nominal shear rate, very significant errors would be made except near open discharge ($\Pi_Q \approx 0.4$ to 0.5). This is true whether transverse flow is included or not. This is an important point, because one is often tempted to *compound* models, i.e., to use several *isolated* modeling ideas together in a manner which may then lead to significant error because, in a sense, the "whole" of the subsequent model turns out to be very different from "the sum of its parts."

An important conclusion to be drawn from the results above is that it is more important to have an accurate rheological characterization of the melt being extruded than to have an exact kinematic model that includes the transverse flow. This places some emphasis on having rheological data over the appropriate range of shear rates experienced within the extruder. It is still a reasonably good rule of thumb to use the nominal shear rate U_z/B to characterize the deformation rates experienced by the fluid. Since the transverse flow increases the *magnitude* of the deformation rate (that is, II_Δ) over the value U_z/B, one should consider U_z/B to be a lower estimate, and if possible require data at shear rates in excess of this value.

To review briefly, we conclude that transverse flow has a small effect on output, not great enough to justify its inclusion in modeling. Transverse flow may have a significant effect on power, especially away from open discharge (significant die resistance) and the more nearly newtonian the melt is. Even so, these transverse flow effects are secondary to the nonnewtonian nature of the melt. An inaccurate rheological characterization will cause much greater modeling errors, in general, than the use of the one-dimensional flow model. Is there any reason, then, to consider transverse flow at all? The answer (yes) lies in the important role of transverse flow as a mechanism of mixing fluid within the screw channel. Figure 6-19 shows a qualitative picture of the helical path followed by an element of fluid as it progresses down the axis of the screw channel. It is the transverse flow which causes the circulatory motion, and this circulation enhances the extent of *mixing* that occurs within the extruder. We leave the word *mixing* undefined here and consider the process of mixing in some detail in Chap. 12. We will find that a realistic mathematical model of mixing in a melt extruder is quite difficult to carry out, and only a few features of this problem have been adequately handled at the present time.

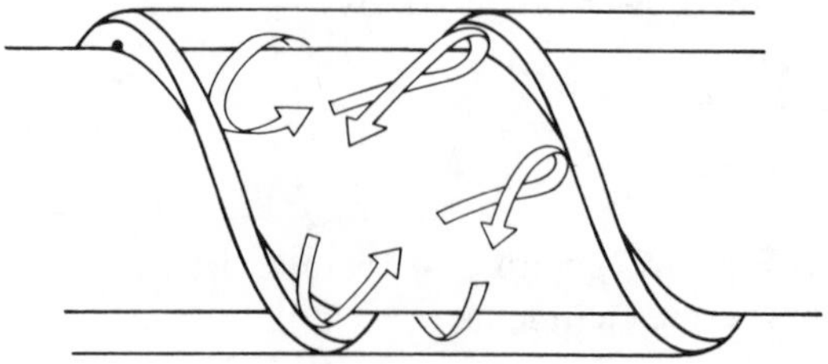

Figure 6-19 Circulatory flow pattern caused by transverse flow.

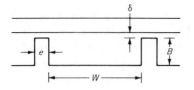

Figure 6-20 Cross section of channel, normal to the flights, showing the clearance δ.

Clearance Flow

In previous models it has been assumed that the screw flights are actually in contact with the barrel, thus forming a *closed* rectangular channel. The true geometry is more like that shown in Fig. 6-20, and we wish to evaluate the importance of flow in the clearance between the "land" of the flight and the barrel. As in the consideration of transverse flow, we examine the role of this geometric simplification with respect to both output and power. As in that case, we find that the power model is much more dependent on the simplifications made than is the output model.

In Fig. 6-21 we show the geometry of a single-flighted screw, with one turn of the screw "unwound," i.e., laid out flat. Consider two points, a and b, on opposite sides of a flight. If a finite clearance under the barrel is allowed for, then we must expect that flow can occur directly between points a and b. There can be both a drag flow and a pressure flow.

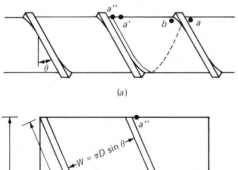

(a)

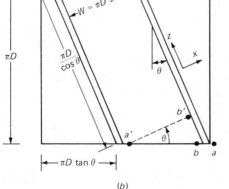

(b)

Figure 6-21 Detail of screw geometry: (a) shows a single-flighted screw; (b) shows the same screw unwound and flattened.

Since δ/e is normally quite small compared to unity, we might approximate the clearance region as a pair of parallel plates. This allows us to use the parallel-plate flow equations already developed in Chap. 5, and in the simplest possible way determine whether clearance effects might be large enough to justify more accurate (and, of course, more tedious) modeling.

With regard to drag flow we know that, at least approximately, $Q_D = \frac{1}{2} U_z A_c$, where A_c is the appropriate cross-sectional area normal to the flow. Drag flow in the clearance is generated by the same U_z as drag flow in the screw channel, and we conclude that the ratio of drag flows in the clearance and channel is given approximately by the area ratio $e\delta/BW$. We can expect this ratio to be quite small, and so drag flow in the clearance cannot be much more than a few percent of the channel drag flow.

The pressure flow is somewhat more difficult to calculate, and we begin with Eq. (5-22) in the form suitable for the clearance geometry:

$$Q_P = \frac{\pi D \delta^3}{12\mu \cos \theta} \left(\frac{\partial p}{\partial x}\right)_c \tag{6-100}$$

We see that the "width" of the clearance region is $\pi D/\cos \theta$ in this case, which is actually the helical length of one turn ($360°$) of the flight. The channel height, of course, is δ, and we denote the appropriate pressure gradient as $(\partial p/\partial x)_c$, which we must now calculate.

For a screw of constant channel geometry we know that the pressure gradient $\partial p/\partial z$ is constant, and we write

$$\frac{\partial p}{\partial z} = \frac{\Delta P}{Z} = \frac{\Delta P}{L} \sin \theta \tag{6-101}$$

We wish to calculate the pressure difference $p_a - p_b$ (see Fig. 6-21) across a flight. We need to calculate the *helical* distance a particle of fluid would travel in order to move from point a to point b along the z axis. We ignore any effect of the transverse flow.

With reference to Fig. 6-21 we note that the points a'' and a' are identical. (We would recover the screw if the unwound sheet were rejoined at points $a'a''$.) In traveling along the helix a particle moving from a to a'' (or a') would travel a distance $\pi D/\cos \theta$. But if we project point a' across to the opposite flight (point b'), which would have the same pressure if transverse flow effects are ignored (that is, $p_{a'} = p_{b'}$), we see that the particle would have actually moved beyond point b by an amount given by $\pi D \tan \theta \sin \theta$. Hence the helical distance from a to b is just $\pi D/\cos \theta - \pi D \tan \theta \sin \theta$.

Then $p_a - p_b$ is just the product of $\partial p/\partial z$ and the distance, *along the helix*, from a to b:

$$p_a - p_b = \left(\frac{\Delta P}{L} \sin \theta\right)\left(\frac{\pi D}{\cos \theta} - \pi D \tan \theta \sin \theta\right) \tag{6-102}$$

For the pressure *gradient* $(\partial p/\partial x)_c$ we use

$$\left(\frac{\partial p}{\partial x}\right)_c = \frac{p_a - p_b}{e} = \frac{\Delta P \, \pi D \cos \theta \sin \theta}{eL} \tag{6-103}$$

From Eq. (6-100), then, we find

$$Q_P = \frac{\pi^2 D^2 \delta^3 \, \Delta P \sin \theta}{12 \mu e L} \tag{6-104}$$

By comparison [Eq. (6-4), with $F_P = 1$, $W = \pi D \sin \theta$, and $Z = L/\sin \theta$] the pressure flow in the channel is

$$Q_P = \frac{\pi D B^3 \, \Delta P \sin^2 \theta}{12 \mu L} \tag{6-105}$$

from which we see that the ratio of the pressure flow across the flights (which we refer to as a *leakage flow*) to the channel pressure flow is just

$$\left(\frac{\delta}{B}\right)^3 \frac{D}{e} \frac{\pi}{\sin \theta}$$

While D/e and $\pi/\sin \theta$ are both greater than unity, the cube of the small factor δ/B ordinarily results in reducing the ratio above to a value considerably less than unity. Under normal circumstances leakage flow is of minor consequence.

The simple analyses above assume isothermal newtonian behavior. The shear rate in the clearance exceeds that in the channel by a factor of the order of B/δ, which is normally in the range of 10 to 20. Viscous heating will be greater in the clearance than in the channel. This factor, along with the usual nonnewtonian viscosity behavior, implies that the viscosity of the fluid in the clearance might be quite a bit less than that in the channel. If this were so, the leakage flow could be considerably in excess of that estimated by the simple analysis just presented.

At this point one might be tempted to account for these factors in a more detailed analysis of flow in the clearance. This may not be justified, however. Fluid moving through the clearance resides in that region for a very short time and spends most of its residence time in the extruder circulating within the channel. It is not at all clear that a steady-state analysis of viscous heating in the clearance region would be realistic. Furthermore, the rheological response of a polymeric melt which rapidly passes into and out of the high-shear clearance region might be expected to show significant transient viscoelastic phenomena. It may be quite misleading to use a power law model (or any *purely viscous* model) to characterize the viscosity reduction due to the high shear rates in the clearance region. For these reasons we avoid the temptation to produce more complicated (but not necessarily more realistic) models of flow in the clearance.

We should, however, examine the role of the clearance flow in affecting the power requirement of the extruder. In the simplest approximation we neglect any pressure flow in the clearance and just consider the power associated with the drag flow. We assume isothermal newtonian behavior, again.

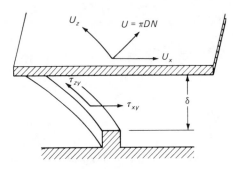

Figure 6-22 Definition sketch for analysis of power expended in the clearance region.

The power associated with the clearance flow is given by

$$\dot{W}_c = \mathbf{F} \cdot \mathbf{U} \tag{6-106}$$

or, in terms of the shear stresses, noting Fig. 6-22,

$$\dot{W}_c = (\tau_{xy} U_x + \tau_{zy} U_z) \frac{eL}{\sin \theta} \tag{6-107}$$

If any pressure flow in the clearance is ignored, the power simplifies to

$$\dot{W}_c = \frac{\mu e L \pi^2 D^2 N^2}{\delta \sin \theta} \tag{6-108}$$

We may compare the contribution of the clearance to the total power requirement most easily by examining the dimensionless power parameter $\Pi_{\dot{W}}$, which now takes the form

$$\Pi_{\dot{W}} = 4 - 6\Pi_Q + 4 \tan^2 \theta + \frac{B}{\delta} \frac{e}{W} \frac{1}{\cos^2 \theta} \tag{6-109}$$

The last term is not ordinarily going to be small compared to unity, in consequence of which we must conclude that the clearance makes a substantial contribution to the power requirement for the extruder. Again, however, we must note that nonnewtonian behavior and nonisothermal conditions would substantially change the contribution of the clearance to the power. In this case we would expect that a more realistic model would show a much smaller contribution of the clearance to the power requirement.

The reservations noted above, regarding the difficulty of producing a realistic model of the rheological and thermal behavior of fluid in the clearance, still hold of course. Other than for its value as an exercise in modeling, it is probably not worthwhile to pursue small modifications (such as the inclusion of power law behavior) of the analyses just presented. Either a complete model should be carried out (which would require a better understanding of the thermal and rheological phenomena than is currently available), or one should conclude from the simple models presented above that the flow in the clearance can be of significance, particularly with regard to power, and keep in mind the sense of the errors that will be made in using these models for design.

6-7 EXTRUSION WITH IMPOSED HEAT TRANSFER

The isothermal and adiabatic cases outlined in previous sections provide limiting models which lead to relatively simple analyses of extruder performance. In the adiabatic case the energy balance is particularly simple: The power input to the fluid appears as an increase in sensible heat [Eq. (6-34)].

Real extruder performance usually lies between these extremes, principally because heat is usually transferred between the fluid and the extruder screw and barrel surfaces. As soon as the adiabatic assumption is lifted, one may no longer write the simple energy balance of Eq. (6-34). Instead, a conduction-convection equation must be introduced. The resultant model contains a large number of parameters, so many that no *compact* presentation of results is possible. It is most instructive if we first examine the problem through dimensional analysis and then look at some typical, though not general, results of a numerical analysis of the model. The reader with no previous background in heat transfer may find it better to skip this section now and come back to it after completing Chap. 13.

The conduction-convection equation has the vector form

$$\rho C_p \, \mathbf{u} \cdot \nabla T = k \, \nabla^2 T + \Phi_v \tag{6-110}$$

where k = thermal conductivity of the melt
  Φ_v = volumetric rate of conversion of mechanical energy into heat

It is possible to write Φ_v in terms of stress τ and the velocity gradient tensor $\nabla \mathbf{u}$ as

$$\Phi_v = \tau : \nabla \mathbf{u} \tag{6-111}$$

If inertial forces are neglected we can write the dynamic equations in the vector form

$$0 = -\nabla p + \nabla \cdot \tau \tag{6-112}$$

The continuity equation is simply

$$\nabla \cdot \mathbf{u} = 0 \tag{6-113}$$

if we continue to neglect possible density variations due to temperature and pressure variations.

We require a constitutive equation, and the power law will be used here in the form

$$\tau = K (\tfrac{1}{2} \mathrm{II}_\Delta)^{(n-1)/2} \, \Delta \tag{6-114}$$

Since we wish to consider temperature effects on viscosity we introduce an *equation of state* for $K(T)$:

$$K = K_b e^{-b(T - T_b)} \tag{6-115}$$

where T_b is some reference temperature at which $K = K_b$.

Boundary conditions on **u** state

$$u_z = \begin{cases} \pi DN \cos \theta & \text{at } y = B \\ 0 & \text{at } y = 0 \end{cases} \qquad \begin{matrix} (6\text{-}116) \\ (6\text{-}117) \end{matrix}$$

$$u_x = \begin{cases} -\pi DN \sin \theta & \text{at } y = B \\ 0 & \text{at } y = 0 \end{cases} \qquad \begin{matrix} (6\text{-}118) \\ (6\text{-}119) \end{matrix}$$

If we neglect the effects of a finite aspect ratio, we can take $u_y = 0$ in the dynamic equations.

Boundary conditions on T might commonly specify a fixed temperature at the barrel surface:

$$T = T_b \qquad \text{at } y = B \qquad (6\text{-}120)$$

At the root of the screw we might assume either of two simple possibilities:

$$T = T_b \qquad \text{at } y = 0 \qquad (6\text{-}121)$$

or

$$\frac{\partial T}{\partial y} = 0 \qquad \text{at } y = 0 \qquad (6\text{-}122)$$

Equation (6-121) specifies the temperature at the screw and asserts that it is the same as that at the barrel. Equation (6-122) specifies that no heat is transferred across the screw boundary. Other cases are possible, but they will not alter the general conclusions to be reached.

Now we will nondimensionalize Eqs. (6-110) to (6-122) in the following way (primes will denote dimensionless variables):

$$(x', y', z') = \frac{(x, y, z)}{B}$$

$$\mathbf{u}' = \frac{\mathbf{u}}{U_z} \qquad \text{where } U_z = \pi DN \cos \theta$$

$$p' = \frac{p}{K_b}\left(\frac{B}{U_z}\right)^n \qquad \tau' = \frac{\tau}{K_b}\left(\frac{B}{U_z}\right)^n$$

$$T' = \frac{T - T_b}{T_b \, \text{Br}} \qquad \text{where } \text{Br} = \frac{K_b U_z^{n+1}}{k T_b B^{n-1}}$$

Then the mathematical model takes the form

$$\text{Pe } \mathbf{u}' \cdot \nabla' T' = \nabla'^2 T' + \tau' : \nabla' \mathbf{u}' \qquad (6\text{-}123)$$

$$0 = -\nabla' p' + \nabla' \cdot \tau' \qquad (6\text{-}124)$$

$$\nabla' \cdot \mathbf{u}' = 0 \qquad (6\text{-}125)$$

$$\tau' = e^{-bT_b \text{Br } T'} \left(\tfrac{1}{2}\text{II}_\Delta'\right)^{(n-1)/2} \Delta' \qquad (6\text{-}126)$$

The boundary conditions state that

On $y' = 1$: $u'_z = 1$ $u'_x = -\tan\theta$ $T' = 0$

On $y' = 0$: $u'_z = 0$ $u'_x = 0$ $\dfrac{\partial T'}{\partial y'} = 0$ [using Eq. (6-122)]

Equations (6-123) to (6-126) and the boundary conditions following are sufficient to determine the variables $\mathbf{u}'$, T', p', and τ', and by inspection we can see that four dimensionless parameters appear: n, θ, Pe, and (bT_b Br). Pe is the Péclet number,

$$\text{Pe} = \frac{\rho C_p B U_z}{k} \tag{6-127}$$

Br is the Brinkman number, but in this formulation it appears only in combination with the parameter b, so that the relevant dimensionless group is a modified Brinkman number defined as

$$\beta = bT_b\ \text{Br} = \frac{bK_b U_z^{n+1}}{kB^{n-1}} \tag{6-128}$$

From this inspection it is an easy matter to show that the screw characteristic will have the functional form

$$\Pi_Q = \Pi_Q(\Pi_P;\ n,\ \theta,\ \text{Pe},\ \beta) \tag{6-129}$$

The model, as formulated above, has been solved by Zamodits and Pearson and by Griffith in papers noted at the end of this chapter.

Figure 6-23 shows the effect of a finite modified Brinkman number, $\beta = 1$. On comparison it can be seen that, at fixed pressure, the effect of heat generation is to increase the output at low pressure (near open discharge) but to decrease the output at high pressure. The *break-even* point is roughly in the neighborhood of $\Pi'_p = 1$. For $\Pi'_p < 1$ the effect of heating is to increase the output; it is otherwise for $\Pi'_p > 1$, and this feature is roughly true over a range of values of n and β.

The strong effect of β, especially on the maximum pressure which may be developed, is shown in Fig. 6-24.

The Péclet number does not appear in the analyses of Griffith or Zamodits and Pearson. They drop the convective terms of Eq. (6-123) (Pe = 0) right from

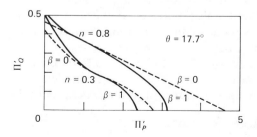

Figure 6-23 Nonisothermal-nonadiabatic screw characteristics for power law fluids for the case $\beta = 1$.

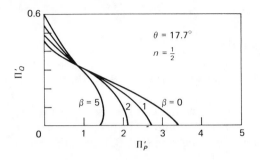

Figure 6-24 Effect of β on screw characteristics for $n = \frac{1}{2}$.

the start. This is in seeming contradiction to the observation that, in fact, the Péclet number is typically† of the order of 10^4. However, both models assume a form of thermal equilibrium, for which T is assumed to be independent of x and z. Thus the only nonzero convective term involves $u_y\, \partial T/\partial y$. Since u_y is neglected in these models it follows that the convective terms disappear identically because of the neglect of u_y and the assumption of thermal equilibrium in the x and z directions.

We should emphasize here the *specific* nature of this model, both in taking Pe = 0 (which neglects thermal convection) and in assuming [through boundary condition (6-122)] that the screw surface is adiabatic. Transverse flow will give rise to significant thermal convection and alter the screw characteristics shown here. Since the metal screw is a considerably better conductor of heat than the polymer melt, it is not very likely to behave in an adiabatic manner, and it is quite likely that there is a temperature distribution along the screw surface down the extruder axis. A rigorous thermal model of extrusion is really beyond our aspirations, and the best that one can do in this regard is carry out a few analyses based on specific *assumptions* in order to get an idea of the effect of gross changes on extruder performance.

6-8 PLASTICATING EXTRUSION

In all cases, so far, the extrusion models have assumed that the screw conveys only a *melt*. In reality, most commercial extruders are fed with *solid* polymer, in the form of small beads or chips, or in powdered form. The process of melting the polymer is usually referred to as *plastication*. The energy required to plasticate or melt the polymer is initially supplied externally across the barrel of the extruder. As melting occurs the mixture of solid and molten polymer is subject to mechanical forces that give rise to significant viscous dissipation of mechanical energy, which further promotes the melting process.

Melting models of the plasticating region of the extruder are fairly complex and are based on experimental observations that are not as yet completely accepted. The lack of acceptance is sometimes with regard to the generality of the

† See R. T. Fenner and J. G. Williams, *J. Mech. Eng. Sci.*, **13**:65 (1971).

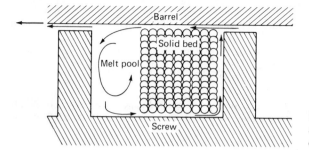

Figure 6-25 Schematic of the melt and solid regions in a plasticating extruder.

observation and sometimes with regard to the implication of the observation. For this reason we will make only the briefest comments on the models currently available and provide in the Bibliography an opportunity for further study of this topic.

Figure 6-25 shows a cross-sectional view of a screw somewhere between the solid-feed and melt-conveying sections of the screw. The general picture, based on observations, is that melting begins with a thin layer of fluid formed at the barrel and, perhaps, along the screw surfaces as well. A *melt pool* forms in which a circulating flow of material exists. The melt pool grows at the expense of the solid bed until eventually all the solid is melted. One of the goals of a plasticating model is the estimation of the distance down the screw axis required to achieve complete melting.

If the solid bed occupies most of the channel cross section over an axial length that is a significant fraction of the screw length, then the melt extruder analysis presented up to this point is inapplicable to the calculation of the output variables of the extruder. Several attempts at developing a model for the plasticating extruder are now available (see the references listed under Sec. 6-8 in the Bibliography) based on a variety of thermal, mechanical, and rheological assumptions. The reader is referred to this literature for further study.

PROBLEMS

6-1 Consider a screw designed with a variable channel depth as in Eq. (6-29). The ratio B_0/B_z is referred to as the *compression* ratio λ. Let Q_{max} be the maximum output (which corresponds to open discharge) for $\lambda = 1$, and let Q'_{max} be the same thing for arbitrary λ. Show that

$$\frac{Q'_{max}}{Q_{max}} = \frac{2\lambda}{1 + \lambda}$$

and

$$\frac{\Delta P'_{max}}{\Delta P_{max}} = \frac{1}{\lambda}$$

where ΔP_{max} is the maximum pressure developed by the screw (which occurs at no output) for $\lambda = 1$, and $\Delta P'_{max}$ is the same at any λ.

Thus we see that a finite compression ratio improves the maximum output but reduces the maximum possible pressure.

6-2 Prove that Eq. (6-48) is valid for the case $N_2 \gg 1 + N_1$ by examining the limit of the right-hand side of

$$(\chi - 1)N_2 - 1 = N_1 \frac{\chi - 1}{\chi \ln \chi} \quad \text{as } \chi \to 1$$

6-3 Consider Example 6-3 involving the tubing die shown in Fig. 6-8. Suppose L is reduced to $\frac{1}{2}$ in. If the maximum permissible pressure is 5000 psi, find the corresponding output Q. What is ΔT at that output? What is the screw speed N?

6-4 In writing Eq. (6-35) it was assumed that the contribution of a term $AN \, \Delta P$ to the power input $\dot{W}$ was unimportant. For Example 6-3 compare the neglected term to the calculated power.

6-5 In Example 6-3 what error would have been made by calculating the output from the isothermal analysis, Eq. (6-19)? Do the same for Example 6-2.

6-6 For the extruder and fluid of Example 6-2 calculate and plot the output (pounds per hour) and the pressure at the die (pounds per square inch) for the case of (a) isothermal flow, (b) adiabatic flow. Assume the die is equivalent to a cylindrical pipe of diameter $\frac{1}{8}$ in and length $\frac{1}{4}$ in. The screw speed N will be in the range 20 to 200 rpm.

6-7 In "Modern Plastics Encyclopedia 1970–1971"† one finds a graph like that in Fig. 6-26. The graph is clearly intended as a "rule of thumb," since the extruder output will depend on the resin, the screw design, screw speed, and die design. As an exercise, let us apply the simple isothermal analysis and determine conditions under which the figure gives reasonable results.

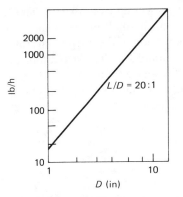

Figure 6-26 Graph for Prob. 6-7.

Assume open discharge in all cases. Assume geometrical factors are held constant and that $\theta = 17.7°$, $D/B = 10$, and $L/D = 20$.

(a) Show that $Q \approx D^3$ under these assumptions, at fixed N. Is that prediction consistent with the graph shown?

(b) What value of N makes the simple isothermal analysis quantitatively consistent with the graph? Is it the same N for any diameter D?

6-8 The data below (see Edwards et al.) were obtained for extrusion of a polyethylene terephthalate melt:

† McGraw-Hill, p. 481,

EXTRUSION **165**

N, rpm	ΔP, psi	M, lb/h
32	490	47.3
32	0	49.1
42.6	20	64.7
52.4	500	78.1
52.4	0	80.3

Extruder dimensions:

$$\theta = 17.7° \qquad D = 2.5 \text{ in}$$
$$L = 24.6 \text{ in} \qquad B = 0.085 \text{ in}$$
$$W = 2.17 \text{ in} \qquad \delta = 0.0055 \text{ in}$$
$$e = 0.21 \text{ in}$$

At the operating temperature of 530°F the melt is nearly newtonian and $\mu = 3500$ P. The melt density is $\rho = 0.042$ lb/in³.

Compare the data with (a) an isothermal theory, (b) an adiabatic theory.

6-9 Rework Example 6-4, using the rheological data given in Fig. 6-27, and compare the results with those given in the example.

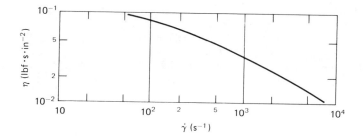

Figure 6-27 Data for Prob. 6-9.

6-10 Zamodits and Pearson present data shown in Fig. 6-28 for extrusion of a natural rubber compound, plotted in terms of Π'_p and Π'_Q. The relevant geometric and rheological parameters are

$$B = 0.2 \text{ in} \qquad T_0 = 100°C$$
$$D = 2.5 \text{ in} \qquad b = 4.2 \times 10^{-3} \text{ °C}^{-1}$$
$$W = 2.15 \text{ in} \qquad n = 0.26$$
$$Z = 82.5 \text{ in} \qquad K = 2.7 \text{ lbf·s}^{0.26}/\text{in}^2$$
$$\theta = 17.7°$$

(a) Compare their results with the one-dimensional isothermal theory (Fig. 6-11).
(b) Do the same using the isothermal two-dimensional theory (Fig. 6-16).
(c) Estimate the adiabatic temperature rise in order to evaluate the isothermal assumption of parts a and b.

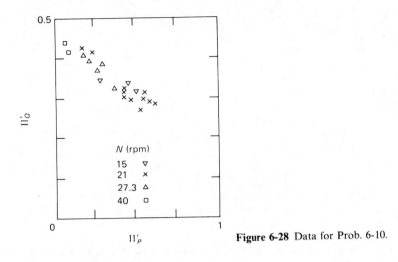

Π'_Q

N (rpm)

15	▽
21	×
27.3	△
40	□

Π'_P

Figure 6-28 Data for Prob. 6-10.

6-11 An extruder is designed with a " throttling die " which imposes a back pressure on the system. Suppose 400 lb/h of a molten acrylic is to be extruded. Find the required die pressure, assuming

(*a*) Isothermal behavior.

(*b*) Adiabatic behavior, giving also the temperature rise.

(*c*) Barrel temperature is $T_b = 375$ F, and adiabatic screw. (Assume Fig. 6-24 is applicable.)

For the extruder: $D = 3.5$ in $L = 20$ in
 $\theta = 17.7$ $N = 120$ rpm
 $B = 0.4$ in

For the acrylic: $K = 4 \text{ lbf·s}^{1/2}/\text{in}^2$ $\rho C_p = 160 \text{ lbf/in}^2 \cdot {}^\circ\text{F}$
 $n = \frac{1}{2}$ $T_0 = 375$ F
 $b = 0.025 \text{ F}^{-1}$ $k = 0.025 \text{ lbf/s} \cdot {}^\circ\text{F}$
 $\rho = 1.12$ g/ml

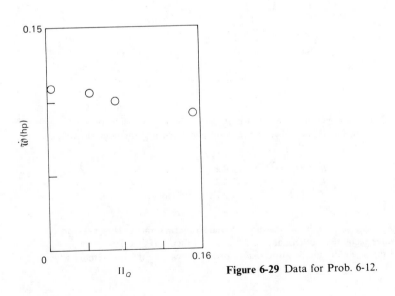

0.15

$\tilde{w}$ (hp)

0

Π_Q

0.16

Figure 6-29 Data for Prob. 6-12.

6-12 Data† on power in a single-flighted extruder are shown in Fig. 6-29. All data were taken at a screw speed of 100 rpm with a variable die resistance. The fluid, a dimethyl silicone polymer, is a liquid at "room" temperature, and is slightly nonnewtonian, with $n = 0.81$ and $K = 0.017$ lbf · s$^{0.81}$/in^2. The extruder dimensions are

$$\theta = 36.6° \qquad D = 1.5 \text{ in} \qquad L = 12 \text{ in}$$

$$\frac{B}{W} = 0.036 \qquad \frac{\delta}{B} = 0.053$$

Compare the data to the predictions of the isothermal and adiabatic analyses.

6-13 Bigg obtained the following data for a polymer solution using the extruder described on page 130 and various dies which allowed control of the pressure. Plot the data in the format Π_Q versus Π_P, and compare to the predictions of the appropriate isothermal analysis. The rheological data are given at the temperature of the experiment (Fig. 6-30), and viscous heating was assumed to be negligible.

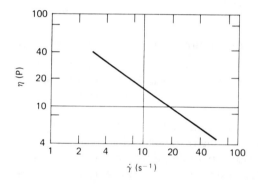

Figure 6-30 Data for Prob. 6-13.

N, rpm	Q, cm^3 min	ΔP, dynes cm^2
20	24	0
60	74	0
100	113	0
20	3.2	125,500
60	7.0	188,000
100	8.8	227,500
20	11	66,100
60	24	127,000
100	35	171,000
20	23	9,650
60	58	34,100
100	86	56,500

6-14 For the extruder of Prob. 6-8, operating at 61.5 rpm, the pressure across the die is 1230 psi, and the observed output is 86.8 lb h. The melt density is $\rho = 0.042$ lb in^3, and the apparent viscosity at the

† Data are from R. T. Fenner and J. G. Williams, *J. Mech. Eng. Sci.*, **13**:65 (1971).

channel shear rate and temperature is 3440 P. An estimate for the viscosity at the clearance shear rate is 2060 P. Assume isothermal newtonian behavior and estimate the required power. Evaluate the isothermal assumption. The observed result is 2.7 hp.†

BIBLIOGRAPHY

6-1 Isothermal newtonian extrusion

The basic development of the simplified extrusion models is in

McKelvey, J. M.: "Polymer Processing," John Wiley & Sons, Inc., New York, 1962.

Another useful source is a collection of papers published as an Industrial and Engineering Chemistry Symposium entitled

Theory of Extrusion, *Ind. Eng. Chem.*, **45**: 970 (1953).

Extension of these ideas to the case of a multisection screw, along with some experimental data, is in

Edwards, R. B., J. E. Fogg, R. R. Kraybill, and J. T. Regan: Flow Rate and Pressure Drop Relationships for Multi-Section Screw Extruders, *Soc. Plast. Eng. J.*, p. 234, March 1964.

6-4 Nonnewtonian isothermal analysis

In dealing with nonnewtonian models one often finds the suggestion to use the simple newtonian theory, with the viscosity replaced by the "apparent" viscosity evaluated at some appropriate shear rate. The basic *failure* of this idea, and illustration of its inaccuracy, is to be found in

Kroesser, F. W., and S. Middleman: The Calculation of Screw Characteristics for the Extrusion of Non-Newtonian Melts, *Polym. Eng. Sci.*, **5**: 1 (1965).

6-7 Extrusion with imposed heat transfer

The first attempt at a "complete" extrusion model which would incorporate nonnewtonian behavior, transverse flow, viscous heat generation, and temperature dependence of viscosity, all within the framework of the basic conservation equations, was given in

Griffith, R. M.: Fully Developed Flow in Screw Extruders, *Ind. Eng. Chem. Fundam.*, **1**: 180 (1962).

A more recent study, which offers some assessment of the assumptions regarding thermal boundary conditions, is

Zamodits, H. J. and J. R. A. Pearson: Flow of Polymer Melts in Extruders. Part I. The Effect of Transverse Flow and of a Superposed Steady Temperature Profile, *Trans. Soc. Rheol.*, **13**: 357 (1969).

---

† The data, and a model for power that is more complicated than the one described here, are presented by R. R. Kraybill, Power for Multi-Section Melt Screw Extruders, *Polym. Eng. Sci.*, **15**: 725 (1975).

Finally we note the comprehensive study of modeling of melt extrusion written by Fenner:

Fenner, R. T.: "Extruder Screw Design," Iliffe Books, London, 1970.

6-8 Plasticating extrusion

A comprehensive book covering a variety of topics is

Tadmor, Z., and I. Klein: "Engineering Principles of Plasticating Extrusion," Van Nostrand Reinhold Company, New York, 1970.

The following papers represent the most significant modeling efforts.

Donovan, R. C.: A Theoretical Melting Model for Plasticating Extruders, *Polym. Eng. Sci.*, **11**: 247 (1971).

————: Pressure Profiles in Plasticating Extruders, *Polym. Eng. Sci.*, **11**: 484 (1971).

Shapiro, J., and J. R. A. Pearson: A Dynamic Model for Melting in Plasticating Extruders, *Imp. Coll. Polym. Sci. Eng. Group Rept.* 5, London, April 1974.

Edmondson, I. R., and R. T. Fenner: Melting of Thermoplastics in Single Screw Extruders, *Polymer*, **16**: 49 (1975).

Lindt, J. T.: A Dynamic Melting Model for a Single-Screw Extruder, *Polym. Eng. Sci.*, **16**: 284 (1976).

SEVEN

CALENDERING

Do not rely on If and Perhaps.

Bahya

Calendering is a process in which molten material is " dragged " through the narrow region between two corotating rolls in such a way as to produce a sheet. Figure 7-1 shows a sketch of the process. In an analysis of the calendering process one seeks the relationships among performance variables, such as sheet thickness, and design and operating variables, such as roll diameters and roll speed. Of course the effect of rheological parameters on these relationships is also sought.

Calenders may also be used to produce a certain surface *finish*, either a degree of smoothness or " gloss " which may be required or perhaps an intentional degree of roughness or pattern. Finish is a result of interaction between the fluid and the roll surface in the region of separation of the two materials. While there is certainly a hydrodynamic aspect to that feature of calendering, the relationship of finish to the usual process variables that are controlled is not understood. We will ignore the surface finish aspect of calendering and treat only the broader hydrodynamic features.

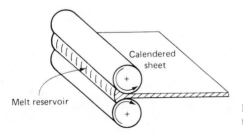

Figure 7-1 Sketch showing calendering between two rolls.

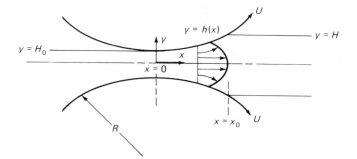

Figure 7-2 Definition sketch for calender flow analysis.

7-1 THE NEWTONIAN MODEL OF CALENDERING

Details of the geometry and the flow field are shown in Fig. 7-2. Assuming that the flow is strictly two-dimensional, so that $\mathbf{u} = [u_x(x, y), u_y(x, y), 0]$, the appropriate equations in the steady state for an isothermal newtonian fluid are

$$\rho\left(u_x \frac{\partial u_x}{\partial x} + u_y \frac{\partial u_x}{\partial y}\right) = -\frac{\partial p}{\partial x} + \mu\left(\frac{\partial^2 u_x}{\partial x^2} + \frac{\partial^2 u_x}{\partial y^2}\right) \tag{7-1}$$

$$\rho\left(u_x \frac{\partial u_y}{\partial x} + u_y \frac{\partial u_y}{\partial y}\right) = -\frac{\partial p}{\partial y} + \mu\left(\frac{\partial^2 u_y}{\partial x^2} + \frac{\partial^2 u_y}{\partial y^2}\right) \tag{7-2}$$

$$\frac{\partial u_x}{\partial x} + \frac{\partial u_y}{\partial y} = 0 \tag{7-3}$$

The inertial terms, of course, make the problem nonlinear, but even if the inertial terms are dropped (with the assumption that viscous forces dominate the process) we still have to deal with a set of *partial* differential equations for which there is no *simple* analytical solution. Our method will be to treat this problem by making some approximations that lead to a formulation involving only a single ordinary differential equation which can be easily solved. It is found that the approximate solution has most of the important features of the full model, and is useful for examination of the relationships among the various parameters that affect the process.

We begin with the plausible argument that the most important dynamic events occur in the region of the minimal roll separation—the *nip* region. In that region, and extending to either side (i.e., in the $\pm x$ direction) by a distance of the order of x_0 (see Fig. 7-2), the roll surfaces are nearly parallel if, as is usually the case, $H_0 \ll R$. Then it is reasonable to assume that the flow is nearly parallel, so that

$$u_y \ll u_x \tag{7-4}$$

and
$$\frac{\partial}{\partial x} \ll \frac{\partial}{\partial y} \qquad (7\text{-}5)$$

If, further, we neglect inertial effects, Eqs. (7-1) and (7-2) become

$$0 = -\frac{\partial p}{\partial x} + \mu \frac{\partial^2 u_x}{\partial y^2} \qquad (7\text{-}6)$$

$$\frac{\partial p}{\partial y} = 0 \qquad (7\text{-}7)$$

From Eq. (7-7) it follows that $p = p(x)$ only, so that u_x satisfies

$$\frac{\partial^2 u_x}{\partial y^2} = \frac{1}{\mu} \frac{dp}{dx} \qquad (7\text{-}8)$$

Equation (7-8) is identical to the *Reynolds lubrication equation*, and the process of transforming this problem from Eqs. (7-1) and (7-2) to Eq. (7-8) is often referred to as *making the usual lubrication approximations*. [The reference to *lubrication* comes from the fact that lubrication problems themselves typically involve a geometry such that Eqs. (7-4) and (7-5) are valid.] Equation (7-8) is easily integrated to give

$$u_x = \frac{1}{2\mu} \frac{dp}{dx} y^2 + ay + b \qquad (7\text{-}9)$$

where a and b are constants of integration.

If we assume that the two rolls are identical and rotate with the same linear speed U, then appropriate boundary conditions are

$$u_x = U \qquad \text{on } y = h(x) \qquad (7\text{-}10)$$

$$\frac{\partial u_x}{\partial y} = 0 \qquad \text{on } y = 0 \qquad (7\text{-}11)$$

and, after evaluating a and b, we may write the velocity as

$$u_x = U + \frac{y^2 - h^2(x)}{2\mu} \frac{dp}{dx} \qquad (7\text{-}12)$$

We note that Eq. (7-12) gives u_x explicitly as a function of y and implicitly as a function of x, through $h(x)$ and through $p(x)$. We note also that since $p(x)$ is unknown at this stage in the analysis, the solution for u_x is really incomplete.

First let us examine $h(x)$, the y distance from the center line to the roll surface. It is easy to verify that

$$h = H_0 + R - (R^2 - x^2)^{1/2} \qquad (7\text{-}13)$$

Introducing Eq. (7-13) into Eq. (7-12) complicates the algebra. Since it is likely that we will confine the analysis to values of x such that $x \ll R$, a good approximation to $h(x)$ is

$$h(x) = H_0\left(1 + \frac{x^2}{2H_0 R}\right) \qquad (7\text{-}14)$$

It is convenient now to define dimensionless variables

$$x' = \frac{x}{\sqrt{2RH_0}} \qquad u'_x = \frac{u_x}{U}$$

$$y' = \frac{y}{H_0} \qquad p' = \frac{pH_0}{\mu U}$$

Then Eq. (7-12) may be written as

$$u'_x = 1 + \sqrt{\frac{H_0}{8R}}\,[y'^2 - (1 + x'^2)^2]\frac{dp'}{dx'} \qquad (7\text{-}15)$$

We may find an expression for the pressure gradient by using a mass balance in the form

$$Q = 2\int_0^h u_x\,dy = 2h(x)\left[U - \frac{h^2(x)}{3\mu}\frac{dp}{dx}\right] \qquad (7\text{-}16)$$

We may solve for dp/dx and find, in dimensionless form,

$$\frac{dp'}{dx'} = \sqrt{\frac{18R}{H_0}}\frac{x'^2 - \lambda^2}{(1 + x'^2)^3} \qquad (7\text{-}17)$$

where a dimensionless flow rate λ has been introduced, defined as

$$\lambda^2 = \frac{Q}{2UH_0} - 1 \qquad (7\text{-}18)$$

Thus we have replaced the unknown pressure gradient with the (still unknown) flow rate λ.

From the physical point of view the events which occur at the point of separation of the sheet from the roll, i.e., at $y = H$ and $x = x_0$, will exert a strong influence on the overall character of the process. We could expect, then, that the assumptions we introduce about the separation point will make themselves felt in the character of the model. With reference to Fig. 7-3 we will introduce two simple assumptions.

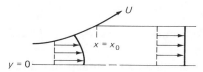

Figure 7-3 Idealization of the separation region.

Downstream of separation, for $x > x_0$, we expect a flat velocity profile, corresponding to rigid motion of the sheet. Upstream, for $x < x_0$ but near the separation point, the velocity will be a function of y according to Eq. (7-15). We will *assume* that right *at* separation, at $x = x_0$, the profile is flat. Since the flat profile corresponds to $\partial u_x / \partial y = 0$ and $\partial^2 u_x / \partial y^2 = 0$, then from Eq. (7-8) we can see that

$$\frac{dp}{dx} = \frac{dp'}{dx'} = 0 \qquad \text{at } x' = x'_0 \tag{7-19}$$

Furthermore, if we ignore the forces acting at the separation boundary (due, e.g., to surface tension), then we expect the stress just inside the fluid to be the same as the ambient pressure. Since the velocity field is assumed flat, the viscous stresses vanish at $x' = x'_0$, and so we can assert that

$$p' = 0 \qquad \text{at } x' = x'_0 \tag{7-20}$$

From Eqs. (7-19) and (7-17) it follows that

$$x'_0 = \lambda \tag{7-21}$$

If we assume that the sheet comes off the roll with the speed U, then a mass balance on the sheet gives

$$Q = 2UH \tag{7-22}$$

from which it follows that

$$\lambda^2 = \frac{H}{H_0} - 1 \tag{7-23}$$

Thus, if H is *measured*, λ may be found and x'_0 follows from Eq. (7-21).

Equation (7-17) may be integrated with respect to x', and if the boundary condition of Eq. (7-20) is imposed the solution for $p'(x')$ is found to be

$$p' = \sqrt{\frac{9R}{32H_0}} \left[\frac{x'^2(1 - 3\lambda^2) - 1 - 5\lambda^2}{(1 + x'^2)^2} x' + (1 - 3\lambda^2)(\tan^{-1} x' - \tan^{-1} \lambda) \right.$$
$$\left. + \frac{1 + 3\lambda^2}{1 + \lambda^2} \lambda \right] \tag{7-24}$$

Figure 7-4 shows the shape of the pressure distribution. The maximum pressure occurs just upstream of the nip at $x' = -\lambda$ and has the value

$$p'_{\max} = \frac{3C(\lambda)}{2} \sqrt{\frac{R}{2H_0}} \tag{7-25}$$

where

$$C(\lambda) = \frac{1 + 3\lambda^2}{1 + \lambda^2} \lambda - (1 - 3\lambda^2) \tan^{-1} \lambda \tag{7-26}$$

From Eq. (7-17), and as suggested in Fig. 7-4, the pressure gradient has three

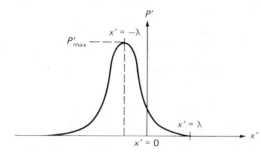

Figure 7-4 Shape of the pressure distribution between the rolls.

extrema: At $x' = \lambda$ the vanishing gradient occurs at $p' = 0$; at $x' = -\lambda$ the maximum pressure occurs. The third value of x' at which the gradient vanishes is seen [from Eq. (7-17)] to be at $x' = -\infty$. The pressure at this "point" is obtained upon setting $x' = -\infty$ in Eq. (7-24). One finds that

$$p'(x' \to -\infty) = \sqrt{\frac{9R}{32H_0}} \left[\frac{1 + 3\lambda^2}{1 + \lambda^2} \lambda - (1 - 3\lambda^2)\left(\frac{\pi}{2} + \tan^{-1} \lambda\right) \right] \quad (7\text{-}27)$$

If an external pressure were imposed far upstream of the nip, then one could obtain λ in terms of that pressure from Eq. (7-27). One does not normally impose such a pressure on a calender. The most reasonable assumption would be that

$$p'(x' \to -\infty) = 0 \quad (7\text{-}28)$$

from which it follows that λ has a specific value, namely,

$$\lambda_0 = 0.475 \quad (7\text{-}29)$$

Subject to the assertion of Eq. (7-28), then, we see that

$$\frac{H}{H_0} = 1 + \lambda_0^2 = 1.226 \quad (7\text{-}30)$$

and that the sheet thickness depends only on H_0.

The problem is essentially solved at this point; we have the velocity distribution and the pressure distribution, and with an assumption on upstream pressure, or a measurement of H, λ may be determined. It is worthwhile to examine the velocity distribution $u_x(x, y)$, which may be written simply as

$$u_x' = \frac{2 + 3\lambda^2(1 - \eta^2) - x'^2(1 - 3\eta^2)}{2(1 + x'^2)} \quad (7\text{-}31)$$

where y is now normalized to $h(x)$ instead of to H_0: $\eta = y/h(x)$.

Figure 7-5 shows the velocity distribution along the path through the calender. Because of the pressure distribution there is a backflow component for $x' < -\lambda$ which is superimposed onto the drag flow component. For $x' < -1.64$ there is a negative flow along the axis, and a circulation pattern develops.

It is clear from Fig. 7-4 that a positive pressure develops in the region between the rolls. Now we must examine the consequences of this pressure field. The most

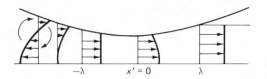

Figure 7-5 Flow pattern in the nip region.

immediate is that "roll bowing" occurs, as shown in Fig. 7-6. The rolls are constrained at their ends by bearings. If the rolls are long enough (a 20-ft width is
found in some operations), then the reactive pressure developed by the flow field
can separate the rolls and cause a nonuniform gap. Obviously this will affect the
uniformity of the sheet thickness.

The roll-separating force can be calculated from

$$\frac{F}{W} = \int_{-\infty}^{x_0} (-T_{yy}) \, dx \tag{7-32}$$

Within the lubrication approximations already made we may ignore the viscous
contribution $(2\mu \, \partial u_y / \partial y)$ that appears in $-T_{yy} = p - \tau_{yy}$ and simply equate $-T_{yy}$
to p. Then we calculate the force from

$$\frac{F}{W} = \int_{-\infty}^{x_0} p(x) \, dx = \mu U \sqrt{\frac{2R}{H_0}} \int_{-\infty}^{\lambda} p' \, dx' \tag{7-33}$$

with the result

$$\frac{F}{W} = \frac{\mu U R}{H_0} G(\lambda) \tag{7-34}$$

where $G(\lambda)$ is a complicated function (see Prob. 7-6) whose value, for the expected
case of $\lambda_0 = 0.475$, is $G(0.475) = 1.22$. Thus we have

$$\frac{F}{W} = 1.22 \frac{\mu U R}{H_0} \tag{7-35}$$

The power transmitted to the fluid by both rolls is calculated from

$$\dot{W} = 2W \int_{-\infty}^{x_0} \tau_{xy} \bigg|_{y=h(x)} U \, dx \tag{7-36}$$

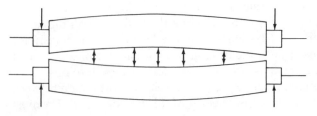

Figure 7-6 An exaggerated view
of the bowing caused by the pressure distribution between the
rolls. It is assumed that forces to
maintain the geometry are imposed only at the bearings, as
shown.

Upon introducing Newton's law for the shear stress, one finds

$$\dot{\mathscr{W}} = 3WU^2\mu\sqrt{\frac{2R}{H_0}}M(\lambda) \tag{7-37}$$

where

$$M(\lambda) = (1 - \lambda^2)\left(\tan^{-1}\lambda + \frac{\pi}{2}\right) - \lambda \tag{7-38}$$

For $\lambda_0 = 0.475$, $M(\lambda)$ is 1.08 and

$$\dot{\mathscr{W}} = 4.58WU^2\mu\sqrt{\frac{R}{H_0}} \tag{7-39}$$

This power input has the potential to raise the temperature of the fluid by an amount which is, at most, given by an *adiabatic* energy balance:

$$\Delta T = \frac{\dot{\mathscr{W}}}{\rho Q C_p} \tag{7-40}$$

Example 7-1 A calender having dimensions $R = 6$ in, $W = 6$ ft, $H_0 = \frac{1}{4}$ in operates on a material of viscosity $\mu = 10^5$ P. The roll speed is $U = 20$ ft/min. Calculate (a) the sheet thickness, (b) maximum pressure, (c) roll-separating force, (d) horsepower delivered to both rolls, and (e) adiabatic temperature rise.

(a) Assuming zero pressure far upstream, we can use Eq. (7-30):

$$H = 1.226H_0 = 0.31 \text{ in}$$

(the sheet *thickness* is twice this).

(b) Using Eq. (7-25), we find

$$p_{\text{max}} = 0.535\frac{\mu U}{H_0}\sqrt{\frac{R}{H_0}} = 61 \text{ psi}$$

(c) From Eq. (7-35) we find

$$\frac{F}{W} = 1.22\frac{\mu UR}{H_0} = 171 \text{ lbf/in}$$

One often sees the roll-separating force reported as force per roll width (F/W) with the units of lbf/in reported as "pounds per linear inch" (pli).

(d) The power input is given by Eq. (7-39):

$$\dot{\mathscr{W}} = 4.58WU^2\mu\sqrt{\frac{R}{H_0}} = 5.35 \text{ hp} = 960 \text{ cal/s}$$

(e) For the adiabatic temperature rise assume $C_p = 0.5$ cal/g·°C and $\rho = 1$ g/cm^3. The volumetric flow rate is

$$Q = 2UHW = 2,893 \text{ cm}^3/\text{s}$$

and we find

$$\Delta T < 0.1°C$$

7-2 THE POWER LAW MODEL OF CALENDERING

Upon introducing the lubrication approximations, one begins the nonnewtonian analysis with

$$0 = -\frac{dp}{dx} + \frac{\partial}{\partial y}\tau_{xy} \tag{7-41}$$

For a power law fluid

$$\tau_{xy} = K \left| \frac{\partial u_x}{\partial y} \right|^{n-1} \frac{\partial u_x}{\partial y} \tag{7-42}$$

where the absolute value sign on the velocity gradient term (which is really $\sqrt{\frac{1}{2}II_\Delta}$ in the lubrication approximation) avoids the problem of taking a fractional root of a negative number.

From the newtonian solution we can see that we should expect two flow regions, one where the velocity gradient is positive ($x' < \lambda$ in the newtonian case) and one where the gradient is negative ($x' > -\lambda$). We shall have to integrate Eq. (7-41) separately in each region. The results are

$$u_x = U + \frac{1}{q}\left(\frac{1}{K}\frac{dp}{dx}\right)^{1/n}[y^q - h^q(x)] \tag{7-43a}$$

$$u_x = U - \frac{1}{q}\left(-\frac{1}{K}\frac{dp}{dx}\right)^{1/n}[y^q - h^q(x)] \tag{7-43b}$$

where $q = n/(1+n)$, and where Eq. (7-43a) is for the region of negative velocity gradient and positive pressure gradient and Eq. (7-43b) is the opposite.

Either of Eqs. (7-43) may then be integrated to give the volumetric flow rate, and as in the newtonian case one may then solve for the pressure gradient. The results may be written as

$$\frac{dp'}{dx'} = -\left(\frac{2n+1}{n}\right)^n \sqrt{\frac{2R}{H_0}} \frac{(\lambda^2 - x'^2)|\lambda^2 - x'^2|^{n-1}}{(1 + x'^2)^{2n+1}} \tag{7-44}$$

and this equation incorporates both cases: For $-\infty < x' < -\lambda$ the pressure gradient is positive, and for $-\lambda < x' < \lambda$ the pressure gradient is negative. As in the newtonian case the extrema of the pressure distribution occur at $x' = -\infty, \pm\lambda$. All dimensionless variables are defined as in the newtonian case except that p' is defined as

$$p' = \frac{p}{K}\left(\frac{H_0}{U}\right)^n \tag{7-45}$$

The pressure distribution is obtained by integration of Eq. (7-44), with the result

$$p' = \left(\frac{2n+1}{n}\right)^n \sqrt{\frac{2R}{H_0}} \int_{-\infty}^{x'} \frac{|\lambda^2 - x'^2|^{n-1}(x'^2 - \lambda^2)}{(1 + x'^2)^{2n+1}} dx' \qquad (7\text{-}46)$$

The integration must be performed numerically. In Eq. (7-46) the boundary condition $p'(x' \to -\infty) = 0$ has been used.

It is somewhat simpler, computationally, to integrate Eq. (7-44) from λ (where $p' = 0$) to x', with the result

$$p' = \left(\frac{2n+1}{n}\right)^n \sqrt{\frac{2R}{H_0}} \int_{x'}^{\lambda} \frac{|\lambda^2 - x'^2|^{n-1}(\lambda^2 - x'^2)}{(1 + x'^2)^{2n+1}} dx' \qquad (7\text{-}47)$$

The maximum pressure occurs at $x' = -\lambda$, so we may write

$$p'_{max} = \left(\frac{2n+1}{n}\right)^n \sqrt{\frac{2R}{H_0}} \int_{-\lambda}^{\lambda} \frac{(\lambda^2 - x'^2)^n}{(1 + x'^2)^{2n+1}} dx' \qquad (7\text{-}48)$$

The dimensionless flow rate parameter λ may be found from Eq. (7-47), since we have *assumed* that p' vanishes as $x' \to -\infty$. Thus λ_0 is found from

$$0 = \int_{-\infty}^{\lambda_0} \frac{|\lambda_0^2 - x'^2|^{n-1}(\lambda_0^2 - x'^2)}{(1 + x'^2)^{2n+1}} dx' \qquad (7\text{-}49)$$

Note that we again use the notation λ_0 to refer to that value of λ corresponding to the inlet condition $p'(-\infty) = 0$. The dependence of λ_0 on n is shown in Fig. 7-7, where it is apparent that the effect of nonnewtonian behavior is to increase the sheet thickness.

Once $\lambda_0(n)$ is known the maximum pressure may be found in the form

$$\frac{p'_{max}}{\sqrt{2R/H_0}} = \left(\frac{2n+1}{n}\right)^n \int_{-\lambda_0}^{\lambda_0} \frac{(\lambda_0^2 - x'^2)^n}{(1 + x'^2)^{1+2n}} dx' = \mathscr{P}(n) \qquad (7\text{-}50)$$

Figure 7-8 shows $\mathscr{P}(n)$.

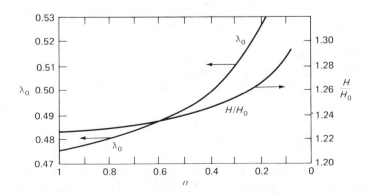

Figure 7-7 Calendered thickness, in terms of λ_0 or H/H_0, as a function of power law index n.

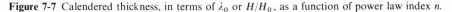

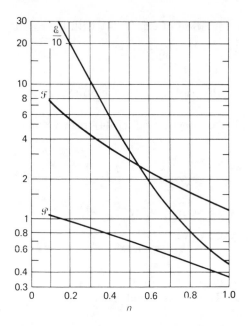

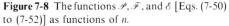

Figure 7-8 The functions $\mathscr{P}$, $\mathscr{F}$, and $\mathscr{E}$ [Eqs. (7-50) to (7-52)] as functions of n.

As in the newtonian case the roll-separating force is obtained by integrating $p'(x')$. It is not too difficult to see that one may write F/W in the form

$$\frac{F}{W} = K\left(\frac{U}{H_0}\right)^n R\mathscr{F}(n) \qquad (7\text{-}51)$$

$\mathscr{F}(n)$ is shown in Fig. 7-8.

The power transmitted to the fluid may be expressed in the format

$$\mathscr{W} = WU^2 K\left(\frac{U}{H_0}\right)^{n-1} \sqrt{\frac{R}{H_0}}\, \mathscr{E}(n) \qquad (7\text{-}52)$$

where $\mathscr{E}(n)$ is given in Fig. 7-8.

According to the model presented here the calendered thickness depends strictly upon the geometry of the system. This follows from the assumption that H_0 is controlled and held fixed. In some systems the roll separation is controlled by "loading" the rolls, i.e., fixing the force acting between the rolls. Then the roll separation adjusts itself so that the reactive force of the calendered fluid just balances the roll loading. If the fluid obeys the power law the separation H_0 may be found from Eq. (7-51):

$$H_0 = U\left[\frac{KR\mathscr{F}(n)}{F/W}\right]^{1/n} \qquad (7\text{-}53)$$

From the point of view of operation of the system the loaded-roll case requires more control than the case of fixed geometry. As Eq. (7-53) shows, the separation H_0, and hence the calendered thickness, depends upon roll speed U and the rheological parameters K and n. Fluctuations in U will produce fluctuations in H_0 in direct proportion. Since $n < 1$ is the usual case, fluctuations in K, due, e.g., to temperature fluctuations, will cause more than proportional changes in H_0. Hence it is more difficult to control calendered thickness in this case.

We can develop a simple model of the sensitivity of calendered thickness to temperature fluctuations. Suppose

$$K = K_0 e^{-b(T-T_0)} \tag{7-54}$$

We write Eq. (7-53) as

$$H_0 = AK^{1/n} = H_0^0 e^{-(b/n)(T-T_0)} \tag{7-55}$$

where H_0^0 is the (half) separation at temperature T_0. Then

$$\frac{dH_0}{dT} = -\frac{b}{n} H_0$$

or

$$\frac{dH_0}{H_0} = -\frac{b \, dT}{n} \tag{7-56}$$

For small changes we may write this as

$$\frac{\Delta H_0}{H_0} = -\frac{b \, \Delta T}{n} \tag{7-57}$$

Suppose we consider a material for which $b = 0.025°\text{F}^{-1}$, and $n = \frac{1}{3}$, subject to temperature fluctuations of magnitude $\Delta T = \pm 3°\text{F}$. Then we find that the fractional change in H_0 (which will be the same as the fractional change in calendered thickness) will be $\Delta H_0 / H_0 = \pm 0.225$. A $3°$ variation in temperature will cause more than a 20 percent variation in calendered thickness!

7-3 CALENDER FED WITH A FINITE SHEET

In the models considered above it was assumed that the calender was fed with a mass of fluid so large that an infinite reservoir of fluid existed upstream from the nip. It is possible, of course, to feed the calender with a *sheet* of fluid, as suggested in Fig. 7-9. In this case a new boundary condition, $p' = 0$ at $x' = -x'_f$, must be introduced, and Eq. (7-46) is replaced by

$$p' = \left(\frac{2n+1}{n}\right)^n \sqrt{\frac{2R}{H_0}} \int_{-x'_f}^{x'} \frac{|\lambda^2 - x'^2|^{n-1}(x'^2 - \lambda^2)}{(1 + x'^2)^{2n+1}} dx' \tag{7-58}$$

and Eq. (7-49) becomes

$$0 = \int_{-x'_f}^{\lambda} \frac{|\lambda^2 - x'^2|^{n-1}(\lambda^2 - x'^2)}{(1 + x'^2)^{2n+1}} dx' \tag{7-59}$$

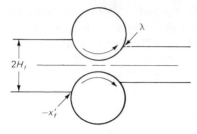

Figure 7-9 Calender fed with a finite sheet.

The thickness of the feedstock enters through the definition

$$x'_f = \left(\frac{H_f}{H_0} - 1\right)^{1/2} \tag{7-60}$$

Equation (7-59) may be solved for λ as a function of n and H_f/H_0, and the results are shown in Fig. 7-10. Several points are of interest.

First, we see that the calendered thickness is reduced somewhat if H_f/H_0 is finite, for all values of n. Next, we see that while $H_f/H_0 = 20$ is large enough that λ is practically equal to λ_0 for the case of the newtonian fluid, a much larger value, say, $H_f/H_0 = 200$, is required for a fluid with $n = 0.25$ to be at nearly its λ_0 value.

More interesting, perhaps, is the observation that if finite H_f/H_0 is accounted for, it is no longer correct to say, without qualification, that the effect of nonnewtonian behavior is to increase the sheet thickness.

Overall, it is apparent that as far as sheet thickness is concerned, there is simply not that much variability in λ with changes in n, or with reasonable variations in H_f/H_0. One could simply state that $H/H_0 = 1.25$ for an expected set of operating conditions and be correct to within a few percent. If *precision* calendering must be done, however, then this few percent uncertainty may be important.

It is important to recognize that the comments of the preceding paragraph hold under the assumption that the system operates at fixed H_0, and refer to λ. If the system operates at fixed roll loading (F/W), as shown at the end of Sec. 7-2,

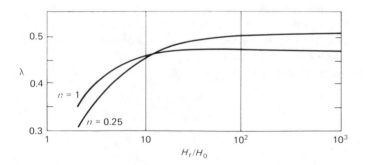

Figure 7-10 Calendered thickness, in terms of λ, as a function of H_f/H_0 for various n. (*Results are from Brazinsky et al.*)

then H_0 may be subject to significant variation if operating conditions change. While one may still state a priori that $H/H_0 = 1.25 \pm 0.03$, the actual value of calendered thickness will vary with the fluctuations in H_0 itself, which may be significant.

7-4 NORMAL STRESSES AND VISCOELASTICITY IN CALENDERING

Let us go back to the dynamic equations for steady two-dimensional flow with negligible inertial effects and examine the way in which normal stresses appear, and subsequently disappear, from the calendering problem. We begin with

$$\frac{\partial \tau_{xx}}{\partial x} + \frac{\partial \tau_{xy}}{\partial y} = \frac{\partial p}{\partial x} \tag{7-61}$$

Instead of nondimensionalizing x and y as before, let us use the characteristic linear dimensions appropriate to the x and y directions and define

$$\tilde{x} = \frac{x}{R} \qquad \tilde{y} = \frac{y}{H_0} \tag{7-62}$$

Then Eq. (7-61) becomes

$$\frac{\partial \tau_{xx}}{\partial \tilde{x}} + \frac{R}{H_0} \frac{\partial \tau_{xy}}{\partial \tilde{y}} = \frac{\partial p}{\partial \tilde{x}} \tag{7-63}$$

If $\partial \tau_{yy}/\partial \tilde{x}$ is subtracted from both sides we find

$$\frac{\partial(\tau_{xx} - \tau_{yy})}{\partial \tilde{x}} + \frac{R}{H_0} \frac{\partial \tau_{xy}}{\partial \tilde{y}} = -\frac{\partial T_{yy}}{\partial \tilde{x}} \tag{7-64}$$

Now let us speculate on the relative magnitudes of the terms in Eq. (7-64). The magnitude of the ratio of the primary normal stress difference to the shear stress is just twice the *recoverable shear* S_R:

$$2S_R = \frac{\tau_{xx} - \tau_{yy}}{\tau_{xy}} \tag{7-65}$$

The recoverable shear is a measure of the elasticity of the fluid. For typical polymer melts one may find $1 < S_R < 10$, but quite commonly $S_R < 1$.

The geometric factor R/H_0 will normally be in the range $100 < R/H_0 < 1000$, which means that the flow region is nearly parallel. This leads to the lubrication approximation that $\partial/\partial x \ll \partial/\partial y$ in a nearly parallel flow field. When x and y are "scaled," i.e., nondimensionalized according to Eqs. (7-62), then at best we might expect that $\partial/\partial \tilde{x}$ is comparable to $\partial/\partial \tilde{y}$. This leads us to conclude that the magnitudes of the two terms on the left-hand side of Eq. (7-64) are given roughly by

$$\frac{\partial(\tau_{xx} - \tau_{yy})}{\partial \tilde{x}} \approx S_R |\tau| \tag{7-66}$$

$$\frac{R}{H_0} \frac{\partial \tau_{xy}}{\partial \tilde{y}} \approx \frac{R}{H_0} |\tau| \tag{7-67}$$

where $|\tau|$ is some approximate measure of the viscous (shear) stresses in the flow, such as $|\tau| = K(U/H_0)^n$.

We see then that the ratio of the normal stress term to the shear stress term is of the order of $S_R/(R/H_0)$. Using the numbers suggested above, we would conclude that the normal stress terms may be dropped from the lubrication approximation for this flow.

Equation (7-64) then becomes

$$\frac{R}{H_0} \frac{\partial \tau_{xy}}{\partial \tilde{y}} = -\frac{\partial T_{yy}}{\partial \tilde{x}} \tag{7-68}$$

Except for notation, this differs from Eq. (7-41) in only one respect: The isotropic pressure p has been replaced by the *total* y-directed stress: $-T_{yy} = p - \tau_{yy}$.

Having just removed the normal stress terms from the analysis in favor of the viscous stresses, one is tempted to introduce a *purely viscous* constitutive equation at this point, such as the power law model of Sec. 7-2. However, the strong lagrangian acceleration associated with this flow suggests the possibility that viscoelastic effects may make an important contribution to the stresses, since fluid is carried from a state of near rest into the high-deformation-rate nip region in a time which may be quite small compared to some appropriate relaxation time for the fluid.

One is faced then with the need to select some appropriate constitutive equation which might be suitable to describe the response of a fluid to a rapidly imposed deformation. A further complication lies in the fact that this particular flow field has an elongational character to it. Indeed, along the line of symmetry for this flow (along $y = 0$) the *shear* components of Δ vanish but elongational components of significant magnitude exist (see Prob. 7-1).

As noted earlier, there is an insufficient experimental base on which to select or recommend a specific constitutive equation suitable to the description of this type of flow. We shall drop this point here but return to it in the next chapter in the context of a closely related flow field which occurs in coating systems.

PROBLEMS

7-1 Combine Eqs. (7-15) and (7-17) to give an expression for $u'_x(x, y)$. The maximum shear rate (at fixed x) occurs along $y = h(x)$. Find $\partial u_x/\partial y$ in the form

$$\frac{\partial u_x}{\partial y}\bigg|_{y=h(x)} = \frac{U}{H_0} g(x')$$

Where does the maximum value of this function occur along x', and what is its value?

Calculate a stretching rate from

$$\frac{\partial u_x}{\partial x}\bigg|_{y=0} = \frac{U}{H_0} e(x')$$

and give the position of maximum stretching. Is the maximum stretching rate comparable in magnitude to the maximum shearing rate?

7-2 With $u_x(x, y)$ known, it is possible to evaluate $u_y(x, y)$ from integration of the continuity equation (7-3). Do so, and discuss the approximation $u_y \ll u_x$. (Is it valid everywhere? How does it depend on R/H_0?)

7-3 Give the value of x' beyond which Eq. (7-14) is a poor approximation to Eq. (7-13). State clearly your definition of "poor."

7-4 Give a practical definition of $x' = -\infty$ by finding the value of $x' < 0$ at which $p'/p'_{max} = 0.01$.

7-5 Suppose a finite pressure p_a were imposed upstream so that Eq. (7-27) had to be replaced by $p'(x' \to -\infty) = p'_a$. Calculate p'_a if we wish to increase λ to $1.1\lambda_0$.

Give the magnitude of p_a if $\mu = 10^4$ P, $U = 100$ ft/min, $R = 6$ in, and $H_0 = \frac{1}{8}$ in, and compare p_a to p_{max}.

7-6 Show that $G(\lambda)$ in Eq. (7-34) is given by

$$\frac{4G(\lambda)}{3} = 1 + \lambda^2 \frac{1 + 3\lambda^2}{1 + \lambda^2} - \lim_{\phi \to -\pi/2} \left\{ \tan \phi \left[(1 - 3\lambda^2)(\phi - \tan^{-1} \lambda) + \lambda \frac{1 + 3\lambda^2}{1 + \lambda^2} \right] \right\}$$

Show that the indicated limit gives a finite value only if $\lambda = 0.475$, and verify that $G(0.475) = 1.22$.

7-7 A calender having dimensions $R = 6$ in and $W = 6$ ft has a net weight of 1000 lb and "floats" on the material being calendered. The roll speed is 20 ft/min, and the material may be considered newtonian with a viscosity of 10^4 P.

Calculate the sheet thickness.

7-8 Derive Eq. (7-44).

7-9 Extend Fig. 7-7 to *dilatant* fluids, defined as power law fluids for which $n > 1$.

7-10 Extend the newtonian theory of calendering to account for (*a*) two rolls with different linear velocities; (*b*) two rolls of different radii but at the same linear speed.

7-11 Suggest a parameter which gives an indication of the length of time that a fluid is subjected to a significant stress during calendering.

7-12 Does a power law fluid exhibit a higher or lower maximum pressure than the corresponding newtonian fluid when calendered according to the models of Secs. 7-1 and 7-2? *Hint:* The answer depends upon your definition of "corresponding newtonian fluid."

7-13 Show that for a newtonian fluid which enters a calender as a finite sheet, the final calendered thickness is given by the solution to

$$\frac{2(1 + \lambda^2)(H_f/H_0 - 1)^{1/2}}{(H_f/H_0)^2} - (1 - 3\lambda^2) \left[\frac{(H_f/H_0 - 1)^{1/2}}{H_f/H_0} + \tan^{-1} \left(\frac{H_f}{H_0} - 1 \right)^{1/2} \right]$$

$$= -\frac{2\lambda}{1 + \lambda^2} + (1 - 3\lambda^2) \left(\frac{\lambda}{1 + \lambda^2} + \tan^{-1} \lambda \right)$$

7-14 (*a*) Find the pressure distribution for the case of calendering a finite sheet of a newtonian fluid.

(*b*) Find the maximum pressure for part *a*, define an appropriate dimensionless pressure, and plot $\tilde{p}_{max}$ as a function of H_f/H_0.

(*c*) Is H_f/H_0 a significant parameter with respect to $\tilde{p}_{max}$?

(*d*) Find the roll-separating force as a function of H_f/H_0.

7-15 Chong presents data, shown in Fig. 7-11, for calendering a high molecular weight cellulose-acetate derivative. Fit Eq. (7-51) to the data and estimate values for K and n at each temperature. Chong gives data on viscosity, at $\dot{\gamma} = 1$ s^{-1}, as a function of temperature. See if your K, n values give a viscosity that is compatible with his data at that shear rate.

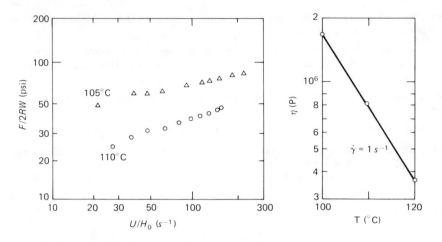

Figure 7-11 Data for Prob. 7-15. (*From Chong.*)

7-16 A calender having the dimensions $R = 4$ in, $W = 40$ in, $H_0 = 4$ mil, operating at a peripheral roll speed of 78.8 ft/min on a material having a viscosity 10^4 P, produces a sheet of thickness 8.6 mil. Calculate the maximum pressure developed in the material, the power required to operate the calender, the roll-separating force, and the average temperature rise of the material, assuming that there is no heat transfer between the material and the rolls.

7-17 The data shown in Fig. 7-12 were obtained by Bergen and Scott, who calendered a plasticized

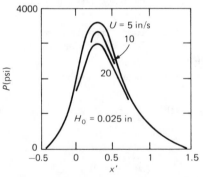

Figure 7-12 Data for Prob. 7-17. (*From Bergen and Scott.*)

thermoplastic resin. The roll diameter was given as 10 in. The only viscosity data were obtained in an instrument which operated at low shear stresses ($\tau \leq 10$ psi) and which indicated newtonian flow with $\mu = 3.2 \times 10^9$ P at 95°F.

  (a) Give the upper limit on shear *rate* for the viscosity data.

  (b) What is the range of nominal shear rates under which calendering was carried out?

  (c) Can the data be explained by the models presented here? Point out any major discrepancies.

  (d) Could viscous heating be significant in these experiments? Support your answer with quantitative arguments.

BIBLIOGRAPHY

7-1 The newtonian model of calendering

The book by McKelvey (see Chap. 6 Bibliography) gives much of the same material presented here. An inaccuracy in one of McKelvey's calculations is corrected in

Ehrmann, G., and J. Vlachopoulos: Determination of Power Consumption in Calendering, *Rheol. Acta*, **14:** 761 (1975).

7-2 The power law model of calendering

The most complete treatment of calendered thickness is given in

Brazinsky, I., et al.: A Theoretical Study of Liquid-Film Spread Heights in the Calendering of Newtonian and Power Law Fluids, *J. Appl. Polym. Sci.*, **14:** 2771 (1970).

A similar analysis, but for a nonnewtonian fluid for which $\eta = A - B \tanh (\dot{\gamma}/k)^n$, is given in

Alston, W. W., Jr., and K. N. Astill: An Analysis for the Calendering of Non-Newtonian Fluids, *J. Appl. Polym. Sci.*, **17:** 3157 (1973).

7-4 Normal stresses and viscoelasticity in calendering

A discussion and development of some features of a model for calendering a viscoelastic fluid is given in

Chong, J. S.: Calendering Thermoplastic Materials, *J. Appl. Polym. Sci.*, **12:** 191 (1968).

Some data on calendering plastics is given in

Bergen, J. T., and G. W. Scott, Jr.: Pressure Distribution in the Calendering of Plastic Materials, *J. Appl. Mech.*, **18:** 101 (1951).

No rheological data were presented for the materials that were calendered, so it is impossible to offer a good test of the models presented here. Problem 7-17 considers several aspects of Bergen and Scott's data.

EIGHT

COATING

I'll give you a definite maybe.

Goldwyn

Coating is a process in which a liquid is applied continuously to a moving sheet in order to produce a uniform application of the fluid onto and/or within the sheet. The term *web* is often used to denote the sheet to be coated, and webs may be of a variety of materials, including:

- Paper and paperboard (to which an adhesive might be coated)
- Cellulosic films (as in photographic emulsion coating)
- Plastic films (as in magnetic surface coating for recording tape)
- Textile fibers and fabrics (to which finishes or backings might be applied)
- Metal foils (to which a polymer might be coated to produce a laminated capacitor foil)

Because of the great variety of coating purposes there is a corresponding variety of coating methods. Several of the more important will be mentioned here in an introductory way. Then analyses of several coating systems will be considered.

Figure 8-1 shows, schematically, a roll coater. The lower roll picks up liquid from a bath and delivers it to a second roll, or directly to the moving web. The web is "squeezed" between two rolls, and the amount of coating applied depends upon fluid properties and the spacing between the rolls in the region through which the

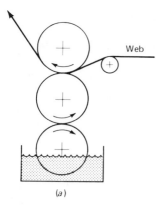

(a)

Figure 8-1 Roll coater.

web moves. In some applications the nip separation is controlled; in others the nip pressure is controlled.

Similar to the roll coater is the "kiss" coater, shown in Fig. 8-2, in which the web is run over the roll without any backup roll on the other side. The amount of wrap around the roll, the tension in the web, and fluid properties control the amount of coating applied to the web. This type of system often has a metering or smoothing device downstream to provide more control over the coating thickness.

Figure 8-3 shows the reverse-roll configuration in which the metering roll, the roll that makes the final delivery of fluid to the web, moves in the opposite direction to the web motion. The geometry of roll coating is very similar to that of calendering, and we shall see similarities in the analyses of the two systems.

Figure 8-4 shows blade coating, which can follow a kiss-coating application or which can directly meter fluid to the moving web from a "pond" of the coating fluid. The blade is usually flexible and acts as a spring. Pressure developed in the flow under the blade determines the precise position of the blade relative to the backup roll and so determines the coating thickness.

Withdrawal coating is shown in Fig. 8-5. Such a system is normally used when it is undesirable to contact the coating with another surface such as a blade or roll. The coating thickness is largely controlled by the dynamics of the region where the web leaves the pond surface.

Sometimes a coating is applied directly to the web from an extruder, as in Fig. 8-6. Rolls may be used, in which case this is simply a modified form of roll

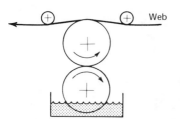

Figure 8-2 Kiss coater.

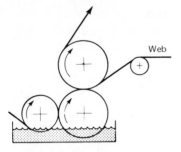

Web

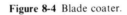
Figure 8-3 Reverse-roll coater.

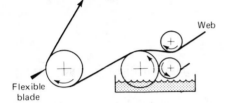

Web

Flexible
blade

Figure 8-4 Blade coater.

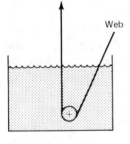

Web

Figure 8-5 Withdrawal coater.

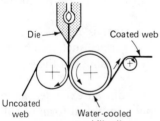

Die

Coated web

Uncoated
web

Water-cooled
chill roll

Figure 8-6 Extrusion coater.

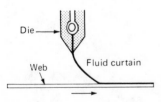

Die

Web

Fluid curtain

Figure 8-7 Curtain coater.

coating, or a "curtain" coating method (Fig. 8-7) may be used. The fluid must be high enough in viscosity that a stable curtain or sheet is produced. Thickness is controlled by precise delivery from the extruder and by control of the web speed. The web is normally moving faster than the curtain, causing the curtain to stretch between the die lips and the web. Some unusual phenomena may occur when viscoelastic fluids are stretched, and the dynamics of the curtain coater are quite complex.

8-1 ROLL COATING

A roll-coating process is similar to calendering in some respects but quite different in others. Figure 8-8 shows a definition sketch for the analyses to follow. Like the calendering flow, there is a converging-diverging character to the kinematics, and we can expect the dynamics to be similar to that described in Chap. 7 for calendering. The major difference is in the character of the separation region, where the fluid splits and adheres to *both* moving surfaces. In the analysis of calendering it is *assumed* that the fluid separates cleanly from the roll; in the analysis of coating it is *assumed* that the fluid evenly wets both the roll and the sheet. The question of whether one or the other of these extremes of behavior occurs must be answered in terms of the rheological characteristics of the fluid, as well as in terms of the fluid-solid and fluid-air surface tensions. In a very loose way we may characterize the calender type of separation region as typical of softened plastics, materials which are of such high viscosity and elasticity that while they are capable of viscous deformation they also have a strong tendency to retain their shape. The coating type of separation region is more typical of relatively low-viscosity materials.

An analysis of roll coating attempts to provide a model which relates coating thickness to the geometric, rheological, and operating variables that characterize a particular system. There are many combinations of the variables. We illustrate the analysis by examining the system whose geometry is as shown in Fig. 8-8. We consider the case where the sheet has the same velocity as the peripheral roll speed and moves in the same direction. Differential speeds, or reverse coating (where the roll motion is opposite that of the sheet), are other commonly encountered cases

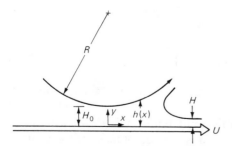

Figure 8-8 Definition sketch for analysis of roll coating.

that can be modeled by similar analyses to the one we will present here. The newtonian fluid will be treated first, in some detail. Nonnewtonian effects are considered subsequently.

Newtonian Coating

It is assumed that a newtonian fluid is being coated in an isothermal steady-state operation. We also assume that viscous forces dominate inertial forces and that the lubrication approximations introduced in Chap. 7 are valid in this flow field. The dynamic equations then reduce to

$$0 = -\frac{\partial p}{\partial x} + \mu \frac{\partial^2 u_x}{\partial y^2} \tag{8-1}$$

It is convenient to introduce the following dimensionless variables:

$$\xi = \frac{x}{(RH_0)^{1/2}} \qquad \eta = \frac{y}{H_0} \qquad u = \frac{u_x}{U} \qquad P = \frac{pH_0^{3/2}}{\mu U R^{1/2}}$$

In terms of these variables Eq. (8-1) becomes

$$\frac{\partial^2 u}{\partial \eta^2} = \frac{\partial P}{\partial \xi} \tag{8-2}$$

Consistent with the lubrication approximations we take P to be independent of η. This allows integration of Eq. (8-2) twice with the result

$$u = \frac{1}{2}\left(\frac{dP}{d\xi}\right)\eta^2 + c_1 \eta + c_2 \tag{8-3}$$

Appropriate boundary conditions include $u = 1$ at $\eta = 0$, from which we find $c_2 = 1$.

As in the calender model, assuming $H_0/R \ll 1$, we take

$$\frac{h(x)}{H_0} \equiv \tilde{h}(\xi) = 1 + \tfrac{1}{2}\xi^2 \tag{8-4}$$

The second boundary condition on u then takes the form $u = 1$ at $\eta = \tilde{h}(\xi)$. Upon evaluating c_1, we may write the velocity field as

$$u = 1 + \frac{1}{2}\frac{dP}{d\xi}(\eta^2 - \tilde{h}\eta) \tag{8-5}$$

which gives u explicitly as a function of η and implicitly as a function of ξ through the functions $\tilde{h}(\xi)$ and $P(\xi)$.

The pressure function is unknown at this point, but its functional form can be determined after introducing the principle of conservation of mass. If we let W be the width of the roll and assume no flow normal to the xy plane, the volumetric flow rate is just

$$Q = W \int_0^{h(x)} u_x \, dy$$

or

$$\lambda \equiv \frac{Q}{UWH_0} = \int_0^{\tilde{h}} u \, d\eta \tag{8-6}$$

The integration may be carried out by using Eq. (8-5), with the result

$$\lambda = \tilde{h} - \frac{1}{12}\left(\frac{dP}{d\xi}\right)\tilde{h}^3 \tag{8-7}$$

While λ is unknown, at this point in the analysis, it is *some* constant, and so one may solve for the pressure gradient and find

$$\frac{dP}{d\xi} = 12\frac{\tilde{h} - \lambda}{\tilde{h}^3} \tag{8-8}$$

Upon integration we find

$$P(\xi) = 12 \int_{-\xi_0}^{\xi} \frac{\tilde{h} - \lambda}{\tilde{h}^3} \, d\xi \tag{8-9}$$

As a boundary condition on pressure we have stated that $P = 0$ at some value of ξ, called $-\xi_0$.

If it is assumed that the liquid splits evenly to coat both the roll and the sheet, then the volumetric flow rate, the coating thickness, and λ are related by

$$Q = 2UH \tag{8-10}$$

and

$$\lambda = \frac{2H}{H_0} \tag{8-11}$$

Note that, as in the analysis of calendering, we use a parameter λ which is related to the coating thickness. The two λ's are quite different however. In the coating analysis λ is defined in such a way that it is directly proportional to H/H_0. But in the calendering analysis the different definition of λ leads to a nonlinear relationship [Eq. (7-23)] between λ and H/H_0.

At this point the analysis is incomplete because λ remains unknown. In order to determine λ it is necessary to consider the geometry and dynamics of the separation region. Figure 8-9 shows a detail of this region. The figure includes the assumption that the film splits evenly, so that the separation point is $(\xi_1, \frac{1}{2}\tilde{h})$.

At the separation point the velocity vanishes,

$$u = 1 - \frac{3}{2}\frac{\tilde{h} - \lambda}{\tilde{h}} = 0 \tag{8-12}$$

from which it follows that

$$\tilde{h}_1 = 1 + \tfrac{1}{2}\xi_1^2 = 3\lambda \tag{8-13}$$

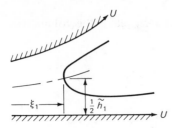

Figure 8-9 Detail of the separation region.

The pressure at the separation point may be found by first integrating Eq. (8-9) (using the condition $-\xi_0 = -\infty$), with the result that

$$P(\xi) = \frac{(6 - \frac{9}{2}\lambda)\xi}{1 + \frac{1}{2}\xi^2} - \frac{3\lambda\xi}{(1 + \frac{1}{2}\xi^2)^2} + \left(\frac{12}{\sqrt{2}} - \frac{9\lambda}{\sqrt{2}}\right) \tan^{-1}\frac{\xi}{\sqrt{2}} + \frac{6\pi}{\sqrt{2}}\left(1 - \frac{3\lambda}{4}\right) \tag{8-14}$$

The simplest dynamic model of the separation region is based on the assertion that the film splits at the point where $u = 0$ *and* $P = 0$. If, in Eq. (8-14), ξ is replaced by $\xi_1(\lambda)$ from Eq. (8-13), and if $P(\xi_1)$ is equated to zero, then Eq. (8-14) becomes a transcendental equation in λ whose solution may be found by numerical methods. The result is

$$\lambda_0 = 1.30 \tag{8-15}$$

A more detailed (but still very crude) model of the separation region has been carried out by Greener and Middleman (1975), which accounts for the effects of viscosity and surface tension on the dynamics of the separation region. Their analysis introduces a dimensionless parameter N_1, defined as

$$N_1 = \frac{\sigma}{\mu U}\left(\frac{H_0}{R}\right)^{1/2} \tag{8-16}$$

It is found that $\lambda_0 = 1.30$ is an asymptotic value to be expected in the limit of vanishing N_1. If N_1 is significantly larger than unity, λ exceeds the value 1.30. In most industrial coating systems N_1 will not exceed unity, and Eq. (8-15) might be expected to provide a useful estimate of λ.

Once λ is known, the pressure distribution follows from Eq. (8-14). The velocity field is then calculated from Eq. (8-5). From this information the stresses in the system may be calculated, from which one may find the roll/sheet-separating force and the forces (and hence the power) required to turn the roll and pull the sheet. The roll/sheet-separating force, for example, is given by

$$F = W\int_{-\infty}^{x_1} p(x)\,dx = \frac{\mu U R W}{H_0}\int_{-\infty}^{\xi_1} P(\xi)\,d\xi \tag{8-17}$$

and for $\lambda_0 = 1.3$ we find

$$\frac{FH_0}{\mu U R W} = 2.6 \tag{8-18}$$

Calculation of the force required to pull the sheet through the coating nip is left as an exercise (Prob. 8-5). Once the pulling force is known it is possible to calculate the tensile stress exerted on the sheet. If this stress is high enough, it is possible that the sheet may undergo tensile deformation during coating with a consequent change in thickness. If a coating is laid down on such a deformed sheet, and if strain relaxation occurs subsequently, it is quite possible that the coating may "buckle" and become uneven. Thus a model for the tensile forces in the sheet itself could be a significant part of a complete coating analysis.

Coating a Power Law Fluid

The power law case is treated in a manner that parallels the newtonian model developed above, and similar complications arise here as do in the power law model of calendering. Consistent with the lubrication approximation the power law is written as

$$\tau_{xy} = K \left[\left(\frac{\partial u_x}{\partial y} \right)^2 \right]^{(n-1)/2} \frac{\partial u_x}{\partial y} \tag{8-19}$$

Other terms that would normally appear in the second invariant II_Δ have been neglected. To avoid taking the fractional root of a negative quantity the velocity gradient is squared before the indicated fractional power is taken.

When nondimensionalized the dynamic equations take the form

$$0 = -\frac{\partial P}{\partial \xi} + \frac{\partial}{\partial \eta} \left\{ \left[\left(\frac{\partial u}{\partial \eta} \right)^2 \right]^{(n-1)/2} \frac{\partial u}{\partial \eta} \right\} \tag{8-20}$$

where the dimensionless variables are defined as in the newtonian case, except that

$$P = \left(\frac{H_0}{U} \right)^n \left(\frac{H_0}{R} \right)^{1/2} \frac{p}{K} \tag{8-21}$$

The solution proceeds as in the newtonian case, but several integrals cannot be explicitly evaluated. The key result, the pressure distribution, is given by

$$P = A \int_{-\infty}^{\xi} \frac{[(\tilde{h} - \lambda)^2]^{(n-1)/2}(\tilde{h} - \lambda)}{\tilde{h}^{1+2n}} \, d\xi \tag{8-22}$$

where

$$A = \left(\frac{1+2n}{n} \right)^n 2^{1+n}$$

As in the newtonian case λ is obtained by asserting that the pressure vanishes at the separation point, at which u also vanishes. Setting $u(\xi_1, \frac{1}{2}\tilde{h}) = 0$ gives

$$\xi_1 = \left[2 \left(\frac{2n+1}{n} \lambda - 1 \right) \right]^{1/2} \tag{8-23}$$

Then the condition $P(\xi_1) = 0$ gives [setting the upper limit at ξ_1 in Eq. (8-22)] an integral equation for λ which must be solved by trial and error. The calculations

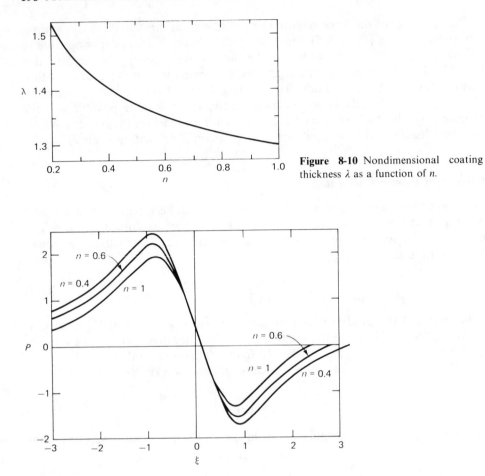

Figure 8-10 Nondimensional coating thickness λ as a function of n.

Figure 8-11 Pressure profiles for newtonian and power law fluids.

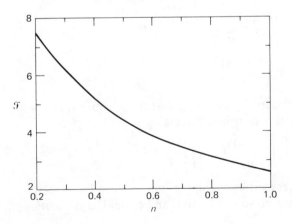

Figure 8-12 Nondimensional roll-separating force as a function of n.

indicate that λ is increased by nonnewtonian shear behavior but that the effect is modest. Figure 8-10 shows $\lambda(n)$.

The effect of n on the pressure distribution is significant, and Fig. 8-11 compares pressure profiles for several n values, including the newtonian case $n = 1$. It is apparent that P is increased by $n < 1$, in comparison to the newtonian case. In making the comparison, however, it is important to keep in mind that it is the nondimensional pressure P being examined. It does not necessarily follow that the actual pressure p, for a nonnewtonian fluid, is always more than that for a newtonian fluid without considering the basis for comparison. Problem 8-7 suggests an investigation of this point.

The roll-separating force F may be calculated from the pressure distribution and written in the form

$$\frac{F}{W} = K\left(\frac{U}{H_0}\right)^n R\mathscr{F}(n) \tag{8-24}$$

Figure 8-12 shows $\mathscr{F}(n)$.

Viscoelastic Effects

If a real viscoelastic fluid is subjected to the flow under consideration here, we might expect three phenomena to occur which would have to be accounted for in any realistic model:

- Nonnewtonian shear behavior
- Normal stress development
- Stress growth and relaxation due to the lagrangian unsteady nature of the kinematics

It is useful to comment here on the problems that arise when one attempts to incorporate these phenomena into a model of roll coating.

We confine our remarks to models based on the lubrication approximations. Otherwise we are discussing the more general and more difficult problem of two-dimensional viscoelastic flow which must be described by simultaneous, coupled, nonlinear partial differential equations. In Chap. 7 it was argued that normal stresses might be expected to enter the lubrication equations only in the total stress term T_{yy}. Thus the equation of motion takes the form

$$\frac{d}{dx}(p - \tau_{yy}) = \frac{\partial \tau_{xy}}{\partial y} \tag{8-25}$$

We can anticipate that in some highly elastic fluids the elastic normal stress τ_{yy} might be comparable to the viscous shear stress τ_{xy}. If these stresses are comparable, then we might argue that the magnitude of $d\tau_{yy}/dx$ is considerably less than that of $\partial \tau_{xy}/\partial y$, since the lubrication assumption itself asserts that for almost parallel flows $\partial/\partial x \ll \partial/\partial y$. Thus we would seek a solution of

$$\frac{dp}{dx} = \frac{\partial \tau_{xy}}{\partial y} \tag{8-26}$$

But this is the same equation as would be solved for the case of a *purely viscous* nonnewtonian fluid. Within the context of the lubrication approximations, then, we find that viscoelasticity does not *explicitly* appear in the equation of motion.

Viscoelasticity may appear explicitly, however, when we write a constitutive equation for the fluid of interest. Thus if we couple Eq. (8-26) with a model of the form

$$\tau + \theta_R \frac{\mathscr{D}\tau}{\mathscr{D}t} = \eta_0 \, \Delta \tag{8-27}$$

where θ_R is a characteristic relaxation time of the fluid, we have incorporated viscoelasticity into the formulation of the problem through the time derivative $\mathscr{D}/\mathscr{D}t$. [The remarks to follow hold equally well, though not in exactly the same format, if we use the Oldroyd derivative $\eth/\eth t$ instead of the Jaumann derivative $\mathscr{D}/\mathscr{D}t$. You may need to review the definitions given in Eqs. (3-126) and (3-127).]

For the shear stress τ_{xy} we have, then,

$$\tau_{xy} + \theta_R \left[u_x \frac{\partial \tau_{xy}}{\partial x} + u_y \frac{\partial \tau_{xy}}{\partial y} - \frac{1}{2}\left(\frac{\partial u_x}{\partial y} - \frac{\partial u_y}{\partial x}\right)(\tau_{yy} - \tau_{xx}) \right] = \eta_0 \left(\frac{\partial u_x}{\partial y} + \frac{\partial u_y}{\partial x}\right) \tag{8-28}$$

Now, if the lubrication approximations, namely, $u_x \gg u_y$ and $\partial/\partial x \ll \partial/\partial y$, are applied to the terms of Eq. (8-28), we find that

$$\tau_{xy} + \theta_R \left[\frac{1}{2}\frac{\partial u_x}{\partial y}(\tau_{xx} - \tau_{yy})\right] = \eta_0 \frac{\partial u_x}{\partial y} \tag{8-29}$$

If, in a similar manner, we write Eq. (8-27) for τ_{xx} and τ_{yy} and then impose the lubrication approximations, we find

$$\tau_{xx} = \theta_R \tau_{xy} \frac{\partial u_x}{\partial y} \tag{8-30}$$

$$\tau_{yy} = -\tau_{xx} \tag{8-31}$$

It follows then that

$$\tau_{xy} = \frac{\eta_0 \, \partial u_x/\partial y}{1 + \theta_R^2(\partial u_x/\partial y)^2} \tag{8-32}$$

and

$$\tau_{yy} = -\theta_R \eta_0 \frac{(\partial u_x/\partial y)^2}{1 + \theta_R^2(\partial u_x/\partial y)^2} \tag{8-33}$$

But Eqs. (8-32) and (8-33) are identical to what one obtains for steady simple shear flow. Hence the lubrication approximations remove those terms of the constitutive equation associated with the lagrangian unsteady character of the flow.

One must ask the question, then, "Is there such a thing as a lubrication theory for a viscoelastic fluid?" The answer is yes, but in a very limited sense.

From Eq. (8-26) we see that any nonlinear *shear* behavior can be accommodated within the theory. However, the nonlinear *acceleration* phenomena associated with the converging-diverging character of the kinematics are lost in the lubrication approximation. The *normal stresses* enter the model only in the calculation of the roll-separating force, using

$$\frac{F}{W} = \int_{-\infty}^{x_1} [-T_{yy}]_{y=h(x)} \, dx = \int_{-\infty}^{x_1} [p - \tau_{yy}]_{y=h(x)} \, dx \qquad (8\text{-}34)$$

Of course, if the nip separation is not fixed but the force F is held constant, then the normal stress will affect the nip separation which will, in turn, affect the coating thickness. We note that if Eq. (8-33) is even qualitatively valid, with respect to the *sign* of the stress τ_{yy}, we can conclude that the roll-separating force tends to be increased by the appearance of normal stresses. However, we would have to examine the effect of nonnewtonian shear behavior on p before concluding anything about the change in total stress T_{yy}, which governs F.

8-2 BLADE COATING

An analytical model of blade coating may be carried out if we assume the blade angle is sufficiently small that the lubrication approximations hold. This is not often the case, since typical operations involve a blade angle in the neighborhood of 45° to the sheet. Nevertheless, the model is of some utility as a first approximation to expected behavior, and we shall therefore illustrate the development of a model for coating thickness and pressure distribution.

Figure 8-13 shows the geometry of the analysis. The blade is flat and set at an angle such that

$$\tan \theta = \frac{H_1 - H_0}{L} \qquad (8\text{-}35)$$

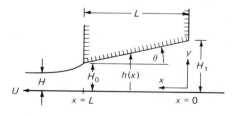

Figure 8-13 Definition sketch for analysis of blade coating.

The volumetric flow rate Q is related to the final coating thickness H by

$$\frac{Q}{W} = UH \tag{8-36}$$

where U is the speed of the web, and W is its width.

Newtonian Coating

We begin with a newtonian analysis. If inertial terms are neglected, the lubrication approximations lead us to dynamic equations in the form

$$0 = -\frac{\partial p}{\partial x} + \mu \frac{\partial^2 u_x}{\partial y^2} \tag{8-37}$$

$$0 = -\frac{\partial p}{\partial y} \tag{8-38}$$

Thus we find $p = p(x)$ only, and Eq. (8-37) may be integrated twice with respect to y to give

$$u_x = U\left(1 - \frac{y}{h}\right) - \frac{h^2}{2\mu} \frac{dp}{dx}\left(\frac{y}{h} - \frac{y^2}{h^2}\right) \tag{8-39}$$

We have used boundary conditions on u_x in the form

$$u_x = \begin{cases} U & \text{on } y = 0 \\ 0 & \text{on } y = h(x) \end{cases}$$

to evaluate the constants of integration.

The volumetric flow rate is found from

$$\frac{Q}{W} = \int_0^h u_x \, dy \tag{8-40}$$

When Eq. (8-39) is integrated an expression is obtained which may be solved for dp/dx with the result

$$\frac{dp}{dx} = 6\mu U \frac{h - 2H}{h^3} \tag{8-41}$$

To integrate for the pressure distribution it is now necessary to introduce an expression for $h(x)$. For a flat blade,

$$h(x) = H_1 - \frac{H_1 - H_0}{L} x \tag{8-42}$$

The integration of Eq. (8-41) may be carried out analytically, and the result for $p(x)$ contains two unknown parameters: an integration constant and the coating thickness H. We shall introduce two boundary conditions on p:

$$p = 0 \qquad \text{at } x = 0$$

$$p = 0 \qquad \text{at } x = L$$

These boundary conditions reflect the notion that the regions $x < 0$ and $x > L$ are exposed to the atmosphere at ambient pressure $p = 0$ and that no significant entrance or exit phenomena occur to alter the pressure from its ambient value. Inertial effects, neglected here, would probably cause an " entrance " loss similar to that at the entrance to a pipe attached to a large reservoir. One might expect exit effects to be associated with surface tension and the dynamic meniscus that is formed in the region near $x \gtrsim L$. We will carry out the analysis using the boundary conditions given above.

With these conditions, $p(x)$ may be written as

$$p(x) = \frac{6\mu U x[h(x) - H_0]}{h^2(x)(H_1 + H_0)} \tag{8-43}$$

and H is found, in the course of applying the boundary conditions, to be

$$H = \frac{H_1 H_0}{H_1 + H_0} \tag{8-44}$$

The maximum in the pressure distribution occurs at a point x^* where

$$h(x^*) = 2H \tag{8-45}$$

It follows easily that

$$p_{\text{max}} = \frac{3\mu U L}{2H_1 H_0} \frac{H_1 - H_0}{H_1 + H_0} = \frac{3\mu U L}{2H_0^2} \frac{\kappa - 1}{\kappa(\kappa + 1)} \tag{8-46}$$

where $\kappa = H_1/H_0$. In terms of κ, H and x^* may be written as

$$H = H_0 \frac{\kappa}{1 + \kappa} \qquad \text{and} \qquad x^* = L \frac{\kappa}{1 + \kappa} \tag{8-47}$$

Since $\kappa > 1$, we see that $H < H_0$ and in fact lies in the range

$$\frac{1}{2} < \frac{H}{H_0} < 1 \qquad \text{for } 1 < \kappa < \infty \tag{8-48}$$

We also note that the dependence of p_{max} on κ is such as to yield a maximum value of the maximum pressure. This maximum occurs at a value of κ of

$$\kappa^* = 2.414 \tag{8-49}$$

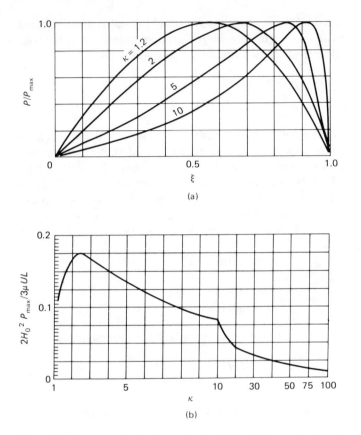

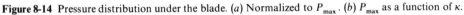

Figure 8-14 Pressure distribution under the blade. (a) Normalized to P_{max}. (b) P_{max} as a function of κ.

at which the maximum pressure is

$$p_{max}^* = 0.258 \frac{\mu U L}{H_0^2} \tag{8-50}$$

Figure 8-14 shows the pressure distribution under the blade with κ as a parameter. We see that the maximum occurs downstream of the center of the blade, and as κ increases the peak approaches the blade tip. The pressure profile is quite sharp for large values of κ. It is not clear a priori how accurate the lubrication model is for values of κ much removed from unity. To answer this question it would be necessary to solve the dynamic equations numerically and compare the results to the simple analytical model given above. (But see Sec. 8-4.)

We note at this point that the coating thickness is predicted to depend only on geometric parameters according to Eq. (8-47). Within the context of this model this prediction is true only if H_1 and H_0 are constrained, independent of operating conditions. In fact, however, most blades are flexible, and H_0 is established by the pressure distribution under the blade.

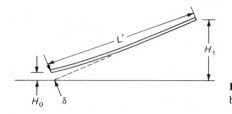

Figure 8-15 Definition sketch for analysis of flexible blade.

The effect of the pressure distribution is to load the blade and cause it to deflect. If the blade were rigid, as assumed above, the *blade loading* $\mathscr{L}$ would be calculated as

$$\mathscr{L} = \int_0^L p(x)\, dx = \frac{6\mu U}{\tan^2 \theta}\left(\ln \kappa - 2\frac{\kappa - 1}{\kappa + 1}\right) \tag{8-51}$$

The flexible blade case is quite complex, but we may model it through the use of some fairly simple approximations.

Figure 8-15 shows a blade under load. Equation (8-51) is no longer strictly valid since θ is not a constant and $h(x)$ is no longer a linear function of x. The format of this simple model will be maintained by taking the angle θ to be defined as the angle to the web made by the *chord* of length L' connecting $y = H_1$ and $y = H_0$. Then we find that

$$\theta = \sin^{-1}\frac{H_1 - H_0}{L'} \tag{8-52}$$

and we take L' to be the undeflected blade length.

The equation for loading is then rewritten in the form

$$\mathscr{L} = 6\mu U f(\kappa, \beta) \tag{8-53}$$

where

$$\beta \equiv \frac{H_1}{L'}$$

and

$$f(\kappa, \beta) = \left[\frac{1}{\beta^2} - \frac{(\kappa - 1)^2}{\kappa^2}\right]\left[\frac{\kappa^2 \ln \kappa}{(\kappa - 1)^2} - \frac{2\kappa^2}{(\kappa - 1)(\kappa + 1)}\right] \tag{8-54}$$

We will consider the blade to deflect as if it were a cantilever beam loaded at a point x_L by a force $\mathscr{L}W$ (where W is the blade width). Figure 8-16 shows the static force diagram. Any elementary text on statics or mechanics will develop the

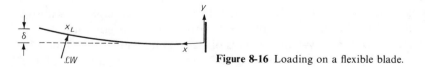

Figure 8-16 Loading on a flexible blade.

solution of this problem. The result of interest is the deflection δ, which is found to be

$$\delta = \frac{\mathscr{L}Wx_L^3}{3EI}\left[1 + \frac{3(L' - x_L)}{2x_L}\right] \tag{8-55}$$

where E is Young's modulus for the blade, and I is the moment of inertia of the cross section, given by

$$I = \frac{WB^3}{12} \tag{8-56}$$

where B is the blade thickness.

We will identify x_L with x^*, the point at which the pressure is a maximum. They are not equal, but related by the angle θ, and as an approximation we take

$$\frac{x^*}{x_L} = \cos\theta = \left[1 - \frac{\beta^2(\kappa - 1)^2}{\kappa^2}\right]^{1/2} \tag{8-57}$$

If the value of H_0 (call it H_{00}) is known in the unloaded case (one might fix H_0 before starting the system), then one would calculate the loaded H_0 from

$$H_0 = H_{00} + \delta\cos\theta \tag{8-58}$$

or

$$\delta = \frac{H_1}{\cos\theta}\left(\frac{1}{\kappa} - \frac{H_{00}}{H_1}\right) \tag{8-59}$$

A scheme for calculating H may be put together as follows:

1. We assume that Eq. (8-47) still holds, but we write it in the form

$$H = \frac{H_1}{1 + \kappa} \quad \text{and} \quad x^* = L'\frac{\kappa}{1 + \kappa}\cos\theta \tag{8-60}$$

2. We assume the blade geometry is known (L', β, W, B) and that an *unloaded setting* of the blade is specified, giving H_1 and H_{00}.
3. We assume H_1 does not change under deflection.
4. We pick values of κ and plot $\mathscr{L}$ versus κ using Eq. (8-53).
5. For each κ we can calculate θ and x_L [from Eqs. (8-57) and (8-60)] and δ [from Eq. (8-59)].
6. We rewrite Eq. (8-55) in the form

$$\mathscr{L} = \frac{3H_1\,EI}{Wx_L^3\cos\theta}\left(\frac{1}{\kappa} - \frac{H_{00}}{H_1}\right)\left[1 + \frac{3(L' - x_L)}{2x_L}\right]^{-1} \tag{8-61}$$

Everything on the right-hand side may be calculated from the equations above for any value of κ. Hence we may plot $\mathscr{L}$ versus κ from this equation.
7. The intersection of the $\mathscr{L}$ versus κ curves of steps 4 and 6 gives the value of κ consistent with the approximations. H follows from Eq. (8-60) in step 1 above.

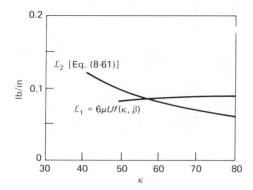

Figure 8-17 Graphical solution of Example 8-1.

Example 8-1 Find the coating thickness developed by a flexible blade coater designed and operated as noted:

$$L = 0.25 \text{ in} \qquad B = 0.005 \text{ in} \qquad U = 100 \text{ cm/s}$$

$$E = 10^7 \text{ psi} \qquad \mu = 10 \text{ P}$$

Before the coating operation begins, the blade is set so that it just touches the sheet $(H_{00} = 0)$ at an angle of $45°$. Assume the blade holder is rigid so that H_1 does not change during coating.

We find $H_1 = 0.177$ in, and $\beta = H_1/L = 0.707$. We have enough information to calculate $\mathscr{L}$ versus κ using the procedure outlined above. The results are shown in Fig. 8-17, and we find

$$\kappa = 58$$

$$H_0 = 0.0031 \text{ in}$$

$$H = 0.003 \text{ in}$$

In this case the blade deflection is quite small, about 3 mil, approximately the same as the blade thickness itself. The maximum pressure developed under the blade is only about 2.6 psi.

Example 8-2 In a paper-coating operation a clay suspension containing 50 percent solids (by weight) is deposited in such a way that the observed coating, on a dry-weight basis, is $\frac{1}{2}$ lb/ream. (A ream is 3300 ft².) The paper speed is 2400 ft/min, the suspension viscosity is 1 P, and its density is 1.52 g/cm³.

The blade is flat, and $L = 0.3$ mm. The blade angle is $4°$.

Find the blade loading and the maximum pressure.

The coating weight per unit area is related to the thickness by

$$w = \rho H$$

and the dry weight is half this (for 50 percent solids loading):

$$w_d = \frac{1}{2}\rho H$$

From the information presented above we find

$$H = \frac{2w_d}{\rho} = \frac{2(0.5)(\frac{454}{3300})(30.5)^2}{1.52} = 0.97 \ \mu m$$

We can find H_1 and H_0 from the pair of equations

$$H = H_0 \frac{\kappa}{1 + \kappa} = 0.97 \times 10^{-4} \ cm$$

$$\sin \theta = 0.07 = \frac{H_1 - H_0}{L'} = \frac{H_0(\kappa - 1)}{0.03}$$

The results are

$$H_0 = 10^{-4} \ cm = 1 \ \mu m \qquad H_1 = 24.5 \ \mu m$$

$$p_{max} = 1186 \ psi \qquad \mathscr{L} = 9 \ pli$$

Note that the value of H_0 is extremely small. If the solids are not submicrometric in size, the application of this model would be very tenuous.

The Effect of Viscoelasticity

Since many coatings are nonnewtonian, we are motivated to examine the blade-coating problem once again but with a constitutive equation suitable for more complex fluids. If we paralleled our treatment of earlier problems, we would probably carry out the analysis for the power law fluid next. Let us depart from that approach, now, in order to illustrate another method of obtaining analytical solutions of complex problems.

We continue to work within the lubrication approximations. Our goal is to gain some insight into the effects of nonlinear viscosity and elastic stresses on the dynamics of blade coating. We begin by recalling the comments on lubrication models of viscoelastic flows, made with respect to the roll-coating analysis, presented in the subsection "Viscoelastic Effects" in Sec. 8-1. Those comments are equally valid for the blade-coating problem. In particular, we recall that the shear stress and normal stress effects are separated in the analysis and so may be considered independently.

We begin with the selection of a constitutive equation which allows for nonnewtonian viscosity and elastic stresses. We choose the Maxwell model with the Jaumann derivative:

$$\tau + \theta_R \frac{\mathscr{D}\tau}{\mathscr{D}t} = \eta_0 \ \Delta \tag{8-62}$$

where θ_R is a relaxation time of the fluid. As in the roll-coating case this gives, because of the lubrication approximations,

$$\tau_{xy} = \frac{\eta_0 \ \partial u_x/\partial y}{1 + (\theta_R \ \partial u_x/\partial y)^2} \tag{8-63}$$

$$\tau_{yy} = -\theta_R \frac{\partial u_x}{\partial y} \tau_{xy} \tag{8-64}$$

The virtues and failures of this constitutive model were discussed in Chap. 3. We note here only that we do not expect the model to be valid for large deformation rates.

With the lubrication approximations the dynamic equations reduce to the form

$$\frac{dp}{dx} = \frac{\partial \tau_{xy}}{\partial y} \tag{8-65}$$

Upon integration with respect to y we find

$$\tau_{xy} = \frac{dp}{dx} y + c \tag{8-66}$$

where c is an integration constant.

When Eq. (8-63) is introduced for τ_{xy}, we have a nonlinear differential equation of the form

$$\frac{\partial u_x/\partial y}{1 + (\theta_R \, \partial u_x/\partial y)^2} = \frac{1}{\eta_0} \left(y \frac{dp}{dx} + c \right) \tag{8-67}$$

It is convenient to introduce the following dimensionless variables:

$$\frac{u_x}{U} = \varphi \qquad \frac{y}{H_0} = \eta \qquad \frac{x}{L} = \xi \qquad \frac{H_0^2 p}{\eta_0 UL} = P$$

$$\frac{\theta_R U}{H_0} = \text{Ws} = \text{Weissenberg number}$$

With these definitions, Eq. (8-67) becomes

$$\frac{\partial \varphi/\partial \eta}{1 + \epsilon(\partial \varphi/\partial \eta)^2} = \eta P' + C \tag{8-68}$$

where we replace $(\text{Ws})^2$ by the parameter ϵ, and where $P' = dP/d\xi$ and C is a dimensionless integration constant. Equation (8-68) will be solved by a *perturbation method*, which is based on the following ideas:

1. We should restrict the solution to small values of ϵ, since we know that Eq. (8-63) has restricted validity.
2. In the limit of $\epsilon \to 0$ we expect to obtain the newtonian solution.
3. For small ϵ we might expect that the functions $\varphi(\xi, \eta; \epsilon)$ and $P(\xi, \epsilon)$ depart from their limiting (newtonian) behavior in a way that depends "smoothly" on

ϵ. An example of "smooth" dependence on ϵ would be a polynomial in ϵ, such as

$$\varphi = \varphi_0 + \sum_{k=1}^{\infty} \epsilon^k \varphi_k \qquad (8\text{-}69)$$

where φ_0 is the newtonian solution.

To carry out the perturbation solution, then, we write

$$\varphi = \varphi_0 + \epsilon\varphi_1 + \epsilon^2\varphi_2 + \cdots \qquad (8\text{-}70)$$

$$P = P_0 + \epsilon P_1 + \epsilon^2 P_2 + \cdots \qquad (8\text{-}71)$$

$$C = C_0 + \epsilon C_1 + \epsilon^2 C_2 + \cdots \qquad (8\text{-}72)$$

We then substitute these expressions for φ, P, and C in Eq. (8-68), clear the denominator on the left-hand side by bringing it across to the right-hand side as a factor, and then we group terms by factoring like powers of ϵ. Thus we find

$$\frac{\partial \varphi_0}{\partial \eta} + \epsilon \frac{\partial \varphi_1}{\partial \eta} + \epsilon^2 \frac{\partial \varphi_2}{\partial \eta} + \cdots = \left[1 + \epsilon \left(\frac{\partial \varphi_0}{\partial \eta} + \epsilon \frac{\partial \varphi_1}{\partial \eta} + \cdots \right)^2 \right]$$

$$\times \left[\eta P_0' + \epsilon\eta P_1' + \epsilon^2 \eta P_2' + C_0 + \epsilon C_1 + \epsilon^2 C_2 + \cdots \right] \qquad (8\text{-}73)$$

Now we equate terms with common factors of ϵ:

Zeroth power:
$$\frac{\partial \varphi_0}{\partial \eta} = \eta P_0' + C_0 \qquad (8\text{-}74)$$

First power:
$$\frac{\partial \varphi_1}{\partial \eta} = \eta P_1' + C_1 + \left(\frac{\partial \varphi_0}{\partial \eta} \right)^2 (\eta P_0' + C_0) \qquad (8\text{-}75)$$

and so on. Boundary conditions require that

$$\varphi = \begin{cases} 0 & \text{on } \eta = \dfrac{h(x)}{H_0} = \kappa + (1 - \kappa)\xi = \tilde{h}(\xi) \\ 1 & \text{on } \eta = 0 \end{cases}$$

Since these conditions must hold for all values of ϵ, it follows that

$$\varphi_0 = \begin{cases} 0 & \text{on } \eta = \kappa + (1 - \kappa)\xi \\ 1 & \text{on } \eta = 0 \end{cases}$$

and all $\varphi_k = 0$ on both boundaries for $k > 0$. Now we can solve Eqs. (8-74) and (8-75) in turn.

From Eq. (8-74) we find

$$\varphi_0 = \tfrac{1}{2}\eta^2 P_0' + \eta C_0 + C_{01} \qquad (8\text{-}76)$$

Using the given boundary conditions on φ_0, we find

$$\varphi_0 = \tfrac{1}{2}(\eta^2 - \tilde{h}\eta)P_0' + 1 - \frac{\eta}{\tilde{h}} \tag{8-77}$$

To find the pressure gradient we must impose a condition of mass conservation in the form

$$\frac{Q}{W} = \int_0^{h(x)} u_x \, dy \qquad \text{or} \qquad \frac{H}{H_0} \equiv \lambda = \int_0^{\tilde{h}(\xi)} \varphi(\xi, \eta) \, d\eta \tag{8-78}$$

[We have used Eq. (8-36) to relate λ to coating thickness.] Consistent with Eq. (8-70) we see that λ itself should be written in the form

$$\lambda = \lambda_0 + \epsilon\lambda_1 + \epsilon^2\lambda_2 + \cdots \tag{8-79}$$

and it follows that

$$\lambda_k = \int_0^{\tilde{h}} \varphi_k \, d\eta \qquad k = 0, 1, \ldots \tag{8-80}$$

Imposing Eq. (8-80) on the zeroth-order (newtonian) solution (8-77) gives P_0' as

$$P_0' = \frac{6}{\tilde{h}^2} - \frac{12\lambda_0}{\tilde{h}^3} \tag{8-81}$$

When this is substituted into Eq. (8-77), we have φ_0 as a function of η and of ξ through $\tilde{h}(\xi) = \kappa + (1 - \kappa)\xi$, and as a function of the (still unspecified) parameter λ_0. To find λ_0 we must integrate Eq. (8-81) with respect to ξ in order to obtain $P_0(\xi)$. On imposing the boundary conditions $P_0(0) = P_0(1) = 0$ (ambient pressure at the entrance and exit of the blade), we find λ_0 to be

$$\lambda_0 = \frac{\kappa}{1 + \kappa} \tag{8-82}$$

These results can be seen to be consistent with the solutions presented in the subsection "Newtonian Coating" in this section.

Once the zeroth-order solution is completed we may go back to Eq. (8-75) and solve for φ_1 and P_1 (and λ_1). This process can be repeated to the exhaustion of the investigator. Algebraic exhaustion sets in at about 2 hours, which is sufficient time to find the correct forms for φ_1, P_1, and λ_1. (Incorrect forms are obtained much more quickly. The 2 hours noted above is required to find and correct the inevitable algebraic errors that occur in five to six pages of hand calculations of this type.)

Equation (8-75) may be written [with the aid of Eq. (8-74)] in the form

$$\frac{\partial\varphi_1}{\partial\eta} = \eta P_1' + C_1 + \left(\frac{\partial\varphi_0}{\partial\eta}\right)^3 \tag{8-83}$$

The solution procedure follows that for the zeroth-order solution:

1. Integrate with respect to η to find φ_1, making use of boundary conditions $\varphi_1 = 0$ on both $\eta = 0$ and $\eta = \tilde{h}$.
2. Impose mass conservation $\int_0^{\tilde{h}} \varphi_1 \, d\eta = \lambda_1$, which gives $P'_1(\xi, \lambda_1)$.
3. Integrate P'_1 with respect to ξ. The conditions $P_1 = 0$ at $\xi = 0$ and 1 give $\lambda_1(\kappa)$.

The results are as follows:

$$(1 - \kappa)P_1 = 6\lambda_1(\tilde{h}^{-2} - \kappa^{-2}) - 43.2\lambda_1^3(\tilde{h}^{-6} - \kappa^{-6}) + 77.76\lambda_0^2(\tilde{h}^{-5} - \kappa^{-5})$$
$$- 57.6\lambda_0(\tilde{h}^{-4} - \kappa^{-4}) + 16.8(\tilde{h}^{-3} - \kappa^{-3}) \tag{8-84}$$

[Keep in mind that $\tilde{h} = \kappa + (1 - \kappa)\xi$.]

$$(1 - \kappa^{-2})\lambda_1 = 7.2(1 - \kappa^{-6})\lambda_0^3 - 12.96(1 - \kappa^{-5})\lambda_0^2 + 9.6(1 - \kappa^{-4})\lambda_0$$
$$- 2.8(1 - \kappa^{-3}) \tag{8-85}$$

We can show that $P_1 < 0$, thus indicating that the effect of nonnewtonian viscosity is a reduction in pressure under the blade. We can also show that $0 < \lambda_1 < 1$, which indicates that the effect of nonnewtonian viscosity is to increase the coating thickness.

The roll-separating force requires calculation of the total stress $-T_{yy}$ exerted on the blade, and we begin with

$$-T_{yy} = p - \tau_{yy} \tag{8-86}$$

The constitutive assumption regarding normal stress [Eq. (8-64)] can be introduced, and we may write the total stress exerted normal to the blade in dimensionless form, and up to linear terms in ϵ we find

$$-\frac{H_0^2 T_{yy}}{\eta_0 UL} = P_0 + \frac{H_0}{L} \epsilon^{1/2} \left(\frac{\partial \varphi_0}{\partial \eta}\right)_{\tilde{h}}^2 + \epsilon P_1 \tag{8-87}$$

The interesting result is that while the viscous effect on stress (pressure) is of first order in ϵ, the elastic effect comes in more strongly, being half order in ϵ. The relative magnitude of the two terms beyond P_0 depends on the geometry (H_0/L), but it is conceivable, *according to this model*, that the positive elastic term could outweigh the negative viscous term (the ϵP_1 term). Hence we cannot generalize and assert that the effect of viscoelasticity is to increase (or decrease) the blade loading. We can show, for example, that in a fluid which is newtonian in shear but which exhibits finite normal stresses, the blade loading is increased relative to the corresponding newtonian fluid. In a fluid which is purely viscous, the loading is decreased.

One must avoid the temptation to investigate the consequences of this model in detail. Because of the restriction of the first-order solution to very small values of $\epsilon = Ws^2$, the model is probably not applicable to realistic viscoelastic fluid-coating operations. Since the constitutive equation has limited validity, there is no justification for going beyond the first-order analysis. It is worthwhile, however, to

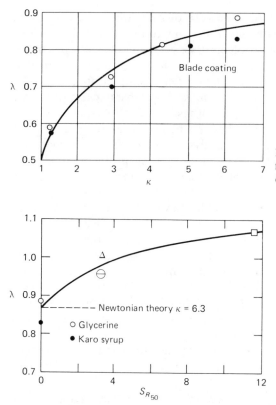

Figure 8-18 Data on $\lambda(\kappa)$ for newtonian fluids, compared to theory. ○ Glycerol; ● Karo Syrup.

Figure 8-19 Data on λ at $\kappa = 6.3$ for aqueous polyacrylamide solutions. ⊖ 1% Polyhall; △ 1.5% Polyhall; □ 1% Polyhall in 50/50 glycerol-water.

examine some data obtained in a laboratory coater to determine if the foregoing models, including the newtonian model, have any capability to predict reality.

Figure 8-18 shows data obtained with two newtonian fluids. Over the full range of κ studied, the data agree well with the theory.

Figure 8-19 shows data obtained with three viscoelastic solutions at a fixed value of κ. The fluids are characterized by their values of recoverable shear S_R measured at 50 s^{-1}. We see, first of all, that λ lies above the newtonian value, as predicted by the theory. However, if S_R and Ws are taken as roughly comparable (see Chap. 3, page 61) we find that the theory outlined above grossly overestimates the expected value of λ. Since the fluids studied have larger values of Ws than those for which the perturbation theory is expected to be valid, the quantitative failure of the data to agree with theory is not unexpected.

8-3 FREE COATING

In the free-coating process the surface to be coated is initially immersed in the coating fluid and then withdrawn. A layer of liquid remains on the surface of the object, the amount depending on the viscosity and surface tension of the coating fluid and on the speed of withdrawal. Gravity, of course, will cause the fluid to tend to drain off the object. Normally some type of postwithdrawal operation, such as drying or freezing of the coated layer, is required to stabilize the coating.

Two of the simplest free-coating problems arise when the object to be coated is either a continuous web or belt of material (such as in film coating) or a cylindrical filament (such as in fiber coating). Geometric simplifications, and the continuous steady-state nature of the process, make it feasible to carry out some models of the process that show some degree of resemblance to experience.

The principal complicating feature of these problems is the strong role played by the free surface in controlling the coating dynamics in the region where the object leaves the surface of the fluid. In previous coating problems (the wire-coating die of Chap. 5 and the roll- and blade-coating problems of this chapter), the coating flow is basically a bounded flow where the free surface enters the problem, if at all, only by (possibly) perturbing the flow near the exit of the system. To a first, and useful, approximation one could ignore surface tension. In the free-coating problem the role of surface tension is central to the dynamics.

The method of analysis is somewhat complicated, and the chain of logic seems more convoluted than in previous models. Assumptions will be made which appear to be, and are, quite arbitrary. But the analyses are interesting and provide a good demonstration of the adage "Nothing ventured, nothing gained." We will be able to compare some of the models discussed with experimental data to assess how much has been gained, and how much is left to do.

We will illustrate the method of analysis with the simplest problem of free coating: the vertical withdrawal of a flat, continuous, wide film from a stationary bath of newtonian fluid.

Figure 8-20 shows the geometry of the model which we will develop for this system. A flat sheet is withdrawn vertically at a speed U from a liquid bath whose free surface is at $x = 0$. Somewhere above the free surface (region 1) a dynamic equilibrium between the effect of gravity and the drag of the sheet leads to a constant coating thickness H_∞. In region 3, nearest the free surface, we assume that the effects of surface tension are dominant, in comparison to viscous and gravitational forces. In region 2 the free surface is the result of the interaction among viscous, gravitational, and interfacial forces. In all regions inertial effects are neglected.

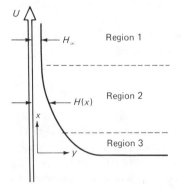

Figure 8-20 Definition sketch for model of free coating onto a flat sheet.

Region 1

The dynamic equations for a newtonian liquid reduce to

$$\mu \frac{d^2 u}{dy^2} - \rho g = 0 \tag{8-88}$$

and the appropriate boundary conditions are

$$u = U \qquad \text{on } y = 0 \qquad \text{(no slip)}$$

$$\mu \frac{du}{dy} = 0 \qquad \text{on } y = H_\infty \qquad \text{(no shear exerted by the air)}$$

We find $u(y)$ to be

$$u = U + \frac{\rho g}{\mu} \left(\frac{y^2}{2} - H_\infty y \right) \tag{8-89}$$

Since H_∞ is unknown a priori the solution is not complete. It will be necessary to find $u(y)$ for region 2, and in "matching" the solutions so that they are continuous between the two regions, an additional condition on H_∞ will appear.

We note at this stage that the liquid film is *not* in plug flow; there is a distribution of velocity. Near the web the fluid is moving upward with velocity U. At the surface the fluid has a lower velocity due to the effect of gravity.

Region 2

Here the dynamic equations include pressure and take the form

$$\mu \frac{\partial^2 u}{\partial y^2} - \rho g - \frac{\partial p}{\partial x} = 0 \tag{8-90}$$

$$-\frac{\partial p}{\partial y} = 0 \tag{8-91}$$

subject to boundary conditions

On $y = 0$: $u = U$ (no slip)

On $y = H$: $\tau \cdot s = 0$ (no shear)

On $y = H$: $p - \tau \cdot n = \dfrac{\sigma}{R}$ (continuity of normal stress)

The two stress boundary conditions are written as shown because the free surface $H(x)$ is not a coordinate surface of the cartesian system with respect to which **u** and τ are measured. **s** and **n** are unit vectors tangential and normal to the free surface, respectively, and $\tau \cdot s$ and $\tau \cdot n$ are the shear and normal components of stress.

We are, in effect, introducing the lubrication approximations here, with the thought that the upward component of velocity dominates the flow and that derivatives of u in the y direction are more important than those in the x direction. Thus we take the flow in this region to be almost parallel to the sheet.

The free surface can be described by the (unknown) function $H(x)$, and the radius of curvature of the free surface is given by

$$\frac{1}{R(x)} = \frac{-d^2H/dx^2}{[1 + (dH/dx)^2]^{3/2}} \tag{8-92}$$

In region 2 we will take the curvature to be sufficiently small that $|dH/dx| \ll 1$, in which case the following approximations may be used in the boundary conditions:

$$\boldsymbol{\tau} \cdot \mathbf{s} = \tau_{xy} = 0 \tag{i}$$

$$\boldsymbol{\tau} \cdot \mathbf{n} = \tau_{yy} = -\tau_{xx} \tag{ii}$$

$$\frac{1}{R} = -\frac{d^2H}{dx^2}$$

The boundary conditions reflect the following ideas:

(i) For small $|dH/dx|$ the free surface is nearly vertical, and the shear stress is well approximated by τ_{xy}.
(ii) By the same token $\boldsymbol{\tau} \cdot \mathbf{n}$ is approximated by τ_{yy}. Since we appear to neglect a y component of velocity, there is a temptation to set $\tau_{yy} = 0$. However, we must note that for a newtonian fluid in a two-dimensional flow, $\tau_{yy} = -\tau_{xx}$ is a consequence of the continuity equation alone.

Thus we must consider a finite normal stress, which may be calculated from

$$\tau_{yy} = -2\mu \frac{\partial u}{\partial x} = -\tau_{xx} \tag{8-93}$$

The second boundary condition then takes the form

$$p + 2\mu \frac{\partial u}{\partial x} = -\sigma \frac{d^2H}{dx^2} \qquad \text{at } y = H(x) \tag{8-94}$$

Equation (8-90) may be integrated twice with respect to y to give the velocity profile as

$$u = U + \frac{\rho g + dp/dx}{\mu}\left(\frac{y^2}{2} - Hy\right) \tag{8-95}$$

which satisfies the no-slip condition at $y = 0$ and the no-shear stress condition at $y = H(x)$.

At this point it is useful to introduce the *entrainment* q (per unit width), defined as

$$q = \int_0^H u \, dy \tag{8-96}$$

If Eq. (8-95) is integrated and we solve for the pressure gradient we find

$$\frac{dp}{dx} = \frac{3\mu}{H^3}\left(UH - q - \frac{\rho g H^3}{3\mu}\right) \tag{8-97}$$

Note that since $H(x)$ is unknown we cannot integrate to find $p(x)$, nor do we need to.

We are now in a position to derive a differential equation for $H(x)$. Equation (8-94) is differentiated with respect to x. Equation (8-97) is substituted for dp/dx. Equation (8-95) is used to eliminate d^2u/dx^2. The result is

$$0 = \sigma\frac{d^3H}{dx^3} - 3\mu q\frac{d}{dx}\left(\frac{1}{H^2}\frac{dH}{dx}\right) + \frac{3\mu}{H^3}\left(UH - q - \frac{\rho g H^3}{3\mu}\right) \tag{8-98}$$

Before dealing further with this equation we note two points: It is nonlinear in H, and it contains the unknown constant q. As a matter of convenience we now introduce the following dimensionless variables and parameters:

$$L = \frac{H}{H_x} \tag{8-99}$$

$$\xi = \frac{x}{H_x}\left(\frac{3\mu U}{\sigma}\right)^{1/3} \tag{8-100}$$

the *capillary number*

$$Ca = \frac{\mu U}{\sigma} \tag{8-101}$$

and a dimensionless equilibrium coating thickness

$$T_x = H_x\left(\frac{\rho g}{\mu U}\right)^{1/2} \tag{8-102}$$

Equation (8-98) then takes the form

$$\frac{d^3L}{d\xi^3} - (3Ca)^{2/3}\left(1 - \frac{T_x^2}{3}\right)\frac{d}{d\xi}\left(\frac{1}{L^2}\frac{dL}{d\xi}\right) + \frac{1}{L^3}[L - 1 + \tfrac{1}{3}T_x^2(1 - L^3)] = 0 \tag{8-103}$$

The unknown parameter q has been replaced by H_x by making use of the solution in region 1, Eq. (8-89), from which it follows that

$$q = UH_x\left(1 - \frac{\rho g H_x^2}{3\mu U}\right) \tag{8-104}$$

Thus Eq. (8-103) contains an unknown parameter T_x [see Eq. (8-102)], and so it is necessary to have *four* conditions imposed on the solution to Eq. (8-103): three boundary conditions for the third-order equation and a fourth condition that will give T_x.

The four conditions are "matching" conditions. We require the region 2 solution to approach the region 1 solution as ξ gets large. In terms of L, region 1 is characterized by

$$L = 1 \tag{8-105}$$

$$\frac{dL}{d\xi} = 0 \tag{8-106}$$

$$\frac{d^2L}{d\xi^2} = 0 \tag{8-107}$$

Equation (8-105) follows directly from the definitions of L and H_∞. Equations (8-106) and (8-107) reflect the fact that the film thickness is uniform and the interface is flat.

The fourth condition must match the curvature of region 2 with region 3. We must inquire first whether Eq. (8-103) will allow imposition of these boundary conditions: Are they compatible with the form of the equation? To answer this we set $L = 1$ in Eq. (8-103) and observe that the vanishing of all derivatives is compatible with the equation. Hence Eq. (8-103) has a solution which asympotically approaches the limiting value of $L = 1$.

The more difficult question is whether we *can* match the solution in regions 2 and 3 somewhere. First we examine region 3.

Region 3

In region 3 we assume that the fluid motion has a negligible effect on the shape of the meniscus, so that $H(x)$ is determined by the *static* balance between gravitational and interfacial forces:

$$\frac{\sigma \, d^2H/dx^2}{[1 + (dH/dx)^2]^{3/2}} = \rho g x \tag{8-108}$$

It is possible to integrate this equation and find the slope in the form

$$\frac{dH/dx}{[1 + (dH/dx)^2]^{1/2}} = \frac{\rho g x^2}{2\sigma} - 1 \tag{8-109}$$

(The boundary condition $dH/dx \to -\infty$ at $x = 0$ has been used.)

We see that in the static case the slope dH/dx vanishes at a height x_∞ above the surface, with x_∞ given by

$$x_\infty = \left(\frac{2\sigma}{\rho g} \right)^{1/2} \tag{8-110}$$

At that point the curvature is given by

$$\left. \frac{d^2H}{dx^2} \right|_{x_\infty} = \left(\frac{2\rho g}{\sigma} \right)^{1/2} \tag{8-111}$$

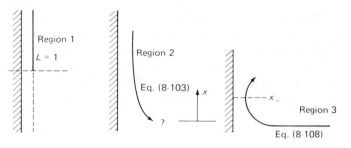

Figure 8-21 Mathematical behavior of the free surface in each region.

In terms of dimensionless variables this is

$$\frac{d^2L}{d\xi^2}\bigg|_{\xi_\infty} = \frac{\sqrt{2}}{3^{2/3}} \mathrm{Ca}^{-1/6} T_\infty \qquad (8\text{-}112)$$

At this point, then, we have the following situation, which may be visualized with reference to Fig. 8-21.

Region 1. The "solution" is trivial: $L = 1$.
Region 2. The solution has proper behavior for $x \to \infty$.
Region 3. The solution has proper behavior for $x \to 0$, but "turns around" at x_∞.

The key to the problem of matching solutions in regions 2 and 3 lies in the behavior of Eq. (8-103) for small x (large L). Since we cannot integrate the differential equation analytically, the behavior of $L(\xi)$ is not obvious. It is useful to rewrite the equation in the following way. We multiply through Eq. (8-103) by $S = d^2L/d\xi^2$ and note that

$$S\frac{d^3L}{d\xi^3} = \frac{d^2L}{d\xi^2}\frac{d^3L}{d\xi^3} = \frac{1}{2}\frac{d}{d\xi}\left(\frac{d^2L}{d\xi^2}\right)^2 = \frac{1}{2}\frac{dS^2}{d\xi} \qquad (8\text{-}113)$$

Equation (8-103) then takes the form

$$\frac{1}{2}\frac{dS^2}{d\xi} - C\left(\frac{S^2}{L^2} - \frac{2S}{L^3}\frac{dL}{d\xi}\right) + \frac{S}{L^3}[L - 1 + \tfrac{1}{3}T_\infty^2(1 - L^3)] = 0 \qquad (8\text{-}114)$$

where $C = (3\,\mathrm{Ca})^{2/3}(1 - T_\infty^2/3)$. For small C and large L this takes the form

$$\frac{1}{2}\frac{dS_0^2}{d\xi} - \frac{S_0 T_\infty^2}{3} = 0 \qquad (8\text{-}115)$$

which has the solution

$$S_0 = \tfrac{1}{3}T_\infty^2\,\xi = \frac{d^2L}{d\xi^2} \qquad (8\text{-}116)$$

(We use the notation S_0 to remind us that this approximation is valid only for $C = 0$, $L \to \infty$, $\xi \to 0$.) Although Eq. (8-116) shows that the curvature S_0 is getting small as $\xi \to 0$, that the film is becoming *flat*, it also shows, upon solving Eq. (8-116) for $L(\xi)$, that L is not becoming large in this limit as $\xi \to 0$. Hence we conclude that Eq. (8-103) does not have the desired behavior for small ξ. This should not be surprising since the development of the model for region 2 implies almost parallel flow. (It is a lubrication approximation.) We should not expect a good model toward the bottom of the film where L gets large.

Suppose we introduce the idea that L is near unity, say, $L = 1 + \epsilon$ where ϵ is small. Then the third term of Eq. (8-114) becomes approximated, after some modest algebra, by

$$\frac{S}{L^3}\{\epsilon + \tfrac{1}{3}T_\infty^2[1 - (1 + 3\epsilon)]\} = (1 - T_\infty^2)\frac{S(L-1)}{L^3} \tag{8-117}$$

Now we examine

$$\frac{1}{2}\frac{dS^2}{d\xi} - C\left(\frac{S^2}{L^2} - \frac{2S}{L^3}\frac{dL}{d\xi}\right) + \frac{(1 - T_\infty^2)S(L-1)}{L^3} = 0 \tag{8-118}$$

While the third term is valid only near $L = 1$, we nevertheless examine the behavior of Eq. (8-118) for *large L*. If $C = 0$ we have

$$\frac{dS^2}{d\xi} = 0 \qquad \text{as } L \gg 1 \tag{8-119}$$

or $S = $ constant as $\xi \to 0$. Hence the curvature becomes constant. The film does not become horizontal necessarily, even if the curvature vanishes, since

$$L = L_0 - \alpha\xi \qquad L_0 \gg 1 \tag{8-120}$$

is compatible with both Eqs. (8-119) and (8-118). For finite C the same conclusions hold, though not so clearly as when $C = 0$. Now it is possible to devise a matching procedure to complete the analysis of the problem.

The Method of Landau and Levich

Their model (see Levich) neglects the effect of gravity in the entrainment region and leads to

$$\frac{d^3L}{d\xi^3} + \frac{L-1}{L^3} = 0 \tag{8-121}$$

Equation (8-121) is solved subject to

$$L = 1$$

$$\frac{dL}{d\xi} = 0 \qquad \text{at } \xi \to \infty$$

$$\frac{d^2L}{d\xi^2} = 0$$

Upon numerical solution of Eq. (8-121) it is found that the curvature at $\xi = 0$ is a finite value,

$$\frac{d^2L}{d\xi^2}\bigg|_{\xi=0} = S_0 \tag{8-122}$$

Landau and Levich match the region 2 and 3 solutions by requiring [see Eq. (8-112)]

$$S_0 = \frac{\sqrt{2}}{3^{2/3}} \, Ca^{-1/6} \, T_\infty \tag{8-123}$$

As a consequence they find

$$T_\infty = \frac{3^{2/3}}{\sqrt{2}} S_0 \, Ca^{1/6} = 0.944 \, Ca^{1/6} \tag{8-124}$$

The numerical coefficient follows upon introducing the value of S_0 found by Landau and Levich. We note that the curvature match is somewhat disjointed, in that Eq. (8-112) holds at $\xi_\infty \neq 0$ whereas Eq. (8-122) uses S_0 at $\xi = 0$. No direct examination of the effect of this aspect of the matching procedure has been given. Since other elements of the model involve various approximations, it is more sensible to examine the comparison of experimental data with Eq. (8-124). We shall do this after examining some other matching methods.

Gravity-corrected Theory

White and Tallmadge (1965) give a gravity-corrected theory which is identical to the development above, except that in the normal stress boundary condition [Eq. (8-94)] the viscous term was neglected. The result is that they solve

$$\frac{d^3L}{d\xi^3} + (1 - T_\infty^2)\frac{L-1}{L^3} = 0 \tag{8-125}$$

and using the Landau-Levich matching procedure they find

$$\frac{T_\infty}{(1 - T_\infty^2)^{2/3}} = 0.944 \, Ca^{1/6} \tag{8-126}$$

For small Ca this is a good approximation to Eq. (8-124); for large Ca one predicts that $T_\infty \to 1$.

The Method of Wilkinson

Wilkinson et al. (see Spiers et al., 1974) correct the normal stress boundary condition to the form given above as Eq. (8-94), and "linearize" the third term of Eq. (8-103) [as in working with Eq. (8-117)]. Thus they solve

$$\frac{d^3L}{d\xi^3} - (3\,Ca)^{2/3}\left(1 - \frac{T_\infty^2}{3}\right)\frac{d}{d\xi}\left(\frac{1}{L^2}\frac{dL}{d\xi}\right) + (1 - T_\infty^2)\frac{L-1}{L^3} = 0 \tag{8-127}$$

Inspection of Eq. (8-118) shows that $d^2L/d\xi^2 = S$ approaches a constant value at small ξ (large L). However, we note that S will be found to be a function of Ca and T_∞. By contrast, if Eq. (8-121) is used, S is a pure number.

Wilkinson, then, solves for T_∞ as a function of Ca by solving Eq. (8-127) for a pair of values of (Ca, T_∞), and then requiring that the pair also satisfy

$$T_\infty = \frac{3^{2/3}}{\sqrt{2}} S(T_\infty, \text{Ca}) \, \text{Ca}^{1/6} \qquad (8\text{-}128)$$

A trial-and-error procedure yields $T_\infty(\text{Ca})$.

Comparison to Experimental Data

Figure 8-22 shows the results of an experimental program carried out by Wilkinson and coworkers. Over a range of $3\frac{1}{2}$ orders of magnitude in Ca ($0.004 < \text{Ca} < 10$) the model of Wilkinson seems superior to, and intermediate between, those of Landau-Levich and White-Tallmadge. The principal failure of the model is its inability to predict a limiting value of T_∞ at high capillary numbers. The region of high capillary number, of course, is one where viscous shear effects are large (large μ and U) and surface tension effects are small (small σ). Hence we would expect the assumption that region 3 can be described as a *static* meniscus to be very poor. One may also look at high Ca as corresponding to high-speed coating (since Ca $\sim U$), and this raises the question of the importance of inertial effects on the coating dynamics. In this regard one should take note of a numerical solution of the two-dimensional dynamic equations for this flow, *including* inertial terms, carried out by Esmail and Hummel. They find that T_∞ is no longer a unique function of Ca but that a parameter γ enters the problem, where

$$\gamma = \sigma \left(\frac{\rho}{\mu^4 g} \right)^{1/3} \qquad (8\text{-}129)$$

Since γ arises because of inclusion of the inertial terms, then in some sense it must be interpretable as an inertial parameter.

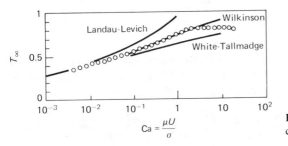

$$\text{Ca} = \frac{\mu U}{\sigma}$$

Figure 8-22 Data of Spiers et al. (1974) compared with several coating models.

Normally we think of a Reynolds number as an appropriate inertial parameter and define it as

$$\text{Re} = \frac{U a \rho}{\mu} \tag{8-130}$$

where a is some appropriate length scale for the flow. Of course H_∞ would be a natural choice for a, but since H_∞ is unknown a priori we prefer to use a different measure. We choose the *capillary length*

$$a \equiv \left(\frac{\sigma}{\rho g} \right)^{1/2} \tag{8-131}$$

which provides a length scale appropriate to surface tension problems. Then a Reynolds number may be defined as

$$\text{Re} = \frac{U}{\mu} \left(\frac{\sigma \rho}{g} \right)^{1/2} \tag{8-132}$$

Now it is possible to show the interrelationship among Re, Ca, and γ, because

$$\text{Re} = \text{Ca } \gamma^{3/2} \tag{8-133}$$

Thus, at fixed γ, large Ca leads to high Reynolds numbers and inertial effects, whereas at fixed Ca, large γ corresponds to high Reynolds numbers.

Figure 8-23 shows the theory of Esmail and Hummel, including a comparison with high Ca data of Soroka and Tallmadge. It would appear that the two-dimensional theory with included inertial terms is essential to the accurate description of high Ca data.

Several attempts at purely viscous nonnewtonian models have been made. Agreement with experiment has been poor, and there is evidence of significant viscoelastic effects. For these reasons we will not discuss nonnewtonian modeling of this particular problem, but the Bibliography at the end of this chapter notes several references to recent papers.

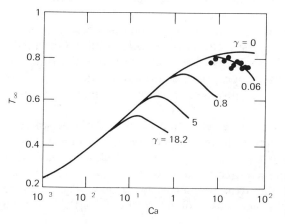

Figure 8-23 Two-dimensional model of Esmail and Hummel. Data of Soroka and Tallmadge correspond to $\gamma = 0.06$.

We turn, instead, to comment on the wire-coating problem corresponding to continuous withdrawal of a cylinder from a free surface. Except for the change in geometry of the moving solid surface, the general approach to this problem follows that for the flat film. We anticipate immediately that a new dimensionless group will enter the problem, since we would expect the cylinder radius to be of importance. The usual choice is in terms of the *Goucher number*,

$$\text{Go} = \frac{R}{(2\sigma/\rho g)^{1/2}} \tag{8-134}$$

At the time of writing, the best available theory is the gravity-corrected theory of White and Tallmadge (1966, 1967), shown plotted in Fig. 8-24. Instead of T_∞ [Eq. (8-102)] we use another dimensionless coating thickness D_∞, defined by

$$D_\infty = H_\infty \left(\frac{\rho g}{\sigma}\right)^{1/2} = T_\infty \, \text{Ca}^{1/2} \tag{8-135}$$

We find D_∞ somewhat more convenient to work with because it does not include the viscosity of the fluid. Thus, in dealing with nonnewtonian fluids we do not have to modify the definition of D_∞. (The capillary number, of course, would have to be redefined for nonnewtonian fluids.)

The gravity-corrected theory is limited to capillary numbers below unity. With most viscous materials we would expect to encounter significantly higher capillary numbers. Shown on Fig. 8-24 are data obtained with several newtonian and viscoelastic fluids, mostly at large capillary numbers. The newtonian data (which cover a viscosity range of 0.5 to 10 P) correlate reasonably well and extrapolate smoothly from the theory. The data show the tendency (noted in the flat sheet data of Fig. 8-22) to level off at high Ca.

The interesting observation is with respect to the viscoelastic polymer solutions studied. The polyacrylamide solutions, which are quite elastic, depart markedly from the newtonian theory and show an asymptotic value of coating thickness that is an order of magnitude below that of the newtonian fluids studied. Clearly the coating dynamics of viscoelastic fluids are much more complex than that allowed for by the simple models offered to date.

Rheological data are available for the polymer solutions studied, but a straightforward assessment of viscoelastic effects is difficult to offer. One might like, for example, to calculate a Weissenberg or Deborah number for this flow, but we have no simple estimate for an appropriate deformation rate or "process time." Let us make an estimate, anyway, just to see what sort of numbers we might be talking about.

The polyacrylamide solutions all have relaxation times of the order of 10 s, whereas the polyethylene oxide solutions have values of the order of unity. Let us estimate a Deborah number (refer back to Chap. 3) in the following way. As a process time we calculate the time required for fluid to accelerate from the relatively quiescent bath through the dynamic meniscus region. The distance traversed, roughly the height of the meniscus, is of the order of $(\sigma/\rho g)^{1/2}$. The time

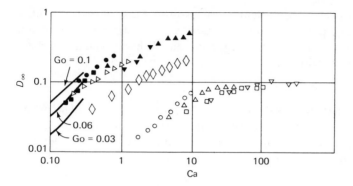

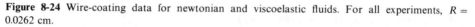

Figure 8-24 Wire-coating data for newtonian and viscoelastic fluids. For all experiments, $R = 0.0262$ cm.

●	Paraffin oil	$\mu = 0.46$ P	Go = 0.1
▼	Glycerol	10	0.08
▲	Glycerol	8.6	0.08
■	80% glycerol	0.61	0.08
○	0.10% polyacrylamide	$\eta_0 = 10$ P	0.07
△	0.25% polyacrylamide	48	0.07
□	0.50% polyacrylamide	160	0.07
▽	0.75% polyacrylamide	350	0.07
▷	0.50% polyethylene oxide	1.1	0.07
◇	0.75% polyethylene oxide	6.6	0.07

Solid lines show the gravity-corrected theory of White and Tallmadge (1966, 1967) for three Goucher numbers. For the viscoelastic fluids the zero shear viscosity is used in the capillary number.

to move this distance would be in the neighborhood of $(\sigma/\rho g)^{1/2}/U$. We define a Deborah number as

$$\text{De} = \frac{\lambda U}{(\sigma/\rho g)^{1/2}} \tag{8-136}$$

For the data of Fig. 8-24 typical values of U were in the range 10 to 50 cm/s, and σ was roughly 60 dynes/cm for all fluids. Thus we find De = 400 to 2000 for the polyacrylamides, and 40 to 200 for the polyethylene oxides. As crude as this estimate is, it does suggest that we are looking at a large Deborah number flow (at least for the data shown). The data are consistent with this suggestion in several respects:

● The 0.5% polyethylene oxide solution has small but measurable normal stresses, and Deborah numbers much smaller than those for the polycrylamide solutions. It shows coating behavior intermediate between the newtonian fluids and the very elastic polyacrylamide fluids.

● The Deborah numbers must be quite high for the polyacrylamide solutions. High Deborah number flows are dominated by elastic effects; i.e., the fluid responds more like an elastic solid than a viscous fluid. This suggests that little viscous entrainment of liquid occurs, and indeed the elastic response would tend to draw entrained fluid back to the meniscus. While speculative, these comments are consistent with the observed reduction in coating thickness of an order of magnitude.

Thus we conclude that this particular type of coating process, when carried out with polymeric fluids, may well be dominated by strong elastic effects that are outside the scope of the models presented earlier. The development of an adequate viscoelastic theory is left as a 3-year exercise for the reader.

8-4 SOME COMMENTS ON THE LUBRICATION APPROXIMATIONS

The lubrication approximations have played a central role in analytical solutions to problems in calendering (Chap. 7) and coating. In some problems we have no alternative if we seek an analytical solution, and the use of the lubrication approximations is justified on the grounds that we are unwilling to attempt a numerical solution to the problem at hand. It is possible, however, to offer some evaluation of the errors associated with these approximations, and to develop thereby a sense of when we might hope to take advantage of these simplifications without throwing away too much accuracy.

Let us consider the problem of pressure-driven flow through a two-dimensional duct with *nonparallel* walls, as shown in Fig. 8-25. If the usual lubrication approximations are imposed we solve

$$\frac{dp}{dx} = \mu \frac{\partial^2 u}{\partial y^2} \tag{8-137}$$

subject to

$$u(x, \pm h) = 0$$

$$\frac{\partial u}{\partial y} = 0 \qquad \text{at } y = 0$$

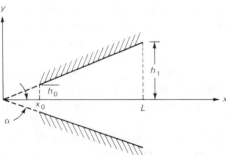

Figure 8-25 Definition sketch for model of pressure flow between nonparallel planes.

We take the pressure to be $p = \Delta P$ at $x = L$, and $p = 0$ at $x = x_0$. We consider here the case of converging flow.

The boundaries are described by

$$h = h_0 + (x - x_0) \tan \alpha \tag{8-138}$$

The solution of this problem may be easily found in the form (see Prob. 8-27)

$$\Delta P = \frac{3\mu Q}{4h_0^2 \tan \alpha} \frac{\kappa^2 - 1}{\kappa^2} \tag{8-139}$$

where

$$\kappa = \frac{h_1}{h_0} \tag{8-140}$$

An analytical solution to this problem may be found without recourse to the lubrication approximations. We set up a polar coordinate system where the velocity field is *assumed* to be strictly radial:

$$\mathbf{u} = [u(r, \theta), 0, 0] \tag{8-141}$$

The continuity equation, in polar coordinates, then gives

$$u = \frac{f(\theta)}{r} \tag{8-142}$$

The radial and angular components of the dynamic equations for a newtonian fluid become (neglecting inertia)

$$\frac{\partial p}{\partial r} = \frac{\mu}{r^2} \frac{\partial^2 u}{\partial \theta^2} = \frac{\mu f''}{r^3} \tag{8-143}$$

$$\frac{1}{r} \frac{\partial p}{\partial \theta} = \frac{2\mu}{r^2} \frac{\partial u}{\partial \theta} = \frac{2\mu}{r^3} f' \tag{8-144}$$

where $' = d/d\theta$.

It is not difficult to find the solutions to be (Prob. 8-28)

$$u(r, \theta) = -\frac{Q}{2\alpha - \tan 2\alpha} \frac{1}{r} \left(1 - \frac{\cos 2\theta}{\cos 2\alpha} \right) \tag{8-145}$$

$$p(r, \theta) = \frac{2\mu Q}{(2\alpha - \tan 2\alpha) \cos 2\alpha} \left[\frac{\cos 2\theta}{r^2} - \left(\frac{\tan \alpha}{h_0} \right)^2 \right] \tag{8-146}$$

If we take

$$p = \Delta P \qquad \text{at } r = x_0 + L = \frac{h_1}{\tan \alpha}$$

and

$$\theta = 0$$

we may eventually find

$$\Delta P = \frac{2\mu Q}{h_0^2 \cos 2\alpha (2\alpha - \tan 2\alpha)} \frac{1 - \kappa^2}{\kappa^2} \tag{8-147}$$

We may evaluate the error in the lubrication approximation by considering the ratio of flow rates at fixed ΔP, as given by Eqs. (8-139) and (8-147):

$$\epsilon = \frac{8 \tan^3 \alpha}{3 \cos 2\alpha (\tan 2\alpha - 2\alpha)} \tag{8-148}$$

We find that the error does not exceed 10 percent until the *half-angle* α exceeds 15°. Crudely speaking, then, we can conclude that the lubrication approximations are quite reasonable even when the nonparallelism is about 30°.

PROBLEMS

8-1 Carry out an analysis, parallel to that of Sec. 8-1, for a roll-coating system operating as shown in Fig. 8-26. Assume the pressure at the separation point is zero, that is, $N_1 = 0$.

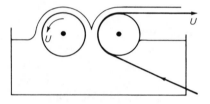

Figure 8-26 Two-roll coating system.

8-2 Analyze the reverse-roll coater, as shown in Fig. 8-27. Assume H_1 is fixed by a blade, as shown. Assume that the pressure at the separation point is zero.

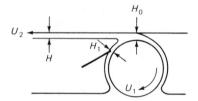

Figure 8-27 Reverse-roll coating.

8-3 A roll coater operates as suggested in Sec. 8-1. Find the coating thickness for the following conditions:

$$R = 6 \text{ in} \qquad H_0 = 10 \text{ mil} \qquad U = 100 \text{ ft/min}$$

The fluid is newtonian, with $\mu = 1$ P and surface tension $= 60$ dynes/cm.

8-4 The fluid whose viscosity is given in Prob. 6-9 is to be coated onto a sheet moving at 200 ft/min by using a roll coater for which $R = 6$ in. The desired coating thickness is 5 mil. Find H_0 and the maximum pressure (pounds per square inch).

8-5 In a sheet/roll-coating system such as discussed and modeled in Sec. 8-1, find the shear stress acting on the sheet and the tensile force required to pull it through the system.

Answer: Tensile force $= 5.4\mu \dfrac{U}{H_0} \sqrt{RH_0}$ per unit width.

8-6 A film 4 mil thick and 2 in wide is roll coated with a newtonian fluid under the conditions listed below. Find the tensile stress in the *film*.

$$R = 4 \text{ in} \qquad H_0 = 0.01 \text{ in} \qquad U = 6 \text{ ft/s} \qquad \mu = 2000 \text{ P}$$

8-7 Consider the observation of Fig. 8-11 that $|P(\xi, n)| > |P(\xi, 1)|$. Does this necessarily imply that $|p(\xi, n)| > |p(\xi, 1)|$?

(a) Show that if the comparison is on the basis of equal apparent viscosities evaluated at a shear rate U/H_0, the answer is yes.

(b) Is the same answer given if the basis is equal apparent viscosities evaluated at a shear rate $2U/H_0$?

8-8 For the case of newtonian roll coating give an analytical expression for $\partial u_x/\partial y$. Is U/H_0 a reasonable value to use as a nominal shear rate?

8-9 The data shown in Fig. 8-28 were extracted from Pitts and Greiller and from Hintermaier and

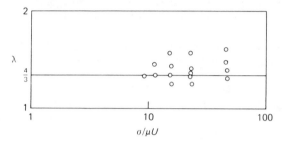

Figure 8-28 Data obtained in a two-roll coating system. (*From Pitts and Greiller and from Hintermaier and White.*)

White. The experiments were performed in a two-roll system which was similar to the design suggested in Prob. 8-1 but without the sheet. The analysis of Prob. 8-1 should be valid. Compare the data with the theory.

8-10 For Example 8-2 calculate an appropriate measure of the nominal shear rate under the blade. Define a Reynolds number for this flow, and give its numerical value. Do you think inertial effects are negligible for this example?

8-11 Derive Eqs. (8-84) and (8-85).

8-12 Plot λ_1 versus κ from Eq. (8-85).

8-13 If Ws = 0.1, find λ/λ_0 according to the first-order theory, using Eqs. (8-79) and (8-85).

8-14 Find limiting values of λ_1 for $\kappa \to 1$ and $\kappa \to \infty$.

8-15 One of the data points in Fig. 8-18 corresponds to the following conditions:

$$\kappa = 3 \qquad L = 8 \text{ cm} \qquad H_0 = 0.15 \text{ cm}$$
$$\mu = 9.3 \text{ P} \qquad U = 9.6 \text{ cm/s}$$

(a) Evaluate inertial effects by calculating a Reynolds number for this flow.

(b) Using Eqs. (8-77) and (8-81), evaluate the assumption that $u_y \ll u_x$.

(c) Assuming adiabatic conditions, estimate the temperature rise of the fluid passing under the blade.

8-16 A photographic film is withdrawn from a coating bath at a speed of 15 in/s. The coating fluid is newtonian, with a viscosity of 25 cP, a density of 60 lb/ft³, and a surface tension of 35 dynes/cm. Find the coating thickness H_∞ and the entrainment rate q. Give the latter in units of pounds per hour per inch width of film.

8-17 Show that at the surface in a film-coating system, as in Fig. 8-20,

$$\boldsymbol{\tau} \cdot \mathbf{s} = \tfrac{1}{2}(\tau_{xx} - \tau_{yy}) \sin 2\theta + \tau_{xy} \cos 2\theta$$
$$\boldsymbol{\tau} \cdot \mathbf{n} = \tau_{xx} \sin^2 \theta + \tau_{yy} \cos^2 \theta + \tau_{xy} \sin 2\theta$$

where θ is the angle between $\mathbf{n}$ and the y axis.

8-18 A parameter with units of length that is often useful in considering problems of free surfaces subject to surface tension is $(\sigma/\rho g)^{1/2}$. What form does Eq. (8-98) take if

$$L = \frac{H}{(\sigma/\rho g)^{1/2}} \qquad \xi = \frac{x}{(\sigma/\rho g)^{1/2}}$$

and

$$D_\infty = \frac{H_\infty}{(\sigma/\rho g)^{1/2}}$$

8-19 From Eq. (8-109) solve explicitly for dH/dx and plot dH/dx versus ξ (as defined in Prob. 8-18).

8-20 Solve for $L(\xi)$ from the result of Prob. 8-19 as an indefinite integral. Give a numerical solution. Use L and ξ as defined in Prob. 8-18.

8-21 Integrate Eq. (8-121) numerically, and find the value of S_0. Compare your S_0 with the value given by Levich, obtainable by inspection of Eq. (8-123).

8-22 Show that if the Landau-Levich solution is known, the gravity-corrected theory [Eq. (8-126)] follows by inspection, with no need for further computation. [*Hint:* Define a new variable $\chi = (1 - T_\infty^2)^{1/3}\xi$.]

8-23 Consider the problem of sensitivity of coating thickness in a withdrawal-coating system (for a flat web) to variations in web speed. To maintain ± 1 percent tolerance in H_∞, what is the allowable variation in U?

8-24 For withdrawal coating onto a flat web, give the sensitivity of H_∞ to variations in σ and μ.

8-25 White and Tallmadge (1966, 1967) give the coating thickness, for Go $\ll 1$ (the small-wire theory) in the form

$$T_\infty = \frac{1.33 \text{ Ca}^{2/3}}{1 - 1.33 \text{ Ca}^{2/3}}$$

A fiber of diameter 0.018 in is withdrawn from a bath of "finish" solution at a speed of 10 ft/s. The finish has a viscosity of 2 cP and a surface tension of 30 dynes/cm. Find the coating thickness using the small-wire theory, and compare it to the value obtained using Fig. 8-23.

8-26 For large cylinders (Go $\gg 1$) the flat-film theory may be used to describe coating by withdrawal. Find the coating thickness and the volumetric entrainment rate for a $\frac{1}{16}$-in wire withdrawn at 1 ft/s from a 1-P fluid with $\sigma = 60$ dynes/cm. Rework for the case $U = 10$ ft/s.

8-27 Derive Eq. (8-139).

8-28 Derives Eqs. (8-145) to (8-147).

8-29 Plot ϵ versus α from Eq. (8-148).

8-30 Derive the analog of Eq. (8-148) for a power law fluid. Is the lubrication approximation more, or less, accurate for nonnewtonian fluids?

8-31 The data tabulated below were obtained in a roll-coating system which operates in the manner described in Sec. 8-1.

(a) Compare the data to the theory.

(b) It was observed that the fluid which was fed to the roll sometimes dammed up behind the roll to a height of the order of the roll radius. Could the hydrostatic pressure due to this "dam" exert a significant influence on the coating thickness?

For all data, $R = 3.64$ cm, $U = 9.3$ cm/s.

Fluid	μ, P	σ, dynes/cm	H_0, mil	H, mil
A	10	64	56	41 ± 3
B	5	65	55	39 ± 3
C	30	75	94	64 ± 3
D	10	64	94	91 ± 8
E	1.6	66	56	70 ± 8

8-32 Rework the models given in Secs. 8-1 and 8-2 for roll coating and blade coating, and account for the effect on coating thickness of a finite pressure at the entrance to the coater ($x = -\infty$ for roll coating, and $x = 0$ for blade coating).

BIBLIOGRAPHY

8-1 Roll coating

The simplest model of flow between two rolls is given in

Hopkins, M. R.: *Br. J. Appl. Phys.*, **8**: 442 (1957).

Experiments are described in

Pitts, E., and J. Greiller: The Flow of Thin Liquid Films between Rollers, *J. Fluid Mech.*, **11**:33 (1961).

This paper establishes the presence of an instability which superimposes uniform waves on the coating. The instability sets in when $\mu U R/\sigma H_0 > 60$. A theory of this, and related instabilities is given in

Pearson, J. R. A.: The Instability of Uniform Viscous Flow under Rollers and Spreaders, *J. Fluid Mech.*, **7**: 481 (1960).

Other data are given in

Hintermaier, J. C., and R. E. White: The Splitting of a Water Film between Rotating Rolls, *20th Eng. Conf.—TAPPI*, 123 (1966).

A theory of coating from a roll to a sheet is given in

Greener, J., and S. Middleman: A Theory of Roll Coating of Viscous and Viscoelastic Fluids, *Polym. Eng. Sci.*, **15**: 1 (1975).

An error in Eq. (35) and fig. 10 of that paper are corrected here by Eq. (8-22) and Fig. 8-11.

8-2 Blade coating

Some aspects of blade coating are described in Pearson above. The analyses given here follow those in

Greener, J., and S. Middleman: Blade Coating of a Viscoelastic Fluid, *Polym. Eng. Sci.*, **14**: 791 (1974).

8-3 Free coating

A general review of a variety of free-coating processes, with an extensive bibliography, is in

Tallmadge, J. A., and C. Gutfinger: Entrainment of Liquid Films, *Ind. Eng. Chem.*, **59**: 19 (1967).

The Landau-Levich approach is presented in

Levich, V. G.: "Physicochemical Hydrodynamics," Prentice-Hall, Inc., Englewood Cliffs, N.J., 1962.

The gravity-corrected theory is found in

White, D. A., and J. A. Tallmadge: Theory of Drag Out of Liquids on Flat Plates, *Chem. Eng. Sci.*, **20**: 33 (1965).

Wilkinson's work is found in

Spiers, R. P., C. V. Subbaraman, and W. L. Wilkinson: Free Coating of a Newtonian Liquid onto a Vertical Surface, *Chem. Eng. Sci.*, **29**: 389 (1974).

The corresponding analysis for the nonnewtonian fluid is given by the same authors in

———, ———, and ———: Free Coating of Non-Newtonian Liquids onto a Vertical Surface, *Chem. Eng. Sci.*, **30**: 379 (1975).

An exchange of views regarding the validity and consistency of the Wilkinson approach is to be found in

Esmail, M. N., and R. L. Hummel: A Note on Linear Solutions to Free Coating onto a Vertical Surface, *Chem. Eng. Sci.*, **30**: 1195 (1975).

The numerical solution including inertial effects appears as

——— and ———: Nonlinear Theory of Free Coating onto a Vertical Surface, *AIChE J.*, **21**: 958 (1975).

This theory is tested against some high-capillary-number data found in

Soroka, A. J., and J. A. Tallmadge: A Test of the Inertial Theory for Plate Withdrawal, *AIChE J.*, **17**: 505 (1971).

Other papers of interest are

Gutfinger, C., and J. A. Tallmadge: Films of Non-Newtonian Fluids Adhering to Flat Plates, *AIChE J.*, **11**: 403 (1965).
Groenveld, P., and R. A. Van Dortmond: The Shape of the Air Interface During the Formation of Viscous Liquid Films by Withdrawal, *Chem. Eng. Sci.*, **25**: 1571 (1970).

Theory and experiments for wire coating are found in

White, D. A., and J. A. Tallmadge: A Theory of Withdrawal of Cylinders from Liquid Baths, *AIChE J.*, **12**: 333 (1966).

and, by the same authors,

——— and ———: A Gravity Corrected Theory for Cylinder Withdrawal, *AIChE J.*, **13**: 745 (1967).

NINE

FIBER SPINNING

An expert is a person who avoids the small errors as he sweeps on to the grand fallacy.

Stolberg

Fiber spinning is a process in which fluid is continuously extruded through an orifice to form an extrudate of, usually, circular cross section. Somewhere downstream of the orifice the extrudate is contacted in such a way that the filament can be pulled and conveyed to further processing steps. Figure 9-1 shows the basic features of a spinning system.

Between the spinneret and the first take-up roll, various events occur which transform the extruded liquid into a fiber. These events may be physical and/or chemical. For example, molten nylon may be extruded and cooled (quenched) before the take-up point to form a filament. If the linear speed at the take-up point exceeds the speed of the filament at the extrusion point, the filament is said to be *drawn*. If the speeds are different, then conservation of mass requires a change in cross-sectional area. Thus a drawn fiber is of smaller diameter than the orifice from which it is extruded. A further consequence of drawing, in many polymers, is the development of morphological features which may depend on the extent and rate of drawing and which may significantly alter the mechanical properties of the fiber. Events occurring in the drawing region are generally the most significant in determining the ultimate properties of the fiber.

If fiber is spun from a polymeric *solution*, then it will be necessary to remove solvent from the fluid before a mechanically coherent fiber structure is attained. This mass transfer process occurs in the drawing zone and interacts with the drawing process in a most complex manner. Some solutions do not form fibers

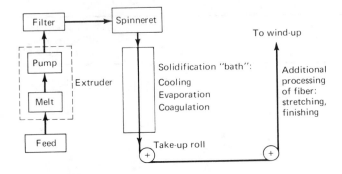

Figure 9-1 Schematic of elements of a fiber-spinning system.

simply by removal of solvent. Often a coagulation step is required which involves contact of the extruded filament with a coagulating bath. Again a very complex mass transfer process occurs in the postextrusion region.

Determination that a fluid is potentially fiber-forming is a necessary, but not sufficient, condition for development of a spinning process. It is often the case that the rheological properties of the extrudate in the postextrusion region are such that a coherent filament cannot be drawn into the quenching or coagulating region. Usually there is an upper limit to the speed of extrusion or a lower limit to the length of the drawing region beyond which the filament "breaks." The word *break* is used loosely here: The liquid stream may be unstable and simply "break" up into a stream of droplets. A principal problem, as yet incompletely solved, is the establishment of criteria which define conditions under which a fiber may be spun.

Fiber spinning is an odd process to analyze in the sense that *if* a stable process is possible, its mechanical description is relatively simple. The principal modeling effort is devoted to the question of stability of the process. In this section we will consider some problems of stable isothermal melt spinning. In Chap. 13 the problems in which heat and mass transfer interact with the spinning process will be considered. Stability is discussed in Chap. 15.

We will define *melt spinning* as that where the extrudate is a molten polymer or other liquid which does not exchange mass with its surroundings. By contrast, *wet spinning* will be that where the extruded filament enters a liquid bath with which it exchanges material, as in a coagulation process. *Solution dry spinning* will refer to the case where solvent must be removed by evaporation to a surrounding gas phase.

9-1 ISOTHERMAL MELT SPINNING—NEWTONIAN FLUID

Figure 9-2 shows the extrudate in the postextrusion region. A typical observation is the *die swell* just after extrusion. Die swell is associated with the relaxation of (elastic) normal stresses developed within the fluid by its deformation history

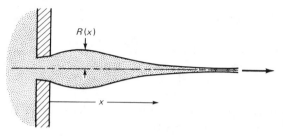

Figure 9-2 Filament in the postextrusion region.

prior to extrusion. The dependence of die swell on rheological properties and on the die inlet geometry is quite complex, and no precise methods exist for accurate a priori prediction of the degree of die swell to be expected under given conditions. This topic is discussed in more detail in Chap. 14, where some methods of estimating die swell are described.

It should suffice here for the reader to recognize that as a consequence of die swell the diameter of the filament near the spinneret is not accurately predictable. The situation is further complicated by the fact that drawing of the filament affects the degree of swelling. Experience suggests that this region of uncertainty lies within a few spinneret diameters of the exit, and that over most of the spinning length the fluid responds to postextrusion, rather than preextrusion, conditions. Analyses of fiber spinning usually take as the initial filament radius the maximum *observed* value and assume that this value is attained right at the spinneret exit.

The simplest analysis neglects any interaction between the filament and the surrounding medium. Thus it is assumed that the filament is isothermal and that no shear or normal stresses act on the filament boundary. Figure 9-3 shows the filament boundary, defined by the local radius $R(x)$ and the unit outward normal vector **n**. The velocity vector is **u**, and the total stress tensor is T.

At the free surface no fluid crosses the boundary, and we may express this in vector form as

$$\mathbf{u} \cdot \mathbf{n} = 0 \qquad \text{at } r = R(x) \tag{9-1}$$

The normal vector **n** has components which are easily found to be

$$n_x = -R'(1 + R'^2)^{-1/2} \tag{9-2}$$

$$n_r = (1 + R'^2)^{-1/2} \tag{9-3}$$

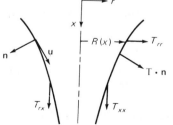

Figure 9-3 Definition sketch for analysis of melt spinning.

where $R' = dR/dx$. At the free surface, then,

$$u_r n_r + u_x n_x = 0$$

or
$$u_r = R'u_x \qquad \text{at } r = R(x) \tag{9-4}$$

The stress boundary condition is slightly more complicated. The stress *vector* normal to the free boundary is $\mathbf{T} \cdot \mathbf{n}$ with components

$$(\mathbf{T} \cdot \mathbf{n})_x = T_{rx} n_r + T_{xx} n_x \tag{9-5}$$

$$(\mathbf{T} \cdot \mathbf{n})_r = T_{rr} n_r + T_{rx} n_x \tag{9-6}$$

If the effect of the ambient fluid is neglected, and if surface tension is ignored, then there is no mechanism for imposing a finite stress *on* the free boundary, and the appropriate stress boundary condition is

$$0 = T_{rx} n_r + T_{xx} n_x \tag{9-7}$$

$$0 = T_{rr} n_r + T_{rx} n_x \tag{9-8}$$

Since the filament of fluid is being drawn in the axial direction, we have reason to expect the existence of a finite axial stress T_{xx}. According to Eq. (9-7), then, there is also a finite shear stress at the free boundary, given by

$$T_{rx} = -\frac{n_x}{n_r} T_{xx} = R'T_{xx} \tag{9-9}$$

This result seems surprising at first, for it states that there is a finite shear stress at the free boundary, even though it appears that above we claimed there is no mechanism for imposing a finite stress at the boundary. The resolution of this point is through a clearer understanding of the *geometry* of the free boundary. The vector $\mathbf{T} \cdot \mathbf{n}$ is normal to the free surface. Referring back to Fig. 9-3 we see that the component T_{rr} is *not* normal to the free surface, and T_{rx} is *not* in the plane of the free surface. Thus, while T_{rx} is a *shear* stress, it is not a shear stress in the *free* surface; it is a shear stress in the cylindrical *coordinate* surface. For a free surface which *is* a cylindrical coordinate surface, i.e., if $R' \equiv 0$, the expected result $T_{rx} = 0$ is obtained.

We are now in position to examine the momentum equations for the filament. In the axial direction we have

$$\rho\left(u_r \frac{\partial u_x}{\partial r} + u_x \frac{\partial u_x}{\partial x}\right) = \frac{1}{r}\frac{\partial}{\partial r}(rT_{rx}) + \frac{\partial T_{xx}}{\partial x} \tag{9-10}$$

We can show a posteriori that a good approximation is $u_x = u_x(x)$, and with this assertion we may integrate each term of Eq. (9-10) across the filament cross section:

$$\int_0^R \rho u_r \frac{\partial u_x}{\partial r} r \, dr = 0 \qquad \text{by assertion that } u_x \neq u_x(r)$$

$$\int_0^R \rho u_x \frac{\partial u_x}{\partial x} r \, dr = \tfrac{1}{2}\rho u_x u_x' R^2 \qquad \text{where } u_x' = \frac{du_x}{dx}$$

$$\int_0^R \frac{1}{r}\left[\frac{\partial}{\partial r}(rT_{rx})\right] r \, dr = RT_{rx}\Big|_R = RR'T_{xx} \qquad \text{using Eq. (9-9)}$$

$$\int_0^R \frac{\partial T_{xx}}{\partial x} r \, dr = \frac{d}{dx}\int_0^R T_{xx} r \, dr - T_{xx}RR' = \tfrac{1}{2}T_{xx}' R^2$$

Thus the momentum equation may be written as

$$\rho u_x u_x' = 2\frac{R'}{R} T_{xx} + T_{xx}' \tag{9-11}$$

This equation provides the starting point for subsequent analyses.

In addition to the momentum balance we must also write a mass balance. For an incompressible fluid the mass balance simply states that the volumetric flow rate across any section of the filament normal to the axis is constant:

$$\pi R^2(x)u_x(x) = Q_0 \tag{9-12}$$

To proceed it is necessary to introduce a constitutive equation. We begin with a newtonian fluid, for which

$$T_{xx} = -p + 2\mu \frac{du_x}{dx} \tag{9-13}$$

This introduces the isotropic pressure p into the problem and necessitates a more detailed look at the dynamic equations. An alternative is to introduce an approximation based on the notion that $R' \ll 1$, as is typical in real spinning systems beyond the maximum in the die swell region.

From Eqs. (9-7) and (9-8) it follows that

$$T_{rr} = R'^2 T_{xx} \qquad \text{at } r = R(x) \tag{9-14}$$

Since there is no significant radial flow we might expect no strong radial variation of stress. Hence Eq. (9-14) can be expected to hold over the filament radius, and if R' is small, a reasonable approximation would be

$$T_{rr} = 0 \tag{9-15}$$

Further, since no angular flow exists, we can expect that

$$T_{\theta\theta} = 0 \tag{9-16}$$

For a newtonian fluid we can easily show that the isotropic pressure is the mean normal stress (see Prob. 9-2):

$$p = -\tfrac{1}{3}(T_{xx} + T_{rr} + T_{\theta\theta}) \tag{9-17}$$

It follows then that

$$T_{xx} = 3\mu \frac{du_x}{dx} \tag{9-18}$$

Equation (9-18), along with Eq. (9-12), may be used to convert Eq. (9-11) to an equation containing only u_x as an unknown:

$$(u_x^2)' = \frac{6\mu}{\rho} u_x \left(\frac{u_x'}{u_x}\right)' \tag{9-19}$$

The solution of this equation is

$$u_x = C_1 \left(C_2 e^{-C_1 x} - \frac{\rho}{3\mu}\right)^{-1} \tag{9-20}$$

If the inertial term $[(u_x^2)'$ in Eq. (9-19)] is neglected the solution is found to be

$$u_x = C_3 e^{C_4 x} \tag{9-21}$$

The constants C_i appearing in either solution must be established through boundary conditions. We note that Eq. (9-20) suggests that a newtonian fluid is "spinnable" only over a finite length. The length increases as ρ/μ becomes small.

For subsequent discussions we will work with Eq. (9-21). We must first deal with an appropriate set of boundary conditions. Since die swell is not accounted for in the model, we will have to define an arbitrary origin $x = 0$ as the point where the maximum $R(x)$ occurs. In most cases the distance from the die exit to the point of attainment of maximum $R(x)$ is quite small in comparison to the total spinning length L between the origin and the take-up point. One simple boundary condition, then, is

$$u_x = U_0 \qquad \text{at } x = 0 \tag{9-22}$$

where U_0 is related to the maximum radius R_0 by

$$\pi R_0^2 U_0 = Q_0 \tag{9-23}$$

The second boundary condition is usually specified in terms of the velocity at take-up:

$$u_x = U_L \qquad \text{at } x = L \tag{9-24}$$

It is convenient to introduce as a parameter the *drawdown ratio*

$$D_R = \frac{U_L}{U_0} \tag{9-25}$$

With the boundary conditions above the solutions for u_x and $R(x)$ become

$$u_x = U_0 \exp \frac{x \ln D_R}{L} \qquad (9\text{-}26)$$

$$R(x) = R_0 \exp \left(-\frac{1}{2} \frac{x \ln D_R}{L} \right) \qquad (9\text{-}27)$$

With these results it is possible to calculate various features which characterize a spinning system, assuming the fluid is newtonian.

Example 9-1 A polyamide of viscosity 5000 P is extruded into air under isothermal conditions where it is drawn in such a way that $D_R = 100$ and $L = 300$ cm. The velocity at the take-up point is $U_L = 10^5$ cm/min. The radius at take-up is $R_L = 10^{-3}$ cm.
(a) Calculate the maximum stretching rate imposed on the melt.
(b) Calculate the maximum tensile stress in the melt.
(c) What force is required to draw the melt?
(d) Estimate the relative importance of inertial effects.
From Eq. (9-26) we may find the maximum stretching rate is

$$u_x' \bigg|_{x=L} = \frac{U_0}{L} D_R \ln D_R = \frac{U_L}{L} \ln D_R$$

For the parameters specified, we find

$$u_x' \bigg|_{x=L} = 25.6 \text{ s}^{-1}$$

The maximum tensile stress is

$$T_{xx} \bigg|_{x=L} = 3\mu u_x' \bigg|_{x=L} = 1.152 \times 10^6 \text{ dynes/cm}^2$$

The force at take-up is

$$F = \pi R_L^2 T_{xx} \bigg|_{x=L} = 3.62 \text{ dynes}$$

The relative importance of inertial effects may be estimated with a parameter discussed in Prob. 9-5, which is found to be

$$\frac{\rho U_L L}{3\mu \ln D_R} = 7.23$$

Since this is greater than unity the calculation suggests that inertial effects are of importance. However, the value given is the maximum, which occurs at $x = L$. At the midpoint of the spinning path, where $x/L = \frac{1}{2}$, we find (see Prob. 9-5) that this parameter is 10 times smaller, or just under unity. Thus inertial effects are unimportant only over the first half of the spinning path.

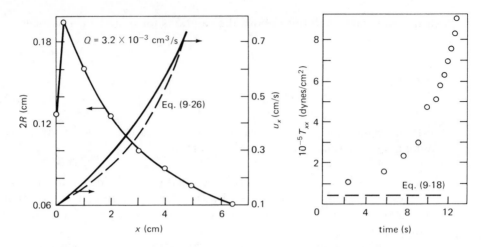

Figure 9-4 Data of Acierno et al. on isothermal spinning of polyethylene compared to new-tonian model.

Example 9-2 Acierno et al. present measurements of filament diameter and axial tensile stress along a polyethylene filament extruded at 160°C into an isothermal chamber. Figure 9-4 shows these measurements. The zero shear viscosity of the melt is 9×10^4 P, and the melt is practically newtonian in shear for shear rates up to 1 s^{-1}.

Evaluate the applicability of the newtonian model to the data.

The filament diameter measurements show clearly the die swell phenomenon mentioned earlier. The ratio of maximum to initial diameter is approximately 1.5. This degree of die swell indicates significant elasticity for this fluid, at least at the shear rates associated with the *die* flow (see Prob. 9-7).

From the measured $R(x)$ and the given Q, we can find the velocity $u_x(x)$ and graphically or numerically differentiate to find $u'_x(x)$. Equation (9-18) then gives T_{xx} along the filament. To avoid the problems of applying the simple model in the die swell region, we take the minimum velocity [at maximum $R(x)$] as U_0, and the value of x at which this occurs as $x = 0$. Figure 9-4 shows some computed results.

The theoretical velocity curve follows from Eq. (9-26) by picking U_0 and D_R so that the curve goes through the observed end points. It is apparent that the velocity is more nearly a linear function of x than is predicted by the newtonian theory.

The tensile stress model gives very poor results in comparison to the measurements. The tension is observed to rise to values much greater than can be explained on the basis of this newtonian model. Clearly the newtonian model is inadequate to explain these data.

9-2 ISOTHERMAL SPINNING OF A POWER LAW FLUID

We begin with the general definition of a power law fluid in the form

$$\mathbf{T} = -p + K(\tfrac{1}{2}\mathrm{II}_\Delta)^{(n-1)/2}\,\Delta \tag{9-28}$$

We can show (see Prob. 9-8) that the second invariant of this spinning flow is

$$\mathrm{II}_\Delta = 6(u'_x)^2 \tag{9-29}$$

from which it follows that

$$T_{xx} = -p + 2K(3)^{(n-1)/2}(u'_x)^n \tag{9-30}$$

On going back to the momentum balance [Eq. (9-11)] and ignoring inertial effects, one finds that (assuming again that $T_{rr} = T_{\theta\theta} = 0$)

$$\frac{u_x}{U_0} = \left[1 + (D_R^q - 1)\frac{x}{L}\right]^{1/q} \tag{9-31}$$

where $q = 1 - 1/n$. It is interesting to see what effect n has on the shape of the u_x versus x curves. Figure 9-5 shows this.

The range $0 < n < 1$ is that usually observed for *shear* viscosity data on polymeric fluids. Whether Eq. (9-28) is applicable to *elongational* flow, and if so, whether n lies only in the range $(0, 1)$, are two unanswered questions at the time of writing. Values of n greater than unity imply an *increasing* viscosity as deformation rate increases. Until reliable elongational viscosity data are available, it is not possible to decide if Eqs. (9-28) and (9-31) are useful.

Available isothermal spinning data are clearly at variance with Fig. 9-5, at least for $n < 1$. Figure 9-5 shows the data of Spearot and Metzner for spinning of a polyethylene melt. The velocity profile is nearly linear, and Eq. (9-31) does not allow for such behavior except in the case $n \gg 1$, for which there is no current experimental justification. This suggests that a purely viscous nonnewtonian model is inadequate to describe spinning dynamics, and we turn then to consideration of a viscoelastic model.

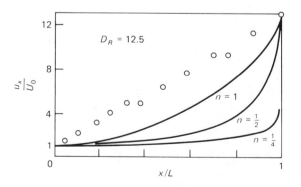

Figure 9-5 Filament velocity according to power law model compared to polyethylene data of Spearot and Metzner.

9-3 ISOTHERMAL VISCOELASTIC SPINNING

Since the elongation rate varies along the axis, spinning is a flow in which the kinematics are unsteady in a lagrangian sense. Hence it is necessary to raise the question as to the importance of viscoelastic phenomena in spinning. A measure of the unsteady nature of the flow is the time rate of change of elongation rate experienced by a "particle" moving with the fluid. This is just

$$\frac{Du'_x}{Dt} = u_x \frac{du'_x}{dx} = u_x u''_x \tag{9-32}$$

If we consider the *relative* rate of change of u'_x, and use the newtonian model as an approximation, we find

$$\frac{u_x u''_x}{u'_x} = \frac{U_0}{L} \tag{9-33}$$

Thus an appropriate Deborah number for this flow would be given by

$$De = \frac{U_0 \lambda}{L} \tag{9-34}$$

where λ is a fluid relaxation time. We should note here that the question has been raised as to whether the relaxation time of a fluid subject to elongation is the same as the (commonly measured) relaxation time in shear. No definitive answer has been given, and the uncertainty adds one more complication to the already difficult study of elongational flows.

Denn and coworkers have solved the problem of isothermal spinning of a fluid which is described by a constitutive equation of the form

$$\tau + \lambda \frac{\delta\tau}{\delta t} = \mu \, \Delta \tag{9-35}$$

where the contravariant Oldroyd *codeformational* time derivative is given, for this particular flow, by

$$\frac{\delta\tau_{xx}}{\delta t} = u_x \frac{d\tau_{xx}}{dx} - 2\tau_{xx} \frac{du_x}{dx} \tag{9-36}$$

$$\frac{\delta\tau_{rr}}{\delta t} = u_x \frac{d\tau_{rr}}{dx} + \tau_{rr} \frac{du_x}{dx} \tag{9-37}$$

in cylindrical coordinates.

The viscosity function μ will be taken to obey a power law, consistent with Eqs. (9-28) to (9-30). The relaxation time is taken to be proportional to the viscosity function, so that

$$\lambda = \frac{\mu}{G} \tag{9-38}$$

where G is a constant *elastic modulus*.

The momentum balance is Eq. (9-11), which, neglecting inertia, may be conveniently written in the format

$$0 = \frac{d}{dx}(R^2 T_{xx})$$ (9-39)

It follows that

$$\pi R^2 T_{xx} = \text{constant} = \frac{T_{xx}}{u_x}$$ (9-40)

the latter part of this equation being a consequence of conservation of mass: $\pi R^2 u_x = \text{constant}$.

It is convenient to introduce the force F exerted at the take-up point, which is given by

$$F = \pi R_L^2 T_{xx}\Big|_L$$ (9-41)

The volumetric flow rate may also be introduced as

$$Q = \pi R_L^2 u_L$$ (9-42)

The constant in Eq. (9-40) may then be evaluated, and the axial stress becomes

$$T_{xx} = \frac{F u_x}{Q}$$ (9-43)

If we again assume that T_{rr} is small [Eqs. (9-14) and (9-15)] then we may put Eq. (9-43) in the form given by Denn and coworkers:

$$T_{xx} - T_{rr} = \tau_{xx} - \tau_{rr} = \frac{F u_x}{Q}$$ (9-44)

Equations (9-35) [with (9-36) and (9-37)] and (9-44) involve the three unknown functions u_x, τ_{xx}, and τ_{rr}. The simplest formulation eliminates the stresses and yields a single equation for u_x, which Denn gives in the form

$$\frac{n\alpha\mu^2 u''}{(u')^{3-n}} + 2\alpha^2 u(u')^{2n-1} - (u')^{n-1}(\alpha u - 3\epsilon) - \frac{u}{u'} = 0$$ (9-45)

where the following dimensionless variables have been introduced:

$$u = \frac{u_x}{U_0} \qquad \alpha = \frac{K 3^{(n-1)/2}}{G}\left(\frac{U_0}{L}\right)^n \qquad \epsilon = \frac{3^{(n-1)/2} KQ}{FL}\left(\frac{U_0}{L}\right)^{n-1}$$

The prime denotes differentiation with respect to $\xi = x/L$. For finite α the equation is nonlinear and must be solved numerically. More to the point, however, is the fact that for finite α the equation is of one higher order than in the newtonian case. This necessitates introduction of an additional boundary condition.

Regardless of the specific constitutive equation used it still is reasonable to specify the kinematic boundary conditions used in the newtonian case [Eqs. (9-22) and (9-24)], which, in terms of dimensionless variables, take the form

$$u = \begin{cases} 1 & \text{at } \xi = 0 \\ D_R & \text{at } \xi = 1 \end{cases} \qquad\qquad\qquad (9\text{-}46) \\ (9\text{-}47)$$

The third boundary condition specifies the axial stress in the fluid at $x = 0$. It arises physically from the viscoelastic nature of the fluid, which has some degree of "memory" of the stress developed by the flow field just upstream of $x = 0$. We will specify

$$\tau_{xx} = \tau_0 \qquad \text{at } x = 0 \qquad\qquad\qquad (9\text{-}48)$$

or, in dimensionless terms,

$$\frac{\tau_{xx} Q}{U_0 F} \equiv T = T_0 \qquad \text{at } \xi = 0 \qquad\qquad\qquad (9\text{-}49)$$

From Eq. (9-44) we see that if τ_{xx} is known, then τ_{rr} may be found if $F u_x / Q$ is specified.

From a mathematical point of view we note, in the course of deriving Eq. (9-45), that we find (see Prob. 9-10)

$$T = \tfrac{2}{3} u - \frac{\epsilon}{\alpha} + \frac{u}{3\alpha (u')^n} \qquad\qquad\qquad (9\text{-}50)$$

Thus the specification of initial conditions on T and u amounts to specification of an initial condition on u':

$$u' = \left[3\alpha \left(T_0 - \frac{2}{3} + \frac{\epsilon}{\alpha} \right) \right]^{1/n} \qquad \text{at } \xi = 0 \qquad\qquad\qquad (9\text{-}51)$$

Since u' must be positive (we can only "pull" the fiber), we see that the initial stress must satisfy

$$T_0 > \frac{2}{3} - \frac{\epsilon}{\alpha} \qquad\qquad\qquad (9\text{-}52)$$

Before the mathematics proceeds to the point of obscuring the physics of this problem, we note that the need for *three* boundary conditions for Eq. (9-45) (or the need for *two* conditions in the newtonian limit) arises not directly from the *order* of the equation, as written, but from the explicit presence of the force at $x = L$, which is contained in the parameter ϵ. This force is not known a priori as part of the specification of the problem. In fact, the force may be considered as an integration constant arising on going from Eq. (9-39) to Eq. (9-40).

Since it is easiest to solve Eq. (9-45) numerically as an initial value problem, the simplest procedure is to carry out the solution using $u = 1$ at $\xi = 0$ and calculating u' at $\xi = 0$ from Eq. (9-51), for a specified value of T_0. The integration is carried out to $\xi = 1$, at which point the value of $u(1) = D_R$ is calculated. If this value of D_R does not match the specified value of Eq. (9-47), the value of ϵ (and hence F) is changed and the integration is repeated. By trial and error the value of $\epsilon(F)$ compatible with the specified draw ratio is found.

The "only" problem to be dealt with is that we have no good basis for picking or calculating T_0. Prior to leaving the spinneret the fluid is subject to a deformation field which generates stresses in the fluid. Just beyond the exit of the spinneret these stresses are free to relax, and this relaxation produces the tendency toward die swell. If the fluid filament were not drawn, then the stress at the maximum R_0 would be zero. The effect of drawing is to shift the position of maximum R_0 to a point where the stresses are not completely relaxed. We do not have enough information about the stress level developed within the spinneret, or the relaxation process, to make good estimates of T_0.

Denn shows that this is not a major problem. The influence of T_0 is restricted to a region near the spinneret, and the velocity profile down the spinning path is only weakly dependent on the specification of T_0. Figure 9-6 shows theoretical results of Denn and Fisher for the velocity profile, with the viscoelastic parameter α taking on small but finite values. We see that as the degree of viscoelasticity increases a linear velocity variation is achieved. Denn uses $T_0 = 1$ for the curves shown in the figure, except that for the newtonian case ($\alpha = 0$) T_0 cannot be independently fixed (and need not, since the order of the differential equation is reduced by one), and the solution itself fixes T_0 to have the value $\frac{2}{3}$.

The particular values of n and D_R illustrated in Fig. 9-6 correspond to a set of

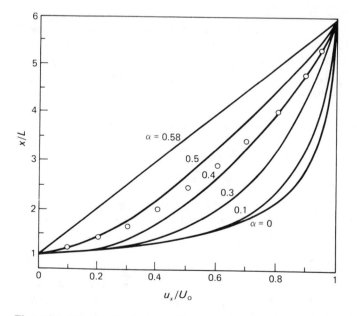

Figure 9-6 Theoretical velocity profiles according to viscoelastic theory of Denn and Fisher for the case $T_0 = 1$, $n = \frac{1}{3}$, $D_R = 5.85$. Data shown are for isothermal spinning of polystyrene.

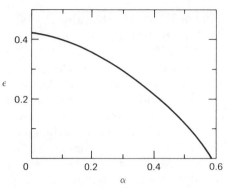

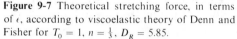

Figure 9-7 Theoretical stretching force, in terms of ϵ, according to viscoelastic theory of Denn and Fisher for $T_0 = 1$, $n = \frac{1}{3}$, $D_R = 5.85$.

experimental data obtained in isothermal spinning of polystyrene. From rheological data the value of the parameter α may be established as somewhere in the range $0.2 < \alpha < 0.3$. Correspondence between these data, which are shown on Fig. 9-6, and the theory would require a larger value of α, in the range $0.4 < \alpha < 0.5$.

From the solution of the equations outlined above it is possible to relate the stretching force, through the parameter ϵ, to the viscoelastic parameter α, using Eq. (9-50). Figure 9-7 shows the theoretical curve for $\epsilon(\alpha)$ for the same conditions as in Fig. 9-6. The polystyrene data referred to above give a stretching force that is not in good agreement with theory. It is again necessary for α to be about 0.5 for the data and theory to be in agreement. (See Prob. 9-14.)

In spite of the discrepancies noted above, it would appear that the viscoelastic model used by Denn and coworkers (which is the White-Metzner generalization of the Oldroyd-Maxwell model) provides the basis for a model of isothermal spinning of viscoelastic fluids that is capable of providing a reasonably good description of observed steady-state behavior. We shall find, in Chap. 15, that this same model provides a basis for description of some significant *transient* observations, as well.

An interesting result of the mathematical analysis is the prediction that, for a given value of the viscoelastic parameter α, there is a maximum draw ratio that can be imposed on the filament. It is given approximately by

$$(D_R)_{\max} = 1 + \alpha^{-1/n} \tag{9-53}$$

The theory does not indicate what would happen if D_R exceeded $(D_R)_{\max}$; it simply implies that an unbounded (and, therefore, physically unattainable) stretching force would be required at draw ratios beyond this limit. As this limit is approached the velocity profile becomes linear.

Figure 9-7 reflects this same result, since we see that $\epsilon \to 0$ (which means $F \to \infty$) at a finite value of α which satisfies Eq. (9-53). If we examine the definition of α, and think of fixing all the operating conditions except for the spinning length L, we may interpret this result in the following way. As we decrease L, the drawing force increases (Fig. 9-7) and the velocity profile becomes nearly linear. There is a minimum value of L over which we can draw down the filament without requiring imposition of an infinite force at the take-up point.

Finally, we note that the limiting linear velocity profile corresponds to a constant stretch rate, given by

$$\frac{du_x}{dx} = \frac{U_0}{L}(D_R - 1) \tag{9-54}$$

Using the definition of α and λ, and imposing Eq. (9-53) as a constraint on D_R, we may show, after some algebraic manipulation, that

$$\lambda \frac{du_x}{dx} = 1 \tag{9-55}$$

Thus the maximum stretch rate, as defined by Eq. (9-54), is such that the draw down is accomplished in one relaxation time.

We end this chapter by noting that commercial fiber spinning is not carried out under isothermal conditions. Thus the models and results presented here cannot be used with any confidence to calculate the forces accompanying non-isothermal spinning, or the velocity profile (and hence the diameter profile) along the spinning path. But the isothermal theory does give considerable physical insight into the process and serves to make clear the essential role of viscoelasticity in the dynamics of fiber spinning. Thus a firm theoretical basis is provided for going on to the problem of nonisothermal modeling, some aspects of which are discussed in Chap. 13.

PROBLEMS

9-1 Derive Eqs. (9-2) and (9-3) by consideration of the geometry of the free boundary as shown in Fig. 9-3.

9-2 Begin with the definition of τ as $\mathsf{T} = -p\,\delta + \tau$ and show that for any incompressible purely viscous fluid the pressure p is the mean normal stress [Eq. (9-17)].

9-3 Using Eq. (9-20) find the maximum spinning length as a relationship among appropriate dimensionless groups.

9-4 Show that the axial stress at the take-up point in an isothermal newtonian spinning system is

$$T_{xx} = \frac{3\mu U_0}{L} D_R \ln D_R$$

9-5 Inertial terms are usually neglected in the spinning analysis. If we consider the ratio $\rho u_x^2 / T_{xx}$ to be a measure of the relative importance of inertial effects to viscous effects, show that the maximum value of this measure is $\rho U_L L / 3\mu \ln D_R$ and that its value at arbitrary x may be written in the form

$$\frac{\rho U_L L D_R^{x/L - 1}}{3\mu \ln D_R}$$

9-6 Show that Eq. (9-26) may be written in the form

$$\frac{u_x}{U_0} = D_R^{x/L}$$

9-7 For Example 9-2 calculate the shear rate in the die, assuming the die is a capillary of diameter 0.128 cm. Are the deformation rates (in terms of the second invariants II_A) higher in the die or along the filament?

9-8 Derive Eq. (9-29).

9-9 Give the derivation of Eq. (9-31).

9-10 Derive Eqs. (9-45) and (9-50).

9-11 Show that the result of Prob. 9-10 behaves properly in the limit of vanishing viscoelasticity:

$$\lim_{\alpha \to 0} T = \tfrac{2}{3}u$$

and that this result is derivable from the newtonian analysis of Sec. 9-1.

9-12 Verify Eqs. (9-36) and (9-37).

9-13 A dimensionless drawing force is defined as

$$\epsilon = \frac{3^{(n-1)/2}K}{FL}\left(\frac{U_0}{L}\right)^{n-1}Q$$

Show that for a purely viscous power law fluid

$$\epsilon = \frac{1}{3[(1/q)(D_R^q - 1)]^n}$$

where $q = 1 - 1/n$. Show that the newtonian result may be obtained from

$$\lim_{n \to 1} \epsilon = \frac{1}{3 \ln D_R}$$

9-14 For an isothermal spinning experiment with polystyrene the following results are available:

$$n = \tfrac{1}{3} \qquad\qquad F = 11{,}700 \text{ dynes}$$

$$K = 4.7 \times 10^4 \text{ P} \cdot \text{s}^{-2/3} \qquad U_0 = 0.29 \text{ cm/s}$$

$$Q = 0.0328 \text{ cm}^3/\text{s} \qquad L = 20 \text{ cm}$$

$$D_R = 5.85$$

Using Fig. 9-7, estimate a value for α.

9-15 Is Eq. (9-11) valid for nonisothermal spinning? Answer for two cases: (a) temperature $= f(r, x)$; (b) temperature $= f(x)$.

BIBLIOGRAPHY

A large and growing literature exists in the area of fiber spinning, which covers a wide range of fundamental as well as technological problems. Many topics have been left untreated in this text because of space limitations. However, nonisothermal behavior will be discussed in Chap. 13, and stability of fiber spinning is discussed in Chap. 15. In this bibliography we will mention a variety of papers covering a wide range of topics which provide opportunities for further study.

One might begin with a book which itself contains a variety of topics in the spinning field and which provides some technological perspective:

Mark, H. F., S. M. Atlas, and E. Cernia (eds.): "Man-Made Fibers: Science and Technology," vol. 1, John Wiley & Sons, Inc., New York, 1967.

The importance of the extensional flow character of spinning is presented in

Weinberger, C. B., and J. D. Goddard: Extensional Flow Behavior of Polymer Solutions and Particle Suspensions in a Spinning Motion, *Multiphase Flow*, **1:** 465 (1974).

Chen, I.-J., et al.: Interpretation of Tensile and Melt Spinning Experiments on Low Density and High Density Polyethylene, *Trans. Soc. Rheol.*, **16:** 473 (1972).

The problem of die swell is discussed in Chap. 14. The interaction between die swell and spinning is studied in

White, J. L., and J. F. Roman: Extrudate Swell During the Melt Spinning of Fiber—Influence of Rheological Properties and Take-up Force, Rept. No. 36, Polymer Science and Engineering Program, University of Tennessee, Knoxville, January 1975.

Crystallization and orientation during spinning are studied in

Nakamura, K., T. Watanabe, T. Amano, and K. Katayama: Some Aspects of Nonisothermal Crystallization of Polymers. III. Crystallization During Melt Spinning, *J. Appl. Polym. Sci.*, **18:** 615 (1974).

Dees, J. R., and J. E. Spruiell: Structure Development During Melt Spinning of Linear Polyethylene Fibers, *J. Appl. Polym. Sci.*, **18:** 1053 (1974).

Dry spinning of fibers from polymer solutions is discussed in

Ohzawa, Y., Y. Nagano, and T. Matsuo: Studies on Dry Spinning. I. Fundamental Equations, *J. Appl. Polym. Sci.*, **13:** 257 (1969).

Griswold, P. D., and J. A. Cuculo: An Experimental Study of the Relationship between Rheological Properties and Spinnability in the Dry Spinning of Cellulose Acetate-Acetone Solutions, *J. Appl. Polym. Sci.*, **18:** 2887 (1974).

Yerushalmi, J., and R. Shinnar: Wet Spinning of Purely Viscous Fluids, *Ind. Eng. Chem. Proc. Des. Dev.*, **10:** 196 (1971).

Paul, D. R.: A Study of Spinnability in the Wet-Spinning of Acrylic Fibers, *J. Appl. Polym. Sci.*, **12:** 2273 (1968).

——— and A. A. Armstrong: The Elastic Stresses Generated During Fiber Formation by Wet-Spinning, *J. Appl. Polym. Sci.*, **17:** 1269 (1973).

9-1 Isothermal melt spinning—newtonian fluid

Development of the basic dynamic equations is available in several places, e.g.:

Matovich, M. A., and J. R. A. Pearson: Spinning a Molten Threadline: Steady State Isothermal Viscous Flows, *Ind. Eng. Chem. Fund.*, **8:** 512 (1969).

——— and T. Matsuo, Studies on Melt Spinning. I. Fundamental Equations on the Dynamics of Melt Spinning, *J. Polym. Sci.*, Pt. A., **3:** 2541 (1965).

Kase, S.: Studies on Melt Spinning. III. Velocity Field Within the Thread, *J. Appl. Polym. Sci.*, **18:** 3267 (1974).

Acierno, D., et al.: Rheological and Heat Transfer Aspects of the Melt Spinning of Monofilament Fibers of Polyethylene and Polystyrene, *J. Appl. Polym. Sci.*, **15:** 2395 (1971).

9-3 Isothermal viscoelastic spinning

The data in Fig. 9-5 are from

Spearot, J. A., and A. B. Metzner: Isothermal Spinning of Molten Polyethylenes, *Trans. Soc. Rheol.*, **16:** 495 (1972).

The spinning mechanics of a viscoelastic fluid described by Eq. (3-135) are given in

Denn, M. M., C. J. S. Petrie, and P. Avenas: Mechanics of Steady Spinning of a Viscoelastic Liquid, *AIChE J.*, **21**: 791 (1975).

The best viscoelastic treatment to date is that in

——— and R. J. Fisher: The Mechanics and Stability of Isothermal Melt Spinning, *AIChE J.*, **22**: 236 (1976).

TUBULAR FILM BLOWING

Experience is a good school, but the fees are high.

Heine

Flat film may be produced by extruding a polymeric melt from a flat die (a "sheeting" die). The resultant film may be drawn or calendered to the desired thickness. Another method may be used to produce flat film, which involves extrusion of the melt from an annular die. Figure 10-1 shows a schematic diagram of the process.

Molten polymer is extruded through an annular die as a thin-walled tube. Air, supplied through the inner mandrel of the die, keeps the tubing inflated, and indeed "blows" the tubing to a larger diameter. Somewhere upstream the "bubble" is cooled, after which the solidified film may be laid flat between rollers which "draw" the tubing between the die and take-up region.

This so-called *blown film process* is quite complex but provides considerable flexibility in producing films of various physical and mechanical properties. In many respects the blown film process is similar to fiber spinning. The kinematics are basically elongational rather than shear, and of course we have in this process, as in fiber spinning, a free boundary flow. The principal distinction, kinematically, to be made in comparing film blowing to fiber spinning lies in the fact that fiber spinning provides a means of *uniaxial* orientation of the filament, whereas film blowing achieves *biaxial* orientation. The two directions of orientation correspond, of course, to axial drawing of the tube and to the circumferential drawing that accompanies the "blow-up" of the tube diameter. Since orientation profoundly affects mechanical properties of film, this particular feature of the blown

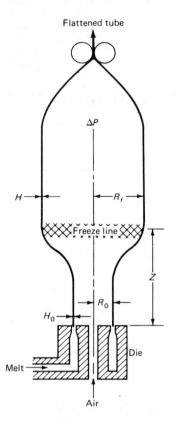

Figure 10-1 Sketch of the film-blowing process.

film process is central to its utility and explains why one might make *flat* film by this more complex route. The *tubular* film is, of course, of commercial utility for such products as sandwich bags or sausage casings.

The two characteristic parameters of the final tubular film are the *blowup ratio*, which is the ratio of the final film diameter to the die diameter, and the *thickness reduction*, calculated as the ratio of thickness as extruded to final film thickness. Alternatively we could use the *machine-direction draw ratio*, defined as the ratio of take-up speed to extrudate speed at the die. For incompressible materials these three ratios are connected through the expression for conservation of mass; the relationship is given in Eq. (10-32). The principal goal of an analysis of this process is development of the relationships among these geometric characteristics and operating conditions such as the bubble pressure, the film speed or tension at take-up, the thermal conditions which "freeze" the film, and the rheological properties of the polymer.

The major complication in developing a model of this system arises from the strong interaction of the heat transfer process between the film and the surrounding air with the temperature-dependent rheological properties of the melt. In addition, crystallization often occurs to an extent, and at a rate, strongly affected by temperature and the degree of orientation of the cooling polymer. Thus the

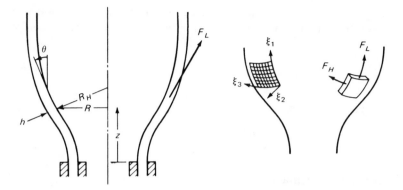

Figure 10-2 Detail for development of the mechanical equations. The ξ_i are coordinates moving with the film.

nonisothermal nature of film blowing cannot be neglected in any analysis that purports to be quantitatively reliable.

Despite this reservation we outline here the simplest newtonian isothermal model of film blowing to show the interaction of the geometric and dynamic parameters and to get an idea of the order of magnitude of some of these interactive phenomena. Extension of the model to the more realistic nonisothermal viscoelastic case is conceptually simple, but computationally quite difficult and tedious.

Figure 10-2 shows a definition sketch for the analysis. Cylindrical polar coordinates are most convenient, and symmetry about the z axis is assumed. The principal geometric assumption is that of a thin film: $h/R \ll 1$. From the kinematic point of view we note that the velocity vector **u** has three nonvanishing components and that, for a thin film, the deformation field is elongational—there are no shear components in this approximation. Such a deformation field gives rise to stresses in the circumferential direction, in the direction of flow (which is at an angle θ to the z axis) and in the direction normal to the film surface.

To calculate the stresses it is most convenient to set up a coordinate system which moves with the fluid. A cartesian system embedded in an element of fluid is shown in Fig. 10-2. While the film thickness will be taken as a small quantity, it is useful to think of the ξ coordinates as embedded in the *inner* surface of the bubble. We let **v** be the velocity vector in the *moving* coordinate system, and **e** is the rate of deformation tensor in that system. **e** has components

$$
e = 2 \begin{pmatrix} \dfrac{\partial v_1}{\partial \xi_1} & 0 & 0 \\[2mm] 0 & \dfrac{\partial v_2}{\partial \xi_2} & 0 \\[2mm] 0 & 0 & \dfrac{\partial v_3}{\partial \xi_3} \end{pmatrix} \tag{10-1}
$$

The velocity v_2 is normal to the bubble surface; it reflects the rate of thinning of the film. At $\xi_2 = 0$ (the inner surface, by definition) we take $v_2 = 0$. At $\xi_2 = h$ we have†

$$v_2 = \frac{dh}{dt} \tag{10-2}$$

and since h is a function of ξ_1,

$$v_2 = \frac{dh}{d\xi_1}\frac{d\xi_1}{dt} = v_1 \frac{dh}{d\xi_1} \tag{10-3}$$

where $v_1 = d\xi_1/dt$ has been defined.

The gradient of velocity in the ξ_2 direction, to a good approximation in a thin film, is just the difference in velocity v_2 across the film divided by the local film thickness. Thus we take

$$e_{22} = \frac{2v_2}{h} = \frac{2}{h}\frac{dh}{d\xi_1}v_1 \tag{10-4}$$

The circumferential velocity v_3 is just the local rate of expansion of the bubble in the circumferential direction:

$$v_3 = 2\pi \frac{dR}{dt} = 2\pi v_1 \frac{dR}{d\xi_1} \tag{10-5}$$

Since we have symmetry about the z axis, the velocity v_3 is uniform about the circumference of the bubble. The stretching rate of deformation, then, is simply the rate of increase of the circumference divided by the local value of the circumference, or

$$e_{33} = \frac{2v_1}{R}\frac{dR}{d\xi_1} \tag{10-6}$$

Since the first invariant of the rate-of-deformation tensor vanishes for an incompressible fluid, we may write

$$e_{11} = -(e_{22} + e_{33}) = -2v_1\left(\frac{1}{h}\frac{dh}{d\xi_1} + \frac{1}{R}\frac{dR}{d\xi_1}\right) \tag{10-7}$$

It is convenient at this point to replace ξ_1 and v_1 by corresponding quantities in the laboratory coordinate system, using

$$d\xi_1 = \frac{1}{\cos\theta}dz \tag{10-8}$$

---

† We are considering a steady-state problem, but in the ξ coordinates h and R appear to be functions of time. ξ is a *moving* coordinate system.

(which follows from the geometry) and

$$v_1 = \frac{Q}{2\pi Rh} \tag{10-9}$$

which is just the overall continuity equation, in terms of the volumetric flow rate Q. Consequently we find

$$e = \frac{Q \cos \theta}{\pi Rh} \begin{pmatrix} -\frac{1}{h}\frac{dh}{dz} - \frac{1}{R}\frac{dR}{dz} & 0 & 0 \\ 0 & \frac{1}{h}\frac{dh}{dz} & 0 \\ 0 & 0 & \frac{1}{R}\frac{dR}{dz} \end{pmatrix} \tag{10-10}$$

Now we are in position to calculate the viscous stresses associated with this deformation. For the newtonian fluid we will have

$$p_{ij} = -p + \mu e_{ij} \tag{10-11}$$

We continue to use the ξ coordinates, so p is the stress tensor in that coordinate system and *not* in the laboratory system.

The stress normal to the free boundary is p_{22}. As a boundary condition we will assume that no external forces act on the bubble and that surface tension forces are insignificant with respect to viscous forces. Then it follows that

$$p_{22} = 0 \tag{10-12}$$

and, from Eq. (10-11), we find the isotropic pressure in the fluid to be

$$p = \mu e_{22} = \frac{Q\mu \cos \theta}{\pi Rh^2} \frac{dh}{dz} \tag{10-13}$$

We may then write the stresses p_{11} and p_{33} as

$$p_{11} = -\frac{\mu Q \cos \theta}{\pi Rh} \left(\frac{2}{h}\frac{dh}{dz} + \frac{1}{R}\frac{dR}{dz} \right) \tag{10-14}$$

$$p_{33} = \frac{\mu Q \cos \theta}{\pi Rh} \left(\frac{1}{R}\frac{dR}{dz} - \frac{1}{h}\frac{dh}{dz} \right) \tag{10-15}$$

At this stage we have calculated the deformation field e and the stress field p. What we desire is a set of equations from which $R(z)$ and $h(z)$ may be determined. These equations follow from simple force balances on the bubble. Referring to Fig. 10-2 we may calculate the forces in the circumferential and longitudinal directions, and we find, for an element of film of dimensions $2\pi R$ by h by $d\xi_1$, that

$$F_L = 2\pi Rh p_{11} \tag{10-16}$$

and

$$dF_H = h\, d\xi_1 p_{33} \tag{10-17}$$

The shape of the bubble is determined by the local balance of forces. The analysis is identical to that carried out for the interface subjected to surface tension which led to Eq. (3-75). Instead of the surface tension we introduce the forces per unit length in the two orthogonal directions in the surface, which are

$$\frac{F_L}{2\pi R} = hp_{11} \tag{10-18}$$

and

$$\frac{dF_H}{d\xi_1} = hp_{33} \tag{10-19}$$

Then Eq. (3-75) would be modified to give the pressure difference across the bubble surface as

$$\Delta P = h\left(\frac{p_{11}}{R_L} + \frac{p_{33}}{R_H}\right) \tag{10-20}$$

where R_L and R_H are the principal radii of curvature in the ξ coordinate system. In terms of laboratory (cylindrical polar) coordinates we have

$$R_H = R \sec \theta \tag{10-21}$$

$$R_L = -\frac{\sec^3 \theta}{d^2 R/dz^2} \tag{10-22}$$

We will define the "freeze line" of the bubble, at $z = Z$, to be the point above which (that is, $z > Z$) the bubble shape does not change. One boundary condition at the freeze line will specify the draw force F_z. Between some arbitrary position z, where the local geometric parameters are $\mathbf{R}$ and θ, and the freeze line Z, an axial force balance takes the form

$$2\pi R \cos \theta \, hp_{11} + \pi(R_f^2 - R^2) \, \Delta P = F_z \tag{10-23}$$

where R_f = radius at the freeze line

ΔP = uniform pressure maintained inside the bubble

When Eqs. (10-20) and (10-23) are manipulated by using Eqs. (10-14), (10-15), (10-21), and (10-22), it is possible to obtain two differential equations: One gives $R(z)$ and the other gives $h(z)$. The equations, in dimensionless form, are

$$2r^2(T + r^2B)r'' = 6r' + r(1 + r'^2)(T - 3r^2B) \tag{10-24}$$

$$\frac{w'}{w} = -\frac{r'}{2r} - \frac{(1 + r'^2)(T + r^2B)}{4} \tag{10-25}$$

where the dependent variables are $r = R/R_0$ and $w = h/R_0$ and the prime (') denotes differentiation with respect to $x = z/R_0$; that is, $(\ )' \equiv d(\ )/dx$. The angle θ is removed from the problem by noting that $r' = \tan \theta$.

Two dimensionless parameters appear in the two differential equations. They are dimensionless pressure

$$B = \frac{\pi R_0^3 \, \Delta P}{\mu Q} \tag{10-26}$$

and a dimensionless stress

$$T = \frac{R_0 F_z}{\mu Q} - B(\text{BUR})^2 \qquad (10\text{-}27)$$

The blowup ratio BUR is simply

$$\text{BUR} = R_f/R_0 \qquad (10\text{-}28)$$

It is convenient to define a dimensionless take-up force as

$$T_z = \frac{R_0 F_z}{\mu Q} \qquad (10\text{-}29)$$

The general procedure requires solution of Eq. (10-24) for $r(x)$, followed by solution of Eq. (10-25) for $w(x)$. Since Eq. (10-24) is second-order we must specify two boundary conditions. These are

$$r = 1 \qquad \text{at } x = 0$$

$$r' = 0 \qquad \text{at } x = X = \frac{Z}{R_0}$$

The second boundary condition on r involves specification of the freeze-line position. The boundary condition on w is simply

$$w = w_0 = \frac{H_0}{R_0} \qquad \text{at } x = 0$$

As in fiber spinning, a die swell phenomenon may cause H_0 to differ from the die lip separation.

If T, B, and X are specified a priori, then solution of the problem gives the thickness reduction at X since, from Eq. (10-25),

$$-\ln \frac{w_z}{w_0} = \ln \frac{H_0}{H} = \ln \sqrt{\text{BUR}} + \int_0^x \frac{(1 + r'^2)(T + r^2 B)}{4} \, dx \qquad (10\text{-}30)$$

We note that T depends upon BUR. Thus it is necessary to specify BUR, solve for $r(x)$ and check the value of $\text{BUR} = r(X)$, and iterate if necessary to find the solution compatible with the choice of BUR. The simplest procedure is to integrate Eq. (10-24) (numerically) from $x = X$, taking $r = \text{BUR}$, $r' = 0$ as *initial* conditions and checking the calculated value of r at $x = 0$. The initial guess on BUR is modified iteratively until a solution produces $r = 1$ at $x = 0$. Then Eq. (10-25) may be integrated to give $w(x)$.

Pearson and Petrie have carried out solutions of this problem using various realistic values of blow ratio, thickness reduction, and freeze-line height. If *one* of the parameters X, B, or T_z is fixed arbitrarily, then solutions of the problem will be curves in the $(\text{BUR}, H_0/H)$ plane, along which the other two parameters are constrained. This can be seen in Fig. 10-3, which shows a set of solutions for fixed freeze-line distance $X = 20$. Any point in the $(\text{BUR}, H_0/H)$ plane, which of course corresponds to specific values of blow ratio and thickness reduction, is the inter-

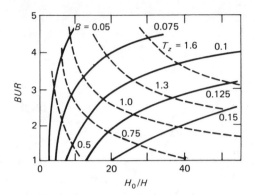

Figure 10-3 Computed results for isothermal newtonian film blowing for the case $X = 20$. (*From Pearson and Petrie, 1970a.*)

section of two curves, one of constant B and the other of constant T_z. The entire graph is for a specified value of X. Hence all the operating conditions are found from the figure.

If, for example, the pressure parameter is arbitrarily specified, then a graph such as Fig. 10-4 is obtained, and one can see the effect, for example, of increasing the freeze-line position at constant T_z. Because of the large number of parameters that appear in this model there is no compact graphical presentation of results. Pearson and Petrie present several sets of curves similar to Figs. 10-3 and 10-4.

The axial force balance [Eq. (10-23)], of necessity, specifies the take-up force F_z as a given operating parameter. It is possible to operate the take-up region in such a way as to control the take-up force. It is also possible to specify and control the take-up speed, or the dimensionless draw ratio

$$D_R = \frac{V}{v_0} \tag{10-31}$$

However, D_R and $T_z(F_z)$ cannot be *independently* specified, since conservation of mass requires that

$$\text{BUR} \frac{V}{v_0} = \frac{H_0}{H} = \text{BUR} \, D_R \tag{10-32}$$

If BUR and H_0/H are independently fixed one finds, as part of the solution, a value for T_z. However, specification of BUR and H_0/H also fixes D_R, through Eq. (10-32). In this sense D_R and T_z are dependent.

> **Example 10-1** Tubular polyethylene film will be extruded in the following manner: The geometric specifications require a thickness of 2 mils and a "lay-flat" width† of 8 in. The extrusion die has $R_0 = 2$ in and $H_0 = 0.05$ in. The production rate is 30 lb/h, and the viscosity is 4.35 lb·s/in². Assume newtonian isothermal flow right up to the freeze line. Take the melt density as $\rho = 56$ lbm/ft³. Determine an acceptable set of operating conditions.

† The lay-flat width is half the perimeter.

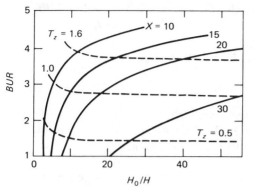

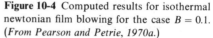

Figure 10-4 Computed results for isothermal newtonian film blowing for the case $B = 0.1$. (*From Pearson and Petrie, 1970a.*)

To solve this problem we will specify values for *one* of the parameters X, T_z, and B and determine the required values for the other two. Let us specify $Z = 40$ in, or $X = 20$. From the geometric specifications we find

$$\frac{H_0}{H} = \frac{0.05}{0.002} = 25$$

and

$$BUR = \frac{8}{\pi R_0} = 1.27$$

From Fig. 10-3 we find

$$B = 0.15 \qquad T_z = 0.70$$

Inverting the definitions of B and T_z we find

$$\Delta P = 0.00675 \text{ psi} \qquad F_z = 0.4 \text{ lbf}$$

The draw ratio is

$$D_R = \frac{H_0}{H \; BUR} = 19.7$$

Either D_R or F_z must be controlled at the given values.

We note that the required ΔP is quite small, and investigate the response of the system to a change in ΔP at fixed X and T_z. Suppose ΔP changed to $\Delta P = 0.0056$ psi so that $B = 0.125$. Then we find from Fig. 10-3 that (assuming T_z is held at 0.7)

$$BUR = 1.5 \qquad \frac{H_0}{H} = 17$$

The major change is the reduction in thickness ratio, which would produce a film of thickness $H = 2.9$ mils, almost 50 percent larger than the desired value. The blow ratio is increased somewhat, with a corresponding increase in lay-flat width from 8 to nearly 9.5 in. Obviously the system cannot tolerate such a large reduction in bubble pressure (-17 percent).

Suppose the draw ratio were controlled (at 19.7) instead of holding $F_z(T_z)$ constant. Then we would find, for $B = 0.125$, $X = 20$, $D_R = 19.7$,

$$\frac{H_0}{H} = 60 \qquad \text{(extrapolating off the figure)}$$

$$\text{BUR} = 3$$

These would be intolerably large changes away from the specifications. This latter result is general and was pointed out by Pearson and Petrie. If possible, it is far better to control take-up *force* than take-up *speed*.

The model presented here is limited in several respects, the most significant of which are the neglect of cooling between the die and the freeze line and the restriction to newtonian flow. Introduction of either of these features complicates the problem but does not introduce any major change in general development of the model. The major difficulty is not so much the *extension* of the theory as its *application*.

The nonisothermal analysis would require some model relating the temperature of the bubble to the rate of loss of heat to the surrounding air. This would introduce at least one dimensionless thermal parameter into the problem. Methods of estimating such parameters are presently quite unreliable, and they would have to be established experimentally. This in itself would be a difficult task.

The flow field in this problem is one of biaxial elongation. To introduce nonnewtonian behavior it is necessary to have a constitutive equation which is realistic in its predictions for elongational flows. This too is an area where no large body of reliable work has been done. (Recall the comments in Chap. 3 on elongational flows.)

Thus we must consider the model outlined here to be very poor, quantitatively, but to provide a basis for developing more realistic models, especially with regard to thermal and rheological phenomena. The Bibliography provides several opportunities for extended reading in these two directions.

PROBLEMS

10-1 Equation (10-13) gives a pressure p which is assumed uniform across the film thickness h. In effect, the pressure ΔP on the inside of the bubble has been neglected.

(a) For the conditions cited in Example 10-1 give the ratio $\Delta P/p$. Show clearly how you estimate dh/dz.

(b) What difficulties ensue if Eq. (10-12) is replaced by

$$p_{22} = \begin{cases} 0 & \text{at } \xi_2 = h \\ -\Delta P & \text{at } \xi_2 = 0 \end{cases}$$

10-2 Give the detailed derivation of Eq. (10-20).

10-3 One boundary condition on r is $r' = 0$ at $x = X$. Show, using Eq. (10-25), that w' is *not* zero at $x = X$, and give its value.

10-4 Show that $r = \text{BUR} = 1$ is a possible solution of Eq. (10-24) for certain values of T and B. Give those values. Find $w(x)$ for this case and plot H_0/H versus B for this case, with X as a parameter.

10-5 Rework Example 10-1 but specify $\Delta P = 0.0045$ psi and solve for Z and F_z. Find the response of the system to ± 10 percent changes in Z, keeping (a) F_z constant or (b) D_R constant.

10-6 In the tubular film-blowing process the film is oriented biaxially. The circumferential orientation is given roughly by the blow ratio BUR, whereas the axial orientation is given by the draw ratio D_R.

Suppose we want to design a process for which $\text{BUR} = D_R$. Specify the operating conditions to achieve $\text{BUR} = 3$ when $X = 20$ for the fluid of Example 10-1. Use the same die geometry and production rate as in that example.

10-7 Find suitable operating conditions for production of nylon hollow-fiber membranes. The die has dimensions $R_0 = 30~\mu\text{m}$ and $H_0 = 15~\mu\text{m}$. The blow ratio is $\text{BUR} = 1$, and take $H_0/H = 3$. The freeze line is at $X = 20$ when the production rate is 1 lb/h total from a block of 100 identical dies. Take nylon to be newtonian with $\mu = 0.5~\text{lbf} \cdot \text{s/in}^2$, $\rho = 0.036~\text{lb/in}^3$.

10-8 Examine the assumption $\Delta P \ll p$ for the operating conditions of Prob. 10-7.

10-9 Calculate a nominal shear rate for the flow in the die of Prob. 10-7. Would you anticipate any problems in operating at this shear rate? Discuss them.

BIBLIOGRAPHY

The material in this chapter follows

Pearson, J. R. A., and C. J. S. Petrie: A Fluid-Mechanical Analysis of the Film-Blowing Process, *Plast. Polym.*, **38**: 85 (1970a).

A more detailed discussion of this problem is given in

———— and ————: The Flow of a Tubular Film. Part 1. Formal Mathematical Representation, *J. Fluid Mech.*, **40**: 1 (1970b).
———— and ————: The Flow of a Tubular Film. Part 2. Interpretation of the Model and Discussion of Solutions, ibid., **42**: 609 (1970c).

The problems associated with considering heat transfer are discussed by Petrie in two papers:

Petrie, C. J. S.: Mathematical Modelling of Heat Transfer in Film Blowing: A Case Study, *Plast. Polym.*, **42**: 259 (1974).
————: A Comparison of Theoretical Predictions with Published Experimental Measurements on the Blown Film Process, *AIChE J.*, **21**: 275 (1975).

Viscoelastic effects are discussed in

————: Memory Effects in a Non-Uniform Flow: A Study of the Behavior of a Tubular Film of Viscoelastic Fluid, *Rheol. Acta*, **12**: 92 (1973).

Blown film extrusion is discussed from the viewpoint of elongational viscosity in

Han, C. D., and J. Y. Park: Studies on Blown Film Extrusion: I. Experimental Determination of Elongational Viscosity, *J. Appl. Polym. Sci.*, **19**: 3257 (1975).

In a companion paper the same authors do a nonisothermal model:

———— and ————: Studies on Blown Film Extrusion: II. Analysis of the Deformation and Heat Transfer Processes, ibid, p. 3277.

ELEVEN

INJECTION MOLDING

The poet is in command of his fantasy, while it is ... the mark of the neurotic that he is possessed by his fantasy.

Trilling

Injection molding is one of the most common operations of the plastics industries. It is used to create finished articles which range from paper clips to automobile front-end assemblies. Because of its versatility, and the significant fraction of the total industrial output of plastics that is injection molded, it is one of the most important of the polymer flow processes.

In terms of basic steps or stages, the injection-molding process may be looked upon as indicated in Fig. 11-1. The solid plastic is melted, and the melt is conveyed to the mold and injected into the mold under high pressure. The mold is cooled to solidify the article, and then the mold is opened and the article is ejected. The mold closes and the cycle repeats.

The simplest molding machine is of the plunger type, illustrated in Fig. 11-2. The plastic is simply pushed forward by a plunger through a heated region.

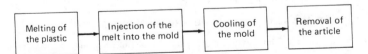

Figure 11-1 Stages in the injection-molding process.

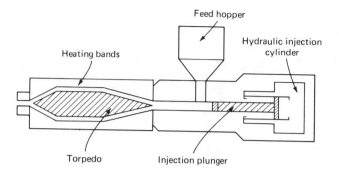

Figure 11-2 A plunger-type injection-molding machine.

Because the high viscosity of melt prevents any significant convective heat transfer, it is necessary to spread the molten material in a thin layer to contact the heated surfaces. One of the more common spreaders is the "torpedo" shown in this example, which simply diverts the material so that it moves through a thin annular region. After melting, the material converges and flows through a nozzle which delivers it to the mold.

In more common use currently is the reciprocating-screw injection-molding machine, shown schematically in Fig. 11-3. In this system the screw function is principally to melt and mix the feed material. For injection the entire screw moves forward as a plunger. A special valve prevents backflow.

The injection-molding machine is quite complex. The mechanical design of the systems which clamp the mold together and then release and eject the solidified article is a major topic in itself. Here we consider only the flow process associated with the filling of the mold. A good general reference to the full range of problems that must be considered in designing an integrated injection-molding process is the book by Rubin.

Before considering some isolated aspects of the molding process, let us follow an element of fluid as it moves from the nozzle toward the mold, and outline some of the flow problems of interest. Transfer of the melt to the mold is simply a problem in creative plumbing. Figure 11-4 shows the typical elements of the

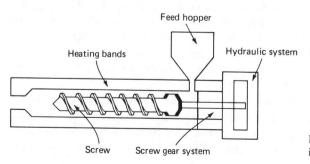

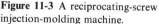

Figure 11-3 A reciprocating-screw injection-molding machine.

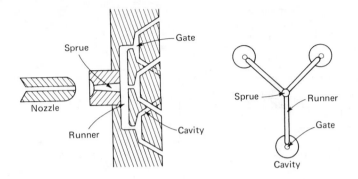

Figure 11-4 Schematic of a three-cavity mold.

transfer system. Material is transferred from the injection-molding machine to the mold block through the nozzle, which is directly coupled to the block with a *sprue bushing*. In a multicavity mold the hot melt is conveyed to each individual mold by a *runner*. Each runner connects to the cavity it feeds by a *gate*, which is simply a restriction in the flow path.

The runners and cavities are normally empty at the beginning of each fill cycle. Hence there is a transient element to the flow while the lines are filling. If the injection rate of material is constant (this is normally the case) then the pressure at the nozzle gradually rises as filling proceeds. We shall find that the transient associated with runner filling is of minor importance in comparison to that associated with the filling of the cavity itself.

A significant complicating feature is the nonisothermal character of mold filling. There are three thermal phenomena which contribute to and complicate the flow analysis. In the first place the mold surfaces, which are the flow boundaries, are not usually at the same temperature as the melt. This follows principally from the difficulty of uniform control of temperature throughout the mold block. In addition to *spatial* variations of temperature, the cyclic nature of the process involves time-varying temperatures within the mold block. Thus the dynamics of heat transfer in the mold block contributes an unsteady nonisothermal character to the mold-filling process.

A consequence of the nonisothermal boundaries is the freezing of the polymer at the flow boundaries if they are sufficiently cold. The principal effect of boundary solidification is the constriction of the flow path, which leads to large pressure increases. Of course, if the problem is not properly accounted for, it is possible to "freeze off" a runner and prevent complete filling of a cavity. Incomplete filling is referred to as a *short shot*.

A third (potential) nonisothermal feature is due to viscous heating. Under some conditions the combination of high viscosity and small flow channels can lead to significant temperature rises of the melt as it proceeds toward the cavity. With some polymers this creates the possibility of thermal degradation. In any

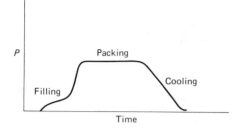

Figure 11-5 Pressure during the molding cycle.

event significant temperature rises will strongly affect the pressure drop–flow rate relationship. Because injection molding can involve pressures as high as several thousand atmospheres, it is necessary to keep in mind that the viscosity of molten polymers is known to depend on pressure. As is the case with viscous heat generation, the pressure-dependent viscosity effect is often ignored in the first stage of process modeling, without justification. We will examine some aspects of this problem briefly in this chapter.

That part of the molding cycle associated with filling is usually short by comparison to the total cycle time. After the cavity is filled the cooling cycle begins. As the polymer solidifies, its density increases slightly, and if the mass of material in the cavity were constant, the volume would decrease. This would correspond to shrinkage of the article and subsequent loss of the geometry of the piece relative to the mold. If the shrinkage were uniform this factor could be easily accommodated by designing the mold a little oversize so as to compensate for shrinkage. Unfortunately there is usually some temperature distribution within the material in the cavity, as well as a distribution of cavity surface temperature. In addition, except with the simplest of moldings, regions of different thickness in the molded material will cool at different rates, leading to different degrees of shrinkage, and this can lead to warpage of the article. To minimize dimensional changes (shrinkage) and shape changes (warpage) a very high pressure is maintained on the cavity during the cooling cycle. As the density of the cooling polymer increases, more melt flows into the cavity to maintain constant volume. This can continue until the gate freezes solid.

The molding cycle is usually visualized graphically as shown in Fig. 11-5. The three basic stages are *filling, packing,* and *cooling.* The pressure rises at a relatively slow rate during the filling cycle. The packing stage is the one in which the shrinkage is offset by maintenance of very high pressure. Relatively little flow occurs at this point. Finally, during the cooling stage, the pressure in the mold relaxes.

The cooling stage generally controls the total cycle time and so depends principally on the thickness of the molded piece, since heat transfer through the low-conductivity polymer is the ultimate resistance to cooling. Cycle times are typically in the range of 10 to 100 s (except for unusually small or large pieces).

After the melt has filled each cavity, and the cooling cycle solidifies the plastic, the mold is opened and the solid plastic piece is ejected. The individual molded

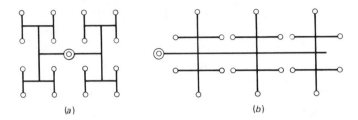

Figure 11-6 (*a*) A symmetrical 16-Chupchik mold. (*b*) An unsymmetrical 18-Chupchik mold.

pieces are usually still connected to each other by the solidified runner material, since the parting line of the mold is usually in the plane of the runners. The desired piece is subsequently separated from the runner at the gate. The gate size and placement is designed to simplify separation and minimize the difficulty of smoothing off the defect that results at the break-off point. The separated runner pieces constitute waste which may, in many cases, be recycled to the feed.

Runner design represents a compromise of many factors. The runner must be large enough to facilitate rapid filling of the cavity but not so large as to significantly increase the time required to freeze the material in the runner. The ideal shape would be circular in cross section, but this is the most difficult to machine, and so trapezoidal runners are often used. One of the most important features of runner designs is *balancing* to ensure that each cavity fills at the same rate. If the runner system is symmetrical, as suggested in Fig. 11-6*a*, then balancing depends principally on the precision of the machining of the mold. Some molds cannot be laid out symmetrically, as in Fig. 11-6*b*, in which case the cross-sectional areas of each runner section must be adjusted to balance the pressure drops from the sprue to each cavity.

We shall outline in this chapter some simple (mostly isothermal) analyses of the mold-filling process. Nonisothermal phenomena cannot generally be neglected, and we shall subsequently have to modify or reformulate the isothermal models. By beginning with the simpler isothermal problems we will find it easier to examine and interpret the role of nonisothermal effects when they are introduced subsequently.

11-1 ISOTHERMAL NEWTONIAN FLOW INTO A CAVITY

The geometry of a simple disk mold is shown in Fig. 11-7. We consider the transient filling of such a mold with a newtonian melt under isothermal conditions. There are two possible cases: constant flow rate (the more likely production case) and constant pressure at the nozzle. In either case we assume that the runner has a small volume by comparison to that of the cavity itself. This allows the assumption that the runner is always filled and that the transient is associated only with the filling of the cavity.

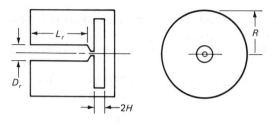

Figure 11-7 A center-gated disk mold.

Figure 11-8 shows a definition sketch for the flow analysis. It is assumed that the velocity vector is given by

$$\mathbf{u} = [u_r(r, z, t), 0, 0] \tag{11-1}$$

for

$$\tfrac{1}{2}D_r \leq r \leq R \qquad -H \leq z \leq H$$

The continuity equation takes the form

$$\frac{1}{r}\frac{\partial}{\partial r}ru_r = 0 \tag{11-2}$$

in the same region.

The position of the interface R^* will be defined in terms of the volumetric flow rate as

$$Q = 4\pi HR^*\frac{dR^*}{dt} \tag{11-3}$$

The radial component of the dynamic equation is

$$\rho\left(\frac{\partial u_r}{\partial t} + u_r\frac{\partial u_r}{\partial r}\right) = -\frac{\partial p}{\partial r} + \frac{\partial}{\partial z}\tau_{rz} + \frac{1}{r}\frac{\partial}{\partial r}r\tau_{rr} - \frac{\tau_{\theta\theta}}{r} \tag{11-4}$$

and for the newtonian fluid the only nonzero stresses are

$$\tau_{rz} = \mu\frac{\partial u_r}{\partial z} \qquad \tau_{\theta\theta} = 2\mu\frac{u_r}{r} \qquad \tau_{rr} = 2\mu\frac{\partial u_r}{\partial r} \tag{11-5}$$

The terms involving τ_{rr} and $\tau_{\theta\theta}$ vanish collectively, for this flow, by virtue of the continuity equation (see Prob. 11-1). As a consequence we must solve

$$\rho\left(\frac{\partial u_r}{\partial t} + u_r\frac{\partial u_r}{\partial r}\right) = -\frac{\partial p}{\partial r} + \mu\frac{\partial^2 u_r}{\partial z^2} \tag{11-6}$$

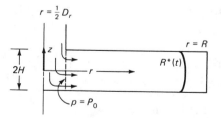

Figure 11-8 Definition sketch for analysis of flow into a disk-shaped cavity.

This is a nonlinear partial differential equation whose solution would require a fairly tedious numerical method. We shall solve it analytically by neglecting the inertial terms [the left-hand side of Eq. (11-6)] and justifying the approximation a posteriori.

Let us begin by noting that the continuity equation is satisfied by a velocity u_r of the form

$$u_r = \frac{1}{r} c(z, t) \tag{11-7}$$

where c is not a function of r but *may* depend on z and t. On neglecting the inertial terms and introducing Eq. (11-7) into Eq. (11-6), we find

$$0 = -\frac{\partial p}{\partial r} + \frac{\mu}{r} \frac{\partial^2 c}{\partial z^2} \tag{11-8}$$

Consistent with the assumptions already made, we take p to be independent of z. This requires that the functions p and c satisfy

$$\frac{r}{\mu} \frac{\partial p}{\partial r} = \frac{\partial^2 c}{\partial z^2} = A(t) \tag{11-9}$$

since $p \neq p(z)$ and $c \neq c(r)$.

It follows then that

$$c = -\frac{AH^2}{2} \left[1 - \left(\frac{z}{H}\right)^2 \right] \tag{11-10}$$

and

$$p - P_0 = A\mu \ln \frac{2r}{D_r} \tag{11-11}$$

The following boundary conditions have been used in obtaining these results:

$$u_r = 0 \qquad \text{at} \quad z = \pm H$$

$$\frac{\partial u_r}{\partial z} = 0 \qquad \text{at} \quad z = 0 \tag{11-12}$$

$$p = P_0 \qquad \text{at} \quad r = \tfrac{1}{2}D_r$$

The function $A(t)$ may be determined in terms of the volumetric flow rate Q from

$$Q = 4\pi \int_0^H r u_r \, dz \tag{11-13}$$

The result is

$$A = -\frac{3Q}{4\pi H^3} \tag{11-14}$$

We note that A is independent of time only if Q is constant.

Before examining some of the features of this solution let us evaluate the assumption by which the inertial terms were neglected. The viscous term is given by

$$\mu \frac{\partial^2 u_r}{\partial z^2} = -\frac{3\mu Q}{4\pi H^3} \frac{1}{r} \tag{11-15}$$

The unsteady-state term is

$$\rho \frac{\partial u_r}{\partial t} = \frac{3\rho}{8\pi} \frac{\dot{Q}}{H} \frac{1}{r} \left[1 - \left(\frac{z}{H} \right)^2 \right] \tag{11-16}$$

where $\dot{Q}$ is dQ/dt. The convective (radial) acceleration is

$$\rho u_r \frac{\partial u_r}{\partial r} = -\frac{9\rho Q^2}{64\pi^2 H^2} \frac{1}{r^3} \left[1 - \left(\frac{z}{H} \right)^2 \right]^2 \tag{11-17}$$

To evaluate the neglect of the latter two terms we estimate the magnitude of each relative to the viscous term.

It is easily seen that the neglect of the transient term depends on the magnitude of a dimensionless group given by

$$\Pi_{tr} = \frac{\rho H^2}{\mu} \frac{\dot{Q}}{Q} \tag{11-18}$$

Of course for the case of constant injection rate, $\dot{Q} = 0$ and the unsteady term vanishes identically. The term $\rho H^2/\mu$ may be considered to be a viscous relaxation time, while $Q/\dot{Q}$ is a process time scale. A typical value of the viscous relaxation time would be (taking $\rho = 1$, $\mu = 10^4$, and $H = 1$, all in cgs units) of the order of 10^{-4} s. The process time scale would be reasonably approximated by the fill time, which would certainly be no less than 1 s. Thus we would always expect Π_{tr} to be less than 10^{-4} and conclude that unsteady effects are unimportant.

The neglect of the inertial term may be seen to be expressible in the form of a parameter

$$\Pi_{in} = \frac{3}{16\pi} \text{Re} \left(\frac{H}{r} \right)^2 \tag{11-19}$$

where

$$\text{Re} = \frac{\rho Q}{\mu H} \tag{11-20}$$

is a Reynolds number of this process. Any reasonable set of values for Q, μ, and H will give a value for Π_{in} that is considerably less than unity. Thus the convective term is safely neglected, and the solution presented is quite reasonable, subject to the assumption of isothermal newtonian flow and the neglect of entrance effects in the neighborhood of $r = \frac{1}{2}D_r$.

Let us consider, now, the two cases of interest: constant pressure or constant flow rate.

Constant pressure By constant pressure we imply that P_0 is held constant. The flow rate would be expected to decrease as the cavity fills. An additional boundary condition on pressure equates the pressure at the advancing interface to atmospheric pressure, or [setting atmospheric pressure at zero in Eq. (11-11)]

$$-P_0 = A\mu \ln \frac{2R^*}{D_r} \tag{11-21}$$

From the relationship of A to Q [Eq. (11-14)] and the definition of R^* [Eq. (11-3)] we find, after a simple integration, that the solution for $R^*(t)$ may be written implicitly as

$$\varphi^2 \ln \varphi - \tfrac{1}{2}(\varphi^2 - 1) = \frac{2Bt}{a^2} \tag{11-22}$$

where

$$a = \frac{D_r}{2R}$$

$$B = \frac{H^2 P_0}{3\mu R^2}$$

and

$$\varphi = \frac{2R^*}{D_r}$$

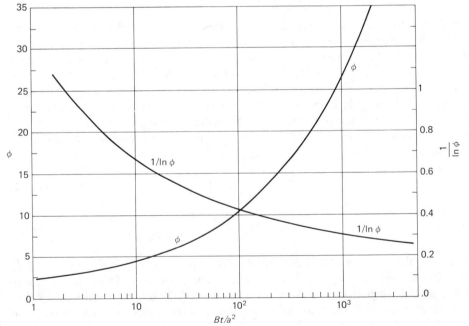

Figure 11-9 Position of the advancing front as a function of time [Eq. (11-22)] and volume flow rate as a function of time [Eq. (11-24)]. Newtonian isothermal filling of a disk.

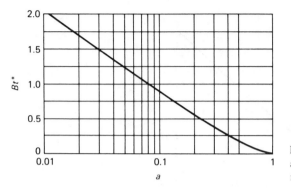

Figure 11-10 Fill time as a function of applied pressure: newtonian isothermal filling of a disk.

The fill time occurs when $R^* = R$ or $\varphi = 1/a$, so that, if t^* denotes the fill time, we find

$$t^* = \frac{1}{2B}[-\ln a - \tfrac{1}{2}(1 - a^2)] \tag{11-23}$$

Figure 11-9 shows the position of the advancing front as a function of time, and Fig. 11-10 gives the fill time as a function of applied pressure.

The volumetric flow rate as a function of time follows most easily from the relationship of A to Q [Eq. (11-14)] and inversion of Eq. (11-21), with the result that

$$\frac{3\mu Q}{4\pi H^3 P_0} = \frac{1}{\ln \varphi} = f\left(\frac{Bt}{a^2}\right) \tag{11-24}$$

This result is also shown on Fig. 11-9.

Equation (11-24) shows an artifact at zero time, when φ is unity, by predicting an infinite value for Q. The maximum Q would be that in the runner itself. This model neglects the runner dynamics on the grounds that, as soon as the cavity begins to fill to a significant extent, the dynamics of the flow in the cavity dominate the process. When R^* is nearly $\tfrac{1}{2}D_r$, entrance effects probably weaken the validity of the model, which assumes only a u_r component to velocity. It does not seem worthwhile to attempt to remove the artifact at short time. Instead we recognize the problem and do not apply the analysis for $R^* < 3D_r$, an arbitrary but reasonable precaution.

Constant flow rate If the flow rate is constant then the fill time is given simply as the ratio of cavity volume to volumetric flow rate:

$$t^* = \frac{2\pi H}{Q}\left(R^2 - \frac{D_r^2}{4}\right) \tag{11-25}$$

The pressure P_0 is no longer a constant but is still given by Eq. (11-21). From the definition of R^* we have, for constant Q [Eq. (11-3)],

$$R^*\dot{R}^* = \frac{Q}{4\pi H} \tag{11-26}$$

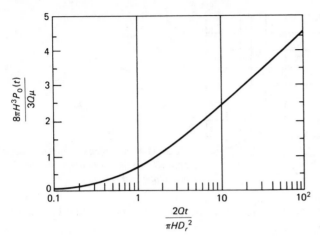

Figure 11-11 Pressure rise: newtonian isothermal filling of a disk at constant flow rate.

or
$$R^{*2} = \frac{Q}{2\pi H}t + \tfrac{1}{4}D_r^2 \qquad (11\text{-}27)$$

The pressure P_0 follows from Eq. (11-21) and increases with time according to

$$P_0 = \frac{3Q\mu}{8\pi H^3}\ln\left(1 + \frac{2Q}{\pi H D_r^2}t\right) \qquad (11\text{-}28)$$

Figure 11-11 shows the pressure rise for this case.

Example 11-1 We will analyze the behavior of a simple disk mold, as illustrated in Fig. 11-8, choosing the following parameters:

$$L_r = 1 \text{ cm} \qquad R = 9 \text{ cm}$$
$$D_r = 1 \text{ cm} \qquad 2H = 0.316 \text{ cm}$$
$$\mu = 10^5 \text{ P} \qquad \rho = 1 \text{ g/cm}^3$$

Two cases will be considered:

(a) Constant Q, with a fill time specified to be 1 s.
(b) Constant P_0, with a fill time specified to be 1 s.

Case (a): Constant Q. From Eq. (11-25),

$$Q = \frac{2\pi H}{t^*}\left(R^2 - \frac{Dr^2}{4}\right)$$

From Eq. (11-28),

$$P_0 = \frac{3Q\mu}{8\pi H^3}\ln\left(1 + \frac{2Q}{\pi H D_r^2}t\right)$$

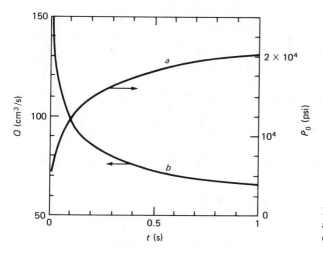

Figure 11-12 Solutions of Example 11-1: (a) $P_0(t)$ at constant Q; (b) $Q(t)$ at constant P_0.

With the parameters given we find the pressure buildup illustrated in Fig. 11-12a.

Let us examine the pressure drop to be expected *in the runner* under these conditions. From Poiseuille's law we find

$$\Delta P_r = \frac{128\mu L_r Q}{\pi D_r^4}$$

which gives a pressure drop of about 5000 psi. This constant pressure must be added to P_0 in order to overcome the resistance of the runner. It is not difficult to see that a runner of much smaller diameter than 1 cm will give a pressure drop quite a bit larger than that associated with the cavity flow itself, for this particular cavity.

Case (b): Constant P_0. Here we use Fig. 11-10. For $a = 0.055$ we find $Bt^* = 1.2$. Since we want $t^* = 1$ s, this gives

$$B = \frac{H^2 P_0}{3\mu R^2} = 1.2$$

from which it follows that $P_0 = 17,200$ psi. Figure 11-12b shows $Q(t)$ for this case, using Fig. 11-9.

11-2 AN EVALUATION OF VISCOUS HEATING IN A RUNNER

We shall find that the flow in the runner is often at such high shear rates that significant viscous heating is generated. Since the temperature dependence of viscosity is so strong, this factor must normally be accounted for. The simplest model of viscous heating assumes an adiabatic flow, for which the first law of thermodynamics leads to an expression for the temperature rise in the form

$$\rho Q C p \, dT = Q \, dp \tag{11-29}$$

Equation (11-29) simply equates the rate of increase of thermal energy to the rate at which pressure is doing work in moving the fluid through the runner at the rate Q. The temperature that appears in this equation is the *flow average*, or *cup-mixing* temperature.

We will assume that the pressure *gradient* in the runner is given by a local form of Poiseuille's law:

$$-\frac{dp}{dz} = \frac{128\mu Q}{\pi D^4} \tag{11-30}$$

where the viscosity μ may be a function of z. We will take the viscosity to depend on temperature according to

$$\frac{\mu}{\mu_0} = e^{-b(T-T_0)} \tag{11-31}$$

In the subsequent analysis we take T_0 to be the melt temperature at the entrance to the runner, and μ_0 to be the viscosity at that temperature.

If the pressure gradient is replaced by the corresponding function of temperature, using the last two equations, we obtain a differential equation for $T(z)$ of the form

$$\frac{dT}{dz} = \frac{\Delta P_0}{\rho C_P L} e^{-b(T-T_0)} \tag{11-32}$$

where ΔP_0 is the pressure drop that would exist under isothermal conditions at $T = T_0$:

$$\Delta P_0 = \frac{128\mu_0 LQ}{\pi D^4} \tag{11-33}$$

Equation (11-32) is easily integrated with the result

$$\chi = 1 + \frac{b \, \Delta P}{\rho C_P} \frac{z}{L} \tag{11-34}$$

where

$$\chi = e^{b(T-T_0)} \tag{11-35}$$

(Recall the appearance of χ in Chap. 6 in the analysis of adiabatic extrusion.) The temperature rise over the length L is given by

$$\chi_L = 1 + \frac{b \, \Delta P_0}{\rho C_P} \tag{11-36}$$

Figure 11-13 shows this result in a simple dimensionless format.

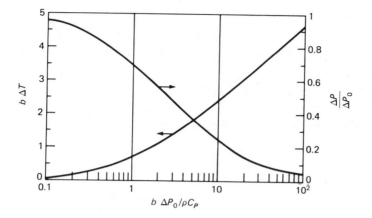

Figure 11-13 Effect of viscous heat generation, in adiabatic capillary flow, on temperature rise and pressure drop.

To find the actual pressure drop we must go back to Eq. (11-30), substitute $\mu[T(z)]$, and integrate. The result is easily found and is conveniently expressed in the form

$$\frac{\Delta P}{\Delta P_0} = \frac{\ln \chi_L}{\chi_L - 1} \tag{11-37}$$

Figure 11-13 shows this result.

Example 11-2 An ABS melt is being injection molded and enters a circular cross-sectional runner of diameter 0.4 in and length 4 in at an inlet melt temperature of 415°F. Estimate the pressure drop and the outlet temperature as a function of injection rate in the range $0.5 < Q < 20$ in^3/s.

Physical properties for the ABS polymer, at the inlet temperature, are

$$\rho = 1.12 \text{ g/cm}^3 \qquad n = \tfrac{1}{3}$$

$$C_p = 0.4 \text{ cal/g} \cdot \text{K} \qquad K = 2.6 \times 10^5 \text{ dyne} \cdot \text{s}^{1/3} \text{ cm}^2$$

$$b = 0.026 \text{ K}^{-1}$$

We will solve the problem using the *newtonian adiabatic* analysis with a viscosity appropriate to each injection rate.

We will need the following results established earlier in Chap. 5:

Nominal shear rate:
$$\dot\gamma = \frac{8(1 + 3n)Q}{n\pi D^3}$$

Poiseuille's law:
$$\Delta P_0 = \frac{128QL}{\pi D^4} \frac{3n + 1}{4n} \eta$$

Power law:
$$\eta = K\dot\gamma^{n-1}$$

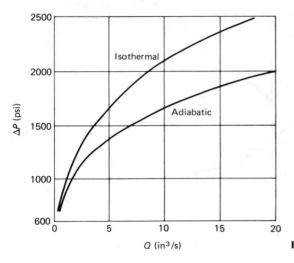

Figure 11-14 Solutions of Example 11-2.

At $Q = 0.5 \text{ in}^3/\text{s}$

$\dot{\gamma} = 119 \text{ s}^{-1}$

$\eta = 1.06 \times 10^4 \text{ P at } 415°\text{F}$

$\Delta P_0 = 744 \text{ psi}$

$\chi_L = 1 + (b \, \Delta P_0/\rho C_p) = 1.1$ (remember to convert the thermal units of C_p to mechanical units: $1 \text{ cal} = 4.2 \times 10^7 \text{ dyne·cm} = 37 \text{ in·lbf}$)

$\Delta T = 5°\text{F}$

From Eq. (11-37) we find $\Delta P = 695$ psi.

Continuing this procedure for flow rates up to 20 in³/s, we obtain results which are presented in Fig. 11-14. For comparison, the pressure drop to be expected in the absence of viscous heating is shown, using the exact power law solution. It is apparent that viscous heating is responsible for significant reductions in the pressure requirement at the higher injection rates.

We will subsequently consider viscous heating effects in the cavity. In some systems the runner flow controls the filling process, and viscous heating in the cavity is of no consequence. When this is not the case, of course, the specific geometry of the cavity and the thermal conditions at its walls control the temperature of the melt during the filling process. We will consider one of the simplest cases when we do examine the cavity problem: adiabatic flow into a disk-shaped cavity.

11-3 EFFECT OF PRESSURE-DEPENDENT VISCOSITY

Because of the very high pressures encountered in high-speed injection molding, it is necessary to consider the fact that the viscosity is a function of pressure and to give a simple model for rough evaluation of this effect. Figure 11-15 shows data for a low-density polyethylene. The viscosity is seen to increase by about an order of magnitude as the pressure varies from atmospheric to 15,000 psi. It should be apparent that viscosity data obtained at atmospheric pressure can be quite misleading if applied to a high-pressure process.

Let us begin with one of the simplest problems: fully developed Poiseuille flow. We may begin with Eq. (5-2), but note that this assumes the viscosity μ is independent of r. We will assume that $\mu = \mu(p)$, and if Eq. (5-3) holds, we then expect that Eq. (5-2) is valid as well. Thus we begin with

$$0 = -\frac{\partial p}{\partial z} + \mu \frac{1}{r} \frac{\partial}{\partial r} r \frac{\partial u_z}{\partial r} \tag{11-38}$$

Over some range of pressure we can take the viscosity to have the form

$$\mu = \mu_0(1 + b'p) \tag{11-39}$$

The variables p and u_z may be separated in Eq. (11-38), and we must solve [see (Eq. 5-6)]

$$\frac{1}{1 + b'p} \frac{dp}{dz} = \mu_0 \frac{1}{r} \frac{d}{dr}\left(r \frac{du_z}{dr}\right) = \text{constant} \tag{11-40}$$

The solution for $p(z)$ is found to be

$$\ln(1 + b'p) = \left(1 - \frac{z}{L}\right) \ln(1 + b'\,\Delta P) \tag{11-41}$$

where ΔP is the pressure drop across the length L of the tube. We assume $p = 0$ at $z = L$.

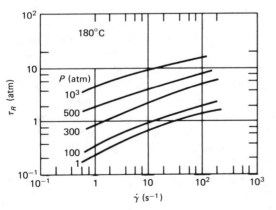

Figure 11-15 Shear stress–shear rate data at various pressures for low-density polyethylene. Data abstracted from K. Ito et al., *J. Appl. Polym. Sci., Appl. Polym. Symp.,* **20**: 109 (1973).

The velocity profile is unchanged in shape by the pressure dependence of μ and is given by

$$u_z = \frac{2Q}{\pi R^2} \left[1 - \left(\frac{r}{R}\right)^2 \right] \tag{11-42}$$

where Q, the volumetric flow rate, is now found to be related to pressure drop ΔP by [see Eq. (5-10)]

$$Q = \frac{\pi R^4 \, \Delta P}{8\mu_0 L} \frac{\ln (1 + b' \, \Delta P)}{b' \, \Delta P} \tag{11-43}$$

We may regard Eq. (11-43) as a modified Hagen-Poiseuille law, and we may easily note that Q is reduced relative to its value for $b' = 0$.

Example 11-3 For a high-molecular-weight polystyrene a value of $b' = 2.9 \times 10^{-3}$ bar^{-1} is given by Penwell et al. Would a significant effect on flow rate be expected for flow through a tube at $\Delta P = 1500$ psi? At $\Delta P = 15,000$ psi?

We may regard Eq. (11-43) to have the form

$$Q = Q_0 \, f(b' \, \Delta P)$$

where Q_0 is the flow rate in the absence of a pressure effect on viscosity. For $\Delta P = 1500$ psi $= 100$ bars, we find $b' \, \Delta P = 0.29$ and $Q/Q_0 = 0.88$. Hence the flow is expected to be reduced by about 12 percent. At 15,000 psi the flow is found to be reduced by about 50 percent.

Now let us examine the problem of flow in a disk-shaped cavity and consider μ to be a function of pressure. Equations (11-1) through (11-5) are still valid, but Eq. (11-6) is not. The normal stress terms do not vanish identically as they do for constant viscosity flow. Instead, a set of terms involving $d\mu/dr$ appears, and if $p = p(r)$ and $\mu = \mu(p)$, then additional terms appear in Eq. (11-6). We still antici-pate that Eq. (11-7) holds (see Prob. 11-18), and we find now that

$$\frac{\partial p}{\partial r} = \frac{\mu}{r} \frac{\partial^2 c}{\partial z^2} - \frac{2c}{r^2} \frac{\partial \mu}{\partial r} \tag{11-44}$$

The variables are no longer separable, and so no simple analytical solution for $p(r)$ and $c(z)$ may be obtained.

Thus, for complex cavity shapes, we are not able to provide a simple model of the effect of pressure-dependent viscosity. This is an unsatisfactory situation, and faced with it, one might be willing to try a model based on removing the offensive term from Eq. (11-44). This allows the separation of the variables $p(r)$ and $c(z)$, and we again have Eq. (11-9) but this time with $\mu = \mu(r)$. Let us again use Eq. (11-39) as a model for $\mu(p)$. Then we may solve for $p(r)$ and find

$$\frac{1 + b'p}{1 + b'P_0} = \left(\frac{2r}{D_r}\right)^{b'\mu_0 A} \tag{11-45}$$

The solution for $c(z)$ [Eq. (11-10)] is still valid, as is the relationship of A to Q [Eq. (11-14)].

Upon introducing the condition $p = 0$ at $r = R^*$, we may write

$$1 + b'P_0 = \left(\frac{2R^*}{D_r}\right)^{2\beta} \tag{11-46}$$

where

$$\beta = \frac{3Q\mu_0 b'}{8\pi H^3} \tag{11-47}$$

R^* is still given by Eq. (11-26), which is simply a material balance.

Now let us consider the constant flow rate case. The pressure will rise with time according to

$$b'P_0 = -1 + \left(1 + \frac{2Q_0}{\pi H D_r^2} t\right)^\beta \tag{11-48}$$

At the fill time [still given by Eq. (11-25)] the pressure is

$$b'P_0^* = -1 + \left(\frac{2R}{D_r}\right)^{2\beta} \tag{11-49}$$

If we define a dimensionless pressure so that

$$\beta \tilde{P}_0^* = b'P_0^*$$

or

$$\tilde{P}_0^* = \frac{8\pi H^3 P_0^*}{3Q_0 \mu_0}$$

then we may rewrite Eq. (11-49) as

$$\tilde{P}_0^* = -\frac{1}{\beta} + \frac{1}{\beta}\left(\frac{2R}{D_r}\right)^{2\beta} \tag{11-50}$$

In the absence of a pressure effect ($\beta = 0$) Eq. (11-28) holds, and at the fill time

$$P_0^* = \frac{3Q\mu}{4\pi H^3} \ln \frac{2R}{D_r}$$

or

$$\tilde{P}_0^* = 2 \ln \frac{2R}{D_r} \tag{11-51}$$

One measure of the effect of pressure-dependent viscosity is the ratio

$$\frac{\tilde{P}_0^*(\beta)}{\tilde{P}_0^*(\beta = 0)} = \frac{(2R/D_r)^{2\beta} - 1}{2\beta \ln (2R/D_r)} \tag{11-52}$$

If we return to Example 11-1 and take $b' = 4 \times 10^{-3}$ bar^{-1}, we may calculate the effect on pressure for that case. We assume $\mu_0 = 10^5$ P.

We find $\beta = 1$ and $\tilde{P}_0^*(\beta) = 56\tilde{P}_0^*(\beta = 0)$. Thus the effect of a pressure-dependent viscosity, to the extent corresponding to the given value of b', is to increase the expected maximum pressure by a substantial factor. If design were based on a viscosity measured at atmospheric pressure, the design could be inadequate with regard to the pressures imposed on the mold.

Let us now examine, a posteriori, the magnitude of the term neglected in Eq. (11-44), relative to the main viscous term. Thus we wish an estimate of

$$\epsilon \equiv \frac{2c/r^2 \, d\mu/dr}{\mu/r \, d^2c/dz^2} \tag{11-53}$$

We may show that

$$\epsilon = \frac{2c(z)}{r^2} \frac{d\mu}{dp} \tag{11-54}$$

If we use $c(0)$ as the maximum value for $c(z)$ and $r = D_r/2$ as the minimum value for r, we find

$$\epsilon = \frac{4|A|H^2}{D_r^2} \frac{d\mu}{dp} \tag{11-55}$$

Using Eq. (11-39) for μ, we find

$$\epsilon_{\max} = 8\beta \left(\frac{H}{D_r}\right)^2 \tag{11-56}$$

as an upper limit on ϵ. The minimum value of ϵ (aside from the fact that ϵ vanishes along $z = \pm H$) would be obtained by using $c(z) = c(0)$ and $r = R$, and we find

$$\epsilon_{\min} = 2\beta \left(\frac{H}{R}\right)^2 \tag{11-57}$$

As expected, small ϵ will correspond to small β. However, even when β is large it is possible for $\epsilon_{\min}$ to be quite small if $H \ll R$, as is often the case. This means only that the approximation to Eq. (11-44) will be valid over the region of the cavity away from the entrance. The analysis will still be incorrect near $r = D_r$ (unless β is very small), and the predicted maximum pressure rise could still be in error. Hence one must exercise caution in using the foregoing analysis.

11-4 RUNNER AND CAVITY COMBINATION

In deriving a simple model for isothermal newtonian flow into a disk-shaped cavity, we considered the pressure P_0 at the cavity entrance $(r = \frac{1}{2}D_r)$ to be a primary variable (either specified or desired from the solution). If we consider the pressure P_0 to be the pressure at the end of a runner which leads to the cavity, then the flow through the runner will affect the value of P_0. We outline here a simple model for a runner and cavity in series.

Actually, the development of the earlier model is still valid down to Eq. (11-11) except that we do not know P_0. If we let P_R be the pressure at the entrance to the runner, then the pressure drop across the runner is just

$$\Delta P_R = P_R - P_0 \tag{11-58}$$

The pressure drop across the cavity is still P_0, and Eq. (11-21) is still valid. We write it in the form [note Eq. (11-14)]

$$P_0 = \frac{3Q\mu}{4\pi H^3} \ln \frac{2R^*}{D_r} \tag{11-59}$$

For ΔP_R we take Poiseuille's law:

$$\Delta P_R = \frac{128\mu L Q_R}{\pi D^4} \tag{11-60}$$

where Q_R is the flow rate through the runner. Thus we find that the pressure P_R (which we take as the sum of the pressure drops ΔP_R and P_0) is given by

$$P_R = \frac{128\mu L Q_R}{\pi D^4} + \frac{3Q\mu}{4\pi H^3} \ln \frac{2R^*}{D_r} \tag{11-61}$$

If we take the simple case where the runner leads only to a single cavity, so that $Q_R = Q$, then we find

$$P_R = \left(\frac{128\mu L}{\pi D^4} + \frac{3\mu}{4\pi H^3} \ln \frac{2R^*}{D_r}\right) Q \tag{11-62}$$

Again we may consider two special cases. If Q is given as a known constant, then Eq. (11-27) holds and P_R is given directly as an explicit function of Q. The more complicated case is that of constant pressure P_R. We may use Eq. (11-26), with which Eq. (11-62) becomes a nonlinear differential equation for R^*:

$$4\pi H \left(\frac{128\mu L}{\pi D^4} + \frac{3\mu}{4\pi H^3} \ln \frac{2R^*}{D_r}\right) R^* \dot{R}^* = P_R \tag{11-63}$$

The solution follows by integration, with the result

$$(g - \tfrac{1}{2})(\varphi^2 - 1) + \varphi^2 \ln \varphi = \tau \tag{11-64}$$

where
$$g = \frac{512 L H^3}{3 D^4} \qquad \varphi = \frac{2R^*}{D_r}$$

and a dimensionless time is defined as

$$\tau = \frac{8H^2 P_R}{3\mu D_r^2} t$$

As expected, Eq. (11-22) is recovered if $g \ll 1$.

11-5 POWER LAW FLOW INTO A CAVITY

For a power law fluid (indeed, for any nonnewtonian viscosity model) the normal stress terms do not vanish from Eq. (11-4), and, as in the case of pressure-dependent viscosity, we are again faced with an inseparable partial differential equation. We will proceed by *assuming* that the shear stress term dominates the flow, and then from the resulting model we will evaluate the neglected terms.

We begin, then, with

$$\frac{\partial p}{\partial r} = \frac{\partial \tau_{rz}}{\partial z} \tag{11-65}$$

and the power law model, written in the form

$$\tau_{rz} = K\left(-\frac{\partial u_r}{\partial z}\right)^n \tag{11-66}$$

Introducing Eq. (11-7), we may separate the variables and write

$$\frac{r^n}{Kn}\frac{dp}{dr} = (-c')^{n-1}c'' = A \tag{11-67}$$

Proceeding as in the newtonian case we find the following:

$$p - P_0 = \frac{AKn}{1-n}[r^{1-n} - (\tfrac{1}{2}D_r)^{1-n}] \tag{11-68}$$

$$c(z) = \frac{(-An)^{1/n}}{1/n+1}(H^{1/n+1} - z^{1/n+1}) \tag{11-69}$$

$$A = -\frac{1}{n}\left(\frac{1+2n}{4\pi nH^{2+1/n}}Q\right)^n \tag{11-70}$$

$$P_0 = \frac{K}{1-n}\left(\frac{1+2n}{4\pi nH^{2+1/n}}\right)^n[R^{*1-n} - (\tfrac{1}{2}D_r)^{1-n}]Q^n \tag{11-71}$$

If Q is constant, for which case Eq. (11-27) gives $R^*(t)$, Eq. (11-71) gives $P_0(t)$ directly. The constant-pressure case is more difficult but still tractable. We use Eq. (11-26) to eliminate Q in favor of R^*, with the result

$$[R^{*1-n} - (\tfrac{1}{2}D_r)^{1-n}]^{1/n}\frac{d}{dt}(R^*)^2 = \text{constant} \tag{11-72}$$

Equation (11-72) may be integrated numerically to give $R^*(t)$. The details are too tedious to lay out here and are left as an exercise (Prob. 11-24). Instead, we consider the nature of the approximation made here.

First, in writing the power law, we have used $\partial u_r/\partial z$ in place of the more exact deformation rate, $(\tfrac{1}{2}II_\Delta)^{1/2}$. Thus we have implied that (see Prob. 11-25)

$$\left(\frac{\partial u_r}{\partial r}\right)^2 \ll \left(\frac{\partial u_r}{\partial z}\right)^2$$

With the solution given above we can evaluate the magnitudes of these terms, and in doing so we find that the square of the stretch rate is smaller than the square of the shear rate by a factor which is, at most, of the order of

$$\epsilon = \left(\frac{H/r}{1/n + 1} \right)^2 \tag{11-73}$$

Since r will normally be larger than H for most cavities, we see that $\epsilon \ll 1$ is a good estimate.

Next, we have ignored the normal stress terms, implying that

$$\frac{1}{r} \frac{\partial}{\partial r} r\tau_{rr} \ll \frac{\partial \tau_{rz}}{\partial z}$$

If we write

$$\tau_{rz} = K \left(\frac{c'}{r} \right)^n \tag{11-74}$$

and

$$\tau_{rr} = 2K \left(\frac{c'}{r} \right)^{n-1} \frac{c}{r^2} \tag{11-75}$$

we find that the normal stress term is smaller than the shear stress term by a factor which is just $(1 + n)\epsilon$. Thus it would appear that the simple model offered here is valid and that the elongational character of radial flow in a disk is of little importance.

These comments are consistent with observations on two other types of radial flow problems that have been studied experimentally. One is a "squeezing flow" created when two parallel disks separated by a fluid are suddenly and steadily set in motion toward each other along their axis. The fluid between the disks is squeezed out, and a radial flow exists which is similar to that in the disk-filling problem under discussion here. Experiments indicate that the neglect of normal stresses in the dynamic equations is valid, and purely viscous nonnewtonian models are adequate to describe the general aspects of this flow under typical operating conditions.

The expected exception to success in modeling this flow with a purely viscous constitutive equation occurs when the rate of squeezing is so high as to create a high Deborah number flow. In this case the Deborah number may be defined as the ratio of the fluid relaxation time to the half-time required for the plates to come together. By analogy, for the disk-filling flow, which is also an unsteady radial flow, one might use the half-time for filling the cavity in defining a Deborah number. One might expect, then, to see significant viscoelastic effects, associated with the transient nature of the flow rather than with elastic stresses per se, under conditions of very high injection rates with highly elastic fluids. No data on cavity filling are available with which one may assess the importance of transient viscoelastic effects. One may, however, gain some insight into this problem by reading the available literature on squeezing flows cited in the Bibliography.

The second radial flow problem of interest here, which has been studied experimentally, is that of steady radial flow in (and out of) the region between fixed parallel disks. Williams and coworkers have examined this problem and find that purely viscous models again are adequate to describe the major features of this *steady* flow. Since the flow is steady, the appropriate criterion for anticipated viscoelastic effects is in terms of a Weissenberg number, which crudely may be taken as the product of the shear rate for the flow and the relaxation time of the fluid. For Weissenberg numbers in excess of unity there is experimental evidence of failure of the theory, with the implication that elastic phenomena become important.

Now let us put some of these results together in a model for filling of a runner-cavity system with a power law fluid. If we add the pressure drops across a circular cross-sectional runner which feeds a circular disk cavity, then we begin with

$$P_0 = \left(8\frac{1+3n'}{n'\pi D_r^3}\right)^{n'}\frac{4K'L}{D_r}Q^{n'} + \left(\frac{2n+1}{4n\pi H^{2+1/n}}\right)^n\frac{K}{1-n}[(R^*)^{1-n} - (\tfrac{1}{2}D_r)^{1-n}]Q^n$$

(11-76)

Note that in the term for the runner we use n' and K' to remind us that if the shear rate in the runner is very different from that in the disk, then different n and K values may be appropriate to each region. In the following development we assume that a single set of values of n and K suffices. The pressure P_0 is now taken to be that at the entrance to the runner.

If Q is constant, R^* is given by Eq. (11-27), and Eq. (11-76) gives $P(t)$ explicitly. We examine the more complex problem of constant pressure filling. To do so we convert Eq. (11-76) to a differential equation for $R^*(t)$ by introducing Eq. (11-3) for Q. The solution may be put in the form

$$\frac{P_0^{1/n}t}{\pi HD_r^2} = \int_1^\varphi (c_1\varphi^n + c_2\varphi)^{1/n}\,d\varphi$$

(11-77)

where

$$\varphi = \frac{2R^*}{D_r}$$

$$c_2 = \frac{K}{1-n}\left(\frac{2n+1}{4n\pi H^{2+1/n}}\right)^n\left(\frac{D_r}{2}\right)^{1-n}$$

$$c_1 = \left(8\frac{1+3n}{n\pi D_r^3}\right)^n\frac{4KL}{D_r} - c_2$$

The integration may be performed analytically only if n is an inverse integer. The fill time is obtained upon setting the upper limit in the integral at $\varphi = 2R/D_r$.

Example 11-4 Find the fill time for a disk-runner system for which the following data are available:

$$D_r = 0.48 \text{ cm} \qquad H = 0.159 \text{ cm}$$

$$L = 2.54 \text{ cm} \qquad R = 9 \text{ cm}$$

Melt properties at 202°C (PVC):

$$n = \tfrac{1}{2} \qquad K = 4 \times 10^5 \text{ dyne} \cdot \text{s}^{1/2}/\text{cm}^2$$

Constant pressure of 15,000 psi at the runner entrance. For the case $n = \tfrac{1}{2}$ Eq. (11-77) may be integrated to give $\varphi(t)$ in the form

$$t = \frac{\pi H D_r^2}{P_0^2} \left[\tfrac{1}{2} c_1^2 (\varphi^2 - 1) + \tfrac{4}{5} c_2 c_1 (\varphi^{2.5} - 1) + \tfrac{1}{3} c_2^2 (\varphi^3 - 1) \right] \qquad (11\text{-}78)$$

Using the specified values of the parameters we find a fill time of 1.25 s.

11-6 VISCOUS HEATING IN A CAVITY FLOW— ADIABATIC ANALYSIS

We can calculate the maximum possible effect of viscous heat generation from an adiabatic analysis, which begins with Eq. (11-29). For a newtonian melt Eq. (11-8) is still valid, and we write it in the form

$$dp = \frac{A\mu \, dr}{r} = -\rho C_p \, dT \qquad (11\text{-}79)$$

If we use Eq. (11-31) as a model for $\mu(T)$, we find

$$\frac{A\mu_0 \, dr}{r} = -\rho C_p e^{b(T-T_0)} \, dT \qquad (11\text{-}80)$$

Integration, with the condition $T = T_0$ at $r = \tfrac{1}{2}D_r$, gives

$$\ln \frac{2r}{D_r} = -\frac{\rho C_p}{A\mu_0 b} \left(e^{b(T-T_0)} - 1 \right) \qquad (11\text{-}81)$$

In effect, we use Eq. (11-81) to give $T(r)$, and thus we find $\mu[T(r)]$. When this is substituted in Eq. (11-79) (the left-hand half of the two equations), we obtain a differential equation for $p(r)$. After a simple integration and some algebra we find

$$p - P_0 = -\frac{\rho C_p}{b} \ln \left(1 + \frac{A\mu_0 b}{\rho C_p} \ln \frac{2r}{D_r} \right) \qquad (11\text{-}82)$$

The solution for A is unchanged from the isothermal case [Eq. (11-14)]. Upon setting $p = 0$ at $r = R^*$, we find

$$P_0 = \frac{\rho C_p}{b} \ln \left(1 - \frac{3Q\mu_0 b}{4\pi H^3 \rho C_p} \ln \frac{2R^*}{D_r} \right) \qquad (11\text{-}83)$$

For constant Q, $R^*(t)$ is given by Eq. (11-27), and so $P_0(t)$ follows explicitly from Eq. (11-83) (see Prob. 11-33).

For constant pressure, Eq. (11-83) becomes a differential equation for $R^*(t)$. The solution takes the form

$$\varphi^2 \ln \varphi - \tfrac{1}{2}(\varphi^2 - 1) = \alpha t \tag{11-84}$$

where

$$\varphi = \frac{2R^*}{D_r} \qquad \alpha = \frac{8H^2\rho C_p}{3\mu_0 D_r^2 b}\left[1 - \exp\left(\frac{bP_0}{\rho C_p}\right)\right] \tag{11-85}$$

We may easily show that Eq. (11-84) reduces to the corresponding isothermal case, Eq. (11-22), in the limit as b vanishes. We may also see that the effect of viscous heating on the fill time is to decrease the fill time by a factor

$$\frac{\alpha(b)}{\alpha(0)} = \frac{\rho C_p}{bP_0}\left[\exp\left(\frac{bP_0}{\rho C_p}\right) - 1\right] \tag{11-86}$$

It is important to keep in mind that the adiabatic analysis gives the maximum possible temperature rise. Adiabatic filling is more nearly achieved with rapid injection rates and with cavity surfaces which are not cooled until after injection. At slower injection rates and with cooled surfaces, there is time for any generated heat to be conducted to the cavity boundaries.

The analysis of the adiabatic temperature rise for flow of a power law fluid into a disk-shaped mold follows the same lines as that just given for the newtonian fluid. We take

$$K = K_0 e^{-b(T - T_0)} \tag{11-87}$$

Though more tedious algebraically, the power law case may be carried through with little difficulty to give

$$P_0 = nAK_0\left(\frac{D_r}{2}\right)^{1-n} \int_1^{2R^*/D_r} \frac{dx}{mx - (1 + m)x^n} \tag{11-88}$$

where

$$m = \frac{AK_0 nb(D_r/2)^{1-n}}{(1 - n)\rho C_p} \tag{11-89}$$

and A is still given by Eq. (11-70).

For the case $n = \tfrac{1}{2}$, for example, we find

$$P_0 = \frac{\rho C_p}{b} \ln\left[-m\left(\frac{2R^*}{D_r}\right)^{1/2} + (1 + m)\right] \tag{11-90}$$

We keep in mind that m depends upon Q. If we consider the constant-pressure case, then Eq. (11-90) may be written as a differential equation for $R^*(t)$, and integrated to give

$$\tfrac{1}{2}(\varphi^2 - 1) - \tfrac{4}{5}(\varphi^{2.5} - 1) + \tfrac{1}{3}(\varphi^3 - 1) = \alpha_{1/2}t \tag{11-91}$$

where

$$\alpha_{1/2} = \left[\frac{1 - \exp{(bP_0/\rho C_p)}}{\dfrac{4K_0\, b(D_r/2)^{3/2}}{\rho C_p H^{3/2}}} \right]^2 \qquad (11\text{-}92)$$

It may be shown that when $b = 0$ Eq. (11-91) is identical to Eq. (11-78). The effect of viscous heating on the fill time is given by the factor

$$\frac{\alpha_{1/2}(b)}{\alpha_{1/2}(0)} = \left[\frac{1 - \exp{(bP_0/\rho C_p)}}{bP_0/\rho C_p} \right]^2 \qquad (11\text{-}93)$$

If this is compared to the corresponding newtonian result, Eq. (11-86), it is seen that the effect of viscous heating is much stronger for nonnewtonian fluids than for newtonian, *at constant pressure*.

11-7 THE EFFECT OF HEAT TRANSFER TO THE CAVITY WALLS—QUALITATIVE COMMENTS

The models presented above all ignore one of the major phenomena occurring in mold filling: heat transfer to the (usually) cold cavity surfaces. In Chap. 13 we establish a basis for quantitative consideration of the effect of heat transfer. Here we offer some brief qualitative remarks.

Figure 11-16 suggests the situation observed in the filling of a rectangular cavity. Heat transfer to the cold walls may cause solidification of the melt before the cavity is filled. In effect this reduces the cross-sectional area available for flow, which would cause a significant increase in pressure buildup if constant flow rate is imposed on the system. (If constant pressure is imposed, then the flow rate will decay as the cavity freezes off.) The thermal " design " of the process must be such as to ensure that the solid polymer does not occlude the cavity (or the runner, as well) before filling is completed. If premature freezing occurs the cavity does not fill completely, giving a short shot.

It should be apparent that a very complex model would be required to describe the more realistic filling problem suggested in Fig. 11-16. Numerical methods involving digital computation are used, and the Bibliography suggests several relevant articles. In general these techniques are fairly successful, and while

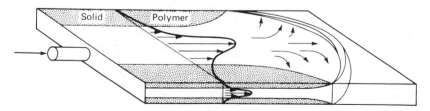

Figure 11-16 Filling of a rectangular cavity with cold walls.

they represent modifications of the simple analyses presented in this text, we will not dwell on the details here. In Chap. 13 we will develop enough information regarding heat transfer that we can assess the range of applicability of the simple models of this chapter.

With reference to Fig. 11-16, we see that the flow field is distorted by strong nonisothermal effects. A form of "channeling" occurs because of the stagnation of the much more viscous cold fluid (or even solid) that forms at the cavity surfaces. The shape of the velocity profile, and to some extent the shape of the advancing front, depends strongly on thermal properties of the melt, such as the temperature coefficient for viscosity. Flow visualization studies referred to in the Bibliography give direct evidence of this strong nonisothermal effect.

We may end our treatment of simple models of the cavity-filling process by examining some data obtained by molding an ABS disk on a 375-ton Cincinnati Milacron reciprocating-screw injection-molding machine with a 32-oz shot size. Figure 11-17 shows the pressure measured as a function of fill rate under various operating conditions. We would like to test the applicability of the simple models outlined here to the description of these data. We need the following information.

The polymer is Monsanto's Lustran ABS Q 714 for which the following physical property data are available:

$$\rho = 1.02 \text{ g/cm}^3 \qquad C_p = 2.43 \times 10^7 \text{ ergs/g} \cdot \text{K}$$

Power law parameters: $n = 0.29 (10^2 < \dot{\gamma} < 10^4 \text{ s}^{-1})$

$$K = 302 \exp \frac{3460}{T_K} \text{ dyne} \cdot \text{s}^{0.29}/\text{cm}^2$$

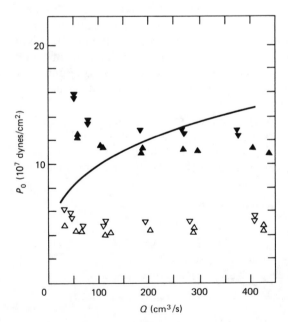

Figure 11-17 Pressure versus flow rate (at the fill time) for a disk cavity with an ABS melt.

	H, cm	T, K
▼	0.127	516
▲	0.127	538
▽	0.203	516
△	0.203	538

Cavity surface temperature is 311 K. The curve is the adiabatic theory for $H = 0.127$ cm, $T = 516$ K. (*Data are from J. F. Stevenson, A. Galskoy, and K. K. Wang, "Injection Molding of Thermoplastics. Part I: Disk Filling Experiments," personal communication.*)

The mold geometry is that of a disk for which $H = 0.127$ cm or $H = 0.203$ cm. The inlet pressure P_0 is measured at a radial position $\frac{1}{2}D_r = 1.9$ cm. (This is not the actual radius of the inlet runner, but we must use, for $\frac{1}{2}D_r$, the actual radial position of pressure measurement.) The pressures plotted are not quite at the fill time but at a time such that R^* is at $r = 10.5$ cm. Thus we use $R = 10.5$ cm in the model.

The inlet temperatures may be varied somewhat, and the mold is cooled by water at 38°C.

We make a calculation, using the adiabatic theory, for the case $H = 0.127$ cm and $T = 243°C = 516$ K. We will use Eq. (11-88), with $R^* = 10.5$ cm and other parameters as noted here. It is first necessary to calculate m [Eq. (11-89)], which requires a value for b. It is not difficult to see that (note Prob. 11-8)

$$b = \frac{\Delta E}{RTT_0} = \frac{3460}{TT_0} = 0.013 \text{ K}^{-1} \qquad \text{(at } T = T_0 = 516 \text{ K)}$$

We then find that $m = -0.006Q^{0.29}$ (Q in cm³/s) which gives values of m much less than unity. Thus we really have the isothermal model to deal with, since viscous heating is apparently negligible in the cavity. From Eq. (11-71) we find

$$P_0 = 2.6 \times 10^7 Q^{0.29} \qquad \text{(P_0 in dynes/cm}^2 \text{ and Q in cm}^3/\text{s)}$$

This curve is plotted on Fig. 11-17. The results are not very good, although at the expenditure of about 10 min we certainly have a rough estimate of the data. At low Q considerable heat transfer to the cavity surfaces occurs, and the observed pressure reflects the strong temperature dependence of viscosity and, possibly, some freezing within the mold. At high Q adiabatic behavior is more nearly achieved, but the model and data are still in poor agreement. Viscous heating in the cavity is still not significant, but it may be in the nozzle-sprue system (see Prob. 11-42). This would tend to lower the model prediction, if accounted for, and increase the deviation between model and reality. However, a compensating effect is the effect of pressure on viscosity which, if accounted for, would raise the theoretical curve. No data are available with which to make a corrected calculation, although estimates could certainly be made (see Prob. 11-43). Thus we conclude that the simple analytical models are useful but must be used with some caution.

11-8 BALANCING OF RUNNERS

Figure 11-18 shows a sprue and runner system from a standard six-cavity telephone-handle molding die. The cavities fed by this system do not fill at equal rates because the system is hydrodynamically unbalanced. Figure 11-19 shows the results of a series of *short shots*, which are injections that are stopped before the entire cavity system is filled. Short shots enable one to see certain features of the filling patterns that cannot be seen from examination of a completed shot. It is obvious, from Fig. 11-19, that the three pairs of cavities do not fill at equal rates,

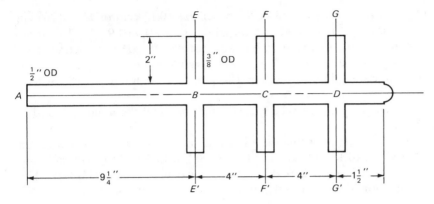

Figure 11-18 Sprue-runner system for a six-cavity telephone-handle molding die.

which reflects the inbalance in the runner system. Quantitative data illustrating the unbalanced-fill situation are shown in Fig. 11-20. A thorough analysis of this specific system is given in the interesting paper of Williams and Lord.

Hydrodynamic balancing of runners is quite a complex problem, although at first glance it appears deceptively simple. With reference to Fig. 11-18, path AB is the common sprue for all six cavities. It is assumed that each *pair* of cavities is balanced, since each is symmetrically placed about the sprue axis. It is necessary

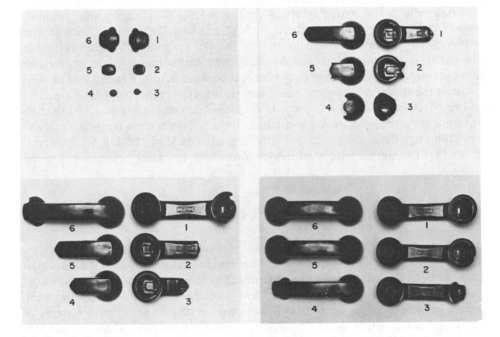

Figure 11-19 Short shots in a telephone-handle molding die. Note the asymmetry, due to unbalanced runners. (*Photo courtesy of G. Williams and H. A. Lord.*)

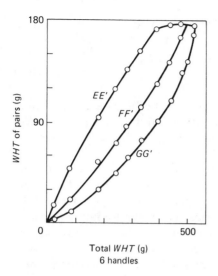

Total *WHT* (g)
6 handles

Figure 11-20 Weights corresponding to short shots as in Fig. 11-19.

to balance the runner paths *BE*, *BF*, and *BG*. This requires that the pressure drops across each of these paths be equal so that the pressure at the entrance of each cavity (points *E*, *F*, *G*) will be the same.

For a power law fluid we can always write the pressure drop–flow rate relationship in the form [Eq. (5-17)]

$$\Delta P = \left(\frac{1+3n}{n\pi R^3}\right)^n \frac{2KL}{R} Q^n \tag{11-94}$$

Let us look at isothermal power law flow and "design" runner *CF* so that the pressure drop over the path from *A* to *E* is the same as that over the path *AF*. Since *AB* is a common path to the cavities at *E* and *F*, we require

$$\Delta P_{BE} = \Delta P_{BCF} \tag{11-95}$$

If we let Q be the total flow rate through the sprue *AB*, then we want the flow rates through all six runners to be one-sixth of Q. Thus we have

$$Q_{BE} = \tfrac{1}{6}Q = Q_{CF} \tag{11-96}$$

However, a simple material balance shows us that

$$Q_{BC} = \tfrac{2}{3}Q \tag{11-97}$$

This lets us write the pressure drops as

$$\Delta P_{BE} = \left(\frac{1+3n}{n\pi R_{BE}^3}\right)^n \frac{2KL_{BE}}{R_{BE}} \left(\frac{Q}{6}\right)^n \tag{11-98}$$

$$\Delta P_{BCF} = \Delta P_{BC} + \Delta P_{CF}$$

$$= \left(\frac{1+3n}{n\pi R_{BC}^3}\right)^n \frac{2KL_{BC}}{R_{BC}} \left(\frac{2Q}{3}\right)^n + \left(\frac{1+3n}{n\pi R_{CF}^3}\right)^n \frac{2KL_{CF}}{R_{CF}} \left(\frac{Q}{6}\right)^n \tag{11-99}$$

If we now equate the pressure drops in these last two equations, we obtain a relationship among the geometric parameters:

$$\frac{L_{BE}}{R_{BE}^{3n+1}} = 4^n \frac{L_{BC}}{R_{BC}^{3n+1}} + \frac{L_{CF}}{R_{CF}^{3n+1}} \tag{11-100}$$

If we take R_{CF} and R_{BC} as the only free parameters, and otherwise use the values shown in Fig. 11-18, we find that there is no unique solution to this problem. We must introduce some *arbitrary* constraint, such as choosing R_{CF} to have a specific value, whereupon a solution for R_{BE} follows.

For example, using the values shown in Fig. 11-18, and choosing $R_{CF} = 0.1$ in, we find (assuming $n = \frac{1}{2}$ as an example) that $R_{BE} = 0.087$ in. These values are reasonably close to each other, which means that the runners will be fairly well balanced thermally as well. (By thermal balancing we imply that heat transfer from the runners occurs at nearly equal rates, so that, for example, the melt entering the two cavities is nearly at the same temperature.)

The fact that n enters the calculations creates a problem, since the runners may no longer be in balance if a different polymer is used, having a different n. For example, if the calculation above is repeated for the case $n = 1$, we find (again choosing $R_{CF} = 0.1$ in) that $R_{BE} = 0.095$ in. Some aspects of this point are picked up in Prob. 11-44.

Other factors which would affect balancing include viscous heating and a pressure-dependent viscosity. If any significant heating occurs over path BC, for example, then the melt would have a different viscosity in runners BE and CF, and an isothermally balanced system would no longer be balanced. Similar comments can be made with regard to a pressure-dependent viscosity. It should be apparent that while the concept of balancing is quite simple, its execution in practice is likely to be difficult.

PROBLEMS

11-1 In deriving Eq. (11-6) it is found that the terms involving the normal stresses collectively vanish. Carry out the derivation in detail.

11-2 For any purely viscous fluid, for which $\tau = \eta(II_\Delta) \, \Delta$, do the normal stress terms drop out of Eq. (11-4)?

11-3 For the purely viscous fluid, Eq. (11-7) still holds for the kinematics of radial flow in a disk. Give the second invariant of Δ for this flow in terms of the function c.

11-4 Derive Eq. (11-22).

11-5 For the runner-cavity system of Example 11-1 plot fill time versus pressure at the nozzle. The fill times of interest would be in the range 0.5 to 3 s. Include the runner pressure drop in your analysis. Assume isothermal flow.

11-6 Rework Prob. 11-5 to account for viscous heating in the runner. Treat the cavity flow as isothermal. Take the thermal properties to be those given in Example 11-2.

11-7 The energy equation for laminar steady tube flow has the form

$$\rho C_p u_z \frac{\partial T}{\partial z} = k \frac{1}{r} \frac{\partial}{\partial r} \left(r \frac{\partial T}{\partial r} \right) + \tau_{rz} \, \Delta_{rz}$$

Derive Eq. (11-29) by integrating each term above over the cross section of the tube. In doing so, point out where the assumption of adiabatic flow enters. Is it necessary to make any assumption about heat conduction in order to arrive at Eq. (11-29)? Show clearly how the average temperature is defined.

11-8 The temperature dependence of viscosity is often expressed in terms of an activation energy ΔE:

$$\frac{\mu}{\mu_0} = \exp \frac{\Delta E}{R}\left(\frac{1}{T} - \frac{1}{T_0}\right)$$

where R is the gas constant and T is an absolute temperature.

(a) Show that the constant b of Eq. (11-31) is related to ΔE by $b = \Delta E/RTT_0$.

(b) Convert the ΔE values of the following table to b values:

		ΔE, kcal/g·mol	ρ, g/cc	Cp, cal/g·K
Polyethylene	Marlex 50	5.8	0.86	0.65
Polystyrene	Styron 700	10	1.15	0.44
Nylon	Zytel 101 NC	21	1.07	0.56
PVC	BF Goodrich	28	1.3	0.45
ABS	Marbon	12	1.12	0.38

Source: This table is abstracted from P. C. Wu, C. F. Huang, and C. G. Gogos, *Polym. Eng. Sci.*, **14:** 223 (1974).

Comment on the manner in which b depends on temperature.

11-9 The temperature dependence of viscosity appears to be an important parameter in the nonisothermal flow analysis. Show, however, that for small temperature rises, $T - T_0$ is independent of b and is given by

$$T - T_0 = \frac{\Delta P_0}{\rho Cp}$$

Give a criterion for application of this result in terms of the value of $b(T - T_0)$.

11-10 Can viscous heating create a maximum in the dependence of ΔP on Q for newtonian tube flow? Investigate the mathematical character of Eq. (11-37) to answer this. Keep in mind that ΔP_0 is a function of Q through Eq. (11-33).

11-11 A particular PVC molding compound is found to degrade (and give an off-color product) if its temperature rises above 495°F. Suppose this melt is injected into a sprue-runner system which can be modeled as a circular tube of diameter $\frac{3}{4}$ in and length 4 in. If the injection temperature is 375°F, what is the maximum tolerable injection rate? (The desired injection rate is in the neighborhood of 5 in³/s.)

Take the thermal properties as given in Prob. 11-8, and assume the PVC is a power law fluid with $K = 4 \times 10^5$ dyne·s$^{1/2}$/cm^2 at 200°C, and $n = \frac{1}{2}$.

11-12 Repeat Example 11-2 for a PVC melt. See Prob. 11-11 for thermal and rheological properties.

11-13 Consider adiabatic newtonian flow through a circular runner. Viscous heating may raise the temperature at the end of the runner above some critical value at which degradation occurs. Is it possible to *reduce* the outlet temperature by *increasing* the injection temperature? Give a general formulation of this problem and obtain the solution.

11-14 Plot the data of Fig. 11-15 and give the best estimate of a parameter b' if the viscosity is taken to vary with pressure according to $1 + b'P$ or $e^{b'P}$.

11-15 Find the velocity profile for steady isothermal fully developed Poiseuille flow in a tube of circular cross section, for a newtonian fluid whose viscosity is a function of pressure according to $\mu = \mu_0 e^{b'P}$.

Show that if the pressure dependence of viscosity were ignored and Poiseuille's law [Eq. (5-10)] used to calculate viscosity, the error in viscosity would be given by the factor

$$\frac{1 - e^{-b' \Delta P}}{b' \Delta P}$$

where ΔP is the imposed pressure drop across the tube.

11-16 Repeat Prob. 11-15 but for a power law fluid. Take $n =$ constant, but let K depend on pressure according to

$$K = K_0 e^{b'P}$$

Give an expression for evaluating the error made in using the case $b' = 0$ to calculate ΔP for a given Q.

11-17 An article is being injection molded of polyethylene. The most important, and rate-determining, part of the filling process involves flow into a capillary of length 1 in and diameter $\frac{1}{16}$ in. If a pressure of 15,000 psi is imposed at the entrance to this capillary, what is the fill time? Use Fig. 11-15 and the table in Prob. 11-8 for physical property data. Include viscous heating and pressure-dependent viscosity.

11-18 In writing Eq. (11-44) it was implied that Eq. (11-7) holds when $\mu = \mu(p)$. Examine the z component of the dynamic equations for this flow, and the continuity equation. Discuss the nature of the approximation that leads to Eq. (11-44).

11-19 Solve Eq. (11-44) for flow into a disk-shaped mold, but assume $\mu = \mu_0 e^{b'p}$. Show that Eq. (11-46) is replaced by

$$e^{-b'P_0} = 1 - 2\beta \ln \frac{2R^*}{D_r}$$

Show that there is an upper limit to R^*. Is this physically realistic? Why does this happen?

11-20 From the solution of Prob. 11-19, find the analog of Eqs. (11-50) and (11-52).

11-21 From the solution of Prob. 11-19, find the analog of Eq. (11-56). Is this relevant to the questions raised in Prob. 11-19?

11-22 Verify Eq. (11-54).

11-23 Derive Eqs. (11-68) through (11-71).

11-24 Derive Eq. (11-72). Show that for $1/n$ an integer Eq. (11-72) may be integrated analytically. Give the solution for $R^*(t)$ for the case $n = \frac{1}{2}$.

11-25 Give II_Δ for disk flow, assuming $u_r = c/r$ and $u_z = 0$.

11-26 Verify Eq. (11-73).

11-27 Derive Eq. (11-77).

11-28 For the system described in Example 11-4, plot filling time versus P_0 for $5000 < p_0 < 15{,}000$ psi.

11-29 For the system described in Example 11-4, estimate the temperature rise at the end of the runner due to viscous heat generation. Assume the melt enters the runner at 202°C, and take the thermal properties given for PVC in the table in Prob. 11-8.

11-30 Work Prob. 11-28 but account for the effect of viscous heating. Begin by evaluating the significance of the effect, and let that guide the complexity of the modeling procedure you choose.

11-31 Verify Eq. (11-78).

11-32 For the system described in Example 11-4, how different are the shear rates in the runner and the cavity? Does your answer depend on flow rate?

11-33 Show, from Eq. (11-83), that if $Q =$ constant is specified, the pressure P_0 becomes infinite at a finite value of R^*. Interpret this result physically.

11-34 Show that Eq. (11-84) reduces to Eq. (11-22) for the case $b = 0$.

11-35 Prove the assertion of Eq. (11-86).

11-36 A melt is being injection molded into a disk-shaped cavity at a pressure of 10,000 psi. Take $\rho C_p = 300$ in·lbf/in^3·K and $b = 0.025$ K^{-1}. Is the fill time significantly reduced by viscous heating?

11-37 Show that Eq. (11-88) may be integrated to give

$$P_0 = \rho \frac{C_p}{b} \ln \left[-m\left(\frac{2R^*}{D_r}\right)^{1-n} + (1+m) \right]$$

11-38 Derive Eqs. (11-91) and (11-92).

11-39 Use the simple model for adiabatic filling of a disk mold to predict P_0 versus Q at fill ($R = 10.5$ cm) for the case $H = 0.203$ cm, shown in Fig. 11-17. Would you expect the model to be more appropriate for the thinner disk, or for the larger disk?

11-40 Show that $P_0 \approx H^{-1-2n}$ in the absence of viscous heating or any other thermal effects for a disk-shaped cavity. Show that if adiabatic heating is the dominant thermal effect, P_0 becomes independent of H_0 as the heating effect increases in magnitude ($|m| \gg 1$).

11-41 Compare the dependence of P_0 on H_0, as exhibited by the data of Fig. 11-17, to the results of Prob. 11-40 and infer from that alone whether viscous heating is significant in these data. Is the inference consistent with the calculated value of m?

11-42 The temperature given with the data of Fig. 11-17 is based on a measurement at the end of the extruder. Between the extruder and the die is a nozzle-sprue system equivalent to a pipe of length of 30.5 cm and diameter 1.27 cm, followed by a tube of length 3.81 cm and diameter 0.63 cm. Could viscous heating cause a significant temperature rise before the cavity?

11-43 Could a pressure-dependent viscosity account for the failure of the adiabatic model to fit the data of Fig. 11-17? No value of b' [in $K = K_0(1 + b'P)$] is available for the ABS polymer. Use the models in this chapter to estimate the probability that this point is worth pursuing. Take $b' = 0.001$ atm^{-1}. If you feel the effort is justified, correct the model and predict the data. If not, justify your reluctance to do the calculation on technical grounds.

11-44 The sprue-runner system of Fig. 11-18 is balanced for a power law fluid ($n = \frac{1}{2}$). Dimensions are as shown on the figure, and $R_{CF} = 0.1$ in and $R_{BE} = 0.087$ in. Suppose the pressure at point B is 10,000 psi. Ignoring cavity resistance, calculate the flow rates at points E and F, and verify the balancing. Take $K = 10^5$ in cgs units.

Suppose the same system were used for a newtonian fluid. To what extent would the system be out of balance?

11-45 Continue the example of Sec. 11-8 and balance the runner DG for a power law fluid ($n = \frac{1}{2}$).

BIBLIOGRAPHY

A comprehensive book covering many practical aspects of injection-molding system design and operation is

Rubin, I. I.: "Injection Molding: Theory and Practice," Wiley-Interscience, New York, 1973.

11-3 Effect of pressure-dependent viscosity

Penwell, R. C., R. S. Porter, and S. Middleman: Determination of the Pressure Coefficient and Pressure Effects in Capillary Flow, *J. Polym. Sci.*, A-2, **9:** 731 (1971).

11-5 Power law flow into a cavity

A flow related to filling of a disk-shaped cavity, which has been studied experimentally, is *squeezing flow*. Several papers shed some light on the role of viscoelasticity in this type of flow field.

Leider, P. J., and R. B. Bird: Squeezing Flow between Parallel Discs, *Ind. Eng. Chem. Fund.*, **13**: 336, 342 (1974).

Kramer, J. M.: Large Deformations of Viscoelastic Squeeze Films, *Appl. Sci. Res.*, **30**: 1 (1974).

Brindley, G., J. M. Davies, and K. Walters: Elastic-Viscous Squeeze Films, *J. Non-Newtonian Fluid Mech.*, **1**: 19 (1976).

For the case of *steady* radial flow one should read

Laurencena, B. R., and M. C. Williams: Radial Flow of Non-Newtonian Fluids between Parallel Plates, *Trans. Soc. Rheol.*, **18**: 331 (1974).

11-7 The effect of heat transfer to the cavity walls

Numerical and experimental studies of injection molding under realistic nonisothermal nonadiabatic conditions include the following:

Harry, D. H., and R. G. Parrott: Numerical Simulation of Injection Mold Filling, *Polym. Eng. Sci.*, **10**: 209 (1970).

Kamal, M. R., and S. Kenig: The Injection Molding of Thermoplastics, *Polym. Eng. Sci.*, **12**: 294, 302 (1972).

Williams, G., and H. A. Lord: Mold Filling Studies for the Injection Molding of Thermoplastic Materials, *SPE 33rd ANTEC*, 307, 318 (1975).

Photographic visualization of mold filling, with interpretations of the observations, may be found in

White, J. L., and H. B. Dee: Flow Visualization for Injection Molding of Polyethylene and Polystyrene Melts, *Polym. Eng. Sci.*, **14**: 212 (1974).

White, J. L.: Fluid Mechanical Analysis of Injection Mold Filling, *Polym. Eng. Sci.*, **15**: 44 (1975).

11-8 Balancing of runners

This material is drawn from the experiments described in Williams and Lord above.

TWELVE

MIXING

He who learns by Finding Out
has sevenfold
The Skill of him who learned by
Being Told.

Guiterman

Hardly any polymer process occurs without mixing playing an important, and often controlling, role. Yet of all the processes which make up an integrated operation, mixing is probably the least understood, and the least amenable to analysis. Why should this be so?

One reason is that the term *mixing* refers to so many different operations having many apparently different ultimate goals. Let us mention some typical mixing processes:

1. It is found that when polystyrene and polybutadiene are intimately mixed in the molten state, the resulting solidified material has superior mechanical properties (relative to either component) when molded into an article. We wish to mix and melt the two solid components just prior to an injection-molding operation.
2. Polystyrene can be foamed by mixing the molten polymer with a "blowing agent," a low molecular weight volatile solvent which, if intimately mixed within the polymer, will form small bubbles when the operating pressure falls below the vapor pressure of the solvent dissolved within the polymer. We wish to perform this mixing operation at some stage upstream of the foaming step.

3. Prior to injection molding an automobile tire, the synthetic (polymer) rubber must be mixed with carbon black particles which impart needed mechanical and thermal properties to the rubber.
4. A commercial PVC coating resin must be compounded with an antioxidant prior to the extrusion-coating operation.
5. Short lengths of synthetic fibers are to be added to a polymer to provide mechanical reinforcement of the final formed article. The fibers must be uniformly distributed.
6. To improve the processibility of polystyrene in a particular operation, zinc stearate is added as a "lubricant" and must be intimately mixed within the polymer melt.
7. Polystyrene is being polymerized in a batch reactor. The reaction is highly exothermic, and if "hot spots" develop in the reactor an undesirable distribution of molecular weights will occur. It is necessary to mix the reactor contents thoroughly so as to maintain a very uniform temperature distribution within the reactor.
8. In a suspension or emulsion polymerization, monomer is charged to a stirred reactor containing an inert liquid in which the monomer is insoluble. An agitator must break the monomer phase into small droplets, each of which is then a miniature batch reactor. These small droplets have a relatively high ratio of interfacial area to volume, and as a consequence the heat of reaction is easily carried off by the suspending phase, giving good control over the course of the reaction.

We could continue like this for several more pages. The appropriate question at this point is: What *common* elements do these processes have? The answer is that in all cases the operative goal is to create and maintain intimate and uniform contact between two or more materials or among the elements of a single material. If the materials are immiscible, as in the case of carbon black and polymer, then a uniform spatial distribution of particles throughout the polymeric matrix is the goal. If the materials are miscible then dissolution must be promoted, followed by promotion of uniform concentration within the solution.

Discussions of mixing must focus on two aspects: There must be some objective measures of the "goodness" of mixing, and there must be some means of evaluating the "efficiency" of mixing in terms of the power required to achieve a certain degree of goodness in a specified time.

12-1 GOODNESS MEASURES

It will be sufficient to introduce mixing concepts for the case of a binary, i.e., two-component, system. Usually one phase or component can be considered the *major* component or continuous phase, while the *minor* component or dispersed phase may be thought of as the material being mixed *into* the continuous phase. These terms are merely semantic crutches in some cases when two streams of

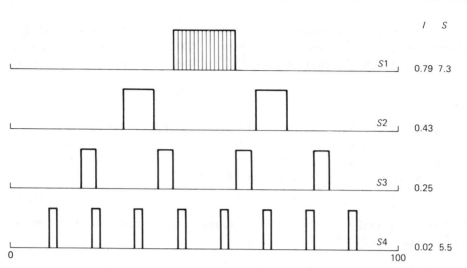

I *S*

S1 0.79 7.3

S2 0.43

S3 0.25

S4 0.02 5.5

0   100

Figure 12-1 A model of a mixing process. Sixteen elements of indivisible height, initially clumped together as in distribution *S1*, are successively subdivided and uniformly redistributed in subsequent mixtures. The intensity of each distribution is shown under *I*, and the scale under *S*.

nearly equal proportions are mixed together. Nevertheless one could arbitrarily define one of the streams as the minor phase, to be mixed "into" the other.

To establish some concepts of goodness of mixing we begin with a simple model of a process involving a one-dimensional space, a line, along which certain elements may be placed. Figure 12-1 shows successive stages in the mixing of this two-component system.

Sixteen elements, each having a unit *mass*, may be placed on any of 100 spaces that the line is divided into. The overall *concentration* of this system is $\phi = 0.16$. We may define the concentration "at a point" by arbitrarily deciding on a *sample space* and measuring the fraction of that space occupied by the unit elements. It should be obvious that the measured concentration depends upon where the measuring points are located, since the distribution of mass is not spatially uniform. The concentration will also depend upon the sample size but presumably will approach ϕ as the sample size increases toward the size of the entire system. (In the following it is to be understood that all concentrations are normalized; they represent the *fraction* of space occupied by elements.)

If we randomly sample the space, and make, say, N measurements of concentration C_i, then the average concentration may be calculated from

$$\bar{C} \equiv \frac{1}{N} \sum_{i=1}^{N} C_i \tag{12-1}$$

In general, $\bar{C}$ will differ from the true concentration ϕ because only a finite number N of samples is averaged.

One measure of the homogeneity of a mixture is the extent to which the concentrations in various regions of the space differ from the mean concentration. The *variance* of the measurements provides such an indication and is defined as

$$s^2 = \frac{1}{N-1} \sum_{i=1}^{N} (C_i - \bar{C})^2 \tag{12-2}$$

A small variance implies that most of the samples yield a concentration C_i that is close to the mean $\bar{C}$ of all samples, thus suggesting a homogeneous system. The maximum variance occurs if the elements are completely segregated, and it can be shown that the maximum variance is given by

$$s_0^2 = \bar{C}(1 - \bar{C}) \tag{12-3}$$

If s^2 is normalized to its maximum value the resulting parameter is called the *intensity*, defined then as

$$I = \frac{s^2}{s_0^2} \tag{12-4}$$

The intensity I ranges from unity (segregation) to zero (homogeneity) and provides one useful measure of the goodness of a distribution. For the mixing process illustrated in Fig. 12-1 the intensity, values of which are given in the figure, continually falls as mixing proceeds, as expected.

A second measure of mixing gives a length scale that is characteristic of the mixture. It is based on the idea of *correlation*. Suppose we measure the concentration "at a point" $C(\mathbf{x})$ by sampling a volume in the neighborhood of $\mathbf{x}$. At the same time let us measure the concentration $C(\mathbf{x}')$ at a nearby point $\mathbf{x}' = \mathbf{x} + \mathbf{r}$. We ask whether there is any relationship between $C(\mathbf{x})$ and $C(\mathbf{x}')$.

We might expect that if $\mathbf{r}$ is small then $C(\mathbf{x})$ and $C(\mathbf{x}')$ would be nearly equal. However, as $\mathbf{r}$ increases, in a well-mixed system, the probability of observing a C greater than the overall $\bar{C}$ would be about the same as that for observing a C less than $\bar{C}$. Another way of saying this is that as $\mathbf{r}$ increases there is less "correlation" between $C(\mathbf{x})$ and $C(\mathbf{x}')$.

One may define a *correlation coefficient* as

$$R(\mathbf{r}) = \frac{1}{s^2 N} \sum_{i=1}^{N} [C_i(\mathbf{x}) - \bar{C}][C_i(\mathbf{x} + \mathbf{r}) - \bar{C}] \tag{12-5}$$

For sufficiently large N [taking note of Eq. (12-2)] we see that $R(0) = 1$, as intended. Once $R(\mathbf{r})$ is calculated it is possible to determine a length scale S, defined as†

$$S = \int_0^\beta R(\mathbf{r}) \, d\mathbf{r} \tag{12-6}$$

---

† β is a measure of the size of the system and is best defined as $\beta = d\mathbf{r}$. If $d\mathbf{r}$ refers to a length then S is a *linear* scale; if $d\mathbf{r}$ is a volume, say, $r^2 \, d\mathbf{r}$, then S is a volumetric scale of mixing.

When S is compared to some characteristic length scale of the system itself (such as β), it is possible to give an idea of whether the mixing is "fine" or "gross."

The scale of mixing is not very useful if and when a system can approach uniform concentration, such that $C(\mathbf{x}) \rightarrow \bar{C}$, as in the case of *dissolution*, as opposed to *dispersion*, of particulates. In that case R approaches unity for any $\mathbf{r}$, and S approaches β, the size of the system. Thus S might go through a minimum as mixing proceeds, and two different states of mixedness might have the same S value.

It is probably useful at this point to illustrate the concepts of scale and intensity with the mixing process of Fig. 12-1. We begin by noting two specific features of this process. The minor component is made up of elements of unit mass. What we mean to illustrate specifically is that in some mixing processes there is a limit to the degree of subdivision possible. Dispersion of particulates provides one such example. Specifically, the elements do not "dissolve"; they can only be dispersed spatially through the space available to them.

The second feature to note is that in Fig. 12-1 we have defined a specific mixing process which successively halves the size of the minor component and uniformly distributes the halves spatially. Thus there is a degree of spatial structure to this mixing process.

Figure 12-2 shows the correlation function. All calculations have been made with a sample "volume" of 10 units, and all averages are based on 20 samples at randomly selected points along the line. The spatial structure of the dispersion shows clearly here, the oscillations arising from the periodicity of the dispersion. The scale decreases as the dispersion process goes on.

By way of contrast, Fig. 12-3 shows a system dispersed to the level of $S4$ but *randomly* distributed along the line. The intensity and scale can be compared and

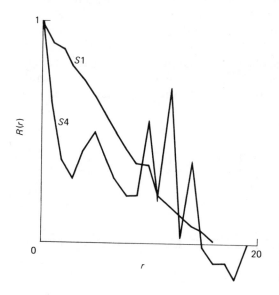

Figure 12-2 Correlation function for the distributions $S1$ and $S4$ of Fig. 12-1. The calculated scales are labeled on Fig. 12-1.

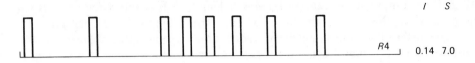

Figure 12-3 Mixture which is dispersed to the same degree as $S4$ but which is randomly distributed along the line. Calculated intensity and scale are labeled.

contrasted for these two cases, and it is interesting to note that the intensity is significantly greater, consistent with our visual observation that $S4$ is better mixed. The calculated scales do not reflect as great a difference.

We might think of the distributions $S1$ through $S4$, and $R4$, as a model of *convective mixing*, where elements of the minor component are separated and redistributed spatially. The distributions $S1$ to $S4$ are the result of an "ordered" convection process, whereas $R4$ could be the result of a random convection process such as occurs in turbulent mixing.

In some systems a "diffusive" mixing may also occur as the minor component dissolves and diffuses through the major component. Figure 12-4 suggests two steps of diffusion, subsequent to convective mixing to the level of $R4$ in Fig. 12-3. The mixtures $D1$ and $D2$ result from $R4$ by permitting the minor component to "diffuse," with the result that the concentration of each element decreases from the unit value previously assigned, and the element is allowed to "spread" in width. The intensity continues to fall, but now we observe the scale going through a minimum and increasing as the system diffuses toward uniformity.

This simplified picture of mixing has been introduced principally to illustrate the concepts of intensity and scale as goodness measures. Because a small volume was "sampled" and only 20 "samples" were taken, the specific values given for intensity and scale will depend on the sampling points selected. Nevertheless, the values cited illustrate the general features expected of a mixing process. To conclude, we state that we can define objective statistical measures of the goodness of a mixing process, the *intensity* and the *scale* of mixing. However, it is generally not practicable to make these measurements, and so they are seldom used.

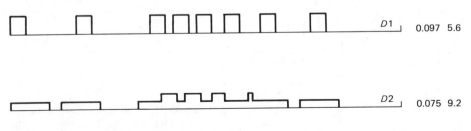

Figure 12-4 Two stages of successive "diffusion" from the distribution $R4$ of Fig. 12-3.

12-2 RESIDENCE TIME DISTRIBUTION AND MIXING

The concept of *mixedness* developed above involves the notion of homogeneity and was discussed in terms of the mixing of two materials, a major and a minor component. In this section we introduce another concept of mixing which is valid in single-component, homogeneous systems. It involves the uniformity of the "history" of the elements of the system, as described by the *residence time distribution.*

Let us begin by considering laminar flow of a newtonian fluid in a long circular pipe: Poiseuille flow. With reference to Fig. 12-5 we consider an element of fluid at radial position r. The element moves at velocity $u(r)$ and requires a time to traverse an axial distance L given by

$$t = \frac{L}{u(r)} \tag{12-7}$$

Fluid elements which all enter the pipe at the same instant of time will leave at different times, in accordance with their radial position, as governed by Eq. (12-7). Hence we speak of a *distribution* of residence times for this flow.

To make this idea quantitative, let us define a function $f(t)$ such that $f(t)\,dt$ is the fraction of the output of the flow which had been in the pipe for a time in the range t to $t + dt$. Then, letting Q be the volumetric flow rate, we may write

$$f(t)\,dt = \frac{dQ}{Q} = \frac{u(r)2\pi r\,dr}{\pi R^2 U} \tag{12-8}$$

where the average velocity U is defined by

$$U = \frac{Q}{\pi R^2} \tag{12-9}$$

From

$$u(r) = 2U\left[1 - \left(\frac{r}{R}\right)^2\right] = \frac{L}{t} \tag{12-10}$$

we find

$$\frac{2r\,dr}{R^2} = \frac{\bar{t}}{2t^2}\,dt \tag{12-11}$$

and

$$\frac{u(r)}{U} = \frac{\bar{t}}{t} \tag{12-12}$$

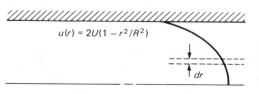

$u(r) = 2U(1 - r^2/R^2)$

Figure 12-5 Definition sketch for consideration of residence time distribution in Poiseuille flow.

We have defined the mean residence time as

$$\bar{t} = \frac{\pi R^2 L}{Q} = \frac{L}{U} \tag{12-13}$$

With these relationships we may write $f(t)$ in the form

$$\bar{t}f(t) = \frac{1}{2}\left(\frac{t}{\bar{t}}\right)^{-3} \tag{12-14}$$

It is more common to describe the residence time distribution through the function $F(t)$ defined by

$$F(t) = \int_{1/2\bar{t}}^{t} f(t)\, dt \tag{12-15}$$

The lower limit is the *first appearance time*, the shortest time any element of fluid resides in the pipe, which of course corresponds to the fluid moving along the pipe axis $(r = 0)$ where the velocity is greatest. For Poiseuille flow we find

$$F(t) = 1 - \frac{1}{4}\left(\frac{t}{\bar{t}}\right)^{-2} \tag{12-16}$$

Figure 12-6 shows $F(t)$, which is called the *cumulative residence-time-distribution function*, or the *F function*, for short.

The interpretation of $F(t)$ is simple: Its value at any time t gives the fraction of fluid with residence times which are less than t. For $t = 5\bar{t}$, for example, we find from Eq. (12-16) that $F = 0.99$. Thus only 1 percent of the fluid has a residence time in excess of $5\bar{t}$. We note, however, that for $t = \bar{t}$ we calculate $F = 0.75$. While we refer to $\bar{t}$ as *the* residence time, in fact 25 percent of the flow has residence times in excess of $\bar{t}$.

The ideal residence time distribution is a unit step function, also shown in Fig. 12-6. It would correspond to "plug flow," for which every element of fluid has exactly the same residence time in the pipe. Because of the no-slip condition at the pipe wall the plug flow condition cannot be approached very closely. However, nonnewtonian behavior of the usual type does give an F function that is closer to plug flow than that for the newtonian fluid.

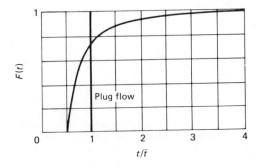

Figure 12-6 Cumulative residence-time-distribution function [Eq. (12-16)] for Poiseuille flow.

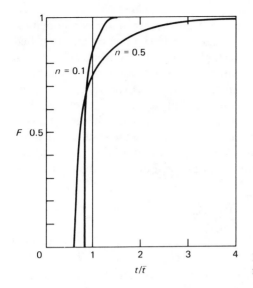

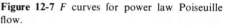

Figure 12-7 F curves for power law Poiseuille flow.

It is not too difficult to show that for a power law fluid the function $f(t)$ is given by

$$\bar{t}f(t) = \frac{2n}{3n+1}\left(\frac{\bar{t}}{t}\right)^3\left(1 - \frac{n+1}{3n+1}\frac{\bar{t}}{t}\right)^{(n-1)/(n+1)} \tag{12-17}$$

Figure 12-7 shows the F function for several values of n. The expected approach to plug flow as the fluid becomes more nonnewtonian is observed.

The F curve is characteristic, to some extent, of the geometry of the flow region. For plane Poiseuille flow, i.e., pressure flow between infinite parallel plates (see Chap. 5), it may be shown that

$$F(t) = \left(1 - \frac{2}{3}\frac{\bar{t}}{t}\right)^{1/2}\left(1 + \frac{1}{3}\frac{\bar{t}}{t}\right) \tag{12-18}$$

This F curve is closer to plug flow than that for a circular pipe.

It is interesting to examine the F curve for flow through a "perfect mixer." With reference to Fig. 12-8 let us consider a flow through a volume V which may be thought of as a stirred tank. A stream of volumetric flow rate Q and volume fraction (concentration) C_0 of some component enters the tank. C_0 is held constant, and the volume flow rate of effluent is Q. Hence the system is at steady state, the contents of the tank are uniform, and the component of interest enters and leaves the system at a rate $C_0 Q$.

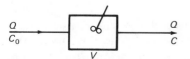

Figure 12-8 Flow through a stirred-tank mixer.

Imagine that at some instant of time $t = 0$ *all* the incoming component is tagged or marked in some way, so that the tagged material can be distinguished from the untagged material already in the tank. The total concentration C_0 of the component of interest entering, residing in, and leaving the system is unchanged. However, the fraction of *tagged* material in the *effluent* will *not* be C_0, since the effluent includes some material which was in the tank prior to the time $t = 0$.

The *tagged* material leaving the tank at some time t has been in the tank for a time less than t. Since $F(t)$ is the fraction of material leaving the system which has residence times less than t, it follows that $F(t)C_0 Q$ is the rate of efflux of tagged material. If we measure the concentration $C(t)$ of tagged material, then a material balance gives

$$QC(t) = F(t)C_0 Q$$

or
$$F(t) = \frac{C(t)}{C_0} \tag{12-19}$$

We conclude that the function $F(t)$ may be obtained as the effluent response to a uniform "step change" in influent concentration of some marker. This provides the simplest measurement of F curves for a flow system.

If the flow field is well defined we may calculate $F(t)$, as was illustrated above for the case of Poiseuille flow. For the case of a perfect mixer, we assume that the contents of the tank are perfectly homogeneous and that the effluent is identical in composition to the contents. A material balance on the tagged component in the tank then takes the differential form

$$V\frac{dC}{dt} = QC_0 - QC \tag{12-20}$$

$$C = 0 \quad \text{at } t \leq 0$$

The initial condition states that no tagged material is in the tank for $t \leq 0$. In Eq. (12-20) C represents the concentration of tagged material in the tank *and* in the effluent, since perfect mixing is assumed. The solution of Eq. (12-20) is

$$\frac{C(t)}{C_0} = 1 - e^{-Qt/V} = F(t) \tag{12-21}$$

If we define the mean residence time as

$$\bar{t} = \frac{V}{Q} \tag{12-22}$$

then we have found that the F curve for a perfect mixer is

$$F = 1 - e^{-t/\bar{t}} \tag{12-23}$$

Figure 12-9 shows this F curve and compares it to plug flow and Poiseuille flow.

One feature to note is that the perfect mixer does not give an F curve with a finite time lag, as in the case of Poiseuille flow (no response before $t/\bar{t} = \frac{1}{2}$) or plug

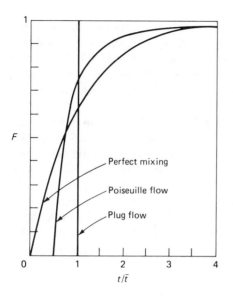

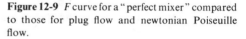

Figure 12-9 F curve for a "perfect mixer" compared to those for plug flow and newtonian Poiseuille flow.

flow (no response before $t = \bar{t}$). The principal distinction of the perfect mixer is the very broad residence time distribution. Since a broad F curve implies heterogeneity with respect to the "history" of material appearing in the output, some caution is necessary in using F curves as a measure of mixing.

Residence time distribution is related to the degree of *backmixing* in a system, which is the mixing that occurs in the primary flow direction. However, homogeneity across a *cross section* of the output is the normally desired mode of mixing, and this is related more to *transverse mixing*, which is the mixing that occurs in the direction normal to the primary flow.

A simple example of this idea is given in Fig. 12-10, which shows a minor component being introduced into the inlet flow of two mixers. One mixer is a long pipe, with the flow being such that a very low Reynolds number is attained. Thus we have laminar flow. Let us suppose, however, that the velocity profile is very flat, so that nearly plug flow is attained. This might be approximated by using a

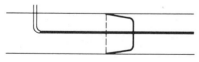

Laminar flow

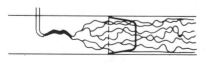

Turbulent flow

Figure 12-10 Comparison of radial mixing in two systems which would show the same F curves.

power law fluid with small n or by having a thin lubricating annular layer of a second fluid surrounding the main fluid component. In any event we assume laminar flow with a very flat velocity profile, leading to a (nearly) plug flow F curve. The second mixer is the same as the first, except that the flow conditions are turbulent. Again a fairly flat velocity profile is attained, and a nearly plug flow F curve is observed.

Mixing in these two systems is distinctly different. Under laminar flow the minor phase simply travels along the pipe axis at the same radial position, and the distribution of minor component across the cross section is essentially unchanged down the length of the pipe. On the other hand, the turbulent flow causes good radial mixing, and the minor phase is dispersed across the radius. Hence the outlet flow is radially homogeneous, and we would declare the minor phase to be well mixed at the outlet. Because the F curve is more strongly affected by backmixing than by radial mixing, the two flows show similar residence time distributions. Because the final state of mixing is more strongly affected by radial mixing, the two flows show very dissimilar *degrees* of mixing.

12-3 LAMINAR SHEAR MIXING

The discussions in the previous section should suggest that transverse mixing plays a greater role in developing good mixedness than does backmixing. In turbulent flows both types of mixing occur as a natural consequence of the dynamics of the flow. In laminar flow the geometry of the flow boundaries plays the major role in determining the extent, if any, of transverse mixing. In fact, the illustration of Fig. 12-10 suggests that *no* mixing occurs in laminar Poiseuille flow. This is not true, and this point again suggests the pitfalls of trying to discuss mixing in general terms. In this section we discuss the type of mixing possible in laminar shear flows and some means of evaluating this type of mixing.

We begin by again referring to a simple shear flow, such as plane Couette flow. Suppose, as illustrated in Fig. 12-11, two streaks of tracer exist at some instant of time, one aligned in the flow direction and the other one aligned transverse to it, in the direction of shear. The streak aligned in the flow direction

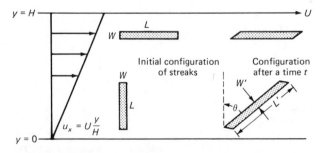

Figure 12-11 "Mixing" of two streaks in a laminar shear flow.

remains so aligned, with only slight changes in the geometry of the ends. If the initial width W were small in comparison to L, then no substantial change in shape would be discernable.

The other streak, initially aligned transverse to the flow direction, is rotated through an angle θ. More important, however, is the fact that the streak is elongated and made thinner. It is easy to calculate the characteristics of the streak after a time t.

Since the velocity field is linear (for this simple example) the upper and lower ends of the streak are displaced, relative to each other, by an amount given by their velocity difference times the duration of motion. In terms of the angle θ we find

$$\theta = \tan^{-1} \dot{\gamma}t \qquad (12\text{-}24)$$

where $\dot{\gamma} = U/H$. The relative increase in length of the streak is

$$\frac{L'}{L} = (\cos \theta)^{-1} = [1 + (\dot{\gamma}t)^2]^{1/2} \qquad (12\text{-}25)$$

while the thickness of the streak decreases according to

$$\frac{W'}{W} = \cos \theta = [1 + (\dot{\gamma}t)^2]^{-1/2} \qquad (12\text{-}26)$$

The thickness W' is referred to as the *striation thickness.*

Within the context of the general goals of mixing the increase of length and decrease in striation thickness both represent an increase of mixedness. The lengthening of the streak increases the interfacial area (or length, in this two-dimensional example) between the two phases or components. Thus there is greater contact, per unit amount of minor component, between the two components. While the decreasing striation thickness is not independent of the increasing length, it can have an independent effect. If diffusion of some component is to be promoted from the minor phase, then the rate of diffusion will be enhanced by creating thinner streaks.

Thus we see that striation thickness can provide a quantitative measure of mixedness in laminar flow systems for which the concept of a striation is meaningful. We see further that we can promote this mixing by introducing the minor component into a shear flow as a streak aligned transverse to the flow direction but in the direction of the velocity gradient.

Let us examine another example of this type of laminar shear mixing. Figure 12-12 shows a cylindrical Couette flow generated by rotation of the inner cylinder.

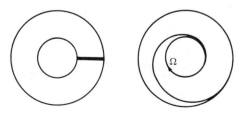

Figure 12-12 Mixing in laminar rotational Couette flow.

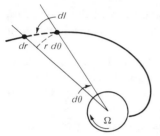

Figure 12-13 Definition sketch for analysis of mixing in a rotational flow.

Suppose that at some initial time an element of fluid is tagged along a radius. After one rotation the element has been drawn into a spiral. With reference to Fig. 12-13 the length of the streak can be calculated.

We assume steady laminar newtonian flow, for which the velocity u_θ is given by [see Eq. (5-58)]

$$u_\theta = r\Omega \frac{1 - (R_0/r)^2}{1 - \varkappa^{-2}} \tag{12-27}$$

(Ω is in radians per time). The angular displacement $\theta(r)$ undergone by a "particle" at a radius r, in a time t, is given by

$$\theta = -\frac{1 - (R_0/r)^2}{1 - \varkappa^{-2}} \Omega t \tag{12-28}$$

(The minus sign arises from our arbitrary choice that clockwise is the direction of positive θ.)

After a single revolution of the inner cylinder, which requires a time $t = 2\pi/\Omega$, two "particles," separated by a radial distance dr, will have different angular positions defined by

$$d\theta = \frac{2\pi}{\varkappa^{-2} - 1} \frac{R_0^2}{r^4} [(r + dr)^2 - r^2] = \frac{4\pi R_0^2}{\varkappa^{-2} - 1} \frac{dr}{r^3} \tag{12-29}$$

The differential length along the direction of the spiral is then

$$(dl)^2 = (dr)^2 + r^2(d\theta)^2 = (dr)^2 \left[1 + \frac{16\pi^2 R_0^4}{r^4(\varkappa^{-2} - 1)^2}\right] \tag{12-30}$$

After one revolution, then, the total spiral length is

$$l = \int dl = \int_{R_i}^{R_0} \left[1 + \frac{16\pi^2 R_0^4}{(\varkappa^{-2} - 1)^2 r^4}\right]^{1/2} dr$$

or

$$\frac{l}{R_0 - R_i} = \frac{1}{1 - \varkappa} \int_{\varkappa}^{1} \left[1 + \frac{16\pi^2}{(\varkappa^{-2} - 1)^2 r'^4}\right]^{1/2} dr' \tag{12-31}$$

where $r' = r/R_0$.

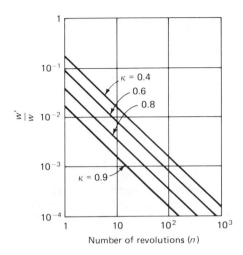

Figure 12-14 Reduction in striation thickness achieved in annular rotational Couette flow.

After n revolutions the relative extension of the streak is

$$\mathscr{S}_n(\varkappa) = \frac{l_n}{R_0 - R_i} = \frac{1}{1 - \varkappa} \int_\varkappa^1 \left[1 + \frac{16n^2\pi^2}{(\varkappa^{-2} - 1)^2 r'^4} \right]^{1/2} dr' \qquad (12\text{-}32)$$

and the relative decrease in striation thickness will be

$$\frac{W'}{W} = [\mathscr{S}_n(\varkappa)]^{-1} \qquad (12\text{-}33)$$

Figure 12-14 shows the reduction in striation thickness as a function of n. Two points are of interest. The first is that the striation thickness decreases inversely as the first power of the number of revolutions. The second is that very significant reductions can be achieved by a small number of revolutions.

Now suppose that we consider an axial annular Poiseuille flow upon which is superimposed the circular annular Couette flow just described. If a single streak of tracer is continuously fed across the radius, it will travel down the axis of the system in a helical path. The analysis of striation thickness given above is valid, since (for a newtonian fluid) the axial and angular flows are independent. It will, however, be necessary to account for the residence time distribution associated with the radial dependence of the axial velocity.

The velocity profile for *axial* annular Poiseuille flow is [see Eqs. (5-29) and (5-30)]

$$u_z = \frac{2Q}{\pi R_0^2} \frac{1 - r'^2 - \dfrac{1 - \varkappa^2}{\ln \varkappa} \ln r'}{1 - \varkappa^4 + \dfrac{(1 - \varkappa^2)^2}{\ln \varkappa}} \qquad (12\text{-}34)$$

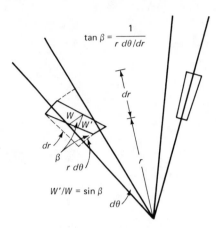

$$\tan \beta = \frac{1}{r\, d\theta/dr}$$

$$W'/W = \sin \beta$$

Figure 12-15 Definition sketch for consideration of striation thickness in combined axial and rotational annular flow.

Over an axial length L an element of fluid will be deformed for a time

$$t(r') = \frac{L}{u_z(r')} \qquad (12\text{-}35)$$

Hence the reduction in striation thickness must account for the fact that each deforming element of the streak is deformed for a time that depends upon its radial position.

With reference to Fig. 12-15 we can easily calculate the desired striation thickness. We see that from purely geometric considerations

$$\frac{W'}{W} = \sin \beta \qquad (12\text{-}36)$$

and β is defined in terms of the angle θ as

$$\tan \beta = \frac{1}{r\, d\theta/dr} \qquad (12\text{-}37)$$

From Eq. (12-28) we find

$$r\frac{d\theta}{dr} = \frac{2\Omega t}{\varkappa^{-2} - 1} r'^{-2} \qquad (12\text{-}38)$$

It follows then that

$$\frac{W'}{W} = \frac{(\varkappa^{-2} - 1)r'^2/2\Omega t}{\left[1 + \dfrac{(\varkappa^{-2} - 1)^2 r'^4}{4\Omega^2 t^2}\right]^{1/2}} \qquad (12\text{-}39)$$

gives the reduction in striation thickness of an element of the streak which is at the radial position r' and which has been deforming for a time t. To complete the calculation we merely have to introduce the residence time associated with each radial position r', using Eqs. (12-35) and (12-34). The result, after introduction of the dimensionless parameter

$$E = \frac{\pi L (R_0^2 - R_i^2)\Omega}{Q} \tag{12-40}$$

gives the reduction in striation thickness at the tube exit as

$$\frac{W'}{W} = E^{-1} \frac{(1 - \varkappa^2)^2}{\varkappa^2} \frac{r'^2 - r'^4 - \dfrac{1 - \varkappa^2}{\ln \varkappa} r'^2 \ln r'}{1 - \varkappa^4 + \dfrac{(1 - \varkappa^2)^2}{\ln \varkappa}} \tag{12-41}$$

In Eq. (12-41) the approximation $E > 10$ has been made. It is not difficult to see that E is just

$$E = \Omega \bar{t} \tag{12-42}$$

where $\bar{t}$ is the mean residence time in the annulus. Under any conditions of practical interest the condition $E > 10$ would be met.

From Eq. (12-41) it can be seen that the striations become exceedingly thin very near the cylindrical boundaries. The maximum striation thickness occurs at a radial position defined by the solution to (see Prob. 12-12)

$$1 - 2r'^2 - \frac{1 - \varkappa^2}{\ln \varkappa}(\ln r' + \tfrac{1}{2}) = 0 \tag{12-43}$$

When this value of $r'(\varkappa)$ is introduced into Eq. (12-41), one finds

$$\left(\frac{W'}{W}\right)_{\text{max}} = E^{-1} f(\varkappa) \tag{12-44}$$

where $f(\varkappa)$ is the term multiplying E^{-1} in Eq. (12-41), evaluated using a numerical solution to Eq. (12-43). Figure 12-16 shows the results of this calculation.

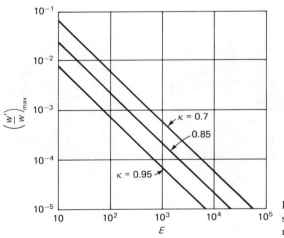

Figure 12-16 Maximum reduction in striation thickness in combined axial and rotational annular flow.

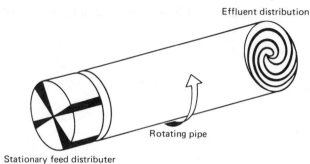

Effluent distribution

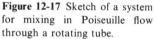

Rotating pipe

Stationary feed distributer

Figure 12-17 Sketch of a system for mixing in Poiseuille flow through a rotating tube.

Thus we have a model with which striation thickness may be estimated as a function of design and operating parameters. Now it is appropriate to address the question: Does this flow provide a viable (i.e., practical) method of mixing, and is the model useful in the calculation of striation thickness?

Alfrey and his coworkers have answered both questions affirmatively and have investigated several rotational mixer configurations both theoretically and experimentally. One of the simplest such mixers is the rotating tube system shown schematically in Fig. 12-17. The feed materials are distributed as alternate circular sectors of each material which then pass in helical Poiseuille flow down the tube axis. The rotation of the tube wall creates sets of spiral striations, and Alfrey shows that the maximum striation thickness is

$$\left(\frac{W'}{W_0}\right)_{max} = (2E)^{-1/2} \tag{12-45}$$

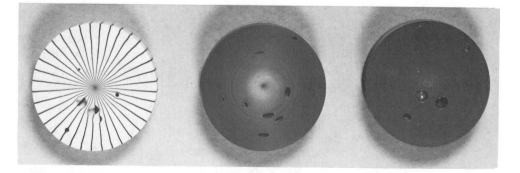

Figure 12-18 Mixing achieved in Poiseuille flow through a rotating tube. (*From Schrenk, Chisholm, and Alfrey.*)

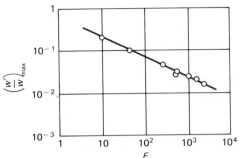

Figure 12-19 Comparison of theory and experiment for mixing in Poiseuille flow through a rotating tube. (*From Schrenk, Chisholm, and Alfrey.*)

where E is again just $\Omega \bar{t}$, and W_0 is defined, for this feed distribution, as†

$$W_0 = \frac{2\pi R_0}{N} \qquad (12\text{-}46)$$

where N is the number of feedports for the minor component.

Experiments were performed by pumping two polystyrene melts to the feed distributor. One was premixed with 2% carbon black; the other was pigmented with 2% TiO_2. Figure 12-18 shows the mixture at the feed and at values of E approximately 100 and 1500. It is apparent that good mixing can be achieved. Measurements of the maximum striation thickness were carried out, and the results are shown in Fig. 12-19. Agreement with the theory is quite good.

Alfrey gives a very simple design equation relating the length and rotational speed of a mixer required to reduce the striations below a specified level at a given volumetric throughput. It takes the form

$$\Omega = \frac{2\pi Q}{LN^2 W_{max}'^2} \qquad (12\text{-}47)$$

and is valid specifically for the feed geometry illustrated here.

It is interesting to note that the rotating annular tube mixer [note Eq. (12-44)] causes the striation thickness to decrease inversely with E to the first power. The rotating circular tube system, however, shows an inverse dependence on the square root of E. Since E is a product of Ω and $\bar{t}$, it may be considered a rough measure of the angular *strain* imposed on the fluid. Thus these two mixers differ significantly in how effectively the shear strain can reduce the striation thickness.

The series of papers by Alfrey and his coworkers is a marvelous example of the interaction of fundamental flow modeling with practical process design. The reader should make the effort to examine this particular series of papers, which are cited in the Bibliography.

---

† Alfrey defines W_0 as the distance between successive "like" interfaces rather than as the actual striation thickness. This is a more generally useful definition for this type of process.

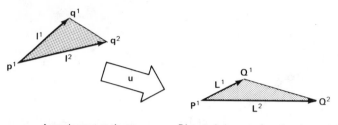

Area element at time t Distorted element after time interval dt

Figure 12-20 Definition sketch for consideration of distortion of an element of area.

Increased Area as a Measure of Mixing

We have seen above that one measure of mixing is the decrease in striation thickness of a streak of "contaminant." In this section we consider a related aspect of mixing: the increase in *area* of an element of contaminant. We find that while striation thickness can be calculated only for fairly simple, well-defined flow fields, consideration of area leads to a more general formulation of "mixing rate," from which several useful generalizations may be drawn.

Figure 12-20 shows a differential element of area defined by two vectors $\mathbf{l}^1$ and $\mathbf{l}^2$. The reader may recall that the area of the triangular element can be obtained as half the magnitude of the cross product $\mathbf{l}^1 \times \mathbf{l}^2$:

$$a^3 = \tfrac{1}{2}|\mathbf{l}^1 \times \mathbf{l}^2| \tag{12-48}$$

Now, suppose that a velocity field $\mathbf{u}$ distorts the area, in a time dt, to one defined by vectors $\mathbf{L}^1$ and $\mathbf{L}^2$. Then the new area is†

$$A^3 = \tfrac{1}{2}|\mathbf{L}^1 \times \mathbf{L}^2| \tag{12-49}$$

The vector $\mathbf{l}^1$ has end points given by position vectors $\mathbf{p}^1$ and $\mathbf{q}^1$. After a time dt the element of material at $\mathbf{p}^1$ moves to a new position $\mathbf{P}^1$ given by

$$\mathbf{P}^1 = \mathbf{p}^1 + \mathbf{u}^p \, dt \tag{12-50}$$

Similarly

$$\mathbf{Q}^1 = \mathbf{q}^1 + \mathbf{u}^q \, dt \tag{12-51}$$

By $\mathbf{u}^p$ and $\mathbf{u}^q$ we mean the velocity vectors at (i.e., in the neighborhood of) points $\mathbf{p}$ and $\mathbf{q}$. The new vector $\mathbf{L}^1$, then, may be written as

$$\mathbf{L}^1 = \mathbf{Q}^1 - \mathbf{P}^1 = \mathbf{l}^1 + (\mathbf{u}^q - \mathbf{u}^p) \, dt \tag{12-52}$$

---

† The superscripts are not exponents; they only indicate which vectors we are considering. The area a^3 may be considered a vector whose *magnitude* is given by Eq. (12-48) and whose *direction* is normal to the plane defined by $\mathbf{l}^1$ and $\mathbf{l}^2$. This is consistent with our earlier treatment of surface elements whose orientation was defined by the unit outward normal.

If the area element is of differential dimensions (and if dt is small as well), then a good approximation to the velocity difference above is found (after thinking and sketching) to be

$$\mathbf{u}^q - \mathbf{u}^p = \mathbf{l}^1 \cdot \nabla\mathbf{u} \tag{12-53}$$

where $\nabla\mathbf{u}$ is the *deformation* gradient tensor with cartesian components

$$\nabla\mathbf{u} = \begin{vmatrix} \dfrac{\partial u_x}{\partial x} & \dfrac{\partial u_x}{\partial y} & \dfrac{\partial u_x}{\partial z} \\[2mm] \dfrac{\partial u_y}{\partial x} & \dfrac{\partial u_y}{\partial y} & \dfrac{\partial u_y}{\partial z} \\[2mm] \dfrac{\partial u_z}{\partial x} & \dfrac{\partial u_z}{\partial y} & \dfrac{\partial u_z}{\partial z} \end{vmatrix} \tag{12-54}$$

The product of the vector $\mathbf{l}^1$ and the tensor $\nabla\mathbf{u}$ is a *vector* with components

$$(\mathbf{l}^1 \cdot \nabla\mathbf{u})_i = l_x^1 \frac{\partial u_i}{\partial x} + l_y^1 \frac{\partial u_i}{\partial y} + l_z^1 \frac{\partial u_i}{\partial z} \qquad i = x,\, y,\, z \tag{12-55}$$

The rate of change of $\mathbf{l}^1$, then, is just

$$\frac{d\mathbf{l}^1}{dt} = \mathbf{l}^1 \cdot \nabla\mathbf{u} \tag{12-56}$$

and the rate of change of area a^3 is, from Eq. (12-48),

$$\frac{da^3}{dt} = \frac{1}{2}\left| \frac{d\mathbf{l}^1}{dt} \times \mathbf{l}^2 + \mathbf{l}^1 \times \frac{d\mathbf{l}^2}{dt} \right| = \tfrac{1}{2}\left| (\mathbf{l}^1 \cdot \nabla\mathbf{u}) \times \mathbf{l}^2 + \mathbf{l}^1 \times (\mathbf{l}^2 \cdot \nabla\mathbf{u}) \right| \tag{12-57}$$

Equation (12-57) provides a general formulation of area increase due to deformation. It shows several features that we have already alluded to in specific examples treated earlier:

1. The rate of change of area clearly depends on the deformation gradients, the components of $\nabla\mathbf{u}$.
2. The dependence on the orientation of a streak relative to the deformation appears explicitly since $(\mathbf{l}^1 \cdot \nabla\mathbf{u}) \times \mathbf{l}^2$ will vanish if $\mathbf{l}^1 \cdot \nabla\mathbf{u}$ is parallel to $\mathbf{l}^2$ ($\mathbf{b} \times \mathbf{c} = 0$ if $\mathbf{b} \parallel \mathbf{c}$).

We can illustrate the application of Eq. (12-57) in two simple flow fields.

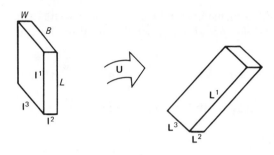

Figure 12-21 Shearing of a parallelepiped.

Simple shear flow Let us calculate the change in area of the streaks shown in Fig. 12-11. For the streak aligned transverse to the flow direction,

$$\mathbf{L} = \mathbf{l}^1 = (0, L, 0) \qquad \nabla\mathbf{u} = \frac{U}{H}\begin{pmatrix} 0 & 1 & 0 \\ 0 & 0 & 0 \\ 0 & 0 & 0 \end{pmatrix}$$

$$\mathbf{W} = \mathbf{l}^2 = (W, 0, 0)$$

$$\mathbf{l}^1 \cdot \nabla\mathbf{u} = \left(\frac{LU}{H}, 0, 0\right) \qquad \mathbf{l}^2 \cdot \nabla\mathbf{u} = 0$$

$$(\mathbf{l}^1 \cdot \nabla\mathbf{u}) \times \mathbf{l}^2 = 0$$

Hence we find that $da^3/dt = 0$. How can this be? Using the criterion of striation thickness reduction, we would say that mixing occurs, since the streak is getting long and thin. The answer is best seen with reference to Fig. 12-21.

It is not the area defined by $\mathbf{l}^1 \times \mathbf{l}^2$ that changes but the one defined by $\mathbf{l}^1 \times \mathbf{l}^3$. Define $\mathbf{l}^3$ by

$$\mathbf{B} = \mathbf{l}^3 = (0, 0, B)$$

Then

$$\frac{da^2}{dt} = \tfrac{1}{2}\left|(\mathbf{l}^1 \cdot \nabla\mathbf{u}) \times \mathbf{l}^3 + \mathbf{l}^1 \times (\mathbf{l}^3 \cdot \nabla\mathbf{u})\right| = \frac{1}{2}\frac{LUB}{H}$$

and

$$\frac{1}{a^2}\frac{da^2}{dt} = \frac{1}{2}\frac{U}{H} = \tfrac{1}{2}\dot{\gamma} \tag{12-58}$$

The *total* area of the streak, initially, is just

$$a_T = 2(WB + LB + LW)$$

The relative rate of change of *total* area is

$$\frac{2}{a_T}\frac{da^2}{dt} = \frac{LB}{WB + LB + LW}\dot{\gamma} \tag{12-59}$$

Figure 12-22 Stretching of a parallelepiped.

When *total* area change is considered this result illustrates another feature of laminar mixing that we have already noted. If the streak is initially distributed so that the area that *will* change is one of the smaller surfaces, then the ratio LB/a_T will be quite small, and the relative change of area will be small. Conversely if the area that *does not* change is the dominant one, no significant area increase occurs. This latter point is relevant to the deformation of the streak aligned with its long direction in the direction of flow in Fig. 12-11.

Simple elongational flow We consider a uniaxial elongation defined by

$$\mathbf{u} = \dot\epsilon(x, \ -\tfrac{1}{2}y, \ -\tfrac{1}{2}z) \tag{12-60}$$

and a streak of material initially aligned as in Fig. 12-22. We may calculate the area changes in a simple manner by using the results given above. We will need the following quantities:

$$\nabla\mathbf{u} = \dot\epsilon \begin{pmatrix} 1 & 0 & 0 \\ 0 & -\tfrac{1}{2} & 0 \\ 0 & 0 & -\tfrac{1}{2} \end{pmatrix}$$

$$\mathbf{L} = \mathbf{l}^1 = (L, 0, 0) \qquad \mathbf{B} = \mathbf{l}^2 = (0, B, 0) \qquad \mathbf{W} = \mathbf{l}^3 = (0, 0, W)$$

$$\mathbf{l}^1 \cdot \nabla\mathbf{u} = \dot\epsilon(L, 0, 0) \qquad \mathbf{l}^2 \cdot \nabla\mathbf{u} = -\tfrac{1}{2}\dot\epsilon(0, B, 0) \qquad \mathbf{l}^3 \cdot \nabla\mathbf{u} = -\tfrac{1}{2}\dot\epsilon(0, 0, W)$$

$$\frac{da^1}{dt} = \frac{1}{2}\left| \frac{d\mathbf{l}^2}{dt} \times \mathbf{l}^3 + \mathbf{l}^2 \times \frac{d\mathbf{l}^3}{dt} \right| = \tfrac{1}{2}BW\dot\epsilon$$

$$\frac{da^2}{dt} = \frac{1}{2}\left| \frac{d\mathbf{l}^1}{dt} \times \mathbf{l}^3 + \mathbf{l}^1 \times \frac{d\mathbf{l}^3}{dt} \right| = \tfrac{1}{2}(LW\dot\epsilon - \tfrac{1}{2}LW\dot\epsilon) = \tfrac{1}{4}LW\dot\epsilon$$

$$\frac{da^3}{dt} = \frac{1}{2}\left| \frac{d\mathbf{l}^1}{dt} \times \mathbf{l}^2 + \mathbf{l}^1 \times \frac{d\mathbf{l}^2}{dt} \right| = \tfrac{1}{2}(BL\dot\epsilon - \tfrac{1}{2}BL\dot\epsilon) = \tfrac{1}{4}BL\dot\epsilon$$

In this deformation all three surfaces change. The total area change is

$$\frac{da_T}{dt} = \dot\epsilon(BW + \tfrac{1}{2}LW + \tfrac{1}{2}BL)$$

and the relative change is

$$\frac{da_T}{a_T \, dt} = \tfrac{1}{2}\dot{\epsilon}\,\frac{BW + \tfrac{1}{2}LW + \tfrac{1}{2}BL}{BW + LW + BL} \tag{12-61}$$

The point of these two examples, in addition to illustrating the application of the analysis given above, is to emphasize the strong role played by the initial distribution of the minor phase which is to be mixed by the main flow. One should also note that *elongational* flows can produce mixing; the relevant parameters are the directions of the **l** vectors relative to the deformation gradient tensor $\nabla\mathbf{u}$.

In complex flow fields neither the **l** vectors nor the $\nabla\mathbf{u}$ tensor may be known. One can conclude, however, that the rate of mixing will depend on the magnitude of $\nabla\mathbf{u}$, and in fact will be linearly proportional to it. Since $\nabla\mathbf{u}$ is related in a simple way to the rate-of-deformation tensor Δ, it follows that the rate of mixing, defined as the relative rate of increase of area, is linearly proportional to $II_\Delta^{1/2}$, where II_Δ is the second invariant of Δ.

If we define a mixing parameter M as the ratio of the area between the major and minor phases to its initial value, say,

$$M = \frac{a_T}{a_{T_0}} \tag{12-62}$$

then we conclude that in laminar mixing

$$\frac{dM}{dt} = kII_\Delta^{1/2} \tag{12-63}$$

where the parameter k is related to the geometry of the mixer and the orientation of the minor phase relative to the flow field.

We end this section by raising a false counterexample of the generalizations developed above. In rationalizing the error of the example we shall again emphasize the care with which it is necessary to examine even simple mixing problems. In the analysis of the rotating tube mixer we find [using Eq. (12-45)]

$$\left(\frac{W'}{W_0}\right)_{\max} = (2\Omega\bar{t})^{-1/2} \tag{12-64}$$

Since striation thickness is inversely proportional to the area separating two components, we may introduce M, defined earlier [Eq. (12-62)], and write

$$M = (2\Omega\bar{t})^{1/2} \tag{12-65}$$

The mixing parameter increases as the residence time increases, consistent with our expectation.

Can we define a mixing *rate* for this system? One interpretation would be the rate of change of M with respect to residence time, which leads to

$$\frac{dM}{d\bar{t}} = \left(\frac{\Omega}{2\bar{t}}\right)^{1/2} \tag{12-66}$$

Is this result consistent with the previous idea that $dM/dt \approx II_\Delta^{1/2}$?

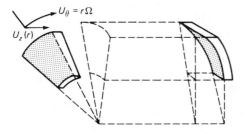

Figure 12-23 Lagrangian distortion of a volume element in a combined rotational-axial flow.

To answer this we note that for this flow

$$\left(\tfrac{1}{2}\mathrm{II}_\Delta\right)^{1/2} = \frac{32Q}{\pi D^3} = \frac{8L}{D\bar{t}} \tag{12-67}$$

Hence it follows that

$$\frac{dM}{dt} = k\mathrm{II}_\Delta^{1/4} \tag{12-68}$$

which clearly is at variance with the general result established earlier. One would conclude that there is something unusual about this particular flow field.

Let us follow a small volume element being deformed by this flow field. Recall that in the rotating *tube*, as opposed to the rotating *annulus*, the angular velocity is $u_\theta = r\Omega$, which is simply a rigid rotation about the axis. Hence the element does not deform in the θ direction. Indeed, if an "observer" were fixed to a "point" in the volume element he could not distinguish the kinematics of this flow from that of ordinary Poiseuille flow, if he confined his observation to the kinematics of the volume element.

The deformation, then, is as illustrated in Fig. 12-23, and only the surface originally normal to the z axis deforms. The relative rate of increase of area of that face is just $\tfrac{1}{2}\dot{\gamma}(r)$, where $\dot{\gamma}(r)$ is du_z/dr. We note, in particular, that Ω does not appear in this analysis of mixing, although it does appear in the previous treatment of this flow. How does one resolve the two analyses?

The resolution lies in the fact that two different "views" of the flow field have been used. If one follows an element of fluid as it moves through the tube the so-called *lagrangian* view is used. Figure 12-23 corresponds to a lagrangian analysis of deformation. On the other hand, if one examines a specific point or plane in space, this is referred to as a *eulerian* view. Figure 12-13 (although for the annular case) or Fig. 12-17 shows the distribution of a streak at a specific cross-sectional plane of the flow. This is a eulerian view. The fluid elements of one of these spirals were not in the same feed segment at the same earlier time.

The eulerian view is the more appropriate if our concern is actually with the cross-sectional distribution of the minor phase, as in the example of tubular extrusion. The lagrangian view leads to the general analysis presented in this section. It is more relevant, in a practical sense, to the calculation of mixedness when the minor phase is introduced as discrete elements. The latter case often occurs when two materials in solid particulate form are brought together to be melted and mixed.

Total Strain as a Measure of Mixedness

An element of material undergoing deformation is mixed at a rate that may be characterized by dM/dt, as calculated in the previous section. In addition to the *rate* of mixing, we are also interested in the *amount* of mixing that a system achieves. Consider the example of Poiseuille flow in a long circular pipe. The shear rate varies across the pipe radius and is easily seen to be expressible as

$$\dot{\gamma}(r) = \frac{32Q}{\pi D^3} \frac{r}{R} \tag{12-69}$$

The mixing rate in this simple flow is then

$$\frac{dM}{dt} = \tfrac{1}{2}\dot{\gamma} = \frac{16Q}{\pi D^3} \frac{r}{R} \tag{12-70}$$

which, obviously, depends upon radial position. An element of material which moves down the axis along the radial position r attains a degree of mixing $M(r)$ given by

$$M = 1 + \frac{16Q}{\pi D^3} \frac{r}{R} t \tag{12-71}$$

where t is the time of exposure to deformation, and $M = 1$ at $t = 0$. We can define the shear strain as $\dot{\gamma}t = \gamma$, and write

$$M = 1 + \tfrac{1}{2}\gamma(r) \tag{12-72}$$

If we take this strain and weight it by the fraction of the flow leaving the pipe which experiences this strain, and sum (integrate) over the pipe cross section, we can define a weighted average total strain (WATS) as

$$\text{WATS} = \bar{\gamma} = \int_{t_0}^{\infty} \frac{32Q}{\pi D^3} \frac{r}{R} t f(t)\, dt \tag{12-73}$$

In this integral we keep in mind that r and t are *dependent* variables in the sense described by Eq. (12-10), and $f(t)$ is the residence time distribution function, given for Poiseuille flow by Eq. (12-14). Using these equations we easily find that

$$\bar{\gamma} = \frac{16L}{3D} = \frac{64Q}{3\pi D^3} \bar{t} \tag{12-74}$$

where $\bar{t}$ is the mean residence time [Eq. (12-13)]. The interesting result is that the WATS depends only on the geometric parameter L/D; increasing the flow rate does not increase the *strain*, although it does increase the *rate* of strain or the rate of mixing.

The value of WATS as a measure of mixing lies in the following idea. In a mixing system for which the geometry of the minor phase (the l vectors of the previous section) is not known but for which the flow field is well defined, $\bar{\gamma}$ provides a quantitative measure of mixedness, since it depends only on the flow field. In terms of our understanding of mixing the concept of $\bar{\gamma}$ provides a useful connection back to the concept of residence time distribution. We shall return to this point shortly in discussing mixing in the extruder.

Effect of Viscosity Ratio on Laminar Shear Mixing

In all the models treated to this point it has been assumed that the minor phase is a fluid with a viscosity identical to that of the major phase. This is not likely to be the case in the mixing of polymers, and would certainly not be the case in the mixing of oils, blowing agents, etc., into molten polymers. In this section we examine a simple model of the effect of viscosity on the mixing of two phases.

We consider a simple shear flow as illustrated in Fig. 12-24. The minor phase, of viscosity μ', is interlayered along the flow direction as shown. For each layer the dynamic equations lead to

$$\frac{d\tau_{xy}}{dy} = 0 \tag{12-75}$$

or

$$\tau_{xy} = C_i \tag{12-76}$$

Furthermore, at the boundary between any two layers, the shear stress must be continuous. Hence all the C_i are equal to the same constant C. Thus in each layer the velocity profile is linear, and

$$u_i = \frac{C}{\mu_i}(y + a_i) \tag{12-77}$$

where a_i is an integration constant.

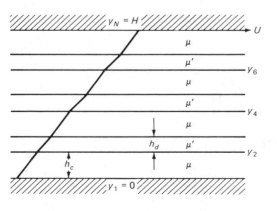

Figure 12-24 Simple shear flow of layers of alternating viscosity.

If we can calculate the constant C we can evaluate the shear strain imposed on the minor phase by the continuous phase. To do so we must evaluate the constants a_i, using the no-slip conditions

$$u_i = u_{i-1} \quad \text{at } y = y_i \tag{12-78}$$

We know that

$$u = U \quad \text{at } y = y_N = H \tag{12-79}$$

from which it follows that

$$C = \frac{\mu U}{H + a_{N-1}} \tag{12-80}$$

(We imply here that the continuous phase, of viscosity μ, wets *both* rigid boundaries.)

From Eqs. (12-78) and (12-77) it follows that

$$a_i = \left(\frac{\mu_i}{\mu_{i-1}} - 1\right)y_i + \frac{\mu_i}{\mu_{i-1}}a_{i-1} \tag{12-81}$$

The no-slip condition at $y = 0 = y_1$ gives

$$a_1 = 0 \tag{12-82}$$

and the recursion relation of Eq. (12-81) gives

$$a_2 = \left(\frac{\mu'}{\mu} - 1\right)h_c$$

$$a_3 = \left(\frac{\mu}{\mu'} - 1\right)h_d$$

and, generally,

$$a_{2n} = \left(\frac{\mu'}{\mu} - 1\right)nh_c$$

$$a_{2n+1} = \left(\frac{\mu}{\mu'} - 1\right)nh_d \tag{12-83}$$

It follows that

$$C = \frac{\mu U}{H + \left(\frac{\mu}{\mu'} - 1\right)\left(\frac{N}{2} - 1\right)h_d} \tag{12-84}$$

In the dispersed phase the shear rate is just

$$\dot{\gamma}_d = \frac{C}{\mu'} = \frac{\mu}{\mu'}\frac{U}{H}\left[1 + \left(\frac{\mu}{\mu'} - 1\right)\left(\frac{N}{2} - 1\right)\frac{h_d}{H}\right]^{-1} \tag{12-85}$$

For the simple geometry assumed here we note that the volume fraction of dispersed phase, ϕ_d, is just

$$\phi_d = \left(\frac{N}{2} - 1\right)\frac{h_d}{H} \tag{12-86}$$

and so we may write the shear rate as

$$\dot{\gamma}_d = \frac{\mu}{\mu'}\frac{U}{H}\left[1 + \left(\frac{\mu}{\mu'} - 1\right)\phi_d\right]^{-1} \tag{12-87}$$

For $\mu = \mu'$ the expected result $\dot{\gamma}_d = U/H$ is recovered.

If $\mu/\mu' \gg 1$,

$$\dot{\gamma}_d = \frac{U}{\phi_d H} > \frac{U}{H} \tag{12-88}$$

Thus if the dispersed phase is very low in viscosity relative to the main phase, then the shear rate in the dispersed phase is greater than the nominal shear rate U/H and is independent of the viscosities, and depends only on the volume fraction ϕ_d.

If $\mu/\mu' \ll 1$,

$$\dot{\gamma}_d = \frac{\mu/\mu'}{1 - \phi_d}\frac{U}{H} \ll \frac{U}{H} \tag{12-89}$$

Thus if the dispersed phase is relatively high in viscosity, it receives a smaller shear rate than U/H, the factor being proportional to μ/μ'.

These results suggest that a high-viscosity dispersed phase is much more difficult to strain (mix) than one of viscosity lower than that of the continuous phase. One must be careful in generalizing this model to real mixing situations since the geometric parameters (orientation of the streaks) may be important.

12-4 THE EXTRUDER AS A MIXER

The extruder is often called upon to perform more than a simple pumping function. Extruders are often fed with a mixture of pellets of two or more polymers, or with a polymer and a second phase (filler, pigment, lubricating agent, etc.), with the expectation that the output of the extruder will be homogeneous. This expectation, of course, is met to varying degrees depending upon design and operating variables.

The ideas developed in this chapter can be used to evaluate the extruder as a mixing device. We make use of the simple parallel plate model of the melt extruder developed earlier in Chap. 6. The details of analysis are quite tedious, and the reader is referred to McKelvey (see Chap. 6 Bibliography) and to Tadmor and Klein for a more complete discussion.

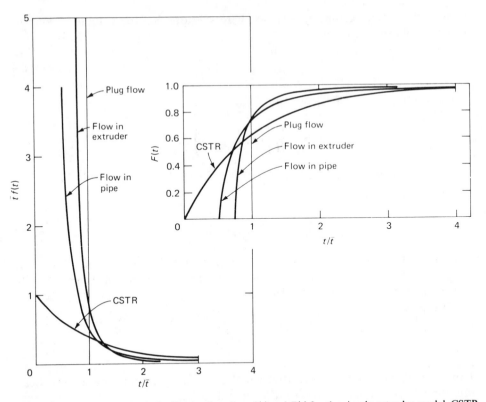

Figure 12-25 Residence-time-distribution functions $f(t)$ and $F(t)$ for the simple extruder model. CSTR is the "perfect" mixer of Fig. 12-8.

The residence-time-distribution functions $f(t)$ and $F(t)$ are shown in Fig. 12-25. In both cases, normalization to the mean residence time $\bar{t}$ removes the dependence of these curves on helix angle and on the ratio of pressure flow to drag flow. The mean residence time is given by

$$\bar{t} = \frac{2L}{U_z \sin \theta(1 + \Phi)} \tag{12-90}$$

where the parameters are as defined in Chap. 6. A new parameter Φ is introduced here, as the ratio of the pressure flow component to the drag flow. For the newtonian fluid this can be shown to be

$$\Phi = -\frac{\Pi_p}{6} \tag{12-91}$$

where Π_p was defined in Chap. 6 as

$$\Pi_p = \frac{\Delta P B^2}{\mu U_z Z} \tag{12-92}$$

Open discharge corresponds to $\Phi = 0$. Closed discharge occurs at $\Phi = -1$. Of course the limiting case of closed discharge gives a value of $\bar{t}$ which approaches infinity.

The F curve for the extruder is more nearly plug flow than that for Poiseuille flow. Thus the extruder should produce an output with a more homogeneous "history" than that of pipe flow, and in this sense might be a good mixer.

Because of the transverse flow there are two components of Δ:

$$\dot{\gamma}_x = \frac{\partial u_x}{\partial y} \qquad \dot{\gamma}_z = \frac{\partial u_z}{\partial y}$$

The magnitude of the strain rate is simply $(\frac{1}{2}II_\Delta)^{1/2}$, which is given by

$$\dot{\gamma} = (\tfrac{1}{2}II_\Delta)^{1/2} = (\dot{\gamma}_x^2 + \dot{\gamma}_z^2)^{1/2} \tag{12-93}$$

The weighted average total strain may be defined as

$$\mathbf{WATS} = \bar{\gamma} = \int_{t0}^{\infty} \dot{\gamma} t f(t)\, dt \tag{12-94}$$

where t_0 is the minimum residence time, and $\bar{\gamma}$ may be calculated by numerical integration. The results of Pinto and Tadmor are shown in Fig. 12-26. It is apparent that in the usual range of helix angles (say, $10° < \theta < 30°$) it is necessary to approach closed discharge in order to raise $\bar{\gamma}$ to high levels.

It is interesting to compare the strain achieved in an extruder to that in a pipe flow. From Fig. 12-26 a typical value of strain would be $\bar{\gamma} = 20L/B$. Writing L/B as $(L/D)(D/B)$ and noting that typical values of these geometric parameters would be $L/D = 25$ and $D/B = 10$, we find an expected value of strain to be $\bar{\gamma} = 5000$. To achieve the same level of strain in a pipe flow we find [from Eq. (12-74)] that an L/D of approximately 1000 would be required. Such a large value would prove impractical.

This model of the extruder as a laminar shear mixer is obviously crude. It does help to suggest the manner in which the relevant parameters enter the process. We see, for example, that the mixing is unaffected by the output, so long as the ratio of pressure flow to drag flow, or Φ, is constant. Thus in scale-up of an extruder from laboratory experiments, Φ would have to be held constant in order to maintain the same degree of mixing, and this would have significant implications with respect to operating conditions.

The extension of Pinto and Tadmor's work to power law fluids has been carried out by Bigg and Middleman with interesting results. Figure 12-27 shows F curves for the case that $n = 0.2$, with Π_P (as defined for a power law fluid) as a parameter. As closed discharge is approached the F curve is nearly that for plug flow.

The strain $\bar{\gamma}$ is shown in Fig. 12-28 as a function of Π_Q, which is related in a nonlinear way to Φ [except for $n = 1$, where $\Pi_Q = \frac{1}{2}(1 + \Phi)$]. We note that the strain axis is logarithmic, and for small Π_Q the nonnewtonian fluid receives significantly *less* strain than the newtonian fluid, at the same value of Π_Q.

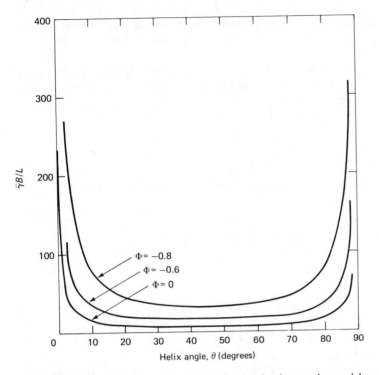

Figure 12-26 Weighted average total strain for the simple extruder model, according to Pinto and Tadmor.

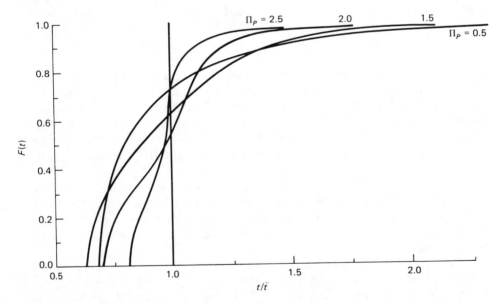

Figure 12-27 F curves for a power law fluid ($n = 0.2$) and for $\theta = 17.7°$

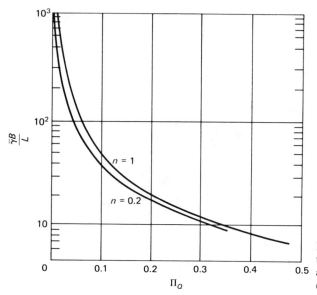

$\dfrac{\bar{\gamma}B}{L}$

$n = 1$

$n = 0.2$

Π_Q

Figure 12-28 Weighted average total strain for a newtonian and a power law fluid ($n = 0.2$), for $\theta = 17.7°$

12-5 MOTIONLESS MIXERS

A class of mixer can be defined that has no moving parts and that achieves mixing by its internal geometric design, which channels fluid being pumped through it in such a way as to cause mixing. Chemical engineers are already familiar with such a motionless mixer: the packed bed. One does not normally encounter packed beds in polymer processes because of the extremely high pressure drops that would be required to achieve a reasonable flow rate through the packing. Nevertheless, the concept of motionless mixing is easily introduced by using the packed bed as an example.

Figure 12-29 shows a section of a packed bed consisting of a tube filled with solid particles. An element of fluid moving through the void spaces of the packing can undergo two events which cause mixing. The deformation field will strain the element, and increase its area, as already discussed in Sec. 12-3. In addition to this action, the packing can also cause "stream splitting" in the neighborhood of stagnation points of the packing. This splitting further reduces striation thickness.

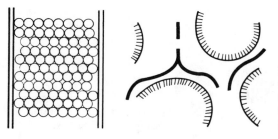

Figure 12-29 A packed bed of particles, viewed as a mixer. The detail shows an element of fluid which is strained and which "splits" in two near a stagnation point in the flow.

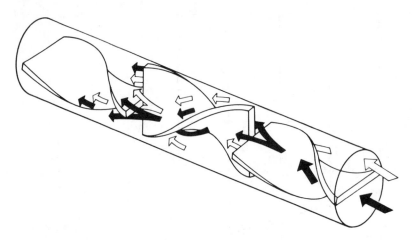

Figure 12-30 Helical elements of a Kenics Static Mixer.

In addition, stream splitting mixes material from one region with that from an adjacent region, thereby providing a degree of radial mixing.

All motionless mixers create strain and cause mixing by that mechanism. The distinguishing feature of commercial motionless mixers is the method whereby stream splitting is achieved. Figure 12-30 shows the design of one popular commercial motionless mixer: the Kenics Static Mixer. This mixer consists of a circular pipe within which are fixed a series of short helical elements of alternating left- and right-hand pitch. The elements are fixed to the pipe wall, and the trailing edge of one element is attached to, and forms a right angle with, the leading edge of the next element.

The helical design of the central element causes a transverse flow to arise in the plane normal to the pipe axis. As a consequence, fluid near the center of the pipe is rotated out toward the circular boundary, and vice versa. Radial mixing is achieved in this manner.

Figure 12-31 suggests what happens to a pair of fluids, initially segregated as they enter a Static Mixer element. To a good approximation the stream is halved each time it passes into a new mixer element. Thus one would anticipate that striation thickness would be reduced by a factor of 2^N, where N is the number of elements in series.

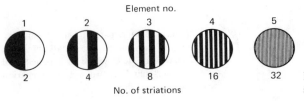

Element no.

Figure 12-31 Idealization of stream splitting in a Static Mixer.

No. of striations

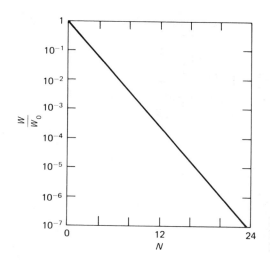

Figure 12-32 Striation thickness reduction based on the idealization of Fig. 12-31 and Eq. (12-96).

A common design provides that the ratio of axial length of a single element to the pipe diameter is about 1.6/1. In terms of the L/D of a pipe containing N elements, then, we find, for

$$\frac{L}{D} = 1.6N \tag{12-95}$$

that

$$\frac{W}{W_0} = 2^{-L/1.6D} \tag{12-96}$$

Figure 12-32 shows the reduction in striation thickness according to Eq. (12-96).

This simple model of flow division is not exactly achieved in practice. Figure 12-33 shows the effect on the helical flow that might be anticipated due to viscous effects, particularly the no-slip conditions which must be met at the solid surfaces. Figure 12-34 shows the results of a mixing experiment in which an epoxy resin, half of which is colored, is fed to a Kenics Static Mixer. After the resin flowed through and filled the mixer, the flow was stopped and the resin polymerized. Sections taken down the length of the mixer show similar striation reductions to those anticipated in the sketch of Fig. 12-33.

The helical element of the mixer effectively reduces the size of the conduit from the diameter of the empty pipe to something like half that diameter. Since pressure drop is such a strong function of the "diameter" (or some appropriate length scale normal to the flow direction), we would expect considerably increased pressure requirements for flow through a Kenics Static Mixer. This turns out to be the case.

Other motionless mixers are on the market, but their geometry is more complex than that of the Kenics device. We choose to examine some features of the Kenics Static Mixer in detail because that system is too complex to yield to the

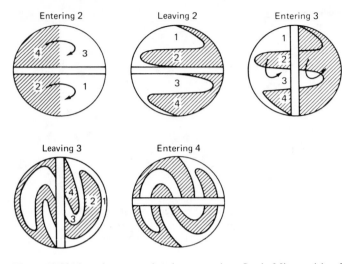

Figure 12-33 Actual cross-sectional patterns in a Static Mixer arising from the transverse flow set up by the helical element.

simplest modeling techniques discussed in earlier chapters but is still amenable to somewhat more sophisticated modeling techniques. Thus the example provides a vehicle for evaluating simple models and illustrates the process of introducing successively more complex models.

We can begin by considering the pressure drop–flow rate relationship for laminar newtonian flow. One of the simplest models would assume that the flow field is identical to that for flow through a long pipe of semicircular cross section.

Figure 12-34 Progressive mixing in a Kenics Static Mixer. The two fluids are initially segregated. Compare with Fig. 12-33. (*Photo courtesy of Kenics Corporation.*)

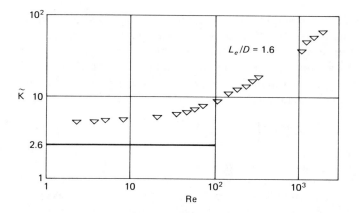

Figure 12-35 Observed $\tilde{K}$ factor for newtonian flow in a Kenics Static Mixer.

Problem 5-12 addresses itself to this flow, and with Fig. 5-9 one can find that

$$\tilde{K}_L = \frac{\Delta P \text{ for Static Mixer}}{\Delta P \text{ for empty pipe of same diameter}} = 2.6 \qquad (12\text{-}97)$$

Figure 12-35 shows experimental data for laminar newtonian flow. It is apparent that this model for $\tilde{K}$ underestimates the actual pressure requirement and fails to predict the observed dependence of $\tilde{K}$ on flow rate. (Note that the Reynolds number for this flow is based on the diameter of the empty pipe, rather than on the hydraulic radius, as discussed in Chap. 5.)

This simple model fails, particularly at higher Re, for several reasons. Primarily it neglects the fact that since each element is only of the order of one diameter in length, the velocity profile is continually developing and rearranging as fluid passes down the pipe axis. This leads to an underestimate of the pressure drop since a developing flow is much more dissipative than the corresponding fully developed flow. Furthermore, the degree of the error due to this aspect of the flow increases with increasing flow rate, since the length required to rearrange the entering profile increases as the flow rate increases.

A second failure of this simple model lies in the fact that the *transverse* flow is not accounted for, since it does not arise in the straight semicircular duct. It seems essential, then, to account for the fact that the velocity vector of this helical flow is a function of all three coordinate positions, and that the components u_z and u_θ both contribute to the friction loss. This leads necessarily to a fairly complex mathematical problem. Before turning to that problem, however, let us examine some simple modifications of the simple model of the straight semicircular duct.

The " path length " in the helical flow is longer than that of the straight flow by a factor

$$\tilde{K}_L = \left(1 + \frac{\pi^2 r^2}{L_e^2}\right)^{1/2} \qquad (12\text{-}98)$$

where L_e is the axial length of a single element, and where r is the radial coordinate of a cylindrical coordinate system whose axis coincides with the pipe axis. A particle traveling along the axis of the central element moves an axial length $L_e(\tilde{K}_L = 1)$, while a particle moving along the outer wall $(r = R)$ travels a helical length $L_R = (L_e^2 + \pi^2 R^2)^{1/2}$, corresponding to $\tilde{K}_L = (1 + \pi^2 R^2/L_e^2)^{1/2}$. Let us define an average $\tilde{K}$, the arithmetic mean of the extreme values of $\tilde{K}$, given by

$$\bar{\tilde{K}}_L = \tfrac{1}{2} + \tfrac{1}{2}\left(1 + \frac{\pi^2 D^2}{4L_e^2}\right)^{1/2} \tag{12-99}$$

We will refer to this as a *path length correction*. For an L_e/D of 1.6 we find $\bar{\tilde{K}} = 1.2$.

A second factor that must be accounted for is the finite area of the helical element in the cross section normal to the axis. If the element is of thickness t, its area is approximately tD, and the presence of the element reduces the area available for flow by a factor

$$\tilde{K}_A = 1 - \frac{4t}{\pi D} \tag{12-100}$$

As a consequence the average velocity in the mixer, at a given volumetric flow rate, is increased by a factor $1/\tilde{K}_A$ relative to the average velocity in an empty pipe. For laminar flow we expect the pressure drop to be proportional to average velocity, and so we might expect a larger pressure drop, by a factor $1/\tilde{K}_A$, relative to the empty pipe, at the same volumetric flow rate. For typical Static Mixers one finds $t/D = 0.1$, for which case $1/\tilde{K}_A = 1.15$.

On the basis of these two factors we might expect a $\tilde{K}$ factor for this simple semicircle model to be

$$\tilde{K} = \frac{2.6\bar{\tilde{K}}_L}{\tilde{K}_A} \tag{12-101}$$

For $L_e/D = 1.6$ and $t/D = 0.1$ we find $\tilde{K} = 3.6$. From Fig. 12-35 the observed value for $\tilde{K}$, in the limit of very small Reynolds numbers, is 5.2. We conclude that these modifications to the straight semicircle model fail to produce a reasonably accurate estimate of $\tilde{K}$, even in the limit of vanishing Reynolds number.

The next level of sophistication examines the effect of the helical flow in producing a transverse flow component. Tung has solved this problem by writing the dynamic equations in a helical coordinate system. In the limit of very low Reynolds numbers, for which the inertial terms may be neglected, and assuming that L_e/D is long enough that the flow is fully developed, Tung finds an analytical solution from which the velocity components may be examined. Figure 12-36 shows the velocity field for the case $L_e/D = 1.6$. In Tung's model L_e/D is the "pitch" of the helix, defined as the axial distance required for the element to twist $180°$. His model assumes that the element *continues* to twist in the same sense; there is no periodic splitting of the flow as in the real system.

A particularly interesting feature of the velocity field is the strength of the transverse flow. From Fig. 12-36 it can be seen that the magnitude of the transverse flow is comparable to that of the axial flow. It is this feature of the Kenics

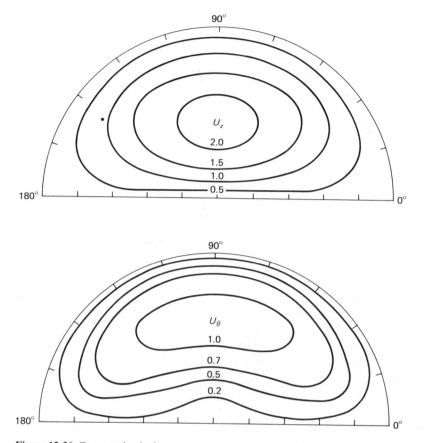

Figure 12-36 Computed velocity components for low Reynolds number flow in a semicircular pipe whose diameter is a continuous helix of pitch $L_e/D = 1.6$

Static Mixer which produces its strong mixing action, even in the limit of very small Reynolds numbers.

For the case $L_e/D = 1.6$ Tung finds that the predicted value of $\tilde{K}$ is 4.7. This is only 10 percent below the observed value. One concludes that, at "vanishingly small" Reynolds numbers, the effect of stream splitting, the periodic redevelopment of the velocity field, is a minor feature of the pressure loss. Of course, the neglect of inertial terms in the helical solution produces a *constant* $\tilde{K}$. The extension of Tung's formulation of this problem to finite Reynolds numbers would require numerical solution of coupled, nonlinear partial differential equations (the full dynamic equations) in a three-dimensional space. The expense of carrying out such a solution would be difficult to justify.

One argument against extension of the solution is the observation that $\tilde{K}$ is constant at Reynolds numbers as high as 10. In most systems involving highly viscous fluids it is not likely that one would exceed that Reynolds number. Figure 12-37 shows data obtained with four molten polymers at shear rates well into the

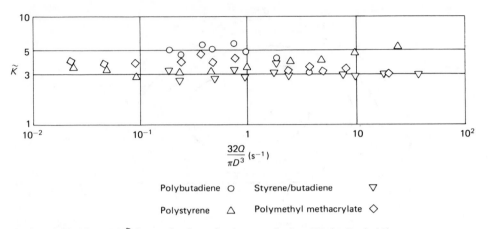

Polybutadiene ○   Styrene/butadiene ▽

Polystyrene △   Polymethyl methacrylate ◇

Figure 12-37 Observed $\tilde{K}$ factors for flow of polymer melts in a Kenics Static Mixer.

region of nonnewtonian behavior. The range of Reynolds numbers (based on apparent viscosity) is approximately 10^{-7} to 10^{-4}. As can be seen, there is some variation from one melt to another, but a constant value of $\tilde{K}$ in the neighborhood of 3 to 5 gives an adequate fit of the data.

It should be noted that the $\tilde{K}$ value is observed to depend *strongly* on both L_e/D and t/D. The data illustrated here, as well as Tung's theoretical work, examined only a limited set of these parameters. Kenics Corporation provides design data for its commercially available systems.

Once the velocity field is calculated other features of interest may be

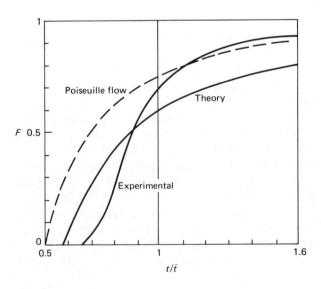

Figure 12-38 F curve for a Kenics Static Mixer. Experimental values were determined with newtonian fluids.

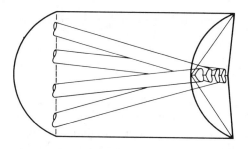

Figure 12-39 The Ross ISG Mixer.

computed. Figure 12-38 shows the observed residence-time-distribution function $F(t)$. The F curve is relatively close to that for plug flow behavior, indicating that such a mixer would be quite good in terms of producing a homogeneous "history" of the elements of fluid leaving the pipe. The F curve resulting from theory is a very poor approximation of the observed data. It is interesting to note that the F curve predicted by the straight semicircular model is practically identical to that derived from the helical model. This suggests that the residence time distribution is strongly affected by the stream-splitting feature of this flow. Tung examines some aspects of this problem, but we will not deal further with it here.

Another motionless mixer that lends itself to some degree of analysis is the ISG Mixer manufactured by Ross. Each element, as shown in Fig. 12-39, consists of a pipe housing through which four cylindrical holes are drilled. At the inlet side the four circular entrances lie along a diameter. At the outlet, the four exits are along a diameter at a right angle to the entrance alignment. The four holes are at oblique angles to the housing axis, with the result that the two holes which enter nearest the pipe wall will exit near the center, and vice versa. This provides a degree of radial mixing. The ends of each element are shaped so that a tetrahedral chamber separates the exit of one element from the entrance of the next.

If two separated streams enter the Ross ISG Mixer in such a way that half of each stream enters *each* of the four channels, then eight layers will emerge into the first tetrahedral chamber. Figure 12-40 shows laminar mixing of a pair of polyester resins. In general, the number of layers will increase by a factor of 4^N, where

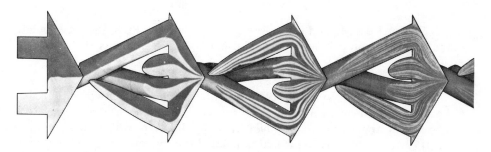

Figure 12-40 Progressive mixing in a Ross ISG Mixer. (*Photo supplied by Ross Corporation.*)

N is the number of elements. Thus the striation thickness can be reduced according to

$$\frac{W}{W_0} = 4^{-N} \tag{12-102}$$

which is considerably better than that achieved with the Kenics Static Mixer. However, this degree of mixing is obtained at a considerable pressure requirement, relative to the Kenics device. The diameter of a single circular channel is about one-fifth the diameter of the pipe into which the element is inserted. For laminar newtonian flow a diameter reduction by this factor would require an increased pressure drop by a factor of $5^4 = 625$. However, each channel carries one-fourth the total flow, so that the pressure increment would be about 160. Ross supplies curves with which the expected pressure drop may be calculated.

Example 12-1 Compare the Ross ISG and the Kenics Static Mixer on the basis of equal reduction of striation thickness at the same throughput and pipe diameter. For the sake of the calculation take $D = 1$ in and $Q = 10$ gal/h, and require that $W/W_0 = 2.5 \times 10^{-4}$.

To achieve this reduction in striation thickness one would require 12 elements of the Kenics Static Mixer and 6 elements of the Ross ISG Mixer. The Kenics Mixer would have a length of about 19 in (taking $L_e/D = 1.6$) while the Ross Mixer would be $6\frac{1}{2}$ in long. Let us assume a viscosity of 100 P.

As a basis, we first calculate the pressure drop through an open pipe 1 in in diameter, 19 in long. From Poiseuille's law,

$$\Delta P = \frac{128\mu LQ}{\pi D^4} = 0.73 \text{ psi}$$

The Reynolds number is

$$\text{Re} = \frac{4Q\rho}{\pi D\mu} = 0.05$$

for which a $\tilde{K}$ factor of $\tilde{K} = 5$ to 6 would apply to the Kenics Mixer. Thus the expected pressure drop in the Kenics Mixer would be approximately

$$\Delta P(\text{Kenics}) = 4 \text{ psi}$$

For the Ross Mixer, Fig. 12-41 is recommended by Ross, and we calculate the pressure drop per *element* as 8.5 psi, and, for six elements,

$$\Delta P(\text{Ross}) = 51 \text{ psi}$$

At these low pressures the advantage of the Kenics Mixer would probably be unimportant and could be offset by the more compact size of the Ross Mixer. For application to very *high*-viscosity fluids, for which the same *ratio* of $\Delta P(\text{Ross})$ to $\Delta P(\text{Kenics})$ would hold, the lower pressure drop of the Kenics Mixer would be a definite advantage.

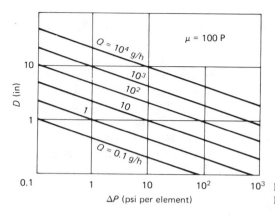

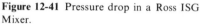

Figure 12-41 Pressure drop in a Ross ISG Mixer.

We had previously made a rough estimate that the Ross ISG Mixer would have a pressure drop approximately 160 times that of an empty pipe of the same length and diameter at the same throughput. The estimated ΔP for the empty pipe in this example was 0.73 psi, and was for a length of 19 in. The Ross ISG Mixer indicated a pressure drop of 51 psi for a length of $6\frac{1}{2}$ in. Adjusting these numbers to an equal-length basis we find

$$\frac{\Delta P(\text{Ross})}{\Delta P(\text{Poiseuille})} = \frac{51}{6.5}\frac{19}{0.73} = 200$$

which is of the order of the expected result. The geometry of the tetrahedral chamber that separates successive elements, and the different lengths of individual channels, makes a more accurate estimate difficult to achieve.

While no theoretical treatment of nonnewtonian flow through motionless mixers is available, and very few experimental observations have been reported, it is possible to anticipate the effect of nonnewtonian behavior on the pressure requirements of the Kenics and Ross mixers. In either case the main factor is the increased shear rate (at a fixed throughput) due to the reduction of "diameter" of the flow channel. The Ross ISG Mixer is the easier to analyze, and we consider it first.

We begin with the power law analog to Poiseuille's law, Eq. (5-17):

$$\Delta P = 2KL\left(\frac{1 + 3n}{n\pi}\right)^n \frac{Q^n}{R^{1 + 3n}} \qquad (12\text{-}103)$$

Suppose we put the same Q through a Ross ISG Mixer of the same outer radius R. Then through each channel we have one-fourth the flow rate. Let us again take the radius of each channel to be reduced by a factor of 5 relative to the empty pipe. Then the pressure ratio would be

$$\frac{\Delta P_{\text{ISG}}}{\Delta P} = (\tfrac{1}{4})^n(5)^{1 + 3n} \qquad (12\text{-}104)$$

Recall that for $n = 1$ this factor is about 160. For small n, say, $n = 0.2$, we find that the pressure factor is only 10, relative to the same length of empty pipe. For $n = 0.5$ the factor is 28.

For the Kenics Static Mixer it is more difficult to use a similar model since the geometry is more complex. In particular the transverse flow is a complicating feature. If we take the newtonian $\tilde{K}$ factor as a basis, we can make an *approximate* correction by considering the viscosity reduction in a Kenics Static Mixer relative to an open pipe at the same flow rate, due to the increased shear rate.

The effective diameter of the Static Mixer is smaller than the corresponding open-pipe diameter by roughly a factor of 2. A suitable number to use would be the ratio of hydraulic diameter [Eq. (5-43)] to pipe diameter:

$$\frac{4r_h}{D} = \frac{4}{D} \frac{\pi D^2/4}{\pi D + 2D} = 0.61 \tag{12-105}$$

The shear rate is increased by the inverse of this factor, and so the viscosity is reduced by a factor of $(0.61)^{1-n}$. This leads us to estimate that the newtonian $\tilde{K}$ factors should be reduced to

$$\tilde{K}_n = \tilde{K}(0.61)^{1-n} \tag{12-106}$$

If we take $\tilde{K} = 5$ in the limit of low Reynolds numbers (note Fig. 12-35), then we estimate

$$\tilde{K}_n = 5(0.61)^{1-n} \tag{12-107}$$

For $n = 0.2$ we find $\tilde{K}_n = 3.4$, while for $n = 0.5$, $\tilde{K}_n = 4$. These numbers are in the range of observed values for molten polymers (as shown in Fig. 12-37), which lends some support to this crude analysis.

If we repeat the comparison of pressure requirements as carried out in Example 12-1 for the case of a highly nonnewtonian fluid, $n = 0.2$, we find that the enormous increase in shear rate in the Ross ISG Mixer is able to offset its pressure disadvantage relative to the Kenics Static Mixer.

If one wishes to mix two liquids under conditions of highly viscous flow, it is not at all clear that the concept of stream splitting that characterizes motionless mixers is truly applicable. The results of Fig. 12-34 certainly suggest the validity of the concept, but this may not be true under *all* possible operating conditions. For example, if the minor component is introduced at a very low volume fraction, it is conceivable that it could travel through a motionless mixer along a tortuous *but coherent* streamline which does not split anywhere. Further, if the viscosities of the two fluids are different by several orders of magnitude, there is a tendency for the low-viscosity fluid to migrate to the solid surfaces where the deformation rates are highest in order to produce a flow with minimum energy dissipation. Thus the low-viscosity fluid "lubricates" the solid boundaries and does not fully participate in the "mixing" character of the flow field. This possibility exists in other "shear" mixing devices such as the extruder and is not restricted to motionless mixers. In general, some preliminary experimental testing is needed to establish feasibility of a specific mixing process.

Many considerations enter into the selection of a mixing process, and many of these considerations involve factors which are not amenable to modeling. (Indeed, many factors are not purely technical.) In a very loose way one must be concerned with the "efficiency" of a mixing process, but this term is very ill defined. Broadly speaking, we would like to get the most mixing for the least energy expenditure, if the energy expenditure is significant. In the mixing of highly viscous materials at high throughputs, energy input is often an important consideration.

We can examine one aspect of this problem here, because it is possible to calculate the power requirements for simple mixers, and it is also possible to calculate some quantitative measure of the extent of mixing achieved by a simple mixer. We illustrate this by comparing a melt extruder to a Kenics Static Mixer.

Let us begin by choosing weighted average total strain (WATS) as the measure of mixing, which we may calculate from Eq. (12-94) if the velocity field is known. The power requirement is also known from the velocity field.

For the simplest model of the melt extruder we may write the power in the dimensionless form [Eq. (6-97)]

$$\frac{\dot{\mathscr{W}} B}{\mu U_z^2 WZ} = 4 - 6\Pi_Q + 4\tan^2\theta$$

Figure 12-26 gives WATS (in the form $\bar{\gamma}B/L$) as a function of θ and $\Phi = 2\Pi_Q - 1$.

For the Kenics Static Mixer we use Tung's theoretical velocity field for helical flow in a semicircular pipe. This theory ignores the effect of the alternating left- and right-hand pitch of the individual elements and so must be considered approximate. As we noted earlier, Tung's theory gives good results for the pressure drop and poor results for the residence-time-distribution function. Since WATS depends upon $f(t)$ through Eq. (12-94), it is likely that the calculation of WATS through Tung's model is in error. Still, it is the best theory available (it is also the *only* theory available), so we pursue the calculation.

Figure 12-42 shows WATS (as $\bar{\gamma}D/L$) as a function of the pitch (L_e/D) of the helix.

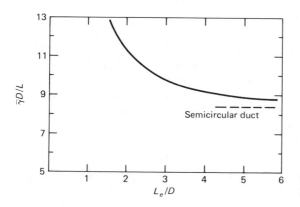

Figure 12-42 WATS for a Kenics Static Mixer.

We compare the systems on the following basis:

$$\text{Flow rate} = 100 \text{ lb/h}$$

$$\text{Viscosity} = 10^5 \text{ P}$$

$$\text{No die resistance}$$

Extruder	Kenics Static Mixer
$D = 2$ in	$D = 1.5$ in
$L/D = 25$	$L/D = 50$
$B/D = 0.1$	$L_e/D = 1.6$
$\theta = 17.7°$	

For the extruder, the isothermal power calculation gives $\dot{W} = 8.5$ hp, and Fig. 12-28 lets us find $\bar{\gamma} = 1750$.

For the Kenics Mixer we find, from Fig. 12-42, a value of $\bar{\gamma} = 640$. We may estimate the pressure drop in the Kenics Mixer (using a $\tilde{K}$ factor of 5) to be about 3500 psi, and using $Q\,\Delta P$ as the power requirement, we find $\dot{W} = 0.4$ hp.

Thus the Kenics Static Mixer achieves only a third of the mixing (using $\bar{\gamma}$ as the measure), but it appears to do so at an expenditure of only 5 percent of the required power for the extruder. At this point one should question whether heat effects in the extruder can reduce the viscosity sufficiently to bring these calculations closer together. This is left for a homework problem (Prob. 12-26).

12-6 MIXING IN STIRRED TANKS

The most common industrial mixing system is the stirred tank. Although there is some limitation to the level of viscosity beyond which good mixing cannot be accomplished in a stirred tank, this level is sufficiently high that a consideration of mixing of polymeric fluids in stirred tanks is in order.

The flow field in a stirred tank defies analytical description. If not turbulent, the flow pattern is at best a complex three-dimensional time-varying laminar flow. On top of this inherent complexity is the fact that an enormous variety of impeller shapes are in use. Consequently one must go from consideration of very general, qualitative features of mixing to very specific observations relevant to (and perhaps *only* to) a particular piece of equipment. Most "design" of stirred-tank mixing processes is based on experience, with only the broadest application of basic principles.

We begin with some general considerations of stirred-tank mixing and then review some experimental results of interest and relevance in consideration of polymeric mixing processes. A useful general reference is the book by Uhl and Gray.

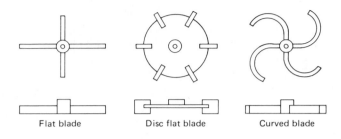

Figure 12-43 Turbine designs.

Mixer Configurations

The mixing vessel itself is generally of cylindrical shape. The shape of the rotating element provides the usual means of classification of stirred-tank systems. While the classification is somewhat arbitrary the most common impellers are usually grouped as *propellers*, *turbines*, or *paddles*.

The propeller is usually similar in design to the marine propeller. It produces a strong *axial* flow and relies on entrainment of the surrounding fluid to give good mixing throughout the vessel. Because entrainment is strongly suppressed by high viscosity, propellers are of little value for mixing fluids whose viscosity exceeds a few thousand centipoise. Hence they are of little interest for polymer processing.

Turbines are usually distinguished by having blades which are at a right angle, or nearly so, to the angular direction of rotation. Figure 12-43 shows some typical turbine designs. In distinction to the propeller, the turbine produces a strong *radial* flow which impacts on the vessel wall, is then directed up or down the wall, and then enters the induced axial flow toward the impeller. The resulting circulation produces good mixing in fluids of viscosities up to the neighborhood of 10^4 to 10^5 cP.

In both the propeller and turbine impellers the maximum velocities are in the central portion of the tank because the moving blades are generally located near the axis of rotation. The flow induced by such impellers relies on strong inertial effects to produce uniform circulation through the vessel. At sufficiently high viscosity inertial flow is suppressed in a relatively short distance from the moving element. As a consequence fluid near the vessel walls, especially at the top and bottom of the vessel, is practically stagnant.

The solution to this problem is to move the main surface of the impeller out toward the periphery of the vessel, thus giving the class of impeller described as a *paddle*. Figure 12-44 shows several types of paddle impellers. Paddle designs such as the "anchor" and the "helical ribbon" provide adequate mixing at viscosities as high as 10^6 cP. Beyond this viscosity stirred tanks are not very useful, and one turns to extruders, motionless mixers, or other high-shear devices.

Another option available regarding the geometric configuration of a stirred tank is the use of baffles, which are usually vertical surfaces rigidly attached to or

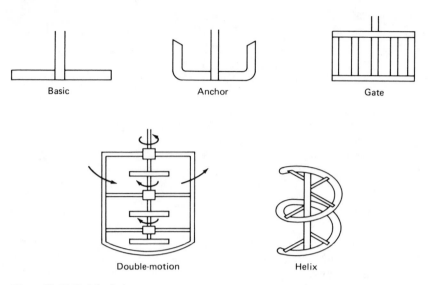

Figure 12-44 Paddle designs.

near the vessel wall. Baffles act to "break up" the flow pattern near the wall. In effect they divert a portion of the slow wall flow radially inward toward the higher shear region of the impeller.

Dimensional Analysis of Stirred-tank Behavior

We may characterize any stirred tank by specifying some appropriate (but arbitrary) linear dimension D and by giving the rotational frequency N as a measure of its operating condition. A set of dimensionless geometric (shape) factors would specify the shape of the mixing system.

Relevant fluid properties would be density ρ and viscosity μ (for a newtonian fluid) or some alternate set of rheological parameters, such as K and n for a power law fluid. Even though most stirred tanks have a free surface, the scale of the system is so large that surface tension effects would be unimportant. The presence of the free surface, however, often leads to vortex formation which affects the flow field. The vortex shape represents a balance between pressure and gravitational forces, and so the gravitational acceleration constant enters the analysis.

There are two main characteristics of a mixing tank for which we desire information. One is related to the power $\dot{W}$ required to operate under a given set of conditions. The other is the time t_m required to achieve a desired degree of mixing.

The power $\dot{W}$ is most easily made dimensionless in the form

$$\dot{W}^* = \frac{\dot{W}}{\rho D^5 N^3} \tag{12-108}$$

[Units, of course, must be consistent. If the units of power include pounds force, while density is in pounds mass, then the conversion factor $g_c = 32.2$ (lbm/lbf) $\times$ (ft·s^{-2}) must be introduced.] One usually refers to $\dot{\mathscr{W}}^*$ as the power number.

A dimensionless mixing time may be defined as

$$t_m^* = Nt_m \tag{12-109}$$

The mixing time is not as precisely defined as the power input. It is a subjective measure based on the approach of the contents of the vessel to some (arbitrarily) defined standard of mixedness. So long as one *consistently* uses an agreed-upon definition of mixedness, then t_m provides a useful relative measure of the time required to mix the contents of the vessel to uniformity.

We may think of $\dot{\mathscr{W}}^*$ and t_m^* as *dependent* variables, and we wish to relate them to the *independent* variables appropriate to the description of stirred-tank design and operation. We already know what these variables are. While the flow field is too complex to yield to theory, we can state that the velocity **u** and pressure p are related through the dynamic equations, along with appropriate boundary conditions. In Chap. 4 we carried out a general dimensional analysis of these equations. We concluded that, for a newtonian fluid, the important dependent variables are the Reynolds number and the Froude number.

Appropriate definitions of these two groups, for a stirred tank, would be

$$\mathrm{Re} = \frac{D^2 N\rho}{\mu} \tag{12-110}$$

and

$$\mathrm{Fr} = \frac{DN^2}{g} \tag{12-111}$$

Thus, for newtonian fluids, we expect that

$$\dot{\mathscr{W}}^* = \dot{\mathscr{W}}^*(\mathrm{Re},\ \mathrm{Fr},\ \text{shape factors}) \tag{12-112}$$

$$t_m^* = t_m^*(\mathrm{Re},\ \mathrm{Fr},\ \text{shape factors}) \tag{12-113}$$

For nonnewtonian fluids we would expect a modified Reynolds number, such as

$$\mathrm{Re}' = \frac{D^2 N^{2-n}\rho}{K} \tag{12-114}$$

for a power law fluid, or we might introduce a Weissenberg number if elastic effects were significant, in the form

$$\mathrm{Ws} = \lambda N \tag{12-115}$$

where λ is some appropriate relaxation time for the fluid.

Let us turn, at this point, to an examination of some typical experimental data and successful methods of correlation of those data. We begin with results for newtonian fluids and then consider correlations of nonnewtonian observations.

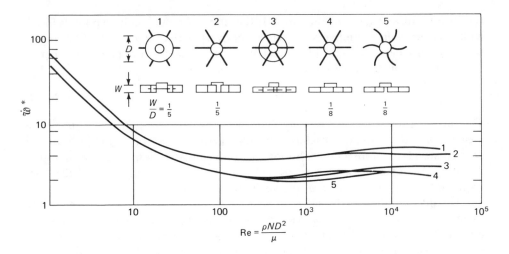

Figure 12-45 Power correlation for turbines. (*After Rushton et al.*)

Correlation of Data

We make no attempt to be comprehensive here or to include the great range of impeller styles and shapes. The Bibliography provides access to a more complete view of this area. Instead we present some data that are generally considered to be reliable and representative of the class of impeller used.

Figure 12-45 shows the curves used to correlate the extensive data of Rushton et al. who used baffled turbine-agitated mixing tanks with newtonian fluids. At high Reynolds numbers the curves flatten out to give nearly constant $\mathscr{W}^*$. This is typical of many forms of impellers but is of minor interest for polymer processing since such high Reynolds numbers are not normally achieved. Instead, we would more likely find ourselves operating in the *low* Reynolds number region where a slope (on log-log coordinates) of -1 is the usual observation. Another point of general interest is that in baffled systems the formation of a significant vortex is suppressed, and as a consequence the Froude number does not enter the correlation of the data.

Six different styles of impeller (all turbines, however) are represented in Fig. 12-45. In the viscous (low Re) region it can be seen that the width W of the blade, relative to the impeller diameter D, is a more significant parameter than the detailed shape of the impeller.

Figure 12-46 shows data obtained by Calderbank and Moo-Young, who studied several nonnewtonian solutions in baffled stirred tanks. Their solutions were representable by the power law, and the level of viscosity, at the level of shear rate in the mixer, was in the range of 10 to 200 P. The most significant observation was that, at least for the systems (fluids and impellers) studied, nonnewtonian data could be superimposed on newtonian data if the right choice of apparent viscosity

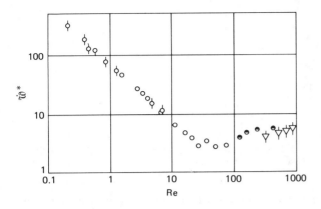

Figure 12-46 Power correlation for four-blade turbines: nonnewtonian fluids. (*After Calderbank and Moo-Young.*)

	$K,$ dyne $\cdot$ s^n/cm^2	n
$\bigcirc$ 4% CMC	112	0.6
φ 5% CMC	430	0.47
$\triangledown$ 55% clay/water	172	0.05
$\ominus$ Kaolin/clay/water	540	0.12

were made. It was found, in agreement with similar studies by Metzner and Otto, that one could define an "equivalent shear rate" for a stirred tank as

$$\dot{\gamma} = 10N \qquad (12\text{-}116)$$

where N is in units of revolutions per time. The coefficient 10 in Eq. (12-116) takes on different values in different studies, depending, apparently, on the style of impeller, but it serves as a useful estimate and we will not belabor its exact value. The implication of this equivalent shear rate is that one may calculate $\dot{\gamma}$ with Eq. (12-116), find the apparent viscosity of the fluid of interest from rheological data in the appropriate shear rate range, and then use a newtonian power correlation. While the approach appears to be reasonably successful, there are some potential problems associated with this view.

In effect, the use of a single apparent viscosity implies that *all* regions of the tank experience the same viscosity. This is clearly not true since the high-shear-rate region is confined to the neighborhood of the impeller. Fluid near the tank walls, or its bottom, is slowly moving, unless the specific impeller design promotes agitation in those regions. Since, for most of the fluids of interest, the low-shear-rate regions will have significantly greater viscosity than the rest of the system, substantially stagnant regions may develop in the tank. One consequence of such behavior may be the development of a "two-compartment" character to a stirred tank.

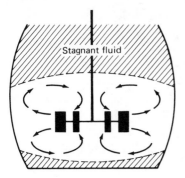

Figure 12-47 Sketch suggesting the development of nearly stagnant regions in mixing of a highly nonnewtonian fluid.

Figure 12-47 suggests a stirred tank in which relatively well-mixed, low-viscosity fluid is surrounded by slowly moving poorly mixed fluid of much greater effective viscosity. As a consequence the system has the appearance of an efficient mixer which occupies only a *portion* of the total vessel volume. The rest of the fluid may exchange very slowly with the well-mixed region. In addition to being a poor mixer, such a system would be quite disadvantageous if thermal effects were important.

Suppose, for example, that the system of Fig. 12-47 were a polymerization reactor, and heat was transferred across the vessel walls in order to maintain a suitable and uniform reaction temperature within the fluid. The stagnant regions would have a different temperature from the well-mixed region since the rate of heat transfer across the vessel wall would be significantly reduced by the adherence of a viscous stagnant layer. Consequently the reaction rate would be highly nonuniform throughout the vessel, and a nonuniform product would result. Other implications of such behavior are equally obvious.

One would, of course, attempt to minimize such problems by using a well-designed impeller, such as a paddle which promotes good peripheral mixing. It is likely, however, that there would still be some degree of segregation. The point is that this kind of problem is aggravated by nonnewtonian behavior, and the use of an effective newtonian viscosity in a power correlation tends to obscure our awareness of other nonnewtonian phenomena.

Mixing-time studies have been carried out by introducing a " contaminant " or " tracer " into a stirred system and measuring, at some point or points within the tank, the concentration of the tracer as a function of time. As mixing proceeds,

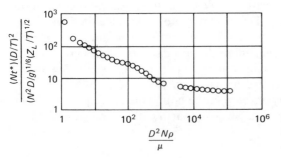

Figure 12-48 Correlation of mixing times for turbines in baffled vessels. (*Data of Norwood and Metzner.*)

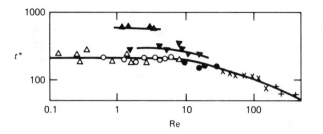

Figure 12-49 Correlation of mixing times for a helical-screw impeller within a draft tube.

		μ, P				$K,$
					n	dyne $\cdot$ s^n $\cdot$ cm^2
×	Glycerine	85				
+	Glycerine	1.7	△	CMC	~0.45	~300
○	Corn syrup	138	▼	PAA	~0.4	~80
●	Corn syrup	26	▲	PAA	~0.25	~270

the tracer concentration fluctuates but eventually approaches a steady value. One defines a mixing time as the time required to reach the steady value, to within some arbitrary standard of deviation. Thus the mixing time is based on an arbitrary definition of mixedness.

Data of Norwood and Metzner are shown in Fig. 12-48 for newtonian fluids in baffled turbine-agitated systems. Reynolds numbers as low as unity were achieved, but extrapolation beyond this point would be dangerous. We note that the ordinate is not simply t^* but includes geometric factors D/T (ratio of impeller diameter to tank diameter) and Z_L/T (ratio of height of liquid in tank to tank diameter). In addition a weak (one-sixth power) dependence on Froude number is noted.

Low Reynolds number mixing studies have been carried out by Chavan et al., and some data are shown in Fig. 12-49. We note that for Re < 10 the dimensionless mixing time flattens out. We also note that the mixing times for highly elastic polyacrylamide solutions are considerably in excess of those for newtonian, or inelastic nonnewtonian, fluids. Ulbrecht and his coworkers have studied flow patterns in the mixing of elastic fluids and suggest that the elasticity significantly alters the kinematics of the circulation patterns within the vessel.

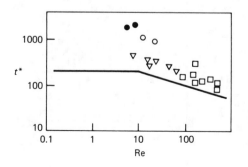

Figure 12-50 Correlation of blending times for a helical-screw impeller within a draft tube. ● 2% PAA; ○ 1% PAA; ▽ 2% CMC; □ 1% CMC.

These "mixing" studies refer to a single fluid within which a tracer or contaminant is mixed. A related study is that of "blending" of two dissimilar (i.e., different viscosity) fluids. Figure 12-50 shows some data of Ford and Ulbrecht on blending of polymer solutions with water. For comparison the corresponding single-fluid mixing-time curve is shown. It is apparent that blending requires more time than mixing. The results are highly dependent on the geometry of the system, the viscosity ratio and volume ratio of the two fluids, and on the elasticity of the fluids. The original references should be consulted.

PROBLEMS

12-1 Determine the scale and intensity of the "mixture" shown in Fig. 12-51. Let the side of the smallest square be the unit of length, so that the total area is 6400 square units. Use a sampling area of 100 square units. State clearly how many samples you take to establish average concentration. For the scale determination calculate the correlation function by sampling at arbitrary points along a diagonal.

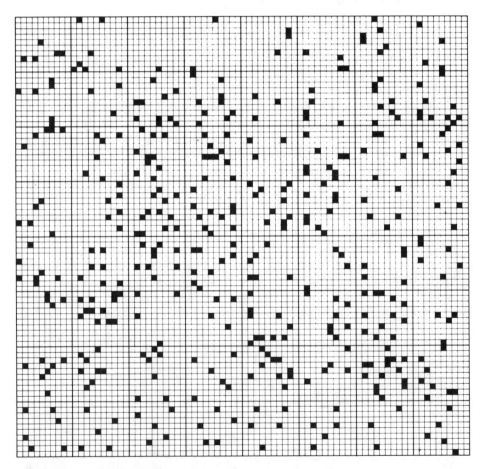

Figure 12-51 "Mixture" for Prob. 12-1.

12-2 Find S for mixtures $S2$ and $S3$ of Fig. 12-1.

12-3 Derive Eq. (12-17).

12-4 Find the F curve for combined pressure and drag flow between infinite parallel plates, for which the velocity profile is

$$u(y) = U\frac{y}{H} - \frac{\Delta P}{2\mu L}y(H - y)$$

Present the result in terms of the parameter $\Pi_P = \Delta PH^2/\mu UL$.

12-5 Show that the area between the plug flow and perfect-mixer F curves is 0.736.

12-6 A quantitative measure of how close a flow system is to the plug flow or perfect-mixer limit is the parameter β, defined as

$$\beta = 1 - \frac{A}{0.736}$$

where A is the area between the F curve of the system and the F curve for the perfect mixer.

Give the value of β for Poiseuille flow.

12-7 The theoretical F curve for a Kenics Static Mixer is shown in Fig. 12-38. Calculate β (see Prob. 12-6).

12-8 Ruthven† calculates the F curve for laminar flow in a circular tube whose axis describes a helix as

$$F = 1 - \frac{1}{4}\left(\frac{t}{\bar{t}}\right)^{-2.81} \qquad \text{for } \frac{t}{\bar{t}} > 0.613$$

Calculate β (see Prob. 12-6).

12-9 Evaluate the F curve for laminar Poiseuille flow with a lubricating wall layer. The sketch (Fig. 12-52) defines this flow as one in which a thin annular layer of low-viscosity fluid surrounds the main phase. Assume the lubricating wall layer is stable and itself is in laminar flow. Define the distribution function for fluid 1 only, i.e., let $f(t)$ be defined such that $f(t)\,dt = dQ_1/Q_1$. Present graphs for $\varkappa = 0.8$ and 0.95, and for $\mu_1/\mu_2 = 10$ and 1000 at each $\varkappa$.

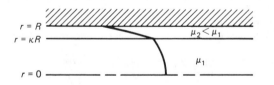

Figure 12-52 Laminar Poiseuille flow with a lubricating layer.

12-10 The F curve may be used to calculate conditions needed to "purge" a system of old material. Suppose a circular pipe has been used to deliver a colored polyethylene to a process. A change in color of the processed material is to be made, and it is planned to simply displace the first material by pumping in the second. The value of $1 - F$ gives the fraction of old material remaining in the pipe at any time after the beginning of the purge.

Suppose we do not wish to start the process using the new color until 99 percent of the old color is purged from the pipe. How much material is wasted in the purging process?

12-11 Derive Eq. (12-41).

12-12 Derive Eq. (12-43).

12-13 Derive Eq. (12-45).

12-14 Derive an equation similar to (12-47) for the *annular* rotational mixer.

† *Chem. Eng. Sci.*, **26**: 1113 (1971).

12-15 Molten polyvinyl chloride (PVC) containing 6.25% carbon black is fed to an annular pipe through which an unpigmented PVC flows at a rate of 25 lb/h. The molten polymers are to pass through a tubing die of 1-in ID and $1\frac{1}{4}$-in OD. The mandrel (the 1-in ID cylinder) is rotated at 60 rpm and is 6 in long. The final carbon black concentration (volume fraction) is to be 0.5%. The annular feedport system (shown in Fig. 12-53) is to be designed so that the black and clear polymers have approximately the same linear velocities at their point of initial contact. The maximum allowable striation thickness is 1 μm.

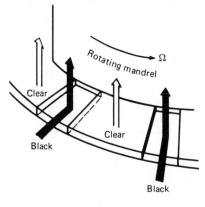

Figure 12-53 Annular feed system.

How many feedports are required for the black melt? What is the angular opening of a feedport segment? What is its width?

12-16 Assuming the pipe produced according to Prob. 12-15 has the same dimensions as the annular die, calculate

(a) How many feet per hour of pipe are produced.

(b) The pressure drop required to pump the melt through the 6-in annular rotating section of the die.

(c) The energy required to rotate the die compared to the pumping energy required.

Assume the melt may be modeled as a power law fluid with $n = \frac{1}{2}$ and $K = 1$ lbf·s$^{1/2}$/in^2.

12-17 Derive an expression for the *number* of striations between any two radial positions in an annular mixer (inner tube rotating). Find the limiting value for the number of striations between the inner and outer radii (the total number of striations) and explain the result. Assume a newtonian fluid.

12-18 For Prob. 12-15 find the number of striations in the pipe cross section that lie in the region 0.505 in $< r <$ 0.620 in.

12-19 Using Eq. (12-41), predict the reduction in striation thickness as a function of axial length at the midpoint between the inner and outer surfaces of an annular tube extruded under the following conditions:

$$D_0 = 4 \text{ in} \qquad Q = 40 \text{ lb/h polystyrene at } 450°F$$

$$D_i = 3.3 \text{ in} \qquad \Omega = 11.7 \text{ rpm}$$

Compare your predictions with the data of Schrenk, Cleereman, and Alfrey shown in Fig. 12-54.

Also shown are data obtained with the *outer* cylinder rotating and the inner mandrel stationary. Should the data be independent of which cylinder rotates, or is this only approximately true?

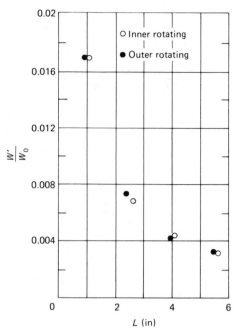

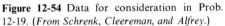

Figure 12-54 Data for consideration in Prob. 12-19. (*From Schrenk, Cleereman, and Alfrey.*)

12-20 Show that the residence time distribution for *plane* Poiseuille flow of a newtonian fluid is

$$f(t) = \frac{1}{3}\frac{\bar{t}^2}{t^3}\left(1 - \frac{2}{3}\frac{\bar{t}}{t}\right)^{-1/2}$$

12-21 Calculate the WATS for *combined* pressure and drag flow between infinite parallel planes. Take the drag flow component to be transverse to the direction of the pressure flow. Present the results as $\bar{\gamma}B/L$ versus some appropriate dimensionless parameter. Consider both positive and negative pressure gradients. You will need the result of Prob. 12-20.

12-22 Repeat the calculation of Prob. 12-21 for the case that the drag and pressure flows are parallel. You will need the results of Prob. 12-4.

12-23 Intuition suggests that the fractional reduction in striation thickness, W'/W_0, should be related in some simple way to the strain imposed on an element of fluid. Calculate WATS for the annular mixer with rotating inner cylinder, and compare WATS to $(W'/W_0)_{\max}$ with regard to the dependence of each on operating and design variables.

12-24 Find $f(\varkappa)$, as defined in Eq. (12-44), in the limit of $\varkappa \to 1$.

12-25 Develop a model for the newtonian kinematics of the combined torsional-radial flow shown in Fig. 12-55.

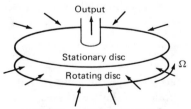

Figure 12-55 Combined torsional-radial flow.

(a) Give the components of **u** and **Δ**.
(b) Calculate II_Δ.
(c) Find the relationship between pressure drop and flow rate.
(d) Find $f(t)$.
(e) Find WATS.

12-26 Rework the example at the end of Sec. 12-5 and account for the effect of heat generation in each system. Use the adiabatic analysis for the estimate.

BIBLIOGRAPHY

A two-volume general reference is

Uhl, V. W., and J. B. Gray (eds.): "Mixing. Theory and Practice," Academic Press, Inc., New York, 1966 (vol. I), 1967 (vol. II).

12-1 Goodness measures

An application of statistical measures of mixedness is given in

Hall, K. R., and J. C. Godfrey: An Experimental and Theoretical Study of Mixing of Highly Viscous Materials, *AIChE–Inst. Chem. Eng. Symp. Ser.*, **10**: 71 (1965).

12-3 Laminar shear mixing

An outstanding series of papers by Alfrey and his coworkers includes the following:

Schrenk, W. J., K. J. Cleereman, and T. Alfrey, Jr.: Continuous Mixing of Very Viscous Fluids in an Annular Channel, *SPE Trans.*, **3**: 192 (1963).
———, D. S. Chisholm, and T. Alfrey, Jr.: Mixing of Viscous Fluids Flowing through a Rotating Tube, *Mod. Plast.*, January 1969, p. 152.
——— and T. Alfrey, Jr.: Coextruding Multilayer Blown Film, *SPE J.*, **29**: 38, 43 (1973).

12-4 The extruder as a mixer

The concept of weighted average total strain is introduced in

Pinto, G., and Z. Tadmor: Mixing and Residence Time Distribution in Melt Screw Extruders, *Polym. Eng. Sci.*, **10**: 279 (1970).

See also

Tadmor, Z., and I. Klein: Mixing in Extruders, chap. 7 in "Engineering Principles of Plasticating Extrusion," Van Nostrand Reinhold Company, New York, 1970.

Application to extrusion of power law fluids, and experimental verification of the theory, is given in

Bigg, D., and S. Middleman: Mixing in a Screw Extruder. A Model for Residence Time Distribution and Strain, *Ind. Eng. Chem. Fund.*, **13**: 66 (1974).

An important mixing topic which we have had to leave out here is that of twin-screw extrusion. An introduction to simple models of these extruder mixers can be gained through the following papers:

Kaplan, A., and Z. Tadmor: Theoretical Model for Non-Intermeshing Twin Screw Extruders, *Polym. Eng. Sci.*, **14:** 58 (1974).

Janssen, L., L. Mulders, and J. Smith: A Model for the Output from the Pump Zone of the Double-Screw Processor or Extruder, *Plast. Polym.*, **43:** 93 (1975).

Wyman, C. E.: Theoretical Model for Intermeshing Twin Screw Extruders: Axial Velocity Profile for Shallow Channels, *Polym. Eng. Sci.*, **15:** 606 (1975).

Todd, D. B.: Residence Time Distribution in Twin Screw Extruders, *Polym. Eng. Sci.*, **15:** 437 (1975).

12-5 Motionless mixers

The theoretical analysis of flow in a Kenics Static Mixer is given in the Ph.D. thesis by Tung:

Tung, T.: "Low Re Entrance Flows: A Study of a Motionless Mixer," Ph.D. thesis, University of Massachusetts, Amherst, 1976.

Chen, S. J., and A. R. MacDonald: Motionless Mixers for Viscous Polymers, *Chem. Eng.*, March 19, 1973, p. 105.

12-6 Mixing in stirred tanks

In addition to the Uhl and Gray reference above, a good general reference is

Holland, F. A., and F. S. Chapman: "Liquid Mixing and Processing in Stirred Tanks," Reinhold Publishing Corporation, New York, 1966.

Unusual agitator designs are discussed in

Ho, F. C., and A. Kwang: A Guide to Designing Special Agitators, *Chem. Eng.*, July 23, 1973, p. 94.

Figure 12-45 is based on the data in

Rushton, J. H., E. W. Costich, and H. J. Everett: Power Characteristics of Mixing Impellers, *Chem. Eng. Prog.*, **46:** 467 (1950).

Figure 12-46 shows a small fraction of the data given in

Calderbank, P. H., and M. Moo-Young: The Prediction of Power Consumption in the Agitation of Non-Newtonian Fluids, *Trans. Inst. Chem. Eng. (London)*, **37:** 26 (1959).

Other power studies in nonnewtonian fluids include

Metzner, A. B., and R. E. Otto: Agitation of Non-Newtonian Fluids, *AIChE J.*, **3:** 3 (1957).

Mixing rates are studied in

Norwood, K. W., and A. B. Metzner: Flow Patterns and Mixing Rates in Agitated Vessels, *AIChE J.*, **6:** 432 (1960).

Chavan, V. V., D. E. Ford, and M. Arumugam: Influence of Fluid Rheology on Circulation, Mixing and Blending, *Can. J. Chem. Eng.*, **53:** 628 (1975).

Ford, D. E., and J. Ulbrecht: Blending of Polymer Solutions with Different Rheological Properties, *AIChE J.*, **21:** 1230 (1975).

Some practical aspects of design of agitated polymerization reactors are discussed in

Beckmann, G.: Design of Large Polymerization Reactors, chap. 3 in *Adv. Chem. Ser. 128*, 37 (1973).

THIRTEEN

HEAT AND MASS TRANSFER

He who, for the sake of learning, lowers himself by exposing his ignorance, will ultimately be elevated.

Ben Azzai

In many polymer processes heat and mass transfer occur within a fluid while it is undergoing some flow process. Most of the problems treated so far have ignored this possibility while isolating and focusing on the fluid dynamics. Exceptions have been those problems where viscous heat generation was included through introduction of a very simple energy balance: the first law of thermodynamics for an adiabatic system.

Now we turn to consideration of flow processes in which heat and mass transfer play a significant role, and for which a more general theoretical foundation must first be established before we can attempt to model such processes. We begin by deriving a *generalized transport equation* which will describe either heat or mass transfer. Then, through a series of examples, we will illustrate the modeling of a variety of important polymer processes in which heat and/or mass transfer either alters the flow process or is itself the dominant process of interest.

13-1 A GENERALIZED TRANSPORT EQUATION

We consider a continuous medium characterized by a velocity vector $\mathbf{u}(\mathbf{x}, t)$, a stress tensor $\mathbf{T}(\mathbf{x}, t)$, and a mass density ρ. For the problems of interest, it will suffice to consider the mass density to be a constant. We wish to allow for the possibility that within the fluid there is a nonuniform temperature distribution $T(\mathbf{x}, t)$ and a composition distribution $c_i(\mathbf{x}, t)$. The subscript i on c refers to each distinct chemical species in the fluid.

The first question to consider is whether the derivation of the continuity equation and the dynamic equations in Chap. 3 must be modified for a nonisothermal multicomponent fluid. The immediate answer is no. So long as we understand that the quantities represented by $\mathbf{u}$, $\mathbf{T}$, and ρ are defined at a point in the fluid which is locally homogeneous, in which composition varies continuously, then the results of Chap. 3 are still valid.

We require here, as well, that the temperature and composition fields $T(\mathbf{x}, t)$ and $c_i(\mathbf{x}, t)$ be well defined and continuous. In particular, we want to introduce the concept of a *density* of internal energy X_T and a *density* of chemical species X_i. We do this in the following way:

$$dX_T = \rho C_v \, dT \qquad \text{or} \qquad X_T = \int \rho C_v \, dT \qquad (13\text{-}1)$$

$$X_i = c_i \qquad (13\text{-}2)$$

Equation (13-1) embodies the notion that internal energy is measured relative to some reference state. Thus we can talk about the change in internal energy, in differential form, or use the integral form with the internal energy measured relative to a standard state at the temperature of the lower limit of the integral.

In Eq. (13-1) C_v is the heat capacity (e.g., in units of Btu/lbm·°F). This gives X_T the units of Btu/ft³ (taking ρ as lbm/ft³), a "density" of internal (thermal) energy. In Eq. (13-2) we choose *molar* concentration units for c_i, that is, moles of species **i** per unit volume.

We note that the mass density ρ and the individual molar densities c_i are not independent but are related by

$$\rho = \sum_{i=1}^{n} M_i c_i \qquad (13\text{-}3)$$

where M_i is the molecular weight of species **i**, and the summation extends over all n separate species in the mixture.

Now let us set up a fixed cartesian coordinate system and consider an element of volume through which fluid passes, as in Fig. 13-1. In a manner similar to that used in Chap. 3 in the derivation of the continuity equation, we can derive an equation for conservation of each chemical species and for conservation of thermal energy. We note that the word *conservation* is something of a misnomer here. While *mass* is conserved, chemical *species* may interconvert through reactions. Thermal energy may be "created" through viscous dissipation (mechanical conversion) or through chemical reaction (exothermic or endothermic processes), or phase change.

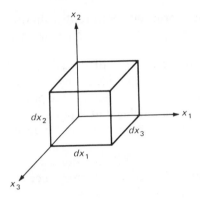

Figure 13-1 Differential control volume for derivation of the transport equation.

Within the volume element the internal energy content, or the amount of any chemical species, changes at a rate which we may write as

$$\frac{\partial}{\partial t} X \, dx_1 \, dx_2 \, dx_3$$

where X may be *either* X_T or X_i.

If fluid crosses a surface of the volume element, then that flow carries internal energy and chemical species in and out of the volume. Any "parcel" of fluid crossing a surface of the control volume has a "density" X (we will not distinguish between heat and chemical species, since the comments and calculations are identical) and crosses the surface with velocity **u**.

As in the derivation of the momentum equation (Chap. 3) we simply multiply the appropriate density by the velocity component *normal* to each face to find the convective *flux* across each face. Thus we find

$$\text{Flow into face } dx_2 \, dx_3 = Xu_1 \, dx_2 \, dx_3$$

$$\text{Flow into face } dx_1 \, dx_3 = Xu_2 \, dx_1 \, dx_3$$

$$\text{Flow into face } dx_1 \, dx_2 = Xu_3 \, dx_1 \, dx_2$$

Note that in each case it is the "density" times the velocity u_i normal to the surface of interest. Since both thermal energy and chemical species concentrations are scalar quantities, X has no direction associated with it. The area factor converts the *flux* Xu_i to a *rate*.

The sum of the three terms above gives the flow rate across the three orthogonal surfaces in the coordinate planes. The other three surfaces experience similar flows, but we must account for the possibility that X varies spatially. Across the parallel surfaces, then, we have, by Taylor's theorem, flow rates given by

$$Xu_i \, dx_j \, dx_k + \frac{\partial}{\partial x_i} (Xu_i \, dx_j \, dx_k) \, dx_i \cdots$$

Consequently the *net* flow, the *convective* flow, across all six faces, is given by [see Eq. (3-33)]

$$C_X = \left(\frac{\partial}{\partial x_1} X u_1 + \frac{\partial}{\partial x_2} X u_2 + \frac{\partial}{\partial x_3} X u_3 \right) dx_1 \, dx_2 \, dx_3 \qquad (13\text{-}4)$$

In addition to convection, both heat and chemical species may cross the control-volume surfaces by a molecular, diffusive, mechanism. In the case of heat transfer we usually refer to this as *conduction*. In speaking generally of X, rather than specifically of X_T or X_i, we will simply use the term *diffusion*. We will *define* a diffusive flux $\mathbf{J}$ so that the rate of transfer of X across each face, by diffusive transport, is

$$J_i \, dx_j \, dx_k \qquad \text{for } i = 1, 2, 3$$

Note that $\mathbf{J}$ is a *vector* quantity, and we multiply the component of the flux vector in *each* i direction by the area normal to the i axis. Similar terms are expressible for the parallel control surfaces, and again the net flow is given, after application of Taylor's theorem, by

$$D = \left(\frac{\partial}{\partial x_1} J_1 + \frac{\partial}{\partial x_2} J_2 + \frac{\partial}{\partial x_3} J_3 \right) dx_1 \, dx_2 \, dx_3 \qquad (13\text{-}5)$$

In considering a balance on thermal energy or chemical species we must account, as noted before, for the possibility of *generation* of heat or species by various physical mechanisms. For the moment, let us simply *define* a generation term, per unit volume, such that $G \, dx_1 \, dx_2 \, dx_3$ is the rate of change of energy or amount of a chemical species due to some mechanism of conversion.

If all these inputs are added we find, after noting that the differential volume terms cancel, an equation of the form

$$\frac{\partial X}{\partial t} = - \frac{\partial}{\partial x_i} (X u_i) - \frac{\partial}{\partial x_i} J_i + G \qquad \text{(sum over } i) \qquad (13\text{-}6)$$

(Our signs arise from the convention that the so-called *net* flows are differences of inputs and outputs: A positive net flow means a greater inflow than outflow. G is defined to be positive if X increases.)

The convective terms may be rewritten in the form

$$\frac{\partial}{\partial x_i} (X u_i) = u_i \frac{\partial X}{\partial x_i} + X \frac{\partial u_i}{\partial x_i} \qquad \text{(sum over } i) \qquad (13\text{-}7)$$

The second term on the right-hand side vanishes by virtue of the continuity equation since we assume the fluid is incompressible. As a consequence we find

$$\frac{DX}{Dt} = \frac{\partial X}{\partial t} + u_i \frac{\partial X}{\partial x_i} = - \frac{\partial}{\partial x_i} J_i + G \qquad \text{(sum over } i) \qquad (13\text{-}8)$$

The diffusive term may be written in vector notation as

$$\frac{\partial}{\partial x_i} J_i = \nabla \cdot \mathbf{J} = \mathrm{div}\ \mathbf{J} \tag{13-9}$$

We may then write Eq. (13-8) in the general format

$$\frac{DX}{Dt} = -\nabla \cdot \mathbf{J} + G \tag{13-10}$$

In cylindrical polar and spherical coordinate systems Eq. (13-10) has the form
Cylindrical:

$$\frac{\partial X}{\partial t} + u_r \frac{\partial X}{\partial r} + \frac{u_\theta}{r} \frac{\partial X}{\partial \theta} + u_z \frac{\partial X}{\partial z} = -\frac{1}{r} \frac{\partial}{\partial r} r J_r - \frac{1}{r} \frac{\partial J_\theta}{\partial \theta} - \frac{\partial J_z}{\partial z} + G \tag{13-11}$$

Spherical:

$$\frac{\partial X}{\partial t} + u_r \frac{\partial X}{\partial r} + \frac{u_\theta}{r} \frac{\partial X}{\partial \theta} + \frac{u_\varphi}{r \sin \theta} \frac{\partial X}{\partial \varphi}$$

$$= -\frac{1}{r^2} \frac{\partial}{\partial r} r^2 J_r - \frac{1}{r \sin \theta} \frac{\partial}{\partial \theta} J_\theta \sin \theta - \frac{1}{r \sin \theta} \frac{\partial J_\varphi}{\partial \varphi} + G \tag{13-12}$$

We will refer to Eq. (13-10) as a *generalized transport equation*. To solve the equation it is first necessary to relate the diffusive flux $\mathbf{J}$ and the generation term G to the dependent variables X and $\mathbf{u}$. This situation is quite similar to what we faced with the dynamic equations: It was necessary to introduce constitutive equations relating $\mathbf{T}$ to $\mathbf{u}$. In the same sense we must now introduce constitutive equations for $\mathbf{J}$ and G.

In the case of internal energy, the variable X_T is usually replaced by temperature, using Eq. (13-1). As a consequence the transport equation for internal energy becomes an equation with T as the dependent variable. We will continue to carry X as the general variable when we do not wish to distinguish between heat or mass transfer, since this will let us economize on space in several subsequent developments. However, we note here that in terms of temperature Eq. (13-10) takes the form

$$\frac{DX_T}{Dt} = \frac{\partial X_T}{\partial T} \frac{DT}{Dt} = \rho C_v \frac{DT}{Dt} = -\nabla \cdot \mathbf{J} + G_T \tag{13-10a}$$

13-2 CONSTITUTIVE EQUATIONS FOR DIFFUSION

Nature has been kind to us! We find that for a great variety of materials having a wide range of properties, the relationship between diffusive flux and density may be expressed in a linear form. In the case of heat conduction Fourier's law is found to hold, which we write in the form

$$\mathbf{J}_T = -k \, \nabla T \tag{13-13}$$

This equation defines the thermal conductivity k.

For molecular diffusion we normally observe Fick's law to hold, expressed in the form

$$\mathbf{J}_i = -\mathscr{D}_i \, \nabla c_i \tag{13-14}$$

This equation defines the species diffusion coefficient $\mathscr{D}_i$ in the fluid. Equation (13-14) is actually an approximation, in the sense that in a multicomponent mixture the flux $\mathbf{J}_i$ depends on the gradients of *all* species. However, virtually no data are available for multicomponent diffusivities of interest to us here, and so Eq. (13-14) is used as a definition of $\mathscr{D}_i$. This point is discussed more fully in chap. 18 of Bird et al. who note that if species **i** is in dilute concentration one may use the diffusivity of species **i** in the material which makes up the bulk of the system. This is not always valid in polymer-solvent systems, and several references are suggested which amplify this point.

Since our interest is in incompressible materials we may define a *generalized diffusivity* which again allows us to write the transport equations for heat and species as a single equation. We begin by writing Eq. (13-14) in the format

$$\mathbf{J}_i = -\alpha_i \, \nabla X_i \tag{13-15}$$

where, obviously, $\alpha_i = \mathscr{D}_i$, and X_i is as defined in Eq. (13-2). We rewrite Fourier's law in the same format, namely,

$$\mathbf{J}_T = -\alpha_T \, \nabla X_T \tag{13-16}$$

by defining a *thermal diffusivity* as

$$\alpha_T = \frac{k}{\rho C_v} = \frac{k}{\rho C_p} \tag{13-17}$$

The latter part of Eq. (13-17) is consistent with the assumption of incompressible fluids, for which $C_v = C_p$.

Thus we may write Eq. (13-10) for *either* heat or mass transfer as

$$\frac{DX}{Dt} = \nabla \cdot \alpha \, \nabla X + G \tag{13-18}$$

It is interesting to note the resemblance of Eq. (13-15) to the model of the purely viscous fluid, written as

$$\tau = \eta \, \Delta = \eta(\nabla \mathbf{u} + \nabla^T \mathbf{u}) \tag{13-19}$$

Here we have used the vector format (in order to display the resemblance) $\nabla \mathbf{u}$ for the tensor whose cartesian components are $\partial u_i / \partial x_j$. $\nabla^T \mathbf{u}$ is the " transpose " of $\nabla \mathbf{u}$ and has components obtained by interchanging the indices on $\nabla \mathbf{u}$; that is, $\nabla^T \mathbf{u} = \partial u_j / \partial x_i$. If we note that $\rho \mathbf{u}$ is the momentum density (it was treated as such in the derivation of the dynamic equations in Chap. 3), then we may write Eq. (13-19) as

$$\tau = v(\nabla \rho \mathbf{u} + \nabla^T \rho \mathbf{u}) \tag{13-20}$$

where $v = \eta/\rho$ is the so-called *kinematic viscosity*. In this form, by analogy to Eq. (13-15), we speak of τ as the *momentum flux tensor*, and v is the *momentum diffusivity*. Note that v, $\mathscr{D}_i$, and $k/\rho C_p$ all have the same units: (length)^2 per time.

While η, and hence v, is a function of the invariants of Δ for most polymeric fluids, there is no strong evidence for a parallel situation with regard to α for heat or mass transfer. Indeed, the amount of data available for the α's of polymers is quite sparse by comparison with that for nonnewtonian viscosity. It does appear that $\mathscr{D}_i$ is quite a strong function of temperature and composition (i.e., of c_i), whereas the thermal conductivity k is a relatively weak function of temperature and composition.

Thermal Transport Coefficients

The thermal diffusivity of polymer melts appears to be a weak function of temperature, independent of molecular weight, and indeed does not vary greatly from polymer to polymer. Excellent data are presented by Shoulberg, a selection of which is presented here. The method of measurement gives the thermal diffusivity directly; individual values of k, ρ, and C_p were not determined. Figure 13-2 shows data for commercial-grade polymers, and includes polymethyl methacrylates, polystyrenes, vinyl copolymers, polyethylene, and polyvinyl chloride. The comments of the paragraph above are obvious and need not be enlarged upon.

Figure 13-3 shows data for polypropylene giving k, ρ, and C_p as functions of temperature. The calculated value of α is also shown. We note the cusp in the C_p data, associated with melting of the polymer, for which the heat of fusion is reported as $\Delta H_m = 41$ cal/g.

Figure 13-4 shows C_p data of Nylon 66. Again the apparent increase in C_p associated with melting, which involves a heat of fusion of $\Delta H_m = 56$ Btu/lb, is seen, suggesting a melting temperature of 262°C. It is noted, however, that when a polymer is heated or cooled at a relatively high rate, as in most processing operations, the phase change does not occur at equilibrium. As a consequence, the

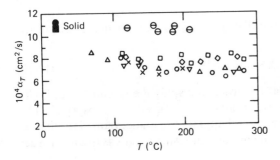

Figure 13-2 Thermal diffusivities of various polymer melts. ◯ Polymethyl methacrylate; □ polystyrene; △ ethyl acrylate–methyl methacrylate; ▽ methyl methacrylate–styrene; ◇ styrene-acrylonitrile; × polyvinyl chloride; ⊖ polyethylene. (*Data from Shoulberg.*)

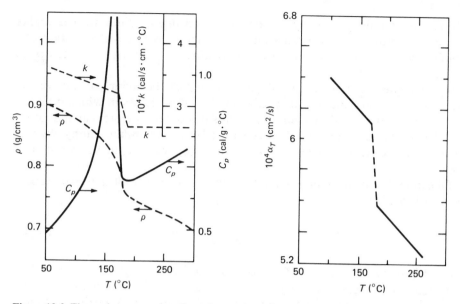

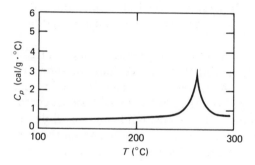

Figure 13-3 Thermal property data for polypropylene. (*Data from Wanger.*)

Figure 13-4 Heat capacity data for Nylon 66.

melting and freezing points can differ considerably and indeed depend on the rate of heat transfer. In making calculations for the freezing of injection-molded nylon parts, one usually takes the freezing point to be about 20°C below the melting point.

Species Diffusion Coefficients

In contrast to observations on *thermal* diffusion, the *species* diffusion coefficient depends strongly on the chemical identity and thermodynamic state of the polymer and the solvent or penetrant. Indeed there is such a wide range of possible behavior that it is useful to begin with a discussion of two limiting situations: the dilute solution, where the polymer concentration is quite small, and the bulk polymer, where the diffusing species is present at a very small concentration. The dilute solution region is of somewhat less interest than that of the bulk polymer, since relatively little processing is carried out on dilute solutions. The higher polymer concentration fluid occurs more commonly, as in devolatilization of a monomer-polymer reactant mixture, or casting of film from concentrated solution.

We will present here some indication of typical behavior so that we can proceed with the presentation of suitable models of processes in which diffusion plays a significant role. More complete discussions of both theoretical and experimental aspects of diffusion in polymers are available in the Bibliography. We also note a general reference, the book " Diffusion in Polymers," edited by Crank and Park.

One of the simplest models of diffusion in liquids is the Stokes-Einstein equation, which predicts

$$\mathcal{D}_{AB} = \frac{\varkappa T}{6\pi\mu_B R_A} \tag{13-21}$$

for the diffusivity of solute A, of "molecular radius" R_A, through a liquid B of viscosity μ_B. The absolute temperature is used, and $\varkappa$ is the Boltzmann constant, which is just the gas constant per molecule, $R/\tilde{N}$, where $\tilde{N}$ is Avogadro's number (6.02×10^{23} molecules/g mol). The Stokes-Einstein equation is found to be a good model for the diffusivity of relatively large spherical molecules through a solvent of smaller molecules in dilute solution where no interaction among diffusing species occurs.

The Stokes-Einstein equation is based on a "hydrodynamic theory"; it treats the diffusing solute molecule as a particle moving through a continuum, subjected to a viscous drag according to Stokes' law. A different approach is the Eyring rate theory, which is based on the "hole theory" of the liquid state. Diffusion, according to this theory, occurs when a molecule shifts its position from one "hole" in the liquid to another. An energy barrier must be overcome to bring about the shift, and absolute rate theory is used to predict the frequency of rearrangement. The result of the rate theory is

$$\mathcal{D}_{AB} = \frac{\varkappa T}{\xi\mu_B}\left(\frac{\tilde{N}}{\tilde{V}_B}\right)^{1/3} \exp \frac{E_{\mu B} - E_{AB}}{RT} \tag{13-22}$$

where ξ = parameter related to the number of nearest neighbors of the diffusing species

$\tilde{V}_B$ = molar volume of the *solvent*

E's = activation energies for viscosity and diffusion

At the other extreme from the dilute solution we must consider the diffusion of a solute molecule through nearly pure polymer. This is the case approached as nearly complete devolatilization is achieved in the removal of solvents from a polymeric solution. It also corresponds to the case of diffusion of solutes through solid polymer, as in plastic films used for packaging to reduce the entry of contaminating species.

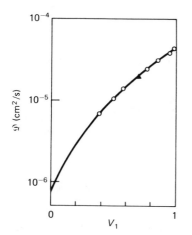

Figure 13-5 Diffusivity of benzene in polyisobutylene-benzene solutions. Data at 70.4°C. (*From Boss, Stejskel, and Ferry.*)

A theory useful for the concentrated region is that due to Doolittle and Fujita. This is basically a "hole theory" written in terms of the "free volume" in a polymer. The free volume is expected to depend strongly on polymer concentration when the fluid is nearly pure polymer. The Fujita-Doolittle theory takes the form

$$\log \frac{\mathscr{D}}{\mathscr{D}_0} = \frac{v_1}{A + Bv_1} \tag{13-23}$$

where v_1 = volume fraction of diluent ($1 - v_1$ is volume fraction of polymer)

A, B = functions of temperature

$\mathscr{D}_0$ = diffusion coefficient in the limit of vanishing v_1 (pure polymer)

With the models above as background, let us examine some examples of experimental data on diffusion coefficients in polymers. We begin with a study, by Ferry and coworkers, of the diffusion of benzene in polyisobutylene-benzene solutions. Figure 13-5 shows diffusion coefficients at 70.4°C, for concentrations in the range $0.385 \leq v_1 \leq 1$. (The point at $v_1 = 1$ is the *self-diffusion* coefficient of benzene.) The polymer has a viscosity average molecular weight of about 1.5×10^6. A sharply fractionated sample, with a molecular weight of 0.5×10^6, was examined (the triangle symbol on the figure) and no effect of molecular weight was observed.

A test of theory is shown in Fig. 13-6. The form of plotting should yield a straight line if the Fujita-Doolittle theory provides a good model; it apparently does for almost all the concentration range studied. If the dilute solution region (v_1 nearly unity) is examined more closely, we find, in Fig. 13-7, that the Fujita-Doolittle theory holds up to about 92% benzene (by volume). For comparison the dilute solution models, which predict an inverse relationship between diffusion coefficient and viscosity, are tested in the form

$$\frac{\mathscr{D}}{\mathscr{D}_s} = \frac{\eta_s}{\eta} = \frac{1}{\eta_R} \tag{13-24}$$

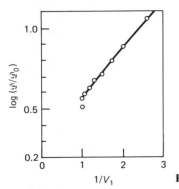

Figure 13-6 Test of Fujita-Doolittle theory for data of Fig. 13-5.

Here η_R is the *relative viscosity*, defined as the ratio of solution to solvent (benzene) viscosity, and $\mathscr{D}_s$ is the self-diffusion coefficient of benzene. We see that, except in very dilute solution ($v_1 > 0.98$), the hydrodynamic theories are not very good.

Ferry's study is basically a solution study at relatively low temperature. It is interesting to look at data obtained at the other extreme, namely, dissolution followed by diffusion of solvent vapor into a polymeric melt. Such data are presented by Duda and Vrentas, who studied the system n-pentane–polystyrene. The results are shown in Fig. 13-8, where the diffusivity of n-pentane at infinite dilution is shown as a function of temperature. Typical Arrhenius behavior is shown, of the form

$$\mathscr{D} = D_0 e^{-E/RT} \tag{13-25}$$

Note that while D_0 has units of diffusivity, it is not the diffusivity at any physically realizable temperature, and should simply be considered an empirical constant.

The unusual feature of the observed behavior is the pair of activation energies, which suggests that a transition of some type may occur in polystyrene around 150°C, which is well above the *glass transition* temperature† of 100°C. Duda and Vrentas discuss this feature, and we will not pause to consider it here. It does raise an important point, however, regardless of its interpretation, and that is with regard to extrapolation of data outside the range of variables covered. Clearly if one had the data of only one branch of Fig. 13-8, and extrapolated to temperatures 20°C past the transition, errors of an order of magnitude could be suffered in the process.

One should also note the order of magnitude of the diffusion coefficient in the limit of pure polymer, ranging from 10^{-8} to 10^{-7} cm²/s, which is considerably below the solution values Ferry shows, which are of the order of 10^{-5} cm²/s. We shall return to some implications of this point subsequently.

---

† If a molten polymer is cooled continuously and if crystallization does not occur, it solidifies to a glassy solid over a narrow range of temperature. The temperature about which this transition takes place is called the glass temperature T_g. The heat capacity and the specific volume exhibit abrupt changes in slope at T_g, and this provides the usual means of measurement.

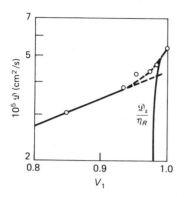

Figure 13-7 Dilute solution region of Figure 13-5.

Duda and Vrentas present data only for *n*-pentane in polystyrene. However, they cite data of Zhurkov and Ryskin for other penetrants into polystyrene and suggest that a simple empirical correlation with which the constants E and D_0 in the temperature range between the glass transition and the second apparent transition at 150°C may be obtained from knowledge of the molar volume of the penetrant.

Figure 13-9 shows a correlation between E and $\tilde{V}_A$ and between D_0 and E whereby one can obtain the constants for the Arrhenius model. Too few data are available for other polymers to allow any suggestion of the generality of such a correlation, but the idea does provide a starting point with which diffusivities may be predicted with minimal information.

The studies cited above have not indicated the role of molecular weight of the polymer in affecting the diffusion coefficient. Diffusion measurements in a series of polystyrenes over a wide range of molecular weights have been carried out by Paul and coworkers. Figure 13-10 shows results for cyclohexanone, with the diffusivity corresponding to concentrated solutions of approximately 30% solvent. The result of interest is the plateau, for $M > 10^5$, where the diffusion coefficient is independent of molecular weight. Paul points out that the plateau occurs in the neighborhood of the *critical molecular weight* M_c for a 30% solution of polystyrene. M_c is defined as the molecular weight above which an individual polymer

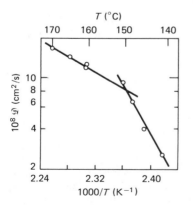

Figure 13-8 Diffusion coefficient versus temperature for *n*-pentane–polystyrene. (*Data from Duda and Vrentas.*)

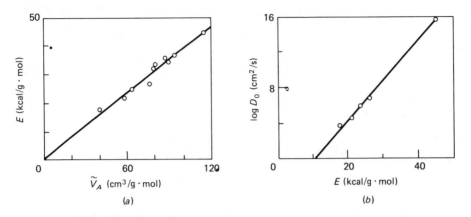

Figure 13-9 (a) Correlation of activation energy with molar volume for various penetrants in polystyrene. [*Data of S. N. Zhurkov and G. Ya. Ryskin, J. Tech. Phys. (USSR),* **24**: 797 (1954).] (b) Correlation of D_0 with activation energy.

chain becomes so entangled among neighboring chains that some of the "individuality" of the chain is lost. It is plausible to suggest that in such an entangled polymer a penetrant diffuses among chain segments of molecules, and the "structure" of this "segment space" no longer depends on molecular weight. In any event, regardless of the interpretation, the observation of Fig. 13-10 seems clear enough.

The experimental technique used by Paul and coworkers allows a determination of the concentration dependence of the diffusion coefficient. Some questions regarding the quantitative reliability of the data are raised and discussed in the paper, but one result emerges that warrants comment. Figure 13-11 shows that the diffusion coefficient may increase, decrease, or be independent as concentration of polymer increases, with molecular weight apparently being the controlling variable. It is the *increase* of diffusion coefficient which seems to be the strange observation. The result appears to be contrary to the usual models of diffusion, as well as contrary to most observations, although Secor makes the same observation in the system dimethylformamide–polyacrylonitrile.

The interesting point to note, apart from the observation itself, is that one can rationalize the result by invoking thermodynamic arguments. Indeed Paul and

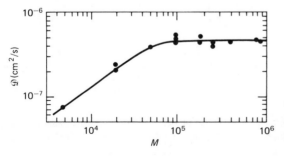

Figure 13-10 Diffusion coefficient of cyclohexanone in various molecular weight polystyrenes at 25°C. Concentrations vary but are approximately 30% polymer. (*From Paul et al.*)

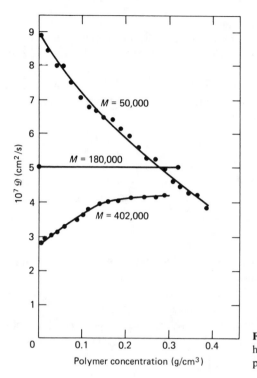

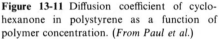

Figure 13-11 Diffusion coefficient of cyclo-hexanone in polystyrene as a function of polymer concentration. (*From Paul et al.*)

coworkers invoke a very simple model of diffusion that introduces a factor which is a difference between a *hydrodynamic* term and a *thermodynamic* term, each of which has a different expected dependence on molecular weight. Depending on the molecular weight it may be possible for this factor to change sign. Since the slope of diffusivity versus concentration is proportional to this factor, the conclusion is drawn that for some polymer-solvent pairs it is possible to find a range of molecular weights over which the observation of Fig. 13-11 would be expected.

Gainer has carried out a theoretical development of diffusion in polymer solutions that allows one to specify necessary conditions under which the diffusion coefficient might increase as polymer concentration increases. The conditions relate to activation energies for diffusion, and Gainer asserts that few systems would satisfy the required conditions. This conclusion would suggest that the results of Paul, shown above, may not be indicative of the usual behavior in some systems.

13-3 CONSTITUTIVE EQUATIONS FOR GENERATION

In order to solve the generalized transport equation (13-18), it is necessary to relate the generation terms to the dependent variables X and $\mathbf{u}$. We will consider the following forms of generation:

1. Heat generation through viscous dissipation.
2. Heat generation through chemical reaction. We use the term *generation* even if the reaction is *endothermic*, although this will not normally be the case in polymerization reactions.
3. Heat generation through phase change. This may include solidification and crystallization of molten polymers or solvent evaporation from solutions.
4. Species generation through chemical reaction.

Viscous Dissipation

It can be shown that the rate (per unit volume) of conversion of mechanical energy into heat (viscous dissipation) is given by a sum of terms involving products of stress components and velocity gradients:

$$G_v = \tau : \nabla \mathbf{u} \tag{13-26}$$

Bird, Stewart, and Lightfoot give the derivation, which will not be repeated here. We use the compact tensor notation, as above, only when it is necessary to display the *format* of the transport equations. For calculations we need G_v in the appropriate coordinate system. Table 13-1 gives the required terms.

Since τ will normally have *mechanical* units, it will be necessary to remember to convert the dissipation terms to *thermal* units when numerical calculations are required:

Multiply	By	To get
ft·lbf	1.285×10^{-3}	Btu
g·cm²·s⁻²	2.39×10^{-8}	cal
g·cm²·s⁻²	9.48×10^{-11}	Btu
ft·lbf	3.24×10^{-1}	cal

Heats of Reaction

We define a molar reaction rate as

$$r_i = \frac{dc_i}{dt} \tag{13-27}$$

where, in this definition, we understand the time derivative to refer only to changes due to chemical reaction. Associated with each reaction we define an enthalpy, or *heat of reaction*, ΔH_i such that the term

$$G_r = -\Delta H_i \, r_i \tag{13-28}$$

gives the rate of change of energy, per unit volume, due to chemical reaction. We adopt the convention that r_i is positive for *production*, rather than *disappearance*, of species **i**, and the heat of reaction is *negative* for *exothermic* (heat-evolving)

Table 13-1 The dissipation function G_v

Cartesian coordinates (x, y, z):

$$G_v = \tau_{xx} \frac{\partial u_x}{\partial x} + \tau_{yy} \frac{\partial u_y}{\partial y} + \tau_{zz} \frac{\partial u_z}{\partial z} + \tau_{xy}\left(\frac{\partial u_x}{\partial y} + \frac{\partial u_y}{\partial x}\right) + \tau_{yz}\left(\frac{\partial u_y}{\partial z} + \frac{\partial u_z}{\partial y}\right) + \tau_{zx}\left(\frac{\partial u_z}{\partial x} + \frac{\partial u_x}{\partial z}\right)$$

Cylindrical coordinates (r, θ, z):

$$G_v = \tau_{rr} \frac{\partial u_r}{\partial r} + \tau_{\theta\theta}\left(\frac{1}{r}\frac{\partial u_\theta}{\partial \theta} + \frac{u_r}{r}\right) + \tau_{zz}\frac{\partial u_z}{\partial z}$$

$$+ \tau_{r\theta}\left[r\frac{\partial}{\partial r}\left(\frac{u_\theta}{r}\right) + \frac{1}{r}\frac{\partial u_r}{\partial \theta}\right] + \tau_{\theta z}\left(\frac{1}{r}\frac{\partial u_z}{\partial \theta} + \frac{\partial u_\theta}{\partial z}\right) + \tau_{rz}\left(\frac{\partial u_z}{\partial r} + \frac{\partial u_r}{\partial z}\right)$$

Spherical coordinates (r, θ, φ):

$$G_v = \tau_{rr}\frac{\partial u_r}{\partial r} + \tau_{\theta\theta}\left(\frac{1}{r}\frac{\partial u_\theta}{\partial \theta} + \frac{u_r}{r}\right) + \tau_{\varphi\varphi}\left(\frac{1}{r\sin\theta}\frac{\partial u_\varphi}{\partial \varphi} + \frac{u_r}{r} + \frac{u_\theta \cot\theta}{r}\right)$$

$$+ \tau_{r\theta}\left(\frac{\partial u_\theta}{\partial r} + \frac{1}{r}\frac{\partial u_r}{\partial \theta} - \frac{u_\theta}{r}\right) + \tau_{r\varphi}\left(\frac{\partial u_\varphi}{\partial r} + \frac{1}{r\sin\theta}\frac{\partial u_r}{\partial \varphi} - \frac{u_\varphi}{r}\right)$$

$$+ \tau_{\theta\varphi}\left(\frac{1}{r}\frac{\partial u_\varphi}{\partial \theta} + \frac{1}{r\sin\theta}\frac{\partial u_\theta}{\partial \varphi} - \frac{\cot\theta}{r}u_\varphi\right)$$

Table 13-2 Heats of reaction—polymerization

Monomer	$-\Delta H$, kcal/mol	Monomer	$-\Delta H$, kcal/mol
Acrylonitrile	17.3	Vinyl chloride	26.0
Methyl acrylate	18.7	Vinylidene chloride	14.4
Methyl methacrylate	13.0	Tetrafluoroethylene	33.0
Styrene	16.4	Isobutene	12.6
α-Methylstyrene	8.4	Acenaphthylene	24.0
Vinyl acetate	21.3	Ethylene	22.0

reactions. G_r, as given by Eq. (13-28), is the appropriate generation term for the thermal energy equation when heats of reaction must be included. If several reactions occur simultaneously, it is necessary to have a term like G_r, above, for *each* reaction that produces significant heat.

Polymerization reactions constitute the main type of chemical reaction of interest to us here. Table 13-2 gives some heats of reaction for cases of importance.

13-4 BOUNDARY CONDITIONS

To solve the transport equations for a specific situation it will be necessary to have appropriate boundary conditions. As in the earlier discussion in Chap. 3 of boundary conditions on **u** and T, most of the required boundary conditions are just mathematical forms of the physical requirement of continuity of the dependent variables.

The continuity of temperature at fluid boundaries is required by simple physical considerations. By the same token the *flux* of heat must be continuous across a fluid boundary, unless a phase change is occurring at the boundary. In that case, the fluxes differ by an amount due to the heat effect of the phase change, such as a latent heat of fusion. We will see this point more clearly in a subsequent example. In some problem formulations a *temperature* can be specified at a boundary. For example, the wall of a pipe may be maintained at a known temperature. In other problems it is the heat *flux* which is constrained, as in the case of heaters on the barrel of an extruder which control the rate of heat transfer.

Boundary conditions on concentration are not quite as simple as those on temperature. We can see this most clearly in a trivial example. Suppose a liquid is in contact, across a free surface, with a gas, and suppose further that the two phases are in thermal and chemical equilibrium. By *equilibrium* we mean specifically that no net transport of heat or mass occurs across the interface. What can we say about the concentration of some particular species which is in both the gas and liquid phases?

We *cannot* simply equate the concentrations c_i in the two phases, because this takes no account of the solubility of the species in the liquid phase. The appropriate condition takes the form

$$c_i^l = \alpha_s c_i^g \tag{13-29}$$

where α_s is a solubility coefficient, defined in fact by Eq. (13-29) as the ratio, *at equilibrium*, of the liquid to gas concentrations. (The same comments hold for two immiscible liquid phases, and we would refer to α_s in that case as a *partition coefficient*.)

When the species of interest is a soluble gas, it is common to write Eq. (13-29) in a form involving partial pressure as the unit of concentration in the gas phase. If the species is dilute in the liquid, for example, one might assume that Henry's law is valid:

$$p_i = Hc_i^l \tag{13-30}$$

Solubility coefficients, partition coefficients, or Henry's law constants are defined only for systems in equilibrium. If mass transfer is occurring, the system is *not* in equilibrium, by definition. Nevertheless, it is common to assume, even in systems which depart from equilibrium, that the boundary between two phases obeys equilibrium relationships among the concentrations of the diffusing species. The justification of the assumption is twofold: Its use does not appear to produce models which are at variance with observation, and (in some ways a more compelling but less comforting justification) we have no viable alternative.

In summary, then, we will use boundary conditions on concentration which are not strictly continuity statements but which account properly for different solubilities across the boundary.

When mass is being transferred across a boundary, and no chemical reaction occurs at the boundary, we expect the mass flux of each species to be continuous. If reaction does occur at the boundary, and is isolated to the boundary, as in the case of a catalytic surface reaction, then the fluxes simply differ by an amount related to the reaction rate.

Let us examine some simple transport problems at this point which illustrate the application of the boundary conditions described above. We will not carry through the solutions here since our goal is to illustrate the process of specifying suitable boundary conditions. Since the problems we select will be quite simple, we will also be illustrating the process of simplifying the general transport equations.

Capillary Flow with Viscous Dissipation

A melt at uniform temperature T_0 enters a long capillary of radius R whose walls are maintained at constant temperature T_R. The fluid obeys the power law. Find the temperature distribution $T(r, z)$, at steady state.

We begin by deciding to use cylindrical coordinates for this problem, and hence we examine Eq. (13-11). For J we use Fourier's law, Eq. (13-13). The generation term will be the viscous dissipation function G_v; we assume no chemical reactions occur which may produce or remove heat.

We will assume that the velocity vector is strictly axial, $\mathbf{u} = (0, 0, u_z)$, and that there is symmetry about the axis. Then Eq. (13-11) becomes

$$\rho C_p u_z \frac{\partial T}{\partial z} = \frac{1}{r} \frac{\partial}{\partial r}\left(kr \frac{\partial T}{\partial r}\right) + \frac{\partial}{\partial z}\left(k \frac{\partial T}{\partial z}\right) + \tau_{rz} \frac{\partial u_z}{\partial r} \tag{13-31}$$

For a power law fluid we may write

$$\tau_{rz} = K\left(-\frac{\partial u_z}{\partial r}\right)^{n-1} \frac{\partial u_z}{\partial r} \tag{13-32}$$

for this flow.

It is usually reasonable to take the *axial* transport of heat in such a flow to be due mainly to convection, and to neglect the axial conduction term by comparison.

Finally, we can often assume that the thermal conductivity k is nearly constant and bring it outside the differentiation. As a result, the thermal energy equation takes the form

$$\rho C_p u_z \frac{\partial T}{\partial z} = k \frac{1}{r} \frac{\partial}{\partial r}\left(r \frac{\partial T}{\partial r}\right) + K\left(-\frac{\partial u_z}{\partial r}\right)^{n-1} \left(\frac{\partial u_z}{\partial r}\right)^2 \tag{13-33}$$

Since u_z appears in this equation we must also write the appropriate dynamic equation, which takes the form

$$0 = -\frac{\Delta P}{L} + \frac{1}{r} \frac{\partial}{\partial r}\left[rK\left(-\frac{\partial u_z}{\partial r}\right)^{n-1} \frac{\partial u_z}{\partial r}\right] \tag{13-34}$$

If we take K to be independent of temperature then we have already solved this problem in Chap. 5, and the solution for $u_z(r)$ may be substituted immediately into Eq. (13-33). Otherwise we must write a model for $K(T)$, such as

$$K = K_0 e^{-b(T-T_0)} \tag{13-35}$$

In that case Eqs. (13-33) and (13-34) are "coupled," since T and u_z appear in both as unknowns.

Turning to consideration of boundary conditions, we impose the following, based on the statement of the problem:

$$u_z = 0 \qquad \text{at } r = R \text{ (no slip at the solid surface)}$$

$$\frac{\partial u_z}{\partial r} = 0 \qquad \text{at } r = 0 \text{ (symmetry)}$$

$$T = \begin{cases} T_0 & \text{at } z = 0 \text{ for } 0 \le r < R \text{ [an entrance condition,} \\ & \text{required since Eq. (13-33) is first order in } z] \\ T_R & \text{at } r = R \text{ (wall temperature specified)} \end{cases}$$

$$\frac{\partial T}{\partial r} = 0 \qquad \text{at } r = 0 \text{ (symmetry)}$$

The latter two conditions hold for all $z > 0$, and provide the required two boundary conditions with respect to r, since Eq. (13-33) is second order in r. The solution of these coupled nonlinear partial differential equations would have to be carried out numerically.

Temperature in a Tubular Blown Film

Consider the process of blowing tubular film, discussed in Chap. 10. Suppose the film is cooled on its outer surface by directing a cold airstream against the film. We make the following assumptions:

Steady state
Axial symmetry
No significant heat loss to the air *inside* the bubble
Air temperature remains constant along the cooling path
Heat transfer from the *outer* surface of the bubble to the air is governed by *Newton's law of cooling*:

$$J_r = h_c(T_s - T_a) \tag{13-36}$$

where h_c = convective heat transfer coefficient
T_a = air temperature
T_s = film temperature at the outer surface

The mechanical equations derived in Chap. 10 are still valid as written there, the only modification being that the viscosity must now be considered a function of position since there will be a temperature variation along the direction of motion. Thus we again introduce an equation for $\mu(T)$, such as

$$\mu = \mu_0 e^{-b(T-T_0)} \tag{13-37}$$

Parallel to the flow analysis, it is convenient to formulate the thermal energy equation in the moving coordinate system ξ used in Chap. 10:

$$\rho C_p v_1 \frac{\partial T}{\partial \xi_1} = k \frac{\partial^2 T}{\partial \xi_2^2} \tag{13-38}$$

Note that we have assumed that k is constant and that *conduction* in the ξ_1 direction is negligible in comparison to *convection* in that direction. Viscous dissipation is not likely to be important in such a process and has been neglected.

Two simple boundary conditions on T are

$$T = T_0 \qquad \text{at } \xi_1 = 0 \text{ (an "initial" condition)}$$

$$\frac{\partial T}{\partial \xi_2} = 0 \qquad \text{at } \xi_2 = 0 \text{ (no heat transfer from the inner surface)}$$

At the outer surface, which is $\xi_2 = h(\xi_1)$, we impose a boundary condition on continuity of the heat flux. The conductive heat flux *to* the surface is just $-k \, \partial T/\partial \xi_2$. The convective heat flux *from* the surface will be taken to be given by Eq. (13-36). Thus, equating the fluxes, we write the third boundary condition as

$$-k \frac{\partial T}{\partial \xi_2} = h_c(T - T_a) \qquad \text{at } \xi_2 = h(\xi_1) \tag{13-39}$$

The complication here is that the film thickness h is a dependent variable itself.

The Newton's law of cooling boundary condition is a very common one to impose because in many processes we do not know the temperature *or* the flux at a boundary. Equation (13-39) specifies neither: It is simply a constraint between the two. We must note, however, that our use of that condition implies that we know the heat transfer coefficient h_c.

The heat transfer coefficient, which is really *defined* by Eq. (13-36), depends principally on the hydrodynamics at the boundary. If the motion of the bubble does not disturb the air which surrounds it, and if no external flow of air is induced, then the convection process is nearly the same as *conduction* into still air. Because air has such a low conductivity this will be a very inefficient mode of heat transfer, which will be reflected in a small heat transfer coefficient. This is the basis of the approximation of no heat flux across the inside surface of the bubble.

By contrast, if the surrounding air is well mixed, either through external means or as induced by the boundary motion itself, then a relatively large value of h_c will reflect the more efficient convective heat transfer process. In any event, the use of the Newton's law of cooling boundary condition implies knowledge of the heat transfer coefficient. We shall return to this point subsequently and examine some theoretical and experimental bases for estimating these coefficients. The major point to make here is that if there is no information from which either the temperature or the temperature gradient may be specified, then Newton's law of cooling is the appropriate boundary condition, and the problem is to find a reasonable estimate for h_c.

Let us continue, then, with examples of formulation of the equations and boundary conditions for heat and mass transfer problems.

Transient Cooling of Runners in a Mold

At the end of injection the runners of a mold are filled with a melt whose temperature is approximately uniform. The cooling cycle brings the mold surfaces very quickly to a lower temperature, and the runners cool and freeze. We wish to find the time required to bring the runners to the freezing temperature.

We make the following assumptions in formulating this problem: At the beginning of the cooling cycle the runner material is at temperature T_0. The mold surfaces are maintained at a uniform temperature T_m. The runner cross section is a semicircle of constant diameter, and the length of the runner is much greater than its diameter. The melt is stationary during this process, so there is no convection within the fluid.

Heat is transferred strictly by conduction, then, and the transport equation reduces to the (vector) form

$$\frac{\partial T}{\partial t} = \alpha_T \ \nabla^2 T \tag{13-40}$$

Note that there is no generation term with which to account for the heat effect associated with the freezing process. Is this a dilemma, or have we formulated the problem improperly?

Let us begin by examining the goal of the analysis. Above we stated that the goal was "to find the time required to bring the runners to the freezing temperature." In mathematical analysis of physical problems one must develop the habit of examining goals. Here we have stated a *false goal*. Assuming that T_m is below the melting temperature of the polymer, we see that the runners are brought to the melting (freezing) temperature *immediately* at the mold surface. As the system cools the amount of solidified polymer increases. Hence there is a melt front, as shown in Fig. 13-12. The hatched region is solidified. The boundary separating solid from melt is the *melt front* and is defined by the closed curve $f_m(r, \theta) = 0$. The appropriate goal, then, is to determine the progress of the melt front so that a

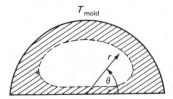

T_{mold}

Figure 13-12 Sketch for analysis of freezing of material in a runner.

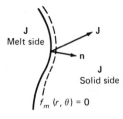

Figure 13-13 Detail of the freezing boundary between solid and molten polymer.

decision can be made as to when the polymer is sufficiently frozen that the mold can be opened.

Heat is conducted through the melt according to Eq. (13-40). There is no generation term in the melt because the molten polymer, by definition, is not undergoing a phase transition. Heat conduction in the solid polymer also obeys the same transport equation, although the thermal diffusivity may differ for a solid and its melt. (In fact, they are not much different, and it would be reasonable to use a single thermal diffusivity if no detailed values of $\alpha_T(T)$ are available on both sides of the melting temperature.) Again, the solid phase does not involve a heat of fusion term.

The heat of fusion enters as a boundary condition connecting the solid and the melt phases, since it is only *at* this boundary that solidification occurs. Let us examine, then, the formulation of this boundary condition with the aid of Fig. 13-13. The *rate* at which heat is conducted to the melt front is simply $(\mathbf{J} \cdot \mathbf{n})r \, d\theta \, dz$. Let us imagine that the solidifying polymer occupies an infinitesimal volume $dV = r \, d\theta \, dz \, dr$. Then the rate of heat absorption due to solidification is just $\Delta H_f \rho \, dV/dt$, where ΔH_f is the heat of fusion (per unit mass) of the polymer. The rate at which heat is conducted *away* from the melt front, toward the mold surface, is also given by a term $(\mathbf{J} \cdot \mathbf{n})r \, d\theta \, dz$.

We must distinguish between $\mathbf{J}$ in the melt and $\mathbf{J}$ in the solid by subscripting, for example, but we must keep in mind that both fluxes are evaluated at the same value of r, defined by $f_m(r, \theta) = 0$.

Now, the appropriate *physical* statement regarding the phase change is that the rate of heat absorption must appear as a difference in the rates of conduction to and from the phase boundary. Consequently we find that the appropriate *mathematical* statement regarding the solidification enters *not* in the transport equation per se but in the boundary condition of the form

$$(\mathbf{J} \cdot \mathbf{n})_{r_{m-}} - (\mathbf{J} \cdot \mathbf{n})_{r_{m+}} = \Delta H_f \rho \, \frac{\partial r_m}{\partial t} \qquad (13\text{-}41)$$

where r_m is defined as the solution to $f_m(r, \theta) = 0$, $r_{m\pm}$ refers to the solid $(+)$ or melt $(-)$ side of the melt front, and $\mathbf{n}$ is the unit normal vector to the curve $f_m(r, \theta) = 0$.

The melt front would be defined as the locus of points along the isothermal $T = T_{\text{melting}}$. Once the relationship $r_m(\theta)$ is found from the solution to $f_m(r, \theta) = 0$, it is a simple matter (in principle) to calculate $\partial r_m / \partial t$ and

$$n_r = \left[1 + \frac{1}{r^2} \left(\frac{dr}{d\theta} \right)^2 \right]^{-1/2} \tag{13-42}$$

$$n_\theta = \left(\frac{1}{r} \frac{dr}{d\theta} \right) n_r \tag{13-43}$$

Assuming that Fourier's law of conduction holds, we find

$$\mathbf{J} \cdot \mathbf{n} = -k \left(n_r \frac{\partial T}{\partial r} + n_\theta \frac{1}{r} \frac{\partial T}{\partial \theta} \right) \tag{13-44}$$

The procedure, then, requires the solution of two conduction equations, one in the melt and one in the solid. The solutions are connected by the heat balance along the melt front, just described above, along with the continuity condition $T_{\text{melting}} = T_{\text{melt}} = T_{\text{solid}}$ along the curve $r_m(\theta)$. Thus the solution procedure must keep track of the melt-front contour as a function of time.

The other (trivial) boundary conditions, based on the original assumptions above, are

$$T = \begin{cases} T_0 & \text{at } t = 0 \text{ everywhere within the polymer} \\ T_m & \text{at the mold surface} \end{cases}$$

For all but the simplest geometries numerical solution would be required.

Solvent Removal from a Solution-cast Film

A polymer solution is cast onto a rotating roll, as shown in Fig. 13-14. A plenum chamber covers a portion of the roll, and air blows across and removes solvent from the "wet" film. The "dry" film is stripped from the roll and passes on to further processing. It is necessary to determine how much solvent is left in the film at the point where the film is taken off the roll.

The following assumptions will be made:

The system is isothermal and at a steady state.
The system is binary, consisting of polymer and a single solvent.
The concentration of solvent in the airstream is practically zero, even at the outlet of the plenum.

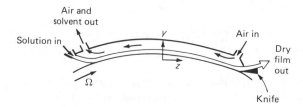

Figure 13-14 Solvent removal from a polymer film cast onto a rotating roll.

The film thickness is unaffected by the loss of solvent and is very small in comparison to the radius R of the roll.

The viscosity is so high that the film is transported as a rigid body at velocity U. Fick's Law holds, with $\mathcal{D}$ a function of solvent concentration c.

With these assumptions the diffusion equation takes the form

$$U\frac{\partial c}{\partial z} = \frac{\partial}{\partial y}\left(\mathcal{D}\frac{\partial c}{\partial y}\right) \tag{13-45}$$

We have neglected the diffusion flux in the z direction in comparison to the convection term. We also assume that no appreciable transfer occurs in the direction of the roll axis. This is likely to be true for a film which is wide compared to its thickness.

Boundary conditions on c are

$c = c_0$ at $z = 0$ (uniform initial solvent concentration)

$\dfrac{\partial c}{\partial y} = 0$ at $y = 0$ (no solvent transfer into the roll)

$c = 0$ at $y = H$ (a high convective mass transfer rate at the free surface is assumed, which reduces the concentration to that of the air)

An additional requirement is a model for $\mathcal{D}(c)$, which we take to be

$$\mathcal{D}(c) = \mathcal{D}_0\, e^{b'(c - c_0)} \tag{13-46}$$

These equations are sufficient to produce a solution for $c(z, y)$. The formulation involves one particularly weak assumption, that of isothermal behavior. In most systems of interest there is a significant heat of vaporization of the solvent, as a result of which there is a significant cooling at, and near, the film surface. This cooling can lead to very large reductions in the diffusion coefficient which will retard the rate of loss of solvent. Hence, in that case, there is a complex interaction between the mass transport and the thermal transport phenomena. If nonisothermal behavior is accounted for, the equations written above are still valid, with the modification

$$\mathcal{D}(c, T) = \mathcal{D}_0\, e^{b(T - T_0)} e^{b'(c - c_0)} \tag{13-47}$$

to account for the temperature dependence of diffusivity. In addition, the heat conduction equation must be introduced, in the form

$$U\frac{\partial T}{\partial z} = \alpha_T \frac{\partial^2 T}{\partial y^2} \tag{13-48}$$

Suitable boundary conditions on temperature include

$$T = \begin{cases} T_0 & \text{at } z = 0 \text{ (an initial condition)} \\ T_R & \text{at } y = 0 \text{ (constant roll temperature)} \end{cases}$$

At the free surface, from which evaporative mass transfer is occurring, the appropriate boundary condition equates the rate at which heat is conducted *to* the surface to the rate at which heat is lost *from* the surface. The heat loss includes both convection and evaporation. The balance gives (taking ΔH_v as the heat of vaporization)

$$-k\frac{\partial T}{\partial y} = h_c(T - T_a) + \Delta H_v\left(-\mathscr{D}\frac{\partial c}{\partial y}\right) \tag{13-49}$$

where T_a is the ambient temperature. All terms are evaluated at the position $y = H$ in this boundary condition.

A finite heat transfer coefficient is assumed, but if (consistent with the assumption earlier that the *mass* transfer coefficient is large so that $c = 0$ at $y = H$) we take $T = T_a$ as the appropriate boundary condition on temperature, Eq. (13-49) still holds [with the term $h_c(T - T_a)$ vanishing, however], and the problem appears overconstrained since two boundary conditions on $y = H$ have then been specified on T.

Another dilemma appears then, and again it must be resolved by clear physical reasoning. We know from physical observation that evaporative cooling accompanies mass transfer if there is a significant heat of vaporization. Any mathematical model must accommodate this effect. Equation (13-49) relates the surface temperature to the mass flux, and solution of the complete problem as formulated above would lead to a surface temperature variation in the direction of sheet travel.

If a high heat transfer coefficient exists, then the film surface and the ambient medium are at the same temperature. But what is that temperature? We can no longer take T_a as known a priori, to be fixed by the inlet conditions on the airstream. Instead, the evaporation process cools the air, and the assumption of a high heat transfer coefficient implies that T_a is the surface temperature. Thus the relevant boundary condition for *large* h_c is

$$-k\frac{\partial T}{\partial y} = \Delta H_v\left(-\mathscr{D}\frac{\partial c}{\partial y}\right) \qquad \text{at } y = H \tag{13-50}$$

which fixes the surface temperature $T(H, z)$ to take on certain values.

In the case of a large throughput of the ambient medium, and for finite h_c, we may specify T_a independently of the rest of the problem formulation because a *large* quantity of the surrounding medium can pick up a finite amount of heat without undergoing a significant temperature change itself. However, if the throughput of the surrounding medium is small, then T_a will vary with z, and an additional energy balance is required. This will take the form of a balance on the ambient fluid:

$$(\dot{m}C_p)_a\frac{dT_a}{dz} = \left[-k\frac{\partial T}{\partial y}\right]_{y=H} = h_c(T - T_a) + \Delta H_v\left[-\mathscr{D}\frac{\partial c}{\partial y}\right]_{y=H} \tag{13-51}$$

Thus if the physical conditions of the problem prevent a priori specification of T_a, then T_a is coupled with the rest of the problem as an unknown, and Eq. (13-51) provides the additional constraint with which it may be obtained.

Coagulation of Wet-spun Fiber

A polymer solution is spun into an acidic bath, and the diffusion of acid into the fiber causes coagulation of the polymer. We wish to find the concentration of coagulated polymer c_{cp} along the spinning path. We make the following assumptions:

The polymer does not diffuse in either its native or coagulated state.
The diffusivity of acid is a function of the degree of coagulation.
The coagulation reaction may be written schematically as polymer + acid $\rightarrow$
    coagulated polymer with rate given by

$$-\frac{dc_a}{dt} = \frac{dc_{cp}}{dt} = k'c_a \qquad (13\text{-}52)$$

The acid bath is stirred, with the result that the concentration of acid at the
    polymer-bath interface may be taken as a constant.

The acid concentration within the fiber obeys a diffusion equation, in cylindrical coordinates, of the form

$$u_z \frac{\partial c_a}{\partial z} = \frac{1}{r} \frac{\partial}{\partial r} \left[r\mathscr{D}(c_{cp}) \frac{\partial c_a}{\partial r} \right] - k'c_a \qquad \text{(acid)} \qquad (13\text{-}53)$$

The concentration of coagulated polymer simply follows the kinetic expression given above, which may be converted to yield

$$\frac{dc_{cp}}{dt} = u_z \frac{\partial c_{cp}}{\partial z} = k'c_a \qquad \text{(coagulated polymer)} \qquad (13\text{-}54)$$

Since c_a is a function of both r and z, this implicitly gives c_{cp} as a function of r.
    A model for the diffusivity is required, such as

$$\mathscr{D} = \mathscr{D}_0 e^{-b''c_{cp}} \qquad (13\text{-}55)$$

Finally the mechanical analysis from which u_z follows must be carried out, as in Chap. 9. It would probably be necessary to modify that simple analysis to include the variation of rheological parameters with c_{cp}. This would couple the momentum equation with the diffusion equation.
    Boundary conditions on c_a would be

$$c_a = 0 \qquad \text{at } z = 0 \text{ (no acid initially present)}$$

$$\frac{\partial c_a}{\partial r} = 0 \qquad \text{at } r = 0 \text{ (symmetry)}$$

$$c_a = c_{a0} \qquad \text{at } r = R(z) \text{ (uniform concentration at the fiber interface)}$$

An initial condition on c_{cp} would be required, such as

$$c_{cp} = 0 \quad \text{at } z = 0 \text{ (no initial coagulated polymer)}$$

The Tubular Flow Reactor

Suppose two fluids, one containing a species A and the other a species B, are mixed just before the entrance of a long tube through which the mixture then flows. A chemical reaction of the form $A + B \rightarrow C$ takes place. The product C is soluble in the mixture and causes a significant increase in the solution viscosity. The reaction is exothermic, and heat must be removed across the tube wall to control the extent of reaction.

The slowest-moving fluid is near the tube wall. Hence it has a longer residence time, which promotes the extent of reaction. This effect is accelerated by the increase in viscosity associated with the increased concentration of product C.

Acting to counter this is the heat transfer which cools the fluid nearest the wall, thereby reducing the reaction rate. In this problem the residence time distribution depends on both the heat transfer and kinetic processes, and a detailed analysis is required to determine the mean concentration of product leaving the tube.

We let a, b, c be the molar concentrations of the corresponding species and assume that the kinetics of the reactions are given by rates

$$r_a = r_b = -r_c = -k'ab \tag{13-56}$$

The following conservation equations are required to determine the dependent variables T, u_z, a, b, and c:

Momentum:
$$0 = -\frac{\Delta P}{L} + \frac{1}{r}\frac{\partial}{\partial r}\left(\mu \frac{\partial u_z}{\partial r}\right) \tag{13-57}$$

Energy:
$$\rho C_p u_z \frac{\partial T}{\partial z} = \frac{k}{r}\frac{\partial}{\partial r}\left(r\frac{\partial T}{\partial r}\right) + \mu\left(\frac{\partial u_z}{\partial r}\right)^2 - \Delta H\, k'ab \tag{13-58}$$

Species A:
$$u_z \frac{\partial a}{\partial z} = \frac{1}{r}\frac{\partial}{\partial r}\left(\mathscr{D}\frac{\partial a}{\partial r}\right) - k'ab \tag{13-59}$$

Species B:
$$u_z \frac{\partial b}{\partial z} = -k'ab \tag{13-60}$$

Species C:
$$u_z \frac{\partial c}{\partial z} = k'ab \tag{13-61}$$

In writing these equations we have assumed:

- Laminar axial flow
- Negligible axial conduction of heat

- Only species A diffuses—B and C have such low diffusivities that convection is the major mode of transport
- Symmetry about the axis

In addition to the conservation equations we will specify the parameters that appear in them as functions of c and T, writing

$$\frac{\mu}{\mu_0} = e^{-b(T-T_0)}e^{b'c} \tag{13-62}$$

$$\frac{\mathscr{D}}{\mathscr{D}_0} = \frac{\mu_0}{\mu} \tag{13-63}$$

$$k' = k'_0 \exp\left(-\frac{\Delta E}{R_g T}\right) \tag{13-64}$$

Now we must write boundary conditions for this set of coupled equations.
On velocity we take the usual conditions

$$u_z = 0 \qquad \text{at } r = R$$

$$\frac{\partial u_z}{\partial r} = 0 \qquad \text{at } r = 0$$

On temperature we need an initial condition, say,

$$T = T_0 \qquad \text{at } z = 0 \qquad \text{for } 0 \le r < R$$

and two conditions with respect to r:

$$T = T_R \qquad \text{at } r = R$$

$$\partial T/\partial r = 0 \qquad \text{at } r = 0$$

On the concentrations we need initial conditions

$$a = a_0 \qquad b = b_0 \qquad c = c_0 \qquad \text{at } z = 0$$

and, for a, two conditions with respect to r:

$$\partial a/\partial r = \begin{cases} 0 & \text{at } r = R \text{ (A does not cross the tube wall)} \\ 0 & \text{at } r = 0 \text{ (symmetry)} \end{cases}$$

Numerical methods would be required to solve this problem.

13-5 DIMENSIONAL ANALYSIS

Under realistic assumptions the transport equations are normally nonlinear, and numerical solution techniques are required. Even linearized models often involve a boundary geometry of such complexity that an analytical solution is impossible. More typically, however, it is the dependence of the transport coefficients on temperature and concentration that introduces mathematical difficulties. In the next section we will consider a variety of very simple models which lead to analytical solutions that are useful for estimating system behavior. In this section we consider the extent to which dimensional analysis provides useful information.

It will be useful to use the more compact vector notation for the purpose of dimensional analysis. We begin, then, with the general transport equation in the form (assuming α is constant)

$$\frac{DX}{Dt} = \alpha \, \nabla^2 X + G \tag{13-65}$$

We assume that the dynamic equations have already been nondimensionalized as in Chap. 4. (The reader should review that material if it is not clearly in mind.) We again introduce a characteristic length L and velocity U and reduce the space and time variables as before. As a consequence the general transport equation takes the form

$$\frac{D'X}{D't'} = \frac{\alpha}{LU} \nabla'^2 X + \frac{L}{U} G \tag{13-66}$$

We see immediately that a dimensionless diffusivity enters the equation. We call

$$\text{Pe} = \frac{LU}{\alpha} \tag{13-67}$$

a Peclet number, in general, whether α is the thermal diffusivity or the species diffusion coefficient [see Eq. (6-127)].

Since a great variety of forms for G are possible there is a whole set of potential dimensionless groups based on specific choices of G. We will examine some of these subsequently. We turn, instead, to examination of groups which may arise from boundary conditions.

Boundary conditions at an interface often equate the diffusive flux *to* the interface to the convective flux *from* the interface, and take the general form

$$-\alpha \, \nabla X = h'(X - X_a) \tag{13-68}$$

where h' = appropriate interfacial (convective) transport coefficient
X_a = ambient value of X

When nondimensionalized this equation becomes

$$-\nabla' X = \frac{h'L}{\alpha}(X - X_a) \tag{13-69}$$

This introduces a generalized *Sherwood number*, defined as

$$\text{Sh} = \frac{h'L}{\alpha} \tag{13-70}$$

When X is chemical species we have

$$\text{Sh} = \frac{h_x L}{\mathscr{D}} \tag{13-71}$$

where h_x is the interfacial mass transfer coefficient, but when heat transfer is considered it is more common to redefine the heat transfer coefficient so that the Sherwood number for heat transfer becomes

$$\text{Sh} = \frac{hL}{k} \tag{13-72}$$

In this form Sh is more commonly called the *Nusselt number* Nu.

Heat and mass transfer coefficients depend strongly on the fluid dynamics in the neighborhood of the transfer surface. Since the dynamic equations, when nondimensionalized, yield dependent variables which depend upon a Reynolds number, the expected and observed result is that the dimensionless transport coefficients (Sherwood numbers) depend upon the Reynolds number too. We shall see that this is the case when we examine some experimental data.

Boundary conditions other than those of the form of Eq. (13-68) usually specify the values of dependent variables, or possibly their gradients. Normally these types of conditions provide natural choices whereby the *dependent* variables are made dimensionless. We can clarify this point most easily by considering some examples. In the course of this we can also examine several types of generation terms and the dimensionless groups that their presence gives rise to.

Example 13-1: Quenching of a melt-spun fiber A melt is spun and drawn through still air. The initial melt temperature is T_0, and the air temperature T_a is constant along the spinning path. We seek the temperature distribution $T(r, z)$ in the fiber. Ignore any effect of the latent heat of fusion on the process.

The thermal energy equation, assuming symmetry about the z axis and neglecting axial conduction, is

$$\rho C_p u_z \frac{\partial T}{\partial z} = k \frac{1}{r} \frac{\partial}{\partial r}\left(r \frac{\partial T}{\partial r} \right) \tag{13-73}$$

and a set of appropriate boundary conditions is

$$T = T_0 \qquad \text{at } z = 0$$

$$-k\frac{\partial T}{\partial r} = h(T - T_a) \qquad \text{at } r = R(z)$$

$$\frac{\partial T}{\partial r} = 0 \qquad \text{at } r = 0$$

The dynamic equation takes the form (refer to Chap. 9)

$$\rho u_z u_z' = 2\frac{R'}{R} T_{zz} + T_{zz}' \tag{9-11}$$

where it was assumed that **u** and **T** are functions only of z, and $' = d/dz$. If newtonian behavior is assumed we may write

$$T_{zz} = 3\mu u_z' \tag{9-18}$$

and Eq. (9-11) becomes (dropping subscripts on u_z)

$$\rho u u' = 6 \frac{R'u'}{R} \mu + 3(\mu u'' + \mu'u') \tag{13-74}$$

Note that for the nonisothermal case μ is a function of z, so that a term μ' appears.

The mass balance takes the form $\pi R^2 u = Q_0$ so that u may be eliminated in terms of R to give

$$R'' - \frac{R'^2}{R} - \frac{\rho Q_0}{3\pi} \frac{R'}{\mu R^2} + \frac{\mu'R'}{\mu} = 0 \tag{13-75}$$

As a model for $\mu(T)$ we use

$$\frac{\mu}{\mu_0} = e^{-b(T-T_0)} \tag{13-76}$$

Equations (13-73), (13-75), and (13-76) constitute three equations in the three unknowns: $T(r, z)$, $R(z)$, and $\mu(z)$. It would be reasonable to assume that for a thin fiber the radial temperature gradient is very small. Thus the viscosity may be evaluated at the radially averaged temperature defined by

$$\bar{T} = \frac{\int_0^R T 2\pi r \, dr}{\pi R^2} \tag{13-77}$$

Since u is assumed independent of r, this definition of $\bar{T}$ is identical with the *cup-mixing* average defined by

$$\langle T \rangle = \frac{\int_0^R uT 2\pi r \, dr}{\int_0^R u 2\pi r \, dr} \tag{13-78}$$

Having introduced the average temperature and the assumption that the radial gradient is small, so that $T(r, z) \approx \bar{T}(z)$, it is useful to rewrite Eq. (13-73) in terms of $\bar{T}$. This is easily done by multiplying both sides of the equation by r and then integrating from $r = 0$ to $r = R$. Using the two boundary conditions on the radial gradients, it is easily found that the average temperature obeys

$$T' + \frac{2R'}{R} T + \frac{2h}{\rho C_p u} \frac{T - T_a}{R} = 0 \qquad T = T_0 \quad \text{at } z = 0 \tag{13-79}$$

where we have dropped the overbar on T for simplicity.

Now let us make this set of equations dimensionless, using as the independent variable $z/R_0 = \tilde{z}$. We have three dependent variables, and we may nondimensionalize two of them in an obvious way:

$$\frac{\mu}{\mu_0} = \tilde{\mu} \tag{13-80}$$

$$\frac{R}{R_0} = \tilde{R} \tag{13-81}$$

There is a very simple way to nondimensionalize T so as to simplify the format of the problem:

$$\tilde{T} = \frac{T - T_a}{T_0 - T_a} \tag{13-82}$$

Introducing these definitions we find

$$\tilde{T}' + \frac{2\tilde{R}'}{\tilde{R}}\,\tilde{T} + \left(\frac{2h}{\rho C_p u}\right)\frac{\tilde{T}}{\tilde{R}} = -\frac{2\tilde{R}'}{\tilde{R}}\,\frac{T_a}{T_0 - T_a} \tag{13-83}$$

$$\tilde{R}'' - \frac{\tilde{R}'^2}{\tilde{R}} - \left(\frac{\rho Q_0}{3\pi\mu_0 R_0}\right)\frac{\tilde{R}'}{\tilde{\mu}\tilde{R}^2} + \frac{\tilde{\mu}'}{\tilde{\mu}}\,\tilde{R}' = 0 \tag{13-84}$$

$$\tilde{\mu} = e^{-b(T_0 - T_a)(\tilde{T} - 1)} \tag{13-85}$$

$$\tilde{R} = \begin{cases} \tilde{T} = \tilde{\mu} = 1 & \text{at } \tilde{z} = 0 \\[2mm] D_R^{-1/2} & \text{at } \tilde{z} = \dfrac{L}{R_0} \end{cases}$$

where D_R is the draw ratio and L is the length of the drawing region. Here we use $' \equiv d/d\tilde{z}$.

The choice of definition of $\tilde{T}$ simplifies the problem in the result that $\tilde{T} = 1$ at $\tilde{z} = 0$, and $\tilde{T}$ approaches zero as thermal equilibrium is achieved. The following dimensionless parameters have now entered the problem:

- The draw ratio D_R and the spinning length L/R_0
- A viscosity-temperature parameter $b(T_0 - T_a)$
- A Reynolds number $\rho Q_0/3\pi\mu_0 R_0 = \text{Re}$
- A temperature factor $T_a/(T_0 - T_a) = \theta_a$
- A heat transfer parameter $2h/\rho C_p u$, which is just the ratio of the Nusselt and Peclet numbers (ignoring the factor of 2), known as the *Stanton number* St.

Thus, we expect to find

$$\tilde{T} = \tilde{T}\left[\tilde{z}; D_R, \frac{L}{R_0}, b(T_0 - T_a), \text{Re}, \theta_a, \text{St}\right] \tag{13-86}$$

This multiparameter character is typical of coupled heat transfer–fluid dynamics problems. Simplifications may result, of course, if some of these parameters play a minor role. For example the Reynolds number may be so small that the term it multiplies in Eq. (13-84) has no effect on the solution. The other parameters are more likely to play a significant role in this particular problem.

Example 13-2: Tubular flow polymerization reactor We examine here the free-radical polymerization of polystyrene in a tubular flow reactor. Styrene monomer is pumped into the inlet of a circular pipe where it is immediately mixed with a second inlet stream of azobisisobutyronitrile (AIBN). AIBN is an "initiator"; it decomposes into free radicals which react with monomer to propagate the growth of the polymer. The simplest view of the reaction scheme is the following:

Initiation:

$$I \xrightarrow{k_I} 2R_I^{\cdot}$$

$$R_I^{\cdot} + M \xrightarrow{fk_I} R_1^{\cdot}$$

Propagation:

$$R_n^{\cdot} + M \xrightarrow{k_p} R_{n+1}^{\cdot}$$

Termination:

$$R_r^{\cdot} + R_s^{\cdot} \xrightarrow{k_T} P_{r+s}$$

Chain transfer:

$$R_n^{\cdot} + M \xrightarrow{k_{trm}} P_n + M^{\cdot}$$

where I = initiator

M = monomer

$R_n^{\cdot}$ = growing radical (polymer) of chain length n

$R_I^{\cdot}$ = radical formed by initiator

P_n = polymer (incapable of further growth) of chain length n

$M^{\cdot}$ = monomer radical capable of growth by propagation

Styrene may polymerize without initiator; this is known as *thermal polymerization*, and its effect is assumed negligible in this simple example.

The reaction proceeds, then, as the mixture flows down the tube. The polymerization is exothermic, with a heat of reaction of -16.7 kcal/g mol. Under most conditions a temperature gradient will be established across the reactor radius. Hence there will be a radial distribution of reaction rates, since the various rate constants are Arrhenius-type functions of temperature. Consequently, at a given axial position, there will be a radial distribution of molecular weights, and the material leaving the reactor will have a broad molecular weight distribution. Good product quality requires some control over the molecular weight distribution.

This can be achieved, in part, by removing heat at the reactor wall so as to flatten the temperature profile across the radius. A uniform temperature profile is promoted by the use of small-diameter tubes, but this strategy requires large numbers of parallel tubes to achieve significant output rates. Hence some compromise must be reached with regard to reactor size.

We outline here a formulation of a model for this process, with the goal of examining the number and type of dimensionless groups that characterize the behavior of the system. We follow the analysis of Wallis et al. The thermal energy equation takes the form

$$\rho C_p u_z \frac{\partial T}{\partial z} = -\Delta H_r R_p + \frac{k}{r} \frac{\partial}{\partial r}\left(r \frac{\partial T}{\partial r}\right) \tag{13-87}$$

The rate of polymerization for this reaction is usually taken to have the form

$$R_p = k_m c_m (f c_I)^{1/2} \tag{13-88}$$

where c_m and c_I are concentrations of monomer and initiator. f is an *initiator efficiency*, with a value of 0.62 cited by Wallis et al. k_m is a rate constant which depends on temperature according to

$$k_m = 1.7 \times 10^{10} e^{-21.6/R_g T} \tag{13-89}$$

(k_m has units of s^{-1}. R_g is the gas constant $= 1.99 \times 10^{-3}$ kcal/g mol · K).

Because of the exponential temperature dependence of k_m (and hence R_p), Eq. (13-87) is nonlinear in temperature.

Appropriate boundary conditions on temperature are

$$T = T_0 \qquad \text{at } z = 0$$

$$\frac{\partial T}{\partial r} = 0 \qquad \text{at } r = 0$$

$$T = T_R \qquad \text{at } r = R$$

The concentrations of initiator and monomer will vary across the tube radius principally because of the thermal profile but also because of diffusion of these species. As the extent of conversion to polymer increases the fluid will become sufficiently viscous to suppress diffusion. In the early stages of conversion diffusion could be significant. Diffusion equations for each species become

$$u_z \frac{\partial c_I}{\partial z} = -k_I c_I + \frac{\mathscr{D}_I}{r} \frac{\partial}{\partial r}\left(r \frac{\partial c_I}{\partial r}\right) \tag{13-90}$$

$$u_z \frac{\partial c_m}{\partial z} = -R_p + \frac{\mathscr{D}_m}{r} \frac{\partial}{\partial r}\left(r \frac{\partial c_m}{\partial r}\right) \tag{13-91}$$

We note that the generation terms for initiator and monomer have been written directly into the species equations. For R_p Eq. (13-88) is used, which couples the two species equations. The conversion of initiator is taken as first order, with a rate constant given by

$$k_I = 1.6 \times 10^{15} e^{-30.8/R_g T} \tag{13-92}$$

with k_I in s^{-1}.

Because of the lack of data, the diffusion coefficient of initiator will be taken to be the same as that of styrene monomer. Wallis et al. find that the diffusion terms make a minor contribution to the results. Appropriate boundary conditions on the concentrations are

$$c_I = c_{I0} \qquad c_m = c_{m0} \qquad \text{at } z = 0$$

and

$$\frac{\partial c}{\partial r} = 0 \qquad \text{at } r = 0, r = R$$

for both c_I and c_m.

These equations are sufficient to allow solutions for $T(r, z)$, $c_I(r, z)$, and $c_m(r, z)$. It can be shown, by methods that are beyond the intended scope of this example, that the chain lengths of the polymer, and various measures of the breadth of the molecular weight distribution, can be determined from knowledge of T, c_I, and c_m. We proceed, then, with a dimensional analysis of the transport equations for this system.

The space variables are made dimensionless by defining $r/R = \tilde{r}$ and $z/R = \tilde{z}$. The dependent variables are normalized to their initial (inlet) values:

$$\frac{T}{T_0} = \tilde{T} \qquad \frac{c_I}{c_{I0}} = \tilde{c}_I \qquad \frac{c_m}{c_{m0}} = \tilde{c}_m$$

The velocity is normalized with its mean value: $u/U = \tilde{u}$.

Upon introducing these definitions the transport equations take the form

$$\tilde{u}\frac{\partial \tilde{T}}{\partial \tilde{z}} = \frac{1}{\mathrm{Pe}_T}\frac{1}{\tilde{r}}\frac{\partial}{\partial \tilde{r}}\left(\tilde{r}\frac{\partial \tilde{T}}{\partial \tilde{r}}\right) + N_1 \tilde{c}_m \tilde{c}_I^{1/2} \tag{13-93}$$

$$\tilde{u}\frac{\partial \tilde{c}_m}{\partial \tilde{z}} = \frac{1}{\mathrm{Pe}_M}\frac{1}{\tilde{r}}\frac{\partial}{\partial \tilde{r}}\left(\tilde{r}\frac{\partial \tilde{c}_m}{\partial \tilde{r}}\right) - N_2 \tilde{c}_m \tilde{c}_I^{1/2} \tag{13-94}$$

$$\tilde{u}\frac{\partial \tilde{c}_I}{\partial \tilde{z}} = \frac{1}{\mathrm{Pe}_M}\frac{1}{\tilde{r}}\frac{\partial}{\partial \tilde{r}}\left(\tilde{r}\frac{\partial \tilde{c}_I}{\partial \tilde{r}}\right) - N_3 \tilde{c}_I \tag{13-95}$$

The Peclet numbers are

$$\mathrm{Pe}_M = \frac{RU}{\mathscr{D}} \qquad \mathrm{Pe}_T = \frac{RU\rho C_p}{k}$$

N_1, N_2, and N_3 are dimensionless *functions* of temperature:

$$N_3 = \frac{k_I R}{U} = \frac{A_I R}{U} e^{-E_I/R_g T}$$

where Eq. (13-92) has been written in the general Arrhenius form in terms of A_I and E_I. On introducing the dimensionless temperature this becomes

$$N_3 = \frac{A_I R}{U} \exp\left(-\frac{E_I}{R_g T_0}\frac{1}{\tilde{T}}\right) \tag{13-96}$$

In the same manner we find

$$N_2 = \frac{A_m R}{U} c_{m0} c_{I0}^{1/2} f^{1/2} \exp\left(-\frac{E_m}{R_g T_0} \frac{1}{\tilde{T}}\right) \tag{13-97}$$

and

$$N_1 = \frac{\Delta H_r}{\rho C_p T_0} N_2 \tag{13-98}$$

The boundary conditions take the form

$$\tilde{T} = \begin{cases} \tilde{c}_I = \tilde{c}_m = 1 & \text{at } \tilde{z} = 0 \\ \dfrac{T_R}{T_0} & \text{at } \tilde{r} = 1 \end{cases}$$

$$\frac{\partial \tilde{c}_I}{\partial \tilde{r}} = \frac{\partial \tilde{c}_m}{\partial \tilde{r}} = 0 \qquad \text{at } \tilde{r} = 0, 1$$

$$\frac{\partial \tilde{T}}{\partial \tilde{r}} = 0 \qquad \text{at } \tilde{r} = 0$$

By inspection the dependent variables must have the functional form

$$\begin{pmatrix} \tilde{T} \\ \tilde{c}_I \\ \tilde{c}_m \end{pmatrix} = \text{some function of}$$

$$\left(\tilde{r}, \tilde{z}; \frac{T_R}{T_0}, \text{Pe}_T, \text{Pe}_M, \frac{\Delta H_r}{\rho C_p T_0}, \frac{A_I R}{U}, \frac{A_m R}{U} c_{m0} c_{I0}^{1/2} f^{1/2}, \frac{E_I}{R_g T_0}, \frac{E_m}{R_g T_0}\right) \tag{13-99}$$

This is quite a collection of dimensionless groups; enough perhaps to discourage a systematic study of the interactions of all these parameters in affecting the behavior of the reactor. Some progress can be made, however, if we constrain somewhat the class of problems we wish to examine.

First, we note that in most cases of interest the Peclet number for mass transfer will be quite large compared to unity. Consequently the diffusion terms in Eqs. (13-94) and (13-95), which are of the order of Pe^{-1}, will be quite small in comparison to the other terms and may be neglected. This gives us mass balances in the form

$$\tilde{u} \frac{\partial \tilde{c}_m}{\partial \tilde{z}} = -N_2 \tilde{c}_m \tilde{c}_I^{1/2} \tag{13-100}$$

$$\tilde{u} \frac{\partial \tilde{c}_I}{\partial \tilde{z}} = -N_3 \tilde{c}_I \tag{13-101}$$

The Peclet number for heat transfer is also quite large, and we might be tempted to throw out the radial conduction term in Eq. (13-93). This would be incorrect, however, since this would leave us with no term to account for the heat transfer to the wall of the tube. This can be seen more clearly if Eq. (13-93) is multiplied by $\tilde{r}$ and integrated across the radius of the tube. The result is

$$\frac{d\langle\tilde{T}\rangle}{d\tilde{z}} = \frac{2}{\text{Pe}_T}\frac{\partial\tilde{T}}{\partial\tilde{r}}\bigg|_{\tilde{r}=1} + 2\int_0^1 N_1\tilde{c}_m\tilde{c}_I^{1/2}\tilde{r}\,d\tilde{r} \qquad (13\text{-}102)$$

where

$$\langle\tilde{T}\rangle = 2\int_0^1 \tilde{u}\tilde{T}\tilde{r}\,d\tilde{r}$$

is the dimensionless cup-mixing temperature. Since N_1 is an exponential function of temperature the integration involving N_1 cannot be performed without further assumptions. We might, for example, assume a flat temperature profile right up to the tube wall:

$$\tilde{T} = \langle\tilde{T}\rangle \qquad \text{for } 0 \le \tilde{r} < 1 \qquad (13\text{-}103)$$

but allow for a finite *gradient* of temperature at $\tilde{r} = 1$. Then we would find

$$\frac{d\tilde{T}}{d\tilde{z}} = \frac{2}{\text{Pe}_T}\frac{\partial\tilde{T}}{\partial\tilde{r}}\bigg|_{\tilde{r}=1} + N_1\tilde{c}_m\tilde{c}_I^{1/2} \qquad (13\text{-}104)$$

with N_1 evaluated at $\tilde{T}$. The temperature gradient would have to be accounted for by introducing some model for heat transfer at the wall in laminar tube flow.

While such an approach is reasonable, and indeed was examined as a special case in the paper of Wallis et al. and found to give fairly good results, the removal of the radial heat transfer term succeeds in "throwing out the baby with the bath," in one respect. A major problem in designing tubular reactors in which there is a significant exotherm is the development of a "hot spot." If heat is generated more rapidly than it can be removed by transfer to the wall, a strong radial gradient of temperature will develop. As a consequence of this uneven temperature distribution, and of the strong dependence of the rate of polymerization on temperature, uneven polymerization rates occur across the tube radius. The result is production of a polymer with a broad molecular weight distribution, which may not have the same product qualities as a polymer with the same degree of conversion (of monomer to polymer) but with a different molecular weight distribution.

If the radial temperature gradient term is removed from the heat conduction equation, then *this* aspect of the problem cannot be modeled completely. Wallis et al. compare predictions of polydispersity, based on the model outlined above, both with and without inclusion of the radial conduction term. The reader should refer to this paper for details that we will not dwell on here. Another paper of related interest is that of Cintron-Cordero et al.

Let us proceed and see what information our dimensional analysis produces if we work with a model based on Eqs. (13-93), (13-100), and (13-101). We have now removed one parameter from the problem by neglecting the diffusion terms. Thus Pe_M no longer appears. We can remove a second parameter by redefining a dimensionless axial variable as

$$\tilde{\tilde{z}} = \frac{A_I R}{U} \tilde{z} = \frac{A_I}{U} z \tag{13-105}$$

After writing out the N functions in full, we find

$$\tilde{u} \frac{\partial \tilde{c}_I}{\partial \tilde{\tilde{z}}} = -\tilde{c}_I \exp\left(-\frac{E_I}{R_g T_0} \frac{1}{\tilde{T}}\right) \tag{13-106}$$

$$\tilde{u} \frac{\partial \tilde{c}_m}{\partial \tilde{\tilde{z}}} = -\left(\frac{A_m}{A_I} f^{1/2} c_{m0} c_{I0}^{1/2}\right) \tilde{c}_m \tilde{c}_I^{1/2} \exp\left(-\frac{E_m}{R_g T_0} \frac{1}{\tilde{T}}\right) \tag{13-107}$$

$$\tilde{u} \frac{\partial \tilde{T}}{\partial \tilde{\tilde{z}}} = \frac{\alpha_T}{R^2 A_I} \nabla^2 \tilde{T} + \left(\frac{\Delta H_r}{\rho C_p T_0}\right) \left(\tilde{u} \frac{\partial \tilde{c}_m}{\partial \tilde{\tilde{z}}}\right) \tag{13-108}$$

where the writing has been simplified by introducing the thermal diffusivity α_T, the vector notation ∇^2, and noting that the right-hand side of Eq. (13-107) appears in Eq. (13-108). The boundary conditions are unchanged. We note that $\tilde{c}_m$ and $\tilde{c}_I$ are implicit functions of $\tilde{r}$ because $\tilde{T}$ is a function of $\tilde{r}$.

Now we may write the list of dimensionless groups that appear in this model:

- The independent variables $\tilde{r}$ and $\tilde{\tilde{z}}$
- The dependent variables $\tilde{u}$, $\tilde{T}$, $\tilde{c}_m$, and $\tilde{c}_I$
- Kinetic parameters $E_I/R_g T_0$, $E_m/R_g T_0$, and $(A_m/A_I)c_{m0} c_{I0}^{1/2} f^{1/2}$
- An exothermic group $-\Delta H_r/\rho C_p T_0$
- A temperature ratio T_R/T_0
- The only parameter related to the size of the reactor $\alpha_T/R^2 A_I$

This is still a large number of parameters, but they are in such a form that we can try to deal rationally with them. First we note that three of the parameters, the first two kinetic parameters and the exothermic group, are not subject to much variation for a particular polymerization reaction. Only T_0 is subject to control in these groups, and since absolute temperatures are used these groups will not vary much over the usual range of operating temperatures.

One is tempted to make the same deduction regarding the temperature ratio T_R/T_0. However, both Cintron-Cordero et al. and Wallis et al. find the temperature profiles down the reactor to be extremely sensitive to this ratio. Hence this is one of the significant variables that must be singled out for study. *It is important to understand that this sensitivity cannot be anticipated on the basis of dimensional analysis.*

The third kinetic parameter contains initial concentrations of initiator and monomer. Significant variations in this parameter are possible, and they are found to exert strong effects on reactor performance. The last parameter listed depends on the reactor radius and on two physical properties which would be nearly constant for a given polymer. The size of the reactor is clearly important from the point of view of heat transfer. One expects that if R is small enough it would not be difficult to transfer all the reaction-generated heat out to the cooled walls and thereby maintain a flat radial profile.

We note that the flow rate (in terms of U) appears only in the axial variable $\tilde{z}$, which is essentially a ratio of a reaction time scale to an average residence time up to the point z. However, if $\tilde{u}$ is other than that for fully developed flow in a conduit of constant cross section, it is likely that $\tilde{u}$ will depend on a Reynolds number, and so $\tilde{T}$ will show a Reynolds number dependence as well. Such might be the case if the reaction were carried out in a Kenics Static Mixer (see Chap. 12) at Reynolds numbers greater than about 10.

On the basis of the comments made above we suggest that the most important dimensionless groups, *for a specific polymerization reaction*, are

- The temperature ratio T_R/T_0
- The size parameter $\alpha_T/R^2 A_I$
- The inlet concentrations, in the form $(A_m/A_I)c_{m0} c_{I0}^{1/2} f^{1/2}$

Mathematical modeling studies, or experimental investigations, should emphasize the behavior of the system with respect to these groups.

To close this example, then, let us suppose that we are interested in the maximum temperature rise that occurs across the tube radius, ΔT_{max}. We expect, at a given reactor length $\tilde{z}$, that

$$\frac{\Delta T_{max}}{T_0} = \text{function of } \left(\frac{T_R}{T_0}, \frac{\alpha_T}{R^2 A_I}, c_{m0} c_{I0}^{1/2} \right) \qquad (13\text{-}109)$$

(Note that since $A_m f^{1/2}/A_I$ is a constant for a given reaction, we need not consider it here.) If the maximum permissible temperature rise is specified a priori, then one may essentially invert this function to the form

$$\frac{\alpha_T}{R^2 A_I} = \text{function of } \left(\frac{T_R}{T_0}, \frac{\Delta T_{max}}{T_0}, c_{m0} c_{I0}^{1/2} \right) \qquad (13\text{-}110)$$

or, ignoring physical properties which are constant,

$$R = \text{function of } \left(\frac{T_R}{T_0}, \frac{\Delta T_{max}}{T_0}, c_{m0} c_{I0}^{1/2} \right) \qquad (13\text{-}111)$$

If ΔT_{max} is specified a priori, Eq. (13-111) gives a constraint on the maximum permissible reactor radius R.

Wallis et al. present just such a result, based on numerical solution of the model outlined here. Figure 13-15 shows their results, calculated for fixed

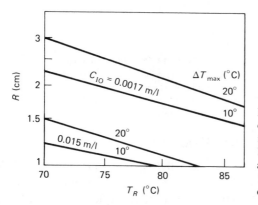

Figure 13-15 Results of Wallis et al. for critical reactor radius which holds the maximum tolerable temperature rise below the specified values (10 or 20°C) as a function of reactor wall temperature T_R. The effect of initial initiator concentration is shown by the parameter C_{I0}.

values of $T_0 = 80°C$, a reactor length of 610 cm, and a fixed but unspecified monomer inlet concentration c_{m0}. The results, consistent with the dimensional analysis, show the relationships among R, T_R, and c_{I0} as functions of the maximum permissible temperature rise.

13-6 CONVECTIVE TRANSPORT COEFFICIENTS

In discussing boundary conditions we saw that a very general format expressing continuity of the flux across an interface was

$$-\alpha \nabla X = h'(X - X_a) \tag{13-112}$$

In the case of heat transfer this led to the definition of a convective heat transfer coefficient h and a Nusselt number Nu defined so that

$$-k \nabla T = h(T - T_a) \tag{13-113}$$

$$\text{Nu} = \frac{hL}{k} \tag{13-114}$$

where L is a characteristic length and k is the thermal conductivity of the medium *from* which convection is occurring. Corresponding equations for mass transfer are

$$-\mathscr{D} \nabla c = h_x(c - c_a) \tag{13-115}$$

$$\text{Sh} = \frac{h_x L}{\mathscr{D}} \tag{13-116}$$

which define the mass transfer coefficient h_x and the Sherwood number.

In this section we examine models with which one may estimate the Nusselt or Sherwood numbers appropriate to a particular process. Specifically, we apply these models to the calculation of transport coefficients for film or fiber extruded into a fluid (often air). *We focus our attention on the dynamics of the external fluid*, assuming that the transport process in the polymeric fluid (the fiber or film) is well defined and may be analyzed by methods to be presented in Sec. 13-7.

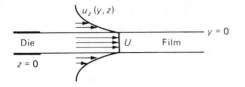

Figure 13-16 Definition sketch for the problem of exterior convective transport from a moving film or sheet.

Transport Coefficients for a Moving Film

We consider the following problem, for which Fig. 13-16 provides a definition sketch. A sheet or film issues from a die at position $z = 0$ and translates at constant speed U through an external fluid. The external fluid is at rest far from the moving sheet, but the motion of the sheet entrains the external fluid and establishes a velocity field $\mathbf{u}(y, z)$.

The dynamics of the *external* flow are constrained to satisfy the continuity and momentum equations:

$$\frac{\partial u_z}{\partial z} + \frac{\partial u_y}{\partial y} = 0 \tag{13-117}$$

$$u_z \frac{\partial u_z}{\partial z} + u_y \frac{\partial u_z}{\partial y} = v_a \frac{\partial^2 u_z}{\partial y^2} \tag{13-118}$$

The temperature or concentration distributions in the external fluid must satisfy an equation of the form

$$u_z \frac{\partial X}{\partial z} + u_y \frac{\partial X}{\partial y} = \alpha_a \frac{\partial^2 X}{\partial y^2} \tag{13-119}$$

We are assuming here that steady state is attained and that the film is wide so that no variations in the x direction occur. We also assume that the effect of the moving film is confined to a thin laminar layer (a boundary layer) normal to the film and that changes in the y direction are much greater than changes along the direction of motion.

If $X = T$ then $\alpha_a = \alpha_T = (k/\rho C_p)_a$; if $X = c_i$ then $\alpha_a = \mathcal{D}_{ia}$. Note that the diffusivities refer to *the external fluid*: $\mathcal{D}_{ia}$ is the diffusion coefficient of some species $\mathbf{i}$ in the *ambient* medium, *not* in the medium of the film. The thermal diffusivity refers to the *ambient* medium. Likewise, the kinematic viscosity v_a in Eq. (13-118) refers to that of the *ambient* medium.

It is important that this change of focus be recognized and kept in mind through the analysis. In previous problems we have written the transport equations for the process occurring within the polymeric fluid. The ambient medium entered only in a boundary condition, such as Eq. (13-112), with the presumption that the convective coefficient h' was known. *Now we focus on the problem of obtaining the coefficient h'*, which naturally requires a detailed analysis of the dynamic and transport equations of the *ambient* medium.

We begin the analysis by eliminating u_y, using the continuity equation. We integrate Eq. (13-117) to find

$$u_y = - \int_0^y \frac{\partial u_z}{\partial z} \, dy \qquad (13\text{-}120)$$

We have imposed the boundary condition $u_y = 0$ at the sheet surface, $y = 0$. This is a no-slip condition, but it is valid only in the absence of a significant mass flux at the sheet surface. If there were a transfer of mass across the surface $y = 0$, then there would be a velocity component u_y proportional to that flux. Models which account for the presence of a high mass flux are discussed by Bird et al. and by Shih and Middleman.

The following boundary conditions will be used:

At $y = 0$: $\qquad\qquad\qquad u_z = U \qquad T = T_0 \qquad c = c_0$

At $z = 0$, $y \to \infty$: $\qquad u_z = 0 \qquad T = T_a \qquad c = c_a$

Again we emphasize the fact that the dependent variables refer to the ambient medium, not to the sheet. However, at $y = 0$ continuity of temperature, for example, requires that the surface of the sheet be at $T = T_0$ for all $z > 0$. Hence the results of the model (the transport coefficients h') may not be directly applicable to the more realistic situation in which U, T_0, and c_0 may change along the z axis. Nevertheless the model will serve the intended purpose: to provide analytical estimates of the appropriate transport coefficients.

The similarity of format of Eqs. (13-118) and (13-119) allows them to be written as a single equation [recall that Eq. (13-119) already stands for two equations] of the form of Eq. (13-119), where X may now refer to u_z, T, or c_i, and α_a may be v_a, α_T, or $\mathscr{D}$, respectively. The boundary conditions are such that a combination of independent variables of the form

$$\eta = \frac{y}{2} \sqrt{\frac{U}{v_a z}} \qquad (13\text{-}121)$$

reduces the transport equation(s) from partial to ordinary differential equations. Using Eq. (13-120) to eliminate u_y, and after quite a bit of exercise with the chain rule for differentiation, the three transport equations may all be written in the form

$$\Pi'' + \Lambda f \Pi' = 0 \qquad (13\text{-}122)$$

subject to boundary conditions

$$\Pi = \begin{cases} 1 & \text{at } \eta = 0 \\ 0 & \text{at } \eta \to \infty \end{cases}$$

The prime refers to differentiation with respect to η. f is defined as

$$f = 2 \int_0^\eta \Pi_u \, d\eta \qquad (13\text{-}123)$$

The dependent variable Π stands for any of the dimensionless profiles

$$\Pi_u = \frac{u_z}{U} \qquad \Pi_T = \frac{T - T_a}{T_0 - T_a} \qquad \Pi_c = \frac{c - c_a}{c_0 - c_a} \tag{13-124}$$

The parameter Λ stands for any of the dimensionless transport numbers

$$\Lambda_T = \frac{v_a}{\alpha_T} = \text{Pr} \qquad \Lambda_c = \frac{v_a}{\mathscr{D}_{ia}} = \text{Sc} \qquad \Lambda_u = \frac{v_a}{v_a} = 1$$

Pr and Sc are the Prandtl and Schmidt numbers.

Equation (13-122) may be solved by integration, and the general solution may be written

$$\Pi = 1 - \frac{\int_0^{\eta} \left[\exp\left(-\int_0^{\eta} \Lambda f \, d\eta\right)\right] d\eta}{\int_0^{\infty} \left[\exp\left(-\int_0^{\eta} \Lambda f \, d\eta\right)\right] d\eta} \tag{13-125}$$

To find $\Pi_u(\eta)$ we set $\Lambda = 1$ in Eq. (13-125). The solution is implicit since Π_u appears in the integrals through f [Eq. (13-123)]. Numerical solution is required to give Π_u. Once Π_u is found f may be calculated (by numerical integration) as a function of η, and then Eq. (13-125) may be solved explicitly for Π_T and Π_c upon setting $\Lambda = \Lambda_T$ and Λ_c, respectively.

Our main interest is not in the profiles Π but in the transport coefficients. The local heat transfer coefficient is defined as

$$-k_a \left[\frac{\partial T}{\partial y}\right]_{y=0} = h(T_0 - T_a) \tag{13-126}$$

and another exercise in algebra and chain rule differentiation confirms that we may define a local Nusselt number as

$$\text{Nu}_z = \frac{hz}{k_a} = -\tfrac{1}{2} \, \text{Re}_z^{1/2} \, \Pi_T'(0) \tag{13-127}$$

Here we have introduced a local Reynolds number

$$\text{Re}_z = \frac{Uz}{v_a} \tag{13-128}$$

We use the word *local* in describing the Nusselt and Reynolds numbers because the characteristic length used in their definitions is the local axial *variable z*. By a similar procedure the Sherwood number is given by

$$\text{Sh}_z = \frac{h_x z}{\mathscr{D}_{ia}} = -\tfrac{1}{2} \, \text{Re}_z^{1/2} \, \Pi_c'(0) \tag{13-129}$$

It is not difficult to find, from Eq. (13-125), that

$$\Pi'(0) = -\left\{\int_0^{\infty} \left[\exp\left(-\int_0^{\eta} \Lambda f \, d\eta\right)\right] d\eta\right\}^{-1} \tag{13-130}$$

Hence, once $\Pi_u(\eta)$ is found the results of interest may be computed.

The most compact format for presentation of these results is Fig. 13-17, based

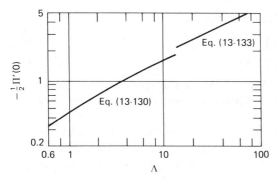

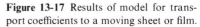

Figure 13-17 Results of model for transport coefficients to a moving sheet or film.

on a numerical solution of the equations presented by Shih and Middleman. The figure may be used for either heat or mass transfer by taking Λ to be the Prandtl or Schmidt number, respectively. In addition, at $\Lambda = 1$, $\Pi'(0)$ gives the dimensionless velocity gradient at the sheet surface, from which the viscous drag exerted by the ambient medium on the sheet may be calculated.

If the ambient medium is a gas (the most common situation) then the Prandtl and Schmidt numbers are both close to unity. If the ambient medium is a liquid the Prandtl and Schmidt numbers are usually larger than unity. In a typical wet-spinning or wet-extrusion process the liquid bath has a viscosity comparable to water, say, $v_a = 0.01$ cm^2/s. Typical thermal diffusivities of such liquids are of the order of 10^{-4} to 10^{-3} cm^2/s, giving Pr $= 10$ to 100. Typical species diffusion coefficients are in the range 10^{-5} to 10^{-4} cm^2/s for simple solutes in water, giving Sc $= 100$ to 1000.

For large Λ an approximate method leads to an analytical solution for $\Pi'(0)$. First, we recognize that large Λ means that the transport coefficient for momentum (v_a) is much greater than the coefficient for heat or mass diffusion. This means that the thermal or species diffusion is confined to a distance from the surface that is small compared to the distance over which the velocity profile falls from unity to (nearly) zero. This, in turn, suggests that the velocity profile $\Pi_u(\eta)$ is nearly flat in the region where Π_T and Π_c fall from unity to zero. As an approximation, then, we write

$$\Pi_u(\eta) = 1 \tag{13-131}$$

For large Λ, then, we may write

$$f = \int_0^\eta 2\Pi_u \, d\eta = 2\eta \tag{13-132}$$

The expression for $\Pi'(0)$ then becomes

$$-\Pi'(0) = \left[\int_0^\infty \exp\left(-\Lambda\eta^2\right) d\eta \right]^{-1}$$

$$= \left[\frac{1}{2} \left(\frac{1}{\Lambda} \right)^{1/2} \Gamma(\tfrac{1}{2}) \right]^{-1} = 1.13\Lambda^{1/2} \tag{13-133}$$

where $\Gamma(\ )$ is the gamma function.

Thus we obtain the simple analytical results

$$\text{Sh}_z = 0.56 \, \text{Re}_z^{1/2} \, \text{Sc}^{1/2}$$ (13-134)

for $\Lambda \gg 1$

$$\text{Nu}_z = 0.56 \, \text{Re}_z^{1/2} \, \text{Pr}^{1/2}$$ (13-135)

We find, then, that we may estimate the convective transport coefficients appropriate to a sheet or film moving in its own plane by using Fig. 13-17 or, for large Λ, Eqs. (13-134) and (13-135).

No data are available with which this model of convective transport may be evaluated. It is instructive, however, to examine an example in which the model is used in order to assess the sensitivity of the final results to the function used for the convective transport coefficient.

Example 13-3: Cooling of a moving film A thin film of polyethylene is extruded into and drawn through still air. Find the film temperature as a function of distance from the die exit. Use the following parameters:

Film speed U	= 1 ft/s	assumed constant
Film thickness W	= 2 mils	
Exit temperature T_0	= 175°C	
Ambient air temperature T_a	= 30°C	
Conductivity k	= 8.0×10^{-4} cal/s·cm·°C	polymer
Diffusivity α	= 1.3×10^{-3} cm²/s	
Kinematic viscosity ν_a	= 0.17 cm²/s	
Conductivity k_a	= 6.5×10^{-5} cal/s·cm·°C	air
Prandtl number Pr	= 0.7	

Because the film is thin (2 mils = 0.005 cm) we will assume that the temperature within the polymer is uniform in the direction normal to its surface. Then, with reference to Fig. 13-18, we may derive a simple energy equation from which the temperature may be found. The assumption of a flat temperature profile across the film can be evaluated as an exercise.

Rate at which heat (per unit width) enters the control volume through the plane at z: $\rho C_p U W T_z$.

Rate at which heat leaves across $z + dz$: $\rho C_p U W T_{z+dz}$.

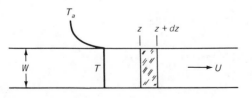

Figure 13-18 Definition sketch for derivation of energy equation for a cooling film.

Rate of convective loss from *both* exposed surfaces:

$$2h \, dz(T - T_a) \text{ per unit width}$$

Dividing by dz, and taking the limit as $dz \to 0$, we find

$$\frac{dT}{dz} = -\frac{2h}{\rho C_p U W}(T - T_a) \qquad (13\text{-}136)$$

Normalizing the temperature we may write this as

$$\frac{d\theta}{\theta} = -\frac{2h}{\rho C_p U W} dz$$

where $\theta = (T - T_a)/(T_0 - T_a)$. An appropriate initial condition is $\theta = 1$ at $z = 0$.

If h were constant the solution would be

$$\theta = e^{-2hz/\rho C_p U W} \qquad (13\text{-}137)$$

For variable h we have

$$\theta = \exp \left(-\int_0^z \frac{2h}{\rho C_p U W} dz \right) \qquad (13\text{-}138)$$

We will use Fig. 13-17 for calculation of $h(z)$. Since $h \sim z^{-1/2}$ the integration above can be done easily, and the result takes the form

$$\theta = e^{-12.8 \times 10^{-3} z^{1/2}} \qquad (13\text{-}139)$$

Figure 13-19 shows h and T as functions of z. It is apparent that the rate of cooling is quite slow, especially in view of the small thickness of the film. The speed of the film is insufficient to generate a large convective coefficient, as a consequence of which the air behaves as a fairly effective insulator in this problem.

With regard to the sensitivity of the cooling rate to the convective coefficient, it should be evident that the dependence is quite strong, since h appears in an exponential function. Hence there is considerable premium to be gained by having an accurate model for h.

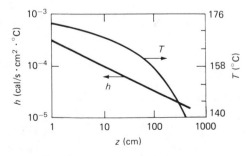

Figure 13-19 Calculated results for Example 13-3.

It would appear that the cooling of extruded film in air through the convection associated with the motion of the film alone is very inefficient. In general, it would be necessary to promote heat (or mass) transfer by external blowing of air either parallel or normal to the film surface.

Transport Coefficients for a Moving Cylinder

In order to calculate transport coefficients appropriate to fiber-spinning or wire-coating operations, we may formulate the transport equations for the fluid region exterior to a cylindrical surface of radius R, translating axially at constant speed $u_z = U$.

The mathematical model parallels the treatment of the previous problem, proper account being taken of the cylindrical symmetry in this case. The continuity equation is

$$\frac{\partial u_z}{\partial z} + \frac{1}{r}\frac{\partial(r u_r)}{\partial r} = 0 \tag{13-140}$$

and the transport equations all take the form

$$u_r \frac{\partial X}{\partial r} + u_z \frac{\partial X}{\partial z} = \frac{\alpha}{r}\frac{\partial}{\partial r}\left(r\frac{\partial X}{\partial r}\right) \tag{13-141}$$

If u_r is taken to vanish at $r = R$, equivalent to the assumption of low mass flux at the cylindrical surface, then we may write u_r in terms of u_z as

$$u_r = -\frac{1}{r}\int_R^r r\frac{\partial u_z}{\partial z}\,dr \tag{13-142}$$

The equation to be solved may be written as

$$\left(-\frac{1}{r}\int_{R_1}^r r\frac{\partial u_z}{\partial z}\,dr\right)\frac{\partial \Pi}{\partial r} + u_z \frac{\partial \Pi}{\partial z} = \frac{\nu}{\Lambda}\frac{1}{r}\frac{\partial}{\partial r}\left(r\frac{\partial \Pi}{\partial r}\right) \tag{13-143}$$

where Π may be any of the dimensionless profiles defined in Eq. (13-124) and Λ is a dimensionless transport number. The boundary conditions are

$$\Pi = \begin{cases} 1 & \text{at } r = R \\ 0 & \text{at } r \to \infty \text{ and } z = 0 \end{cases}$$

If a new independent variable

$$\eta = \frac{U}{4\nu_a}\frac{r^2}{z} \tag{13-144}$$

is introduced, Eq. (13-143) may be written in the form

$$\frac{\partial^2 \Pi}{\partial \eta^2} + \frac{\Lambda}{\eta} f \frac{\partial \Pi}{\partial \eta} = 0 \tag{13-145}$$

with $\quad \Pi = \begin{cases} 1 & \text{at } \eta = \dfrac{UR^2}{4v_a}\dfrac{1}{z} = \eta_0 \\[2mm] 0 & \text{at } \eta = \infty \end{cases}$

The function f is defined as

$$f = \eta_0 + \frac{1}{\Lambda} + \int_{\eta_0}^{\eta} \Pi_u \, d\eta \tag{13-146}$$

It is important to note that η_0 is a *variable*: It is a function of axial coordinate z. Since η_0 appears in the first boundary condition, as well as in the function f, we conclude that $\Pi = \Pi(\eta, z)$, and Eq. (13-145) is still a *partial* differential equation. Thus the variable η does not transform Eq. (13-143) to an ordinary differential equation; η is not a similarity transform.†

If η_0 is fixed at a specific value, then in effect we fix the position z at which Eq. (13-145) and its boundary conditions hold. Since η_0 appears only in the coefficient f and in the first boundary condition, we develop an approximate solution to Eq. (13-145) in the following way. For a fixed value of η_0 Eq. (13-145) is solved as an *ordinary* differential equation for $\Pi(\eta)$, with η_0 as a constant parameter. The solution may be written formally as

$$\Pi = 1 - \frac{\displaystyle\int_{\eta_0}^{\eta} \exp\left(-\int_{\eta_0}^{\eta} \frac{\Lambda f}{\eta} \, d\eta\right) d\eta}{\displaystyle\int_{\eta_0}^{\infty} \exp\left(-\int_{\eta_0}^{\eta} \frac{\Lambda f}{\eta} \, d\eta\right) d\eta} \tag{13-147}$$

As in the problem of film transport, Π_u is found by setting $\Lambda = 1$ and solving iteratively. Once Π_u is found, Eq. (13-147) is an explicit solution for Π_T and Π_c (using $\Lambda_T = \mathrm{Pr}$ and $\Lambda_c = \mathrm{Sc}$, respectively). The gradients at the cylindrical surface are found from

$$\Pi'(0) = \left[\int_{\eta_0}^{\infty} \exp\left(-\int_{\eta_0}^{\eta} \frac{\Lambda f}{\eta} \, d\eta\right) d\eta\right]^{-1} \tag{13-148}$$

The dimensionless transport coefficients may be found from

$$\mathrm{Sh} = 4\eta_0 \, \Pi'(0) \tag{13-149}$$

where for mass transfer we have $\mathrm{Sh} = 2h'_x R/\mathscr{D}$, and for heat transfer the Nusselt number is $\mathrm{Nu} = 2hR/k_a$.

Figure 13-20 shows Sh as a function of η_0 with Λ as a parameter. Keep in mind that since $\eta_0 \sim 1/z$, Fig. 13-20 gives the Sherwood number as a function of position. Near the origin of the cylinder (the die exit) η_0 approaches infinity, and the Sherwood number becomes unbounded. It might be appropriate, then, to examine reasonable ranges of η_0 for problems of interest to us here.

---

† η was incorrectly treated as a similarity transform in a paper by Vasudevan and Middleman. The error was pointed out subsequently in a communication from Fox and Hagin.

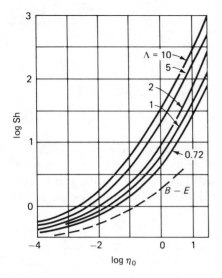

Figure 13-20 Results of model for transport coefficients to a moving cylinder. The curve labeled *B-E* is the solution of Bourne and Elliston for $\Lambda = 0.72$.

Recall that

$$\eta_0 = \frac{UR^2}{4v_a z} \tag{13-150}$$

For fiber spinning (ignoring drawing)† typical values for U and R are of the order of 300 cm/s and 0.02 cm, respectively. For dry spinning we may take $v_a = 0.15$ cm²/s; for wet spinning the kinematic viscosity would be of the order of 0.01 cm²/s.

This gives $\eta_0 \approx 0.2/z$ (dry) or $\eta_0 \approx 3/z$ (wet). The axial length of interest might extend approximately from 1 to 100 cm. Except very close to the die exit, we see that the values of η_0 of interest are quite small compared to unity.

If η_0 were identically zero, then Eq. (13-145) would be independent of z, and η would be a true similarity transform for the problem. Eqs. (13-147) and (13-148) become more nearly exact, then, for small η_0. In the absence of an "exact" solution of Eq. (13-145) it is difficult to assess the accuracy of the approximate solution offered here. We can, however, compare the model to other approximate solutions of this problem and to some experimental data on heat transfer from fibers.

Another method of solution of this problem is based on the classical Karman-Pohlhausen treatment of the laminar boundary layer. The initial application of the technique to the problem of the moving sheet or fiber was by Sakiadis. Subsequently Bourne and coworkers solved the corresponding heat transfer problem and made some comparisons of the predicted results with data on the cooling of fibers.

Figure 13-20 shows a comparison of the solution of Bourne and Elliston, for

---

† Even under conditions of drawing, the product UR^2 would be nearly constant.

Pr = 0.72, with the solutions presented earlier. The solutions differ by about 25 percent at $\eta_0 = 10^{-4}$ and become more nearly equal as η_0 decreases. In the absence of an exact numerical solution to the problem it is not possible to argue which solution is the more accurate. We turn, then, to evaluation of experimental data.

We need a model for the temperature of a fiber as a function of distance from the spinneret. Using the same procedure as in the derivation of the corresponding case for a film [Eq. (13-138)], we find

$$\theta = \exp\left(-\frac{2}{\rho C_p}\int_0^z \frac{h\,dz}{RU}\right) \tag{13-151}$$

For the cylindrical fiber the quantity $\rho R^2 U$ must be constant to satisfy mass conservation. It is convenient, then, to write the exponent so that

$$\theta = \exp\left(-\frac{2}{\rho R^2 U C_p}\int_0^z Rh\,dz\right)$$

$$= \exp\left(-\frac{k_a}{\rho R^2 U C_p}\int_0^z \mathrm{Nu}\,dz\right) \tag{13-152}$$

where $\mathrm{Nu} = 2hR/k_a$ has been introduced. On examining the application of this model in the example that follows it will be seen that Eq. (13-152) is in a convenient format for computation.

Example 13-4: Cooling of a polyester fiber Hill and Cuculo present data on melt-spun polyethylene terephthalate (PET). Figure 13-21 gives the velocity along the spinning path (from measured diameters) and the temperature at the fiber surface.

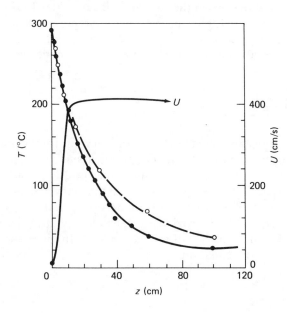

Figure 13-21 Filament velocity and temperature profiles measured by Hill and Cuculo. The model in Example 13-4 gives the values indicated by ○.

We need the following information, taken from the paper by Hill and Cuculo unless otherwise noted.

$$\pi R^2 \rho U = 6.3 \times 10^{-3} \text{ g/s}$$

$$T_0 = 290°C \qquad T_a = 22°C$$

$$2R_0 = 0.0031 \text{ cm}$$

$$\rho = 1.2 \text{ g/cm}^3 \text{ (approx. average over the temperature range)}$$

$$\rho C_p = 0.5 \text{ cal/cm}^3 \cdot °C \text{ (assumed)}$$

$$k_a = 6.5 \times 10^{-5} \text{ cal/cm} \cdot °C \cdot s$$

$$v_a = 0.17 \text{ cm}^2/s \qquad \text{Pr} = 0.72$$

From Fig. 13-21 we represent U by

$$U = \begin{cases} 7 + \dfrac{413z}{15} & z < 15 \\[2mm] 420 & z > 15 \end{cases}$$

For z in the range $2 \le z \le 100$ cm we find η_0 in the range $1.2 \times 10^{-3} \ge \eta_0 \ge 2.4 \times 10^{-5}$. We will use Fig. 13-20 for determination of Nu, extrapolating where necessary to $\eta_0 < 10^{-4}$. The integration in Eq. (13-152) is performed numerically on a hand calculator. The results are shown in Fig. 13-21.

We see that the theory underestimates the rate of heat transfer, since the predicted temperatures are everywhere higher than those observed. The agreement is not bad, however, giving some confidence in the use of the model for cooling of fibers. We note that since the model of Bourne and Elliston predicts smaller values of Nu, it will give an even poorer representation of the data, in this case. Other data can be examined (see Prob. 13-11), and it would appear that Fig. 13-20 can be used with some confidence for $\eta_0 < 10^{-2}$.

13-7 SIMPLE TRANSPORT MODELS

In this section we consider transport models that are relevant to polymer processes and that, by virtue of their simple geometry and the assumed independence of the diffusivity with respect to temperature and concentration, lead to linear equations for which analytical solutions may be found. As in the case of the simple flow models treated in Chap. 5, these transport models provide a basis for extension to more complex cases, and they provide a means of generating quick order-of-magnitude estimates for the behavior of related systems. In all cases we are dealing with the general transport equation.

$$\frac{DX}{Dt} = \alpha \nabla^2 X + G \tag{13-153}$$

under assumptions that guarantee linearity. This requires only that G be, at most, a linear function of X, that α be constant, and that the velocity field $\mathbf{u}$, which appears implicitly in the derivative DX/Dt, be uncoupled from the transport equation. The boundary conditions must, of course, be linear as well.

Transient Diffusion in Solids

In a stationary solid, for which $\mathbf{u} = 0$, we examine solutions of

$$\frac{\partial X}{\partial t} = \alpha \, \nabla^2 X \tag{13-154}$$

or, introducing a characteristic linear scale L,

$$\frac{\partial X}{\partial \tilde{t}} = \tilde{\nabla}^2 X \tag{13-155}$$

where $\tilde{t} = \alpha t / L^2$ and the space variables are normalized to L. As boundary conditions we take

$$X = X_0 \qquad \text{at } \tilde{t} = 0$$

$$-\tilde{\nabla} X = \text{Sh} \, (X - X_a) \qquad \text{at the boundaries}$$

In the boundary conditions, Sh is the generalized Sherwood number introduced earlier in this chapter. For heat transfer $\text{Sh} = \text{Nu} = hL/k$. For mass transfer $\text{Sh} = k_x L / \mathcal{D}$. (See Probs. 13-15 and 13-16.)

We define a dimensionless dependent variable as

$$\tilde{X} = \frac{X - X_a}{X_0 - X_a} \tag{13-156}$$

Then the equations become

$$\frac{\partial \tilde{X}}{\partial \tilde{t}} = \tilde{\nabla}^2 \tilde{X} \tag{13-157}$$

$$\tilde{X} = 1 \qquad \text{at } \tilde{t} = 0$$

$$-\nabla \tilde{X} = \text{Sh} \, \tilde{X} \qquad \text{at the boundaries}$$

Solutions of Eq. (13-157) are easily found by the method of separation of variables for three simple geometries:

1. The slab of thickness $2L$ and infinite area
2. The cylinder of radius L and infinite length
3. The sphere of radius L

In each case we have one-dimensional transport (i.e., there is only *one* dependent space variable of which $\tilde{X}$ is a function), and, by inspection, we know that

$$\tilde{X} = \tilde{X}(\tilde{r}, \tilde{t}, \text{Sh}) \tag{13-158}$$

($\tilde{r}$ is any of the space variables for the three geometries.)

Table 13-3 Solutions of Eq. (13-157)

Slab:

$$\tilde{X} = \sum_{n=1}^{\infty} \frac{4 \sin \lambda_n}{2\lambda_n + \sin 2\lambda_n} \exp\left(-\lambda_n^2 \tilde{t}\right) \cos\left(\lambda_n \tilde{r}\right) \tag{13-158a}$$

where λ_n are the roots of $\lambda \tan \lambda = \text{Sh}$

Cylinder:

$$\tilde{X} = 2 \, \text{Sh} \sum_{n=1}^{\infty} \exp\left(-\lambda_n^2 \tilde{t}\right) \frac{J_0(\lambda_n \tilde{r})}{(\lambda_n^2 + \text{Sh}^2)J_0(\lambda_n)} \tag{13-158b}$$

$$\lambda J_1(\lambda) - \text{Sh} \, J_0(\lambda) = 0$$

Sphere:

$$\tilde{X} = 2 \, \text{Sh} \, \frac{1}{\tilde{r}} \sum_{n=1}^{\infty} \frac{[\lambda_n^2 + (\text{Sh} - 1)^2] \sin \lambda_n}{\lambda_n^2[\lambda_n^2 + \text{Sh}(\text{Sh} - 1)]} \exp\left(-\lambda_n^2 \tilde{t}\right) \sin\left(\lambda_n \tilde{r}\right) \tag{13-158c}$$

$$\lambda \cot \lambda + \text{Sh} - 1 = 0$$

The solutions for each case are presented in Table 13-3. Because the solutions are algebraically cumbersome, it is convenient to have them in a graphical format. Two types of graphs are usually presented. One type corresponds to the case $\text{Sh} = \infty$, equivalent to the boundary condition $X = X_a$ at the exposed boundaries. The plots are in the form $\tilde{X}$ versus $\tilde{r}$ with $\tilde{t}$ as a parameter. For finite Sh one usually finds plots of $\tilde{X}$ versus $\tilde{t}$, with $\tilde{r}$ and Sh as parameters. Figures 13-22 to 13-30 show the solutions in graphical form.

We note that for small Sherwood numbers Figs. 13-25 to 13-30 become inadequate to the task. It turns out that an alternative model, valid in that limit,

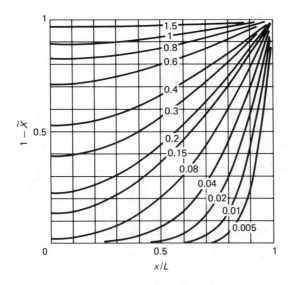

Figure 13-22 $\tilde{X}$ profiles with $\tilde{t}$ as a parameter: slab.

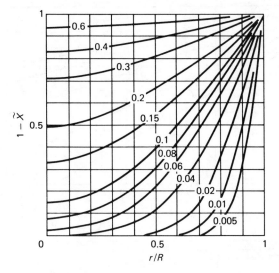

Figure 13-23 $\tilde{X}$ profiles with $\tilde{t}$ as a parameter: cylinder.

gives a particularly simple analytical solution for our use. We derive the model in cartesian coordinates for the case of the slab. The extension to the other two cases is straightforward.

We begin with

$$\frac{\partial \tilde{X}}{\partial \tilde{t}} = \frac{\partial^2 \tilde{X}}{\partial \tilde{r}^2} \tag{13-159}$$

and integrate both sides of the equation over the space $0 \leq \tilde{r} \leq 1$:

$$\int_0^1 \frac{\partial \tilde{X}}{\partial \tilde{t}} \, d\tilde{r} = \int_0^1 \frac{\partial^2 \tilde{X}}{\partial \tilde{r}^2} \, d\tilde{r} \tag{13-160}$$

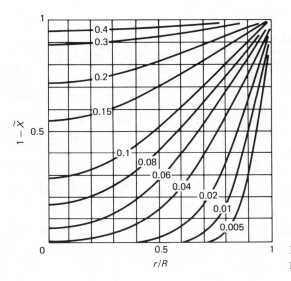

Figure 13-24 $\tilde{X}$ profiles with $\tilde{t}$ as a parameter: sphere.

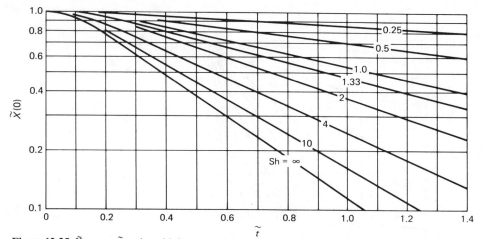

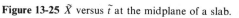

Figure 13-25 $\tilde{X}$ versus $\tilde{t}$ at the midplane of a slab.

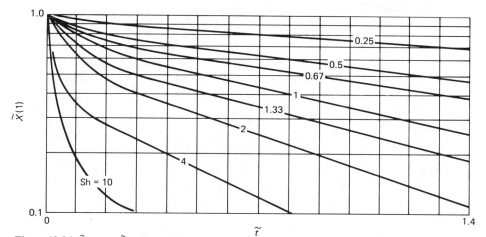

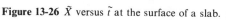

Figure 13-26 $\tilde{X}$ versus $\tilde{t}$ at the surface of a slab.

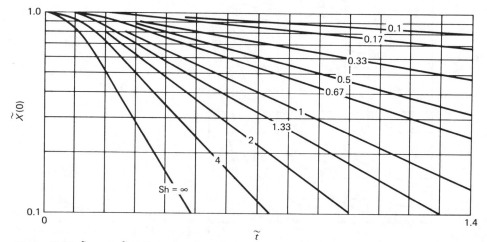

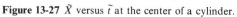

Figure 13-27 $\tilde{X}$ versus $\tilde{t}$ at the center of a cylinder.

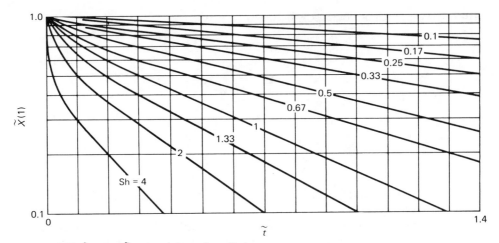

Figure 13-28 $\tilde{X}$ versus $\tilde{t}$ at the surface of a cylinder.

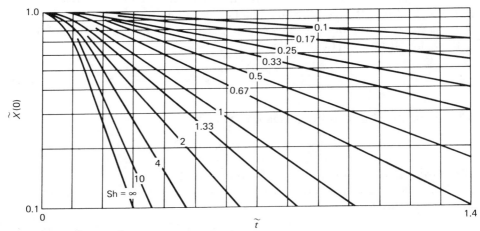

Figure 13-29 $\tilde{X}$ versus $\tilde{t}$ at the center of a sphere.

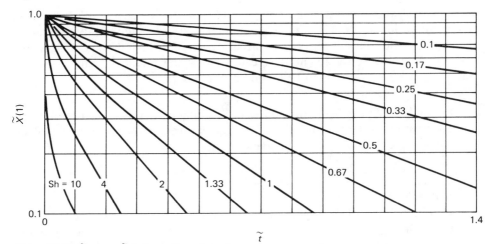

Figure 13-30 $\tilde{X}$ versus $\tilde{t}$ at the surface of a sphere.

We define the average concentration $\bar{\bar{X}}$ as

$$\bar{\bar{X}} = \int_0^1 \tilde{X} \, d\tilde{r} \tag{13-161}$$

and find

$$\frac{d\bar{\bar{X}}}{d\tilde{t}} = \left. \frac{\partial \tilde{X}}{\partial \tilde{r}} \right|_0^1 = -\text{Sh} \ \tilde{X}_1 \tag{13-162}$$

[The boundary conditions have been used in obtaining the right-hand term of Eq. (13-162).]

Now, let us consider what is meant by "small" Sherwood number. Since $\text{Sh} = k_x L/\alpha$, we see that the limit of small Sherwood number implies that the *internal* diffusion process is much more rapid than the *external* transport process $(\alpha/L \gg k_x)$. If this is the case, then we might expect a uniform $\tilde{X}$ profile across $\tilde{r}$. We introduce that assumption, now, by equating $\tilde{X}$ to $\bar{\bar{X}}$ (and using the symbol $\tilde{X}$), so that we find, from Eq. (13-162), that

$$\frac{d\tilde{X}}{d\tilde{t}} = -\text{Sh} \ \tilde{X} \tag{13-163}$$

The solution of this first-order equation is simply

$$\tilde{X} = \exp(-\text{Sh} \ \tilde{t}) \tag{13-164}$$

where the initial condition $\tilde{X}(0) = 1$ has been used.

For the cylinder or sphere similar results are found. In fact, all three solutions may be written as

$$\tilde{X} = \exp(-n \, \text{Sh} \ \tilde{t}) \tag{13-165}$$

with $n = 1, 2$, and 3 for the slab, cylinder, and sphere, respectively. We must keep in mind that this solution is valid only for small Sherwood number. By comparing the approximate solution to the more accurate model (which does not neglect the internal concentration gradients), we can see that the qualifier "small" implies $\text{Sh} < 0.05$ if Eq. (13-165) is to be a good approximation.

Example 13-5: Cooling a mold runner Give the temperature distribution and the position of the solid front in a $\frac{1}{4}$-in diameter runner of an injection-molded nylon piece. Make the following assumptions:

- Melt temperature is initially 564°F
- Runner surface is lowered to 186°F and held at that value
- Solidification of the Nylon does not affect the heat transfer process
- Thermal properties are constant over the temperature range and are $k = 6 \times 10^{-4}$ cal/cm·s·°C, $\rho = 1.1$ g/cm^3, and $C_p = 0.8$ cal/g

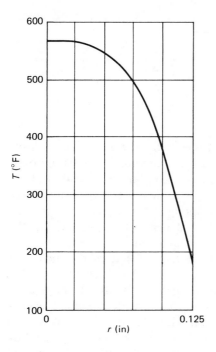

Figure 13-31 Solution to Example 13-5.

Choose $t = 6$ s, for which we find $\alpha t/R^2 = 0.04$. From Fig. 13-23 we can pick off values of $\tilde{X}$ versus $\tilde{r}$. Noting that $T = 378\tilde{X} + 186$ we can prepare Fig. 13-31. If the freezing temperature is taken as 467°F we see that the solid boundary is nearly halfway across the runner, after 6 s.

Example 13-6: Cooling of a polymer coating Polyethylene is blade coated onto a web of cellular plastic which moves under the blade at 20 cm/s. The coating temperature is 400°F, and cooling is achieved by blowing air, at 80°F, across the film. From earlier heat transfer studies it is known that, for the system of air cooling that is used, $h = 0.08$ cal/cm²·s·°C. The coating thickness is 0.1 cm. How far downstream from the blade must the web travel before the surface temperature of the coating falls below 144°F?

If we assume that the coating moves as a rigid body with the web, then we may treat this as a transient solid-cooling problem, with time replaced by exposure time x/U. Assuming constant physical properties we might expect to use Fig. 13-26. Note, however, that the boundary conditions on the slab problem include the notion (implicit in the formulation) that *both* surfaces experience identical thermal conditions. In this coating problem the exposed surface may be taken to satisfy a Newton's law of cooling boundary condition, but the interface with the web certainly does not satisfy the same condition. Since foamed plastic is such a good insulator it is likely that the appropriate boundary condition is one of *no flux* at the web surface. Figure 13-32 shows the expected situation.

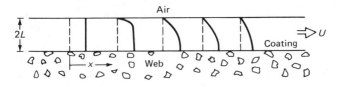

Figure 13-32 Sketch for Example 13-6.

In problems involving the plane slab geometry there is a very simple relationship connecting the solutions of the problem with symmetrical boundary conditions (for which the solutions presented above are valid) and the problem in which one surface is insulated (the problem at hand). In both cases there is a zero-gradient boundary condition along one surface. In the symmetrical problem this is the mathematical statement of symmetry. In the insulated-surface problem this is the no-flux statement.

The difference in the two problems lies in the fact that the distance from the exposed surface to the no-flux surface is L (by definition of L) in the problem with symmetry, and it is $2L$ in the problem with one surface insulated. This means that we can use the solutions already presented, except that everywhere that L appears we replace it with $2L$.

Thus, in the problem at hand, we set $L = 0.1$ cm and use Fig. 13-26. For polyethylene we take $k = 3.3 \times 10^4$ g·cm/s^3·°C $= 7.9 \times 10^{-4}$ cal/s·cm·°C and $\alpha = 1.3 \times 10^{-3}$ cm^2/s. The Nusselt number is $hL/k = 0.08(0.1)/7.9 \times 10^{-4} = 10$.

The desired value of $\tilde{X}$ is

$$\tilde{X} = \frac{144 - 80}{400 - 80} = 0.2$$

From Fig. 13-26 we find $\alpha t/L^2 = 0.08$, or

$$x = \frac{0.08 L^2 U}{\alpha} = \frac{0.08(0.1)^2 20}{1.3 \times 10^{-3}} = 12 \text{ cm}$$

We note that in Example 13-6 the reduced " time " is almost too small to allow use of Fig. 13-26. We could, of course, use the infinite-series solution in Table 13-3, and calculate $\tilde{X}$ for any desired value of $\tilde{t}$. However, the series converges slowly for small $\tilde{t}$, and machine computation would be required. Because of these two factors one seeks a model, valid for "short times," which has computational and/or graphical advantages over the solutions already presented.

Let us begin by considering the *physics* of short-time transport processes before plunging into a mathematical model. Figure 13-33 shows the expected concentration profiles shortly after $\tilde{t} = 0$. The salient feature is that there is some period of time (up to that corresponding to curve 3) before the center of the solid "feels" the effect of the concentration change at the boundary $\tilde{r} = 1$. Up to that time, if the solid were larger, the transport process would be no different because in the region where $\tilde{X} = 1$ it is also true that $\tilde{\nabla}\tilde{X} = 0$. Hence no flux occurs in the

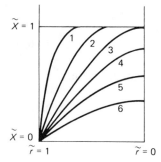

Figure 13-33 $\tilde{X}$ profiles in a slab at short times. The boundary condition $\tilde{X}(1, \tilde{t}) = 0$ (Sh $= \infty$) is shown.

central region at short times, and the *extent* of the central region does not affect the transport process in the boundary region.

This suggests that the finite solid may be replaced by a *geometric* model known as the semi-infinite solid, which has a boundary at, say, $z = 0$, and is *unbounded* in the direction $z > 0$. The transport equation still has the form

$$\frac{\partial X}{\partial t} = \alpha \, \nabla^2 X \tag{13-166}$$

(neglecting generation for simplicity), but now the boundary conditions are†

$$X = X_0 \qquad \text{at} \quad t = 0$$

$$\alpha \, \nabla X = h'(X - X_a) \qquad \text{at} \quad z = 0$$

$$X = X_0 \qquad \text{for } z \to \infty$$

The only change from the previous (bounded-solid) formulation is the shift in space coordinate, so that $z = 0$ is in the free surface, and in the third boundary condition which reflects the physical idea that if the extent of the solid is great enough, the change at the boundary $z = 0$ is never felt. (See Prob. 13-15 for discussion of the use of the notation h' in the second boundary condition above.)

Now we must demonstrate that this change in mathematical formulation, reflecting a change in the physical model, is advantageous with respect to the problem of "short-time solutions." We begin by making the problem dimensionless. $\tilde{X}$ is defined as before. We no longer have a characteristic length scale with which to make z dimensionless because the "slab" is semi-infinite. However, we note that the boundary condition at $z = 0$ may be made dimensionless if $\tilde{z} = h'z/\alpha$. This definition of $\tilde{z}$ puts the transport equation in the form

$$\frac{\partial \tilde{X}}{\partial t} = \frac{h'^2}{\alpha} \frac{\partial^2 \tilde{X}}{\partial \tilde{z}^2} \tag{13-167}$$

---

† The convection boundary condition has a different sign than in Eq. (13-155), because the space coordinate now originates at the surface.

Now it should be apparent that a suitable definition of $\tilde{t}$ is

$$\tilde{t} = \frac{h'^2 t}{\alpha} \tag{13-168}$$

with the result that the equation and boundary conditions take the form

$$\frac{\partial \tilde{X}}{\partial \tilde{t}} = \frac{\partial^2 \tilde{X}}{\partial \tilde{z}^2} \tag{13-169}$$

$$\tilde{X} = 1 \qquad \text{at } \tilde{t} = 0, \tilde{z} \to \infty$$

$$\frac{\partial \tilde{X}}{\partial \tilde{z}} = \tilde{X} \qquad \text{at } \tilde{z} = 0$$

Through these definitions, all parameters are removed from the problem and $\tilde{X} = \tilde{X}(\tilde{z}, \tilde{t})$.

The solution of Eq. (13-169) is most easily found by the use of Laplace transforms, and the result may be written in the form

$$\tilde{X} = \text{erf } \frac{\tilde{z}}{2\tilde{t}^{1/2}} + \exp (\tilde{z} + 4\tilde{t}) \text{ erfc} \left(\frac{\tilde{z}}{2\tilde{t}^{1/2}} + 2\tilde{t}^{1/2} \right) \tag{13-170}$$

where the *complementary error function* is defined by

$$\text{erfc } \phi = 1 - \frac{2}{\sqrt{\pi}} \int_0^\phi e^{-x^2} \, dx = 1 - \text{erf } \phi \tag{13-171}$$

Figure 13-34 shows $\tilde{X}$ as a function of $\tilde{t}$, with $\tilde{z}$ as a parameter in the form $\tilde{z}/2\tilde{t}^{1/2}$. The choice of form for the $\tilde{z}$-dependent parameter follows from the fact

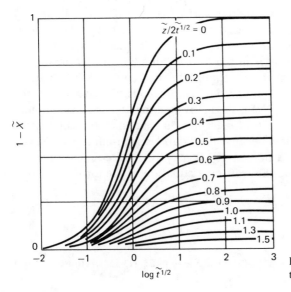

Figure 13-34 $\tilde{X}(\tilde{t})$ in a slab according to the "short-time" model (13-170).

that for large Sh the solution becomes a unique function of $\tilde{z}/2\tilde{t}^{1/2}$ and does not depend on $\tilde{z}$ and $\tilde{t}$ separately. The Sherwood number is easily forced into the definition of the ordinate by noting that

$$\tilde{t} = \frac{h'^2 t}{\alpha} = \left(\frac{h'L}{\alpha}\right)^2 \frac{\alpha t}{L^2} = \text{Sh}^2 \frac{\alpha t}{L^2} \tag{13-172}$$

Recall that this model imposes no linear scale L on the geometry, and note that $\tilde{t}$ does not depend on L. Real systems do, of course, have a finite value of L, and so it is useful to see where L comes into (and out of) the definition of $\tilde{t}$. With the same idea in mind, we see that $\tilde{z}/2\tilde{t}^{1/2}$ may be written as

$$\frac{\tilde{z}}{2\tilde{t}^{1/2}} = \frac{z/L}{2\sqrt{\alpha t/L^2}} \tag{13-173}$$

Now let us examine some implications and applications of this form of the solution. First we examine the surface concentration as a function of time for finite Sherwood numbers. It can be shown that Eq. (13-170) takes the form

$$\tilde{X}(0, \tilde{t}) = \exp 4\tilde{t} \ \text{erfc} \ 2\tilde{t}^{1/2} \tag{13-174}$$

Figure 13-35 shows this result. Note that the ordinate is $\tilde{t}^{1/2}$, which compresses the time scale for convenience.

It is instructive to use Eq. (13-174) to solve Example 13-6 and compare the result with the more exact model. Figure 13-36 shows the surface temperature of the coating as a function of distance from the blade. We note that $T = 144°\text{F}$ at $z = 12$ cm, exactly the same as found with the more exact model upon which Fig. 13-26 was based. From this it would appear that up to 12 cm, which corresponds to an exposure time of about 0.5 s, the process corresponds to short times.

It would be worthwhile to establish some criterion for application of Eq. (13-174), the semi-infinite model—a definition of short time. We do so by

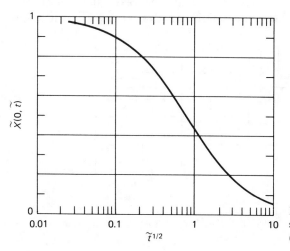

Figure 13-35 $\tilde{X}(0, \tilde{t})$ on the surface of a slab according to the "short-time" model (13-174).

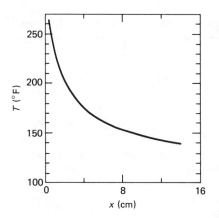

Figure 13-36 Solution to Example 13-6. Surface temperature of the coating as a function of distance downstream of the coating blade.

defining a position $z_{0.9}$ at which $\tilde{X}$ has changed slightly toward the ambient value, say, $\tilde{X} = 0.9$. Figure 13-37 shows $z_{0.9}/2\sqrt{\alpha t} = (z_{0.9}/L)/2\sqrt{\alpha t/L^2}$ as a function of $\tilde{t}^{1/2} = \text{Sh}\sqrt{\alpha t/L^2}$. For a conservative estimate we note that, for large $\tilde{t}$,

$$\frac{z_{0.9}}{2\sqrt{\alpha t}} = 1.2 \quad \text{or} \quad \frac{z_{0.9}}{L} = 2.4\sqrt{\frac{\alpha t}{L^2}} \qquad (13\text{-}175)$$

The idea of the semi-infinite model is that the temperature change does not extend to the center of the slab. To ensure that $z_{0.9}/L$ be less than unity we require $2.4\sqrt{\alpha t/L^2} < 1$ or $\alpha t/L^2 < 0.17$. Thus our definition of "short time" is $t < 0.17L^2/\alpha$. In Example 13-6 the value of $\alpha t/L^2 = 0.08$, so we should expect that the short-time solution would be valid.

Figure 13-37 also defines what we might think of as a *penetration thickness*. At a given time we may calculate a value of $z_{0.9}/L$, which we may interpret as the degree to which the external "concentration" field has "penetrated" the slab. If $z_{0.9}/L$ is small with respect to unity then the process is still in its short-time stage.

This semi-infinite model has been presented for the plane-slab geometry. However, in the case of a cylinder or sphere, if the penetration thickness is quite

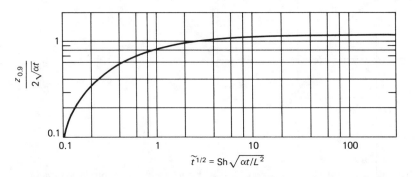

Figure 13-37 Position $z_{0.9}$, within a slab, at which $\tilde{X}$ has changed from its initial value of 1 to a value $\tilde{X} = 0.9$, based on Fig. 13-34.

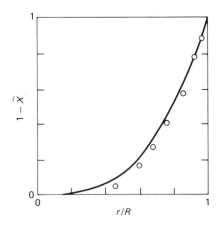

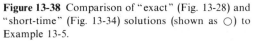

Figure 13-38 Comparison of "exact" (Fig. 13-28) and "short-time" (Fig. 13-34) solutions (shown as $\bigcirc$) to Example 13-5.

small with respect to the radius, then the curvature of the solid is of no consequence and the system behaves, at short times, as if its surface is plane. We may illustrate this point by calculating the temperature distribution for Example 13-5 from the semi-infinite model. At the desired time of 6 s we found $\alpha t/R^2 = 0.04$, small enough to expect the short-time model to hold. We use Fig. 13-34, noting that the assumption that the runner surface is maintained at a fixed value corresponds to large Sherwood number. Hence we use the right-hand asymptotes of the figure, and we obtain Fig. 13-38. The agreement between the exact and short-time solutions is adequate for most purposes.

Simple Convection-diffusion Models

In the previous section we considered transport problems in which there was no internal flow to provide convection; at most, rigid motion occurred. Here we examine diffusion in the presence of very simple kinematics.

There are several systems of industrial interest in which transfer of heat or mass takes place from a thin layer of fluid flowing down an inclined surface. A simple model of such a process can be developed with reference to Fig. 13-39. We assume the existence of a region over which a film of constant thickness H exists. (If this assumption were false we would find that no solution of the dynamic equations could be found compatible with conservation of mass.)

Assuming laminar steady unidirectional flow we can reduce the dynamic equations to the form

$$0 = \frac{\partial}{\partial y} \tau_{zy} + \rho g \tag{13-176}$$

or, integrating once,

$$\tau_{zy} = \rho g(H - y) \tag{13-177}$$

(We have used the boundary condition that no shear stress is exerted on the free surface at $y = H$.)

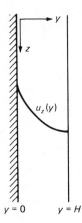

$y = 0$ $y = H$

Figure 13-39 Definition sketch for analysis of diffusion from (or to) a falling film along a vertical surface.

As a constitutive equation we will take the power law model, which leads to an equation for velocity in the form

$$\frac{du_z}{dy} = \left[\frac{\rho g(H - y)}{K}\right]^{1/n} \tag{13-178}$$

With the boundary condition $u_z = 0$ at $y = 0$ we may solve formally for $u_z(y)$ and find

$$u_z = \int_0^y \left[\frac{\rho g(H - y)}{K}\right]^{1/n} dy \tag{13-179}$$

If a transport process is occurring then K will be a function of y, and the velocity field is coupled thereby with the transport equation. If K is taken to be constant the integration can be performed immediately to give

$$u_z = H\left(\frac{\rho g H}{K}\right)^{1/n} \frac{n}{1 + n}\left[1 - \left(1 - \frac{y}{H}\right)^{1/n + 1}\right] \tag{13-180}$$

We will use this simple form of the velocity profile in modeling heat and mass transfer from a falling film.

The transport equation takes the form

$$u_z \frac{\partial X}{\partial z} = \alpha \frac{\partial^2 X}{\partial y^2} \tag{13-181}$$

under the assumptions that α is constant and that no generation occurs. The solution, with $u_z(y)$ given by Eq. (13-180), can be obtained in the form of an infinite series, but it is rather cumbersome for calculational purposes. Instead, we examine some approximate models which are easier to evaluate. The approximations are based in part on a simplification of the velocity profile.

Consider first the problem generated by Eq. (13-181) subject to boundary conditions

$$X = \begin{cases} X_0 & \text{at } z = 0 \\ X_w & \text{at } y = 0, z > 0 \end{cases}$$

A second boundary condition with respect to y is required, dictated by physical conditions at the free surface. We consider first the case that the transfer from (or to) the falling film is principally across the solid-fluid boundary ($y = 0$) rather than across the fluid-air interface ($y = H$). We will examine a solution valid for short contact time, much in the sense of the short-time model of the previous section. We take as the appropriate boundary condition

$$X = X_0 \qquad \text{at } y \to \infty, \, z > 0$$

As described in the previous section, the short-time model corresponds to the idea that the change in X penetrates only a small distance into the fluid layer. As a simplification to the convective term we assume that the velocity in the region close to the wall is linear, with the slope given by the value *at* the wall. From Eq. (13-178) we obtain du_z/dy at $y = 0$, and we write the velocity profile as

$$u_z(y) = \left(\frac{\rho g H}{K}\right)^{1/n} y \qquad \text{for } \frac{y}{H} \ll 1 \tag{13-182}$$

The equation to be solved, then, is

$$\left(\frac{\rho g H}{K}\right)^{1/n} y \frac{\partial X}{\partial z} = \alpha \frac{\partial^2 X}{\partial y^2} \tag{13-183}$$

We may define a dimensionless dependent variable as

$$\tilde{X} = \frac{X - X_0}{X_w - X_0} \tag{13-184}$$

Equation (13-183) may be solved by Laplace transform methods or by introducing a similarity transform. The solution is

$$\tilde{X} = \frac{1}{\Gamma\left(\frac{4}{3}\right)} \int_{\eta}^{\infty} e^{-\eta^3} \, d\eta \tag{13-185}$$

where $\Gamma(\ )$ is the gamma function and η is defined as

$$\eta = \frac{y}{(9\beta z)^{1/3}} \tag{13-186}$$

The parameter β is given by

$$\beta = \alpha \left(\frac{K}{\rho g H}\right)^{1/n} \tag{13-187}$$

The integral in Eq. (13-185) must be evaluated numerically. Figure 13-40 shows $\tilde{X}$ as a function of η.

To calculate the *amount* of heat or mass picked up by the fluid we need the cup-mixing average of $\tilde{X}$, defined by

$$\langle \tilde{X} \rangle = \int_0^H \frac{u_z \tilde{X} \, dy}{Q/W} \tag{13-188}$$

where Q/W is the volumetric flow rate per unit width of film. The result is most conveniently put in the format

$$\langle \tilde{X} \rangle = 0.186 \left(\frac{1 + 2n}{n} \right)^{1/3} \left(\frac{9\alpha}{UH} \frac{z}{H} \right)^{2/3} \tag{13-189}$$

where the average velocity $U = Q/WH$ has been introduced. We see that $\langle \tilde{X} \rangle$ depends upon a Peclet number UH/α, the power law index n, and the normalized vertical distance z/H.

What is the expected region of applicability of this model? In the first place we have assumed that the velocity is linear or, equivalently, that the gradient is constant. We may write the velocity gradient as

$$\frac{du_z/dy}{(\rho g H/K)^{1/n}} = \left(1 - \frac{y}{H} \right)^{1/n} \tag{13-190}$$

For regions near the wall (small y/H) we may write the right-hand side as $1 - y/nH$. We will restrict the solution to $y/H < 0.1n$ in order to stay in the nearly linear region.

It is necessary, then, that the region of penetration of the wall value of $\tilde{X}$ not exceed the region of linear velocity. Hence we require $\tilde{X} < 0.1$ within $y/H < 0.1n$. From Fig. 13-40 we see that $\tilde{X} < 0.1$ requires $\eta > 1$. Putting these constraints together we find the requirement that

$$\left(\frac{\beta}{H^2} \frac{z}{H} \right)^{1/3} < 0.1n$$

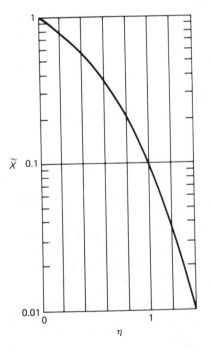

$\tilde{X}$ 0.1

η

Figure 13-40 $\tilde{X}(\eta)$ according to Eq. (13-185).

or
$$\frac{z}{H} < 10^{-4} \frac{H^2 n^3}{\beta} \qquad (13\text{-}191)$$

The dependence on n^3 is quite significant here; there is an order-of-magnitude difference in the region of application of this model between the newtonian fluid and a power law fluid of $n \approx 0.46$ for the same apparent viscosity evaluated at the wall shear rate. The nonnewtonian model is valid over a much smaller region than the newtonian model.

We note that β may also be expressed as

$$\beta = \frac{\alpha H^2}{Q/W} \frac{n}{2n + 1} \qquad (13\text{-}192)$$

from which Eq. (13-191) may be written as

$$\frac{z}{H} < 10^{-4} \frac{Q/W}{\alpha} n^2(2n + 1) \qquad (13\text{-}193)$$

If we make the comparison at constant Q/W we see that the region of applicability is still strongly dependent on n, though slightly less so than on the basis of constant apparent viscosity.

Example 13-7: Falling-film heat exchanger Let us examine a simplified application of these ideas so that we can see the magnitude of the effects that this model can accommodate. We consider a "falling-film" heat exchanger designed as shown in Fig. 13-41. A heated cylindrical tank, 2 ft in inside diameter, receives fluid through a header at its top in such a way that a film of the fluid flows along the wall under the action of gravity. A steam-jacketed

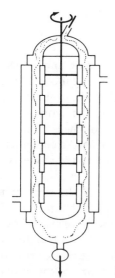

Figure 13-41 Schematic of a falling-film heat exchanger.

shell on the outside of the pipe maintains the inside wall surface at a temperature of 250°F. The fluid enters the system at 80°F. A series of scrapers is fixed to a rotating shaft and effectively mix the fluid film every 6 in along the tank axis. Hence we may think of the process as periodic: Film falls a distance of 6 in, during which time it picks up heat from the wall. The fluid is mixed, and, at a slightly elevated average temperature, falls to the next scraper level. We wish to determine whether such a system can bring the fluid to 200°F over a reasonable axial distance.

Assume the following operating conditions and physical properties:

$$Q = 1500 \text{ lb/h} \qquad \rho = 60 \text{ lb/ft}^3$$

$$\mu = 6700 e^{-0.025(T-80)} \text{ P (newtonian)}$$

$$\alpha = 7 \times 10^{-4} \text{ cm}^2/\text{s (independent of temperature)}$$

We will take the fluid to be isothermal over each 6-in section for the purpose of calculating fluid properties.

The film thickness may be shown to be related to the flow rate by (Prob. 13-22)

$$H = \left(\frac{Q}{W} \frac{3\mu}{\rho g} \right)^{1/3} \tag{13-194}$$

Over the first section this gives (taking $\mu = 6700$ P)

$$H_1 = 2.8 \text{ cm}$$

$$U = 0.37 \text{ cm/s}$$

$$\beta = 18 \times 10^{-4} \text{ cm}^2$$

$$\langle \tilde{X} \rangle = 0.028$$

The average temperature entering the second section will be $\langle X \rangle = (X_w - X_0)\langle \tilde{X} \rangle + X_0 = 84.8°F$.

The process requires a temperature change of 120°F total, and we estimate about a 5°F rise over a 6-in section. Thus we would require more than 24 such sections, with a total length, excluding headers, of about 12 ft. The question of whether this is a practical length is outside the scope of the problem. The point is that one can readily produce a rough estimate for the purpose of determining feasibility of a design. We only need to comment, then, on some weaknesses of the model in order to determine if the model is at all reasonable for such use.

First, we note that, at $z = 6$ in, we have $z/H = 5.4$, while the right-hand side of the inequality of Eq. (13-193) has the value 0.44. Thus the model is in error to some degree, and it would be necessary to have the more exact model (taking account of the actual velocity profile and, possibly, the finite nature of the film) in order to evaluate the error and provide a more accurate estimate of the required length of the heater. Since the estimated size of the system might be on the border of feasibility one would probably seek a more accurate solution to the problem in this case, and investigate alternate designs as well.

We turn, next, to a *mass* transfer example based on the falling-film model. Exposed liquid-film systems are often used for devolatilization of polymers, especially as a last stage or "finishing" step of a polymerization reaction. In that case the primary transport process occurs at the free surface rather than at the polymer-solid boundary. The transport equation

$$u_z \frac{\partial X}{\partial z} = \alpha \frac{\partial^2 X}{\partial y^2} \tag{13-181}$$

still holds, but different boundary conditions would be used. We illustrate this with the falling-film case.

If the extent of penetration is small then a useful approximation would come from recognition that the velocity u_z would be quite flat in the neighborhood of the exposed surface. It would be most convenient to set up the origin of the y coordinate in the free surface. The velocity at the free surface is [from Eq. (13-180)]

$$U_0 = \frac{n}{1+n} H\left(\frac{\rho g H}{K}\right)^{1/n} \tag{13-195}$$

Equation (13-181) then takes the form

$$U_0 \frac{\partial X}{\partial z} = \alpha \frac{\partial^2 X}{\partial y^2} \tag{13-196}$$

Since U_0 is constant we may define a dimensionless exposure time $\tilde{t} = \alpha z / U_0 H^2$ and write Eq. (13-196) in the form

$$\frac{\partial X}{\partial \tilde{t}} = \frac{\partial^2 X}{\partial \tilde{y}^2} \tag{13-197}$$

where $\tilde{y} = y/H$. The problem is thus reduced to a form similar to that considered previously. This is not surprising, since the assumption of constant u_z transforms the problem to one of rigid motion. Let us select a new set of boundary conditions which produce a model with physical features different from those of earlier problems, even though there are some basic mathematical similarities.

Example 13-8: Devolatilization of a falling film Polymeric fluid containing some residual volatile monomer is transferred from the reactor and must undergo devolatilization in order to reduce the concentration of residual monomer. In practice the interaction of heat and mass transfer renders the modeling of this problem quite complex. We will simplify the model to be able to examine some features of the solution which would be obscured in a more exact analysis. In effect, we uncouple the heat and mass transfer phenomena to the point that *analytical* solutions are possible.

The monomer concentration will obey a diffusion equation of the form

$$\frac{\partial \tilde{c}}{\partial \tilde{t}} = \frac{\partial^2 \tilde{c}}{\partial \tilde{y}^2} \qquad (13\text{-}198)$$

$$\tilde{c} = \begin{cases} 1 & \text{at } \tilde{t} = 0, \ \tilde{y} \to \infty \\ 0 & \text{at } \tilde{y} = 0 \end{cases}$$

We have defined the reduced concentration $\tilde{c}$ as

$$\tilde{c} = \frac{c}{c_0} \qquad (13\text{-}199)$$

where c_0 is the initial monomer concentration. We assume that the ambient medium is effectively at zero monomer concentration and that the free surface is brought to zero concentration by efficient convective transfer. We use a short-time assumption so that a semi-infinite geometry is relevant.

We assume that the falling film enters the devolatilizer at the desired temperature T_0, and that the only means of heat transfer from the film is through the effect of vaporization of the monomer at the free surface. The heat conduction equation takes the form

$$\frac{\partial \tilde{T}}{\partial \tilde{t}} = \frac{\alpha_T}{\mathscr{D}} \frac{\partial^2 \tilde{T}}{\partial \tilde{y}^2} \qquad (13\text{-}200)$$

$$\tilde{T} = 1 \qquad \text{at } \tilde{t} = 0, \ \tilde{y} \to \infty$$

where $\tilde{T} = T/T_0$ is the reduced temperature. Note that we will define $\tilde{t}$ using $\alpha = \mathscr{D}$, as a consequence of which the energy equation will have the ratio of thermal to mass diffusivities appearing explicitly.

The third boundary condition on $\tilde{T}$ couples the energy and species diffusion equations by equating the conductive heat flux to the surface to the evaporative heat flux from the surface:

$$-k \frac{\partial T}{\partial y} = -\mathscr{D} \frac{\partial c}{\partial y} (\Delta H_v) \qquad \text{at } y = 0 \qquad (13\text{-}201)$$

In terms of the dimensionless variables already defined this takes the form

$$\frac{\partial \tilde{T}}{\partial \tilde{y}} = \frac{\mathscr{D} c_0 \, \Delta H_v}{k T_0} \frac{\partial \tilde{c}}{\partial \tilde{y}} \qquad \text{at } \tilde{y} = 0 \qquad (13\text{-}202)$$

We will let

$$\Pi_H \equiv \frac{\mathscr{D} c_0 \, \Delta H_v}{k T_0} \qquad (13\text{-}203)$$

be a heat of vaporization parameter for this problem.

If all transport coefficients are assumed constant we can obtain an analytical solution to Eq. (13-198) for $\tilde{c}$. Thus we can explicitly write the boundary

Volatiles out

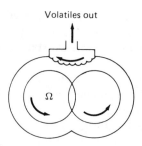

Figure 13-42 Schematic of a twin-screw devolatilizer.

condition on $\tilde{T}$ [Eq. (13-202)] as a function of $\bar{t}$ and obtain the solution to Eq. (13-200) for $\tilde{T}$. The particular feature we wish to examine is the magnitude of the cooling effect of monomer vaporization: Is it large enough that it must be accounted for in designing a devolatilization system?

The solution for $\tilde{T}$ is particularly simple for the uncoupled model formulated here. In fact, $\tilde{T}$ is constant along the surface and is given by

$$\tilde{T}(0) = 1 - \left(\frac{\alpha_T}{\mathscr{D}}\right)^{1/2} \Pi_H \tag{13-204}$$

For a choice of parameters typical of devolatilization of solutions at room temperature one can predict a cooling of the film surface of the order of 25°C. Such a large temperature drop would have a significant retarding effect on the rate of loss of volatiles.

Let us consider one more mass transfer example which uses the simple models developed above. Again the goal is development of simple estimates without recourse to complicated numerical procedures.

Example 13-9: Design equation for a twin-screw devolatilizer A vented twin-screw extruder operated partially filled can be used as a devolatilizer. Figure 13-42 shows an idealization of the geometry. In the region where fluid crosses from one screw to the other a free surface exists which allows evaporation of residual volatiles. We may model this system as a stagewise process in which surface evaporation is followed by mixing and conveying to the next evaporation stage. Figure 13-43 shows a stagewise model of the process. A single stage consists of the screw root and flights rotated through one turn.

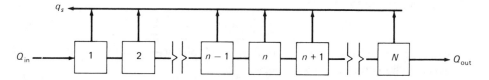

Figure 13-43 Schematic for stagewise analysis of a twin-screw devolatilizer.

In the simplest model we assume that each stage operates in an identical manner to all the others. We will define an exposure time t_e during which evaporation occurs from each stage. The *total* time spent in each stage is just the overall residence time divided by the number of stages: $t_s = t_R/N$. The exposure time can be expected to be some fraction of t_s that depends on the geometry of the screws and the degree of fill. The residence time t_R is just the ratio of the total volume to the volumetric flow rate.

To model the evaporation process we will calculate the loss of volatile material in time t_e, using the semi-infinite transport model for diffusion. If we assume effective volatiles removal at the exposed surface so that the surface concentration is reduced to zero, then the solution for the concentration field, at time t, is (for the nth stage)

$$\frac{c}{c_{n-1}} = \text{erf} \, \frac{y}{\sqrt{4\mathscr{D}t}} = \frac{2}{\sqrt{\pi}} \int_0^{y/\sqrt{4\mathscr{D}t}} e^{-\phi^2} \, d\phi \qquad (13\text{-}205)$$

for $0 \le t \le t_e$. (It should be apparent that we have solved $\partial c/\partial t = \mathscr{D} \, \partial^2 c/\partial y^2$ subject to conditions $c = 0$ at $y = 0$ and $c = c_{n-1}$ at $t = 0$ and $y \to \infty$.) In Eq. (13-205) we use c to denote $c(y, t)$, and c_{n-1} is the average concentration (mixed) entering the nth stage.

The solvent flux at the free surface is just

$$-\mathscr{D} \left. \frac{\partial c}{\partial y} \right|_{y=0} = c_{n-1} \sqrt{\frac{\mathscr{D}}{\pi t}} \qquad (13\text{-}206)$$

and the total loss of volatiles, δm_n, in time t_e is just the integral of the flux over the exposure time multiplied by the area of exposed surface (per stage) a_e. The result is

$$\delta m_n = a_e c_{n-1} \sqrt{\frac{4\mathscr{D}t_e}{\pi}} \qquad (13\text{-}207)$$

We see that the loss from each stage depends upon the average concentration at the end of the previous stage. A simple material balance gives

$$c_{n-1} - c_n = \frac{N}{V_h} \delta m_n = \frac{N a_e c_{n-1} \sqrt{4\mathscr{D}t_e/\pi}}{V_h} \qquad (13\text{-}208)$$

where V_h is the *hold-up volume*, the total volume of solution in the system. We take V_h/N to be the volume per stage and assume it is the same for each stage.

Thus we find that the concentration at each stage is given by a *difference* equation of the form

$$c_n = c_{n-1}(1 - F) \qquad (13\text{-}209)$$

where F combines all the geometric and physical parameters introduced earlier. The solution is easily seen to be

$$c_n = c_0(1 - F)^n \qquad (13\text{-}210)$$

We note that the solution is meaningful only if $F < 1$, corresponding to the idea that this solution is valid for short time.

It is convenient to write F in the form

$$F = v_h^{-1/2} a_e \sqrt{\frac{4\mathscr{D} f_e}{\pi}} Q^{-1/2} \qquad (13\text{-}211)$$

where $v_h = V_h/N$ and a_e are geometric parameters, f_e is defined by $t_e = f_e t_s$, and Q is the volumetric flow rate of material through the system. Thus F is separated into geometric factors independent of the size of the system and factors related to the diffusivity and the volumetric flow rate Q.

The model may also be used to assess the effect of a change in throughput Q on the behavior of the system. Suppose, for the sake of a numerical illustration, that a twin-screw devolatilizer conveys 10,000 lb/h of polymer, containing 4% volatiles, and is able to reduce the volatiles content to 0.5%. We desire to estimate the impact on the volatiles content of an increase in throughput to 15,000 lb/h, with no other change in design.

Let us assume that inspection of the screw design suggests that the number of stages is $N = 20$. From Eq. (13-210) we find $F = 0.1$. The projected change in throughput reduces the residence time, changing F to 0.08. The volatiles content is found to be 0.75% under these conditions.

Heat Transfer in Confined Laminar Flows

In several previous chapters we have encountered flow problems in which the fluid and the confining boundaries might be at different temperatures. A knowledge of the heat transfer process between the fluid and the boundary is necessary to evaluate the temperature change within the fluid and the effect of that change on the flow process. In addition to such problems, which may arise in die flows or in mold filling (to name just two examples), we may also purposefully pump fluid through heated or cooled conduits to change the fluid temperature. Of course in this latter case we are really talking about the conduit as a heat exchanger.

Some relatively simple analyses are possible with which one may estimate thermal behavior in confined laminar flows. We shall outline the formulation of several problems of interest and examine the use of the solutions in some example problems. One of the simplest problems to begin with is that of flow between parallel plates, both of which are maintained at uniform temperature T_w. We take the fluid to obey a power law, with temperature-independent properties. Hence the velocity profile is already given in Chap. 5, and we rewrite it in the format

$$u_z = U \frac{1 + 2n}{1 + n} \left[1 - \left(\frac{2y}{B} \right)^{1 + 1/n} \right] \qquad (13\text{-}212)$$

where U = average velocity
B = separation between plates $(-\tfrac{1}{2}B \le y \le \tfrac{1}{2}B)$

The conduction-convection equation may be written in the dimensionless form (see Prob. 13-26)

$$(1 - \tilde{y}^{1+1/n})\frac{\partial \tilde{T}}{\partial \tilde{z}} = \frac{\partial^2 \tilde{T}}{\partial \tilde{y}^2} + \mathrm{Br}\ \tilde{y}^{1+1/n} \tag{13-213}$$

by defining $\tilde{y} = 2y/B$, $\tilde{T} = (T - T_w)/(T_0 - T_w)$, and

$$\tilde{z} = \frac{4kz}{\rho C_p B^2 U}\frac{1+n}{1+2n} \tag{13-214}$$

The viscous heat generation term is included, and this introduces a nonnewtonian Brinkman number

$$\mathrm{Br} = \left(\frac{1+2n}{n}\right)^{1+n}\frac{KU^{1+n}B^{1-n}}{k(T_0 - T_w)2^{1-n}} \tag{13-215}$$

The boundary conditions are those for uniform inlet temperature, followed by contact with an isothermal boundary for all $\tilde{z} > 0$.

$$\tilde{T} = \begin{cases} 1 & \text{at } \tilde{z} = 0 \\ 0 & \text{at } \tilde{y} = \pm 1 \end{cases}$$

Of principal interest is the cup-mixing temperature $\langle \tilde{T} \rangle$, which can be seen to depend on the following parameters:

$$\langle \tilde{T} \rangle = \langle \tilde{T} \rangle(\tilde{z}, n, \mathrm{Br}) \tag{13-216}$$

Analytical solutions are possible, but the format is cumbersome and it is most convenient to have the results in graphical form. Figure 13-44 shows the cup-mixing temperature as a function of axial distance from the entry plane. The effect of nonnewtonian behavior, at least for $\frac{1}{4} < n < 1$, is not very significant. The results shown are for the case $\mathrm{Br} = 0$ (no viscous heat generation) and are taken

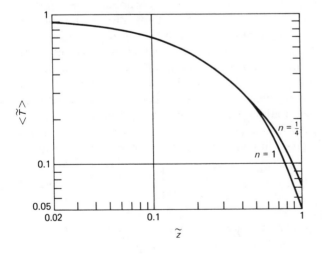

Figure 13-44 Average temperature for power law flow between parallel plates at uniform temperature.

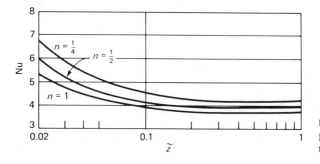

Figure 13-45 Nusselt number for flow of a power law fluid between parallel plates at uniform temperature.

from a *numerical* solution of Vlachopoulos and Keung. Their results show that for Br < 1 the temperature $\langle \tilde{T} \rangle$ is not significantly affected over the range $0 < \tilde{z} < 1$ by viscous heating.

In terms of the heat transfer characteristics of such a flow it is of interest to calculate the heat flux at the solid boundary. This may be obtained most simply from the temperature gradient at the duct wall, and it is convenient to write this in the form (see Prob. 13-27)

$$-\frac{\partial \tilde{T}}{\partial \tilde{y}} = \tfrac{1}{2} \, \mathrm{Nu} \, \langle \tilde{T} \rangle \qquad \text{at } \tilde{y} = 1 \tag{13-217}$$

Equation (13-217) thereby defines the local Nusselt number, and this definition is consistent with our earlier introduction of the Nusselt number as a dimensionless heat transfer coefficient. Figure 13-45 shows Nu versus $\tilde{z}$ for the case Br = 0.

If we have a solution for the temperature distribution along the duct then the Nusselt number provides no *independent* information, since it is derivable from $\tilde{T}(\tilde{y}, \tilde{z})$ through Eq. (13-217). However, in some models of convective heat transfer we cannot solve easily for the detailed temperature field, and the format of the model may involve a heat transfer coefficient which must be established from independent assumptions. We have already seen an example of this: Example 13-2, where Eq. (13-104) provides a possible model for the mean temperature change down the axis of a tubular reactor and requires knowledge of the temperature gradient at the wall. As an approximation, then, one could introduce the Nusselt number from a simpler related model and thereby solve the more complex model. (In the case of Example 13-2 we would use the Nusselt number for *tubular*-flow heat transfer, to be presented below.)

Before considering the tubular geometry we illustrate the application of the parallel plate model just developed.

Example 13-10: Heat transfer in axial annular flow A commercial polyethylene resin is extruded through an annular die of 2-in radius, the separation between the concentric cylindrical surfaces being 0.02 in. The output rate is 100 lb/h. The melt enters the die at a uniform temperature of 375°F, but we wish to extrude at 300°F. Can the melt be brought to the desired extrusion temperature by cooling the die surfaces to a uniform temperature of 270°F?

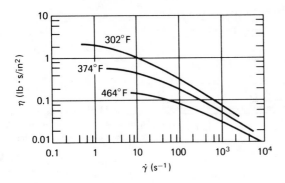

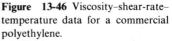

Figure 13-46 Viscosity–shear-rate–temperature data for a commercial polyethylene.

We will need the following physical property data:

$$\text{Density } \rho = 57 \text{ lb/ft}^3$$

$$\text{Thermal diffusivity } \alpha_T = 1.3 \times 10^{-3} \text{ cm}^2/\text{s}$$

$$\text{Thermal conductivity } k = 3.3 \times 10^4 \text{ g} \cdot \text{cm/s}^3 \cdot {}^\circ\text{C}$$

$$\text{Viscosity–shear rate data are shown in Fig. 13-46}$$

First we estimate the nominal shear rate in the die. The average velocity is $U = Q/2\pi RB = 3.4$ in/s, and the nominal shear rate [Eq. (5-25)] is $\dot{\gamma} = 6U/B = 1020 \text{ s}^{-1}$. At this shear rate, and at 375°F, we estimate $n = \frac{2}{3}$, $K = 3.45 \times 10^4 \text{ dyne} \cdot \text{s}^{2/3}/\text{cm}^2$.

The desired value of $\langle \tilde{T} \rangle$ is

$$\langle \tilde{T} \rangle = \frac{300 - 270}{375 - 270} = 0.35$$

and from Fig. 13-44 we estimate a required value of $\tilde{z}$ of approximately 0.4. From Eq. (13-214) this indicates a required die length of $z = 6$ cm.

The solution assumes no significant viscous heating occurs in the die flow. To evaluate this we calculate the Brinkman number from Eq. (13-215) and find Br = 1.57. Judging from the results of Vlachopoulos and Keung it would appear that measurable viscous heating effects occur, which would increase the required cooling length by 10 to 15 percent.

Very few data are available for evaluating the solution presented above; none are known to us for the case of heat transfer to polymeric fluids in conduits that approximate parallel plates. The more common geometry is that for Poiseuille flow in circular tubes or pipes, so we turn to consideration of that problem now.

As in the case of the plane Poiseuille flow, the simplest model assumes that the isothermal velocity profile is still valid in the nonisothermal case. (The alternative, of course, requires solving the conduction-convection equation *coupled* with the appropriate momentum equation.) The mathematical model takes the form, for a power law fluid,

$$(1 - \tilde{r}^{1/n+1})\frac{\partial \tilde{T}}{\partial \zeta} = \frac{1}{\tilde{r}}\frac{\partial}{\partial \tilde{r}}\left(\tilde{r}\frac{\partial \tilde{T}}{\partial \tilde{r}}\right) + \text{Br}\, \tilde{r}^{1/n+1} \tag{13-218}$$

where $\tilde{r} = r/R$ and $\tilde{T} = (T - T_w)/(T_0 - T_w)$. The reduced axial variable is

$$\zeta = \frac{kz}{\rho C_p R^2 U}\frac{1+n}{1+3n} \tag{13-219}$$

and the Brinkman number for this geometry is defined by

$$\text{Br} = \left(\frac{1+3n}{n}\right)^{1+n}\frac{KU^{1+n}R^{1-n}}{k(T_0 - T_w)} \tag{13-220}$$

Appropriate boundary conditions for the problems of interest would be

$$\tilde{T} = \begin{cases} 1 & \text{at } \zeta = 0 \\ 0 & \text{at } \tilde{r} = 1 \end{cases}$$

$$\frac{\partial \tilde{T}}{\partial \tilde{r}} = 0 \qquad \text{at } \tilde{r} = 0$$

corresponding to the case of flow into a tube whose wall is maintained iso-thermally at a different temperature from the inlet condition.

In the case of no viscous heating, $\text{Br} = 0$, the solution to Eq. (13-218) is associated with the name of Graetz, who first solved this problem for the newton-ian fluid. The Graetz solution may be written in the form of an infinite series, and the cup-mixing temperature is given by

$$\langle \tilde{T} \rangle = \sum_{i=1}^{\infty} A_i e^{-a_i \zeta} \tag{13-221}$$

For $\zeta > 0.01$ only the first few terms of the series are required. Table 13-4 gives values of A_i and a_i for the first three terms. The values for the power law case can be obtained from a paper by Lyche and Bird. Figure 13-47 shows $\langle \tilde{T} \rangle$ as a function of ζ for several values of the power law index n.

Table 13-4 Coefficients for use in Eq. (13-221)

i	a_i				A_i			
	$n = 1$	0.5	0.33	0	1	0.5	0.33	0
1	7.31	6.58	6.26	5.78	0.82	0.81	0.81	0.69
2	44.6	39.1	36.4	30.5	0.10	0.11	0.11	0.13
3	114	99.5	92.3	74.9	0.032	0.039	0.046	0.048

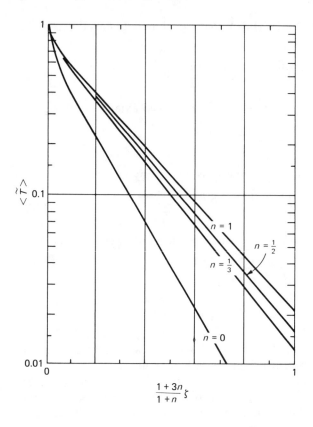

Figure 13-47 Average temperature for power law flow through a pipe with uniform wall temperature.

The temperature distribution $\tilde{T}(\tilde{r}, \zeta)$ is not presented here but of course is required for the calculation of the average $\langle \tilde{T}(\zeta) \rangle$. The paper by Lyche and Bird gives sufficient detail if one desires the radial profiles. In addition, one needs $\tilde{T}(\tilde{r}, \zeta)$ to calculate the local Nusselt number, defined as $\mathrm{Nu} = hD/k$, where $D = 2R$. As in the case of parallel plates, the local heat transfer coefficient is given by

$$-k \left. \frac{\partial T}{\partial r} \right|_{r=R} = h(\langle T \rangle - T_w)$$

or, in dimensionless form,

$$\mathrm{Nu} = \frac{2}{\langle \tilde{T} \rangle} \left[-\frac{\partial \tilde{T}}{\partial \tilde{r}} \right]_{r=1} \tag{13-222}$$

Figure 13-48 shows Nu as a function of axial position for several n values.

For small ζ, say, $\zeta < 0.01$, the series solution for $\tilde{T}$ requires evaluation of many terms, and an alternate solution is more convenient. It is based on the notion, already introduced in several previous analyses, that for short contact times (in this case for small ζ) the region of penetration of heat conduction is quite

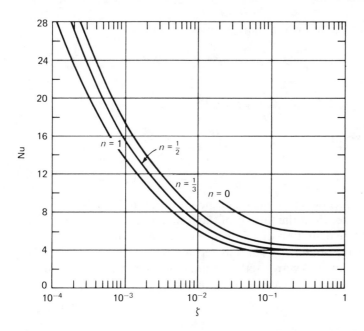

Figure 13-48 Nusselt number for power law flow through a pipe with uniform wall temperature.

small and is confined to the wall region. The name of Leveque is usually attached to the short-contact-time solution.

We begin with the conduction-convection equation in the usual form, namely,

$$\rho C_p u_z \frac{\partial T}{\partial z} = k \frac{1}{r} \frac{\partial}{\partial r}\left(r \frac{\partial T}{\partial r}\right) \tag{13-223}$$

For a power law fluid the velocity profile is

$$u_z = U_{max}\left[1 - \left(\frac{r}{R}\right)^{1/n+1}\right] \tag{13-224}$$

where

$$U_{max} = \frac{3n+1}{n+1} U \tag{13-225}$$

In the wall region we linearize the velocity profile, using

$$u_z = -\frac{du_z}{dr}\bigg|_{r=R}(R-r) \tag{13-226}$$

from which we obtain

$$u_z = \frac{U_{max}}{R}\frac{1+n}{n}(R-r) \tag{13-227}$$

We now introduce the following dimensionless variables:

$$s = 1 - \frac{r}{R} \qquad \tilde{z} = \frac{z}{R} \qquad \tilde{T} = \frac{T - T_w}{T_0 - T_w}$$

$$\text{Pe} = \frac{3n + 1}{n} \frac{UR}{\alpha_T} = \text{a Peclet number}$$

Equation (13-223) takes the form

$$\text{Pe } s \frac{\partial \tilde{T}}{\partial \tilde{z}} = \frac{\partial^2 \tilde{T}}{\partial s^2} - \frac{1}{1 - s} \frac{\partial \tilde{T}}{\partial s} \tag{13-228}$$

We would like to redefine one of the independent variables in such a way as to remove the Peclet number from the format of the problem. This could be done by incorporating Pe into the definition of $\tilde{z}$, in which case the independent variable ζ [Eq. (13-219)] is (nearly) obtained. (The difference lies in a factor involving n.) However, this leaves an awkward set of terms on the right-hand side of Eq. (13-228). An alternative is to change the s variable to

$$\tilde{s} = \text{Pe}^{1/3} s \tag{13-229}$$

If this is done, Eq. (13-228) becomes

$$\tilde{s} \frac{\partial \tilde{T}}{\partial \tilde{z}} = \frac{\partial^2 \tilde{T}}{\partial \tilde{s}^2} - \frac{1}{\text{Pe}^{1/3} - \tilde{s}} \frac{\partial \tilde{T}}{\partial \tilde{s}} \tag{13-230}$$

Now, the notion of short penetration "time" may be thought of in terms of the Peclet number, since a very large Peclet number, corresponding to rapid flow (large U) or poor conductivity (small α_T), would lead to a small degree of "penetration" of heat from the wall into the fluid interior. Thus it is appropriate to make the restriction $\text{Pe} \gg 1$, which reduces Eq. (13-230) to the form

$$\tilde{s} \frac{\partial \tilde{T}}{\partial \tilde{z}} = \frac{\partial^2 T}{\partial \tilde{s}^2} \tag{13-231}$$

Boundary conditions are

$$\tilde{T} = \begin{cases} 1 & \text{at } \tilde{z} = 0 \\ 0 & \text{at } \tilde{s} = 0 \\ 1 & \text{at } \tilde{s} \to \infty \end{cases}$$

The latter boundary condition reflects the idea that if the extent of heat penetration is small we may consider the variable $\tilde{s}$ to range from zero to infinity. The advantage of the semi-infinite boundary condition lies in the fact that it allows a solution of Eq. (13-231) in "closed," i.e., nonseries, form. The solution is found to be

$$\tilde{T} = 1 - \frac{1}{\Gamma(\frac{4}{3})} \int_\eta^\infty e^{-\eta^3} \, d\eta \tag{13-232}$$

where

$$\eta = \frac{\tilde{s}}{(9\tilde{z})^{1/3}} \tag{13-233}$$

This integral appeared in an earlier problem, and it is presented graphically in Fig. 13-40.

Since the solution is valid only for large Peclet number we may evaluate $\tilde{T}$ from Eq. (13-232) only at large η. Consequently the solution only covers the range of $\tilde{T}$ values near $\tilde{T} = 1$. Clearly one could extrapolate Fig. 13-47 into that region with reasonable accuracy. The utility of this solution lies in the accurate calculation of the Nusselt number for small z.

With Eq. (13-222) we find

$$\text{Nu} = \frac{2 \text{ Pe}^{1/3}}{\langle \tilde{T} \rangle} \left[\frac{\partial \tilde{T}}{\partial \tilde{s}} \right]_{\tilde{s}=0} \tag{13-234}$$

For the region of small heat penetration a good approximation is $\langle \tilde{T} \rangle = 1$. We find, then, that

$$\text{Nu} = \frac{2}{9^{1/3}\Gamma(\frac{4}{3})} \left(\frac{3n+1}{n} \frac{UR}{\alpha_T} \frac{R}{z} \right)^{1/3} \tag{13-235}$$

or, in terms of ζ,

$$\text{Nu} = \frac{2}{9^{1/3}\Gamma(\frac{4}{3})} \left(\frac{1+n}{n} \frac{1}{\zeta} \right)^{1/3} \tag{13-236}$$

Figure 13-48 shows Nu as a function of axial position for several values of n. Equation (13-236) was used for $\zeta < 0.01$. For $\zeta > 0.01$ the infinite series solution for $\tilde{T}$, given by Lyche and Bird, was used, as indicated in Eq. (13-222).

As noted before, the principal utility of the *local* Nusselt number, which we can now estimate from Fig. 13-48, is in supplying a convective heat transfer coefficient for use in formulating a particular boundary condition for the energy equation. Such a need arises in problem formulations quite often; Example 13-2 is a case in point. One does not normally measure the local average temperature of the fluid, $\langle T \rangle$, and consequently the local Nusselt number is not usually measured for comparison to theory. The usual experimental study of heat transfer in a confined flow, such as is under discussion here, involves measurement of the average temperature $\langle T \rangle$ at the *outlet* of the heated pipe or duct, say, at $z = L$.

From a theoretical solution for $\tilde{T}(\tilde{r}, \zeta)$ it is, of course, possible to calculate $\langle \tilde{T}(\zeta) \rangle$ and evaluate the theory at a specific axial position corresponding to the pipe exit. While a direct check on the temperature is thus possible, one normally casts the theory and the data into a format involving an average Nusselt number. The analysis is quite simple, and its exposition helps to make clear what we mean by an "average" Nusselt number.

An energy balance on an element of fluid moving down the axis of a circular pipe, assuming that only *thermal* energy need be considered, is simply†

$$dq = -wC_p \, d\langle T \rangle = h(\langle T \rangle - T_w)2\pi R \, dz \tag{13-237}$$

where $w = \rho U \pi R^2$ is the mass flow rate. This is a deceptive equation and is sometimes presented as if h were defined only by *this* equation. While it is true that one *may* define h by Eq. (13-237), it is also true that the h that appears here is the *same* h introduced earlier in Eq. (13-222). This follows upon taking Eq. (13-223) and integrating it across the cross-sectional area of the pipe and making use of Eq. (13-222) as the appropriate boundary condition at $r = R$. The details are left as an exercise.

To integrate Eq. (13-237) we first separate the variables, obtaining

$$-\frac{d\langle T \rangle}{\langle T \rangle - T_w} = \frac{h}{wC_p} 2\pi R \, dz \tag{13-238}$$

After introducing the dimensionless parameters used earlier we may write this as

$$-\frac{d\langle \tilde{T} \rangle}{\langle \tilde{T} \rangle} = \text{Nu} \left(\frac{\alpha_T}{UR} \right) d\left(\frac{z}{R} \right) \tag{13-239}$$

Integration gives

$$\ln \langle \tilde{T} \rangle_L = -\frac{\alpha_T}{UR} \int_0^{L/R} \text{Nu} \, d\left(\frac{z}{R} \right) \tag{13-240}$$

To proceed further one could introduce a model for $\text{Nu}(z)$ and perform the integration. A useful alternative is to define an average Nusselt number as

$$\overline{\text{Nu}} = \frac{R}{L} \int_0^{L/R} \text{Nu} \, d\left(\frac{z}{R} \right) \tag{13-241}$$

in which case we find

$$\ln \langle \tilde{T} \rangle_L = -\frac{\alpha_T L}{UR^2} \overline{\text{Nu}} \tag{13-242}$$

Since $\langle \tilde{T} \rangle_L$ is easily measured (by collection of the output of the pipe at $z = L$), Eq. (13-242) provides a means of calculating experimental values of an average Nusselt number. We note that the dimensionless group $U\pi R^2 / \alpha_T L = wC_p/kL$ is usually referred to as the *Graetz number* Gz. It is, of course, a form of Peclet number.

The overall heat transfer rate q is given by

$$q = -wC_p(\langle T \rangle - T_0) \tag{13-243}$$

---

† Note that we take the heat exchange dq to be *negative* if heat is transferred *to* the fluid and *positive* if heat flows *from* the fluid.

With Eq. (13-242), this allows one to express the Nusselt number as

$$\overline{\text{Nu}} = \frac{-q}{\pi k L} \frac{\ln\left[(\langle T \rangle - T_w)/(T_0 - T_w)\right]}{\langle T \rangle - T_0} \tag{13-244}$$

The more familiar, dimensional, form of this equation is

$$q = \bar{h}(2\pi R L)\,\Delta T_{\text{lm}} \tag{13-245}$$

where the *log-mean temperature difference* is defined as

$$\Delta T_{\text{lm}} = \frac{\langle T \rangle - T_0}{\ln \dfrac{\langle T \rangle - T_w}{T_0 - T_w}} \tag{13-246}$$

The main reason for writing these expressions in alternate forms is to point out that the definition of average Nusselt number, Eq. (13-241), is not universally followed. For example, one can find an average Nusselt number defined as

$$\text{Nu}_a = \frac{q}{\pi k L} \frac{1}{T_a - T_w} \tag{13-247}$$

where

$$T_a = \tfrac{1}{2}(\langle T \rangle + T_0) \tag{13-248}$$

When $\langle T \rangle$ and T_0 are not very different, corresponding to a small extent of heat transfer, it is not difficult to show that

$$\lim_{(\langle T \rangle - T_0) \to 0} \Delta T_{\text{lm}} = T_a - T_w \tag{13-249}$$

and consequently

$$\lim_{(\langle T \rangle - T_0) \to 0} \overline{\text{Nu}} = \text{Nu}_a \tag{13-250}$$

However, for significant temperature changes the two Nusselt numbers are quite different, and one must be aware of this difference when using published correlations for Nusselt numbers. The point is not that Nu_a is *wrong*, but that it is *different* from $\overline{\text{Nu}}$ (see Prob. 13-30).

Example 13-11: Comparison of measured and predicted Nusselt numbers
Griskey and Wiehe present data for heat transfer to molten polyethylene pumped through a $\tfrac{3}{8}$-in heated pipe. They present the data in terms of an "arithmetic average Nusselt number," shown plotted in Figure 13-49. Compare the data with theory.

We begin by constructing the theoretical curve in terms of the average Nusselt number Nu_a. For very small values of $U\pi R^2/\alpha_T L = wC_p/kL$ the Leveque solution holds, and since the extent of heat transfer is not great, we expect that $\text{Nu}_a = \overline{\text{Nu}}$. If Eq. (13-235) is used for the local Nusselt number,

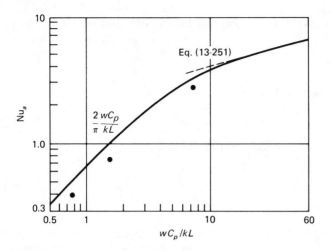

Figure 13-49 Data of Griskey and Wiehe compared to theory [Eq. (13-251) using $n = 0.7$].

and if the integration indicated in Eq. (13-241) is carried out, the result is found to be

$$\mathrm{Nu}_a = \overline{\mathrm{Nu}} = 1.61 \left(\frac{3n + 1}{4n}\right)^{1/3} \left(\frac{4UR^2}{\alpha_T L}\right)^{1/3}$$

$$= 1.75 \left(\frac{3n + 1}{4n}\right)^{1/3} \left(\frac{wC_p}{kL}\right)^{1/3} \tag{13-251}$$

It is much more tedious to carry out the same procedure using the Graetz infinite series solution, and instead we examine the limiting behavior at the extreme where the fluid is almost completely heated to the wall temperature. Under those conditions Eq. (13-243) gives

$$q = -wC_p(\langle T \rangle - T_0) = -wC_p(T_w - T_0)$$

and, from Eqs. (13-247) and (13-248), we find

$$\mathrm{Nu}_a = \frac{2}{\pi} \frac{wC_p}{kL} \tag{13-252}$$

Figure 13-49 shows this asymptotic relation, as well as the Leveque limit [Eq. (13-251)] for $n = 0.7$ (the value noted by Griskey and Wiehe). It is not very difficult to interpolate a smooth curve between the two asymptotic limits.

The data of this example are seen to be in reasonably good agreement with the theory. Other sets of experimental data, obtained with polymer solutions, also bear out the general validity of the models presented above. We must recall, however, that the models are subject to certain assumptions which are not always met. In particular we have assumed:

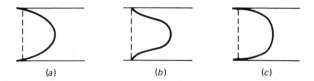

Figure 13-50 Distortion of the velocity profile due to a temperature-dependent viscosity: (a) isothermal profile, (b) cooling at the wall, (c) heating at the wall.

1. The viscosity is independent of temperature.
2. The pipe wall is isothermal.
3. No viscous heat generation occurs.

Let us examine, then, some models which relax each of these assumptions.

The effect of temperature-dependent viscosity Because of the strong temperature dependence of viscosity we might expect that heat transfer models that ignore this effect could be subject to error. The physical phenomenon that arises from the temperature-dependent viscosity is the distortion of the velocity profile, with its subsequent effect on the convection of heat down the pipe axis. Figure 13-50 suggests the distortion to be expected and points out the fact that the change in profile is different for cooling than for heating.

In the case of *cooling*, the fluid near the wall is more viscous than the bulk of the fluid, as a consequence of which the fluid moves more slowly near the wall relative to the isothermal profile. In the case of *heating*, the fluid near the wall is of relatively low viscosity and so moves faster than that of the isothermal case.

It is possible then to speculate on the effect of heating or cooling on the convection of heat. The cup-mixing temperature, defined as

$$\langle T \rangle = \int_0^R \frac{T(r, z)u_z(r)2\pi r \, dr}{\pi R^2 U} \tag{13-253}$$

will be greater if the velocity is reduced by cooling in the wall region because this is the region of maximum temperature reduction. Hence $\langle T \rangle$ is increased somewhat by the effect of cooling, and this would give the appearance of less effective heat transfer. The opposite holds for the effect of heating.

Christiansen and coworkers carried out numerical solutions of the equations of motion and energy for laminar flow of power law fluids in circular pipes of constant wall temperature. They write the temperature dependence of viscosity in such a way that the power law becomes

$$\tau = [K(\sqrt{\tfrac{1}{2}\mathrm{II}_\Delta})^{n-1} \Delta] \exp{(n \, \Delta E/R_g T)} \tag{13-254}$$

Note this is slightly different than the models introduced in previous discussions of temperature-dependent viscosity. The inclusion of n in the term with ΔE simply changes the meaning of ΔE. In the end one obtains parameters such as ΔE from viscosity-temperature data, and the theoretical or physical significance of the parameter is often of no concern.

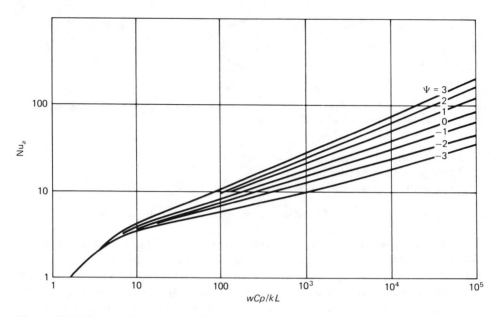

Figure 13-51 Theoretical calculation of the effect of a temperature-dependent viscosity on the average Nusselt number for power law flow ($n = \frac{1}{2}$) in a pipe with isothermal wall. (*After Christiansen et al.*)

Figure 13-51 shows a set of curves for the *arithmetic average* Nusselt number as calculated from the numerical solutions of Christiansen et al. The case $n = 0.5$ is shown; the original paper also gives such curves for $n = 0.1$, 0.3, and 1.0. The parameter Ψ is defined as

$$\Psi = \frac{\Delta E}{R_g T_w} \frac{T_w - T_0}{T_0} \tag{13-255}$$

We note that Ψ is positive for heating and negative for cooling. For $\Delta E / R_g T_w$ in the neighborhood of 10 (which is a typical value) the curves for Nu_a depend only on Ψ, so long as Ψ itself is not too large (say, $\Psi < 3$). As expected from our earlier qualitative interpretation of Fig. 13-50, Nu_a is reduced by cooling.

Figure 13-52 shows data obtained for cooling of a Carbopol solution ($n = 0.46$, $\Delta E = 4720$ cal/mol). The data fall between the appropriate isothermal model, Eq. (13-251), and the numerical solution for $\Psi = -1$ and are well described by the $\Psi = -1$ curve. Nu_a was calculated from Eq. (13-247).

Other boundary conditions All the solutions presented above are based on a boundary condition which assumes that the conduit wall is held at some fixed and uniform temperature. We call this the *isothermal-wall* boundary condition. It is

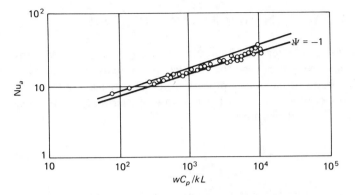

Figure 13-52 Data for cooling of a Carbopol solution. (*From Christiansen et al.*) Upper curve is Eq. (13-251); lower curve is from Fig. 13-51.

possible to operate a heat exchanger in such a way that the heat flux at the wall is fixed, as in the case of electrical heating with controlled power input. This changes the boundary condition at the conduit surface and alters the solutions somewhat.

For example, the Leveque solution for flow in a tube with constant heat flux at the wall gives a local Nusselt number in the form

$$\text{Nu} = 1.41 \left(\frac{3n + 1}{4n}\right)^{1/3} \left(\frac{wC_p}{kz}\right)^{1/3} \tag{13-256}$$

which may be compared to Eq. (13-235). If Eq. (13-235) is put in the same format as Eq. (13-256), we see (Prob. 13-34) that the Nusselt number is higher for the case of prescribed heat flux than for the case of prescribed temperature. This is true of the Graetz solutions, as well.

We will not dwell on this point further here, even though there is a significant difference between the solutions based on these two boundary conditions. Sufficient references are cited for further study if desired.

Viscous heat generation If viscous heat generation occurs while a fluid is exchanging heat across the conduit walls, the average temperature $\langle T \rangle$ will be increased somewhat, whether the fluid is being heated or cooled, relative to the value $\langle T \rangle$ would have in the absence of dissipation. If one were to calculate a Nusselt number from data subject to a dissipation effect, the Nusselt number would be too high in the case of heating and too low in the case of cooling. This point must be kept in mind in evaluating heat transfer data or in carrying out heat transfer calculations.

The viscous heat generation problem is important in its own right, outside the context of the calculation of heat transfer coefficients. For this reason we drop the point here, with the simple qualitative comments offered just above, and treat the viscous dissipation problem more fully in the next section.

13-8 HEAT GENERATION IN NONADIABATIC FLOWS

In several places throughout the text we have introduced the notion of viscous heat generation. In all cases we considered the simplest possible model: adiabatic flow. For flow through a conduit (such as a pipe or annulus) with stationary walls the adiabatic analysis just equates, through the first law of thermodynamics, the power input to the heat rise, and one finds

$$\Delta T = \frac{\Delta P}{\rho C_p} \tag{13-257}$$

Equation (13-257) is of value because it provides a rapid estimate of viscous heating effects. However, it can give a very misleading estimate in some respects. In the first place, it assumes adiabatic flow—no heat exchange across the boundaries of the conduit. If the conduit is insulated, or if the residence time is very short, then the adiabatic assumption may be reasonable. Under many conditions it is not a good approximation, as we shall see subsequently.

The second misleading feature is that ΔT given above is an *average* temperature. It is possible for the maximum temperature to be considerably in excess of this average value, even in adiabatic flow. This latter point is important if thermal degradation must be considered, since some of the fluid may be subjected to very high temperatures while the *average* temperature could be relatively small.

In this section we consider some features of viscous heat generation in more detail than earlier models. We illustrate only the case of laminar flow in a tube of uniform circular cross section. We begin, then, with the energy equation in the form [Eq. (13-33)] for a power law fluid:

$$\rho C_p u_z \frac{\partial T}{\partial z} = \frac{k}{r} \frac{\partial}{\partial r}\left(r \frac{\partial T}{\partial r}\right) + K\left(-\frac{\partial u_z}{\partial r}\right)^{n+1} \tag{13-258}$$

Equation (13-258) assumes that the thermal conductivity k is independent of temperature (note Fig. 13-2).

If K is taken as a function of temperature,

$$K = K_0 e^{-b(T - T_0)} \tag{13-259}$$

then the velocity field, assuming steady fully developed Poiseuille flow, must satisfy

$$\frac{\Delta P}{K_0 L} = \frac{1}{r} \frac{d}{dr}\left[r e^{-b(T - T_0)}\left(-\frac{du_z}{dr}\right)^n\right] \tag{13-260}$$

Equations (13-258) and (13-260) are coupled and not amenable to simple analytical solution. Several special cases have been worked out, however, which can provide useful information.

Fully developed temperature profile with K independent of temperature By *fully developed* we imply that the temperature profile becomes independent of z, so that $T = T(r)$ only. If K is independent of temperature the velocity field is already known from Chap. 5:

$$u_z = \frac{1 + 3n}{1 + n} U \left[1 - \left(\frac{r}{R} \right)^{1 + 1/n} \right] \tag{13-261}$$

Under these assumptions Eq. (13-258) takes the form (after a lot of algebra)

$$0 = \frac{1}{s} \frac{d}{ds} \left(s \frac{d\tilde{T}}{ds} \right) + s^{1 + 1/n} \tag{13-262}$$

where $s = r/R$ and $\tilde{T}$ may be defined in such a way as to remove all parameters from the problem:

$$\tilde{T} = \frac{T - T_0}{T_0 \, \text{Br}} \tag{13-263}$$

The Brinkman number is defined here as

$$\text{Br} = \frac{KR^2}{kT_0} \left(\frac{1 + 3n}{n} \frac{U}{R} \right)^{1 + n} \tag{13-264}$$

We need two boundary conditions on $T(r)$. One simply expresses symmetry about the axis:

$$\frac{d\tilde{T}}{ds} = 0 \qquad \text{at } s = 0$$

For the other boundary condition we assume the tube wall is isothermal, at temperature T_0, so that

$$\tilde{T} = 0 \qquad \text{at } s = 1$$

Equation (13-262) is easily solved to give

$$\tilde{T} = \frac{1 - s^q}{q^2} \tag{13-265}$$

where $q = 3 + 1/n$. Figure 13-53 shows $\tilde{T}(s)$ for $q = 5$ ($n = \frac{1}{2}$). The temperature profile is quite flat over the central core of the pipe, but falls off sharply as the isothermal wall is approached.

The cup-mixing (average) temperature may be calculated from

$$\langle \tilde{T} \rangle = \frac{\int_0^1 \tilde{T} u_z s \, ds}{\int_0^1 u_z s \, ds} \tag{13-266}$$

with the result

$$\langle \tilde{T} \rangle = \frac{1 + 4n}{1 + 5n} \left(\frac{n}{1 + 3n} \right)^2 \tag{13-267}$$

Figure 13-53 compares $\langle \tilde{T} \rangle$ for $n = \frac{1}{2}$ with the corresponding profile.

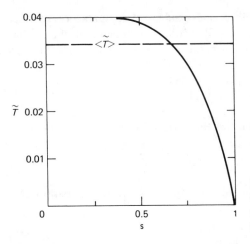

Figure 13-53 Temperature profile for fully developed conditions in power law flow through an isothermal pipe ($n = \frac{1}{2}$). The corresponding average $\langle \tilde{T} \rangle$ is shown as well.

Fully developed temperature profile with $K = K_0 e^{-b(T-T_0)}$ Now we wish to estimate the effect of a temperature-dependent viscosity on the dissipation. Although Eqs. (13-258) and (13-260) are nonlinear and coupled, an analytical solution is possible, as shown by Sukanek. However, the analytical solution is in a format which requires numerical calculations on the computer. Instead of showing the development of the equations, then, we cite Sukanek's paper as reference and show some calculations based on his work. (But see Prob. 13-41.)

Figure 13-54 shows $\langle \tilde{T} \rangle$ as a function of a parameter $\mathscr{B}$, related to the Brinkman number by

$$\mathscr{B} = bT_0 \, \text{Br} \qquad (13\text{-}268)$$

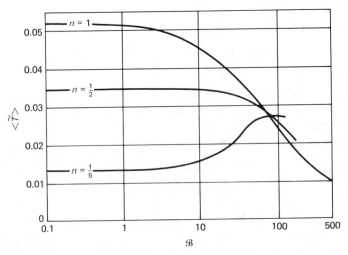

Figure 13-54 $\langle \tilde{T} \rangle$ as a function of $\mathscr{B}$, showing the effect of a temperature-dependent viscosity on the dissipation. Power law flow under fully developed conditions in an isothermal pipe.

We see that $\mathscr{B} \to 0$ for small Brinkman number or small values of bT_0. Thus we expect that the effect of temperature-dependent viscosity will vanish for small $\mathscr{B}$, and this is what is observed in Fig. 13-54. The left-hand asymptotes are identical with the values calculated from Eq. (13-267). Significant deviations from Eq. (13-267) do not occur until $\mathscr{B}$ exceeds unity, especially for the more strongly nonnewtonian fluids.

While Fig. 13-54 gives an indication of the effect of a temperature-dependent viscosity on viscous heat generation, it should be kept in mind that the model assumes a fully developed temperature profile. As in the case of development of the velocity profile, discussed in Chaps. 4 and 5, some finite axial length is required for such development to occur. In the case of the velocity field, with viscous fluids, the entry length was seen to be quite short, perhaps of the order of a few tube diameters at most. In the next section we see that the development of the thermal field requires a long entry length, which may in fact require thousands of tube diameters. In many practical systems the fully developed temperature field is never achieved.

Axial development of the temperature profile with K independent of temperature

The two previous models share the same assumption of fully developed temperature profiles. We know, however, that some finite axial length must be required for this to occur, and the analysis of this problem was given by Bird for the power law fluid. We go back to Eq. (13-258) and introduce the dimensionless radial variable s and again define $\tilde{T}$ by Eq. (13-263) with Brinkman number given by Eq. (13-264). To remove all parameters from the differential equation it is necessary to define an axial variable ζ as

$$\zeta = \frac{k(1+n)}{R^2 \rho C_p U(1+3n)} z \tag{13-269}$$

Equation (13-258) then takes the form

$$(1 - s^\alpha) \frac{\partial \tilde{T}}{\partial \zeta} = \frac{1}{s} \frac{\partial}{\partial s} \left(s \frac{\partial \tilde{T}}{\partial s} \right) + s^\alpha \tag{13-270}$$

where

$$\alpha = 1 + \frac{1}{n}$$

For the isothermal wall case the boundary conditions are, again,

$$\frac{\partial \tilde{T}}{\partial s} = 0 \qquad \text{at } s = 0$$

$$\tilde{T} = 0 \qquad \text{at } s = 1$$

Now an inlet condition on $\tilde{T}$ at $\zeta = 0$ is required, and we choose

$$\tilde{T} = 0 \qquad \text{at } \zeta = 0$$

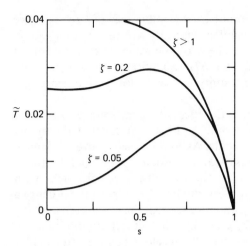

Figure 13-55 Temperature profiles at several axial positions. Power law flow ($n = \frac{1}{2}$) in a pipe with isothermal wall.

which states that the fluid enters the pipe at the temperature of the pipe wall. Since Eq. (13-270) is linear an analytical solution is possible, which Bird obtained as an infinite series.

The format of the solution is cumbersome to work with, and we present some relevant results in graphical form first. Figure 13-55 shows temperature profiles (for the case $n = \frac{1}{2}$) for several values of the axial position ζ. We see that before the profile is fully developed the temperature has a maximum value somewhere between the axis and the wall.

Figure 13-56 shows the cup-mixing average temperature as a function of axial position ζ. We can infer immediately from this figure that the models given earlier for fully developed temperature profiles cannot be valid unless ζ is larger than about 0.5. Thus we may define a *thermal entry length* z^* such that $\zeta^* = 0.5$, or [using Eq. (13-269)]

$$\frac{z^*}{R} = \frac{0.5(1 + 3n)}{1 + n} \frac{UR}{\alpha_T} \tag{13-271}$$

where the thermal diffusivity α_T has been introduced.

It is not difficult to see that under most conditions the thermal entry length is not exceeded. Since α_T for polymer melts is of the order of 10^{-4} cm^2/s, we may write Eq. (13-271), approximately, as

$$\frac{z^*}{R} = O\left(10^4 \frac{Q}{R}\right) \qquad \left(\frac{Q}{R} \text{ in cm}^2/\text{s}\right) \tag{13-272}$$

Since in many processing systems R is normally of the order of 1 cm or less and Q is not normally much less than 10^{-1} cm^3/s in magnitude, it is apparent that the thermal entry length is several orders of magnitude greater than the tube radius. Thus the fully developed solutions [such as Eq. (13-267)], while providing an easily calculated *upper* limit, may be *very* misleading, and Fig. 13-56 provides the more relevant information.

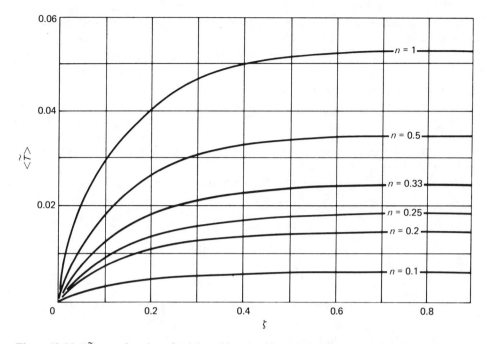

Figure 13-56 $\langle \tilde{T} \rangle$ as a function of axial position ζ, with n as a parameter, for power law flow in a pipe with isothermal wall.

Since the small ζ region is of interest, it may be useful to give the analytical solutions for the temperature profiles. A detailed development of the viscous heating problem, including the effects of *both* a temperature- and pressure-dependent viscosity, for a power law fluid is given in two papers by Takserman-Krozer and her coworkers (see Galili et al.). The details are too lengthy to present here, and even the simplest results occupy a fair amount of space. Unfortunately no calculations are made of the average temperature rise even under simplifying assumptions. We present here the solution for the temperature profile, ignoring, as a first approximation, the effect of viscosity variation. Calculations based on the analytical solutions are suggested in several problems at the end of this chapter.

For the case of the isothermal wall the solution is given as

$$\theta_{(\text{iso})}(s, \zeta) = G\left(\frac{1 - s^q}{q} + \sum_{i=1}^{\infty} B_i e^{-a_i \xi/G} \sum_{k=0}^{\infty} b_{ik} s^k\right) \tag{13-273}$$

where s and q are defined as before ($s = r/R$, $q = 3 + 1/n$). A dimensionless temperature θ is defined as

$$\theta = \frac{T - T_0}{\Delta P/\rho C_p} \tag{13-274}$$

Note that this normalizes the temperature rise to the adiabatic value [Eq. (13-257)].

The parameter G is similar to a Brinkman number in some respects and is defined as

$$G = \frac{\rho C_p Q}{2\pi k L} \tag{13-275}$$

The axial variable ξ is simply

$$\xi = \frac{z}{L} \tag{13-276}$$

The coefficients B_i, a_i, and b_{ik} depend upon n, and are given for the cases $n = 1$ and $n = \frac{1}{2}$ in Table 13-5.

In all cases the b_{ik} are given by the recursion formula

$$b_{ik} = -\frac{2qa_i}{(q-2)k^2}(b_{i,k-2} - b_{i,k-q})$$

$$b_{i0} = 1 \qquad \text{for all } i$$

$$b_{ij} = 0 \qquad \text{if } j < 0$$

$$\sum_{k=0}^{\infty} b_{ik} = 1 \text{ (isothermal case)} \qquad \sum_{k=0}^{\infty} k b_{ik} = 0 \text{ (adiabatic case)}$$

For the adiabatic boundary the solution is

$$\theta_{(\text{adi})} = \xi + G\left[\frac{1}{q-2}\left(\frac{q}{2}s^2 - s^q\right) + \sum_{i=1}^{\infty} B_i e^{-a_i\xi/G} \sum_{k=0}^{\infty} b_{ik} s^k\right] \tag{13-277}$$

Table 13-5 Coefficients for use in Eqs. (13-273) and (13-277)

i	Isothermal wall		Adiabatic wall	
	a_i	B_i	a_i	B_i
		$n = 1$		
1	1.828	−0.338	0	−0.250
2	11.15	0.126	6.42	0.320
3	28.48	−0.0580	20.97	−0.102
4	53.81	0.0331	43.54	0.0498
5	87.14	−0.0210	74.13	−0.0296
		$n = \frac{1}{2}$		
1	1.974	−0.279	0	−0.25
2	11.73	0.117	6.73	0.325
3	29.85	−0.0584	21.93	−0.109
4	56.34	0.0339	45.53	0.0537
5	91.19	−0.0222	77.52	−0.0320

From these basic solutions it is possible to calculate cup-mixing temperatures, and the results for the isothermal case should give Fig. 13-56 when the dimensionless variables are converted. In the entrance region the analytical solutions can be used to generate numerical results with more accuracy than can be obtained from Fig. 13-56. However, for small ξ/G, which is the situation of principal interest here, the series converges slowly, and for accuracy the summations indicated in the solutions must be carried out beyond the first few terms. It is best to simply program a computer to perform the requisite calculations, including those needed to calculate the b_{ik} coefficients. [The summation $\sum\limits_{k=0}^{\infty} b_{ik}\, s^k$ will ordinarily have to be taken out to 20 or 30 terms when s is near unity (the wall region).]

For adiabatic boundaries the cup-mixing value of $\theta_{(adi)}$ should agree with the first law of thermodynamics (see Probs. 13-47 and 13-48). It is of some interest to see how the *maximum* temperature behaves in the adiabatic case, and this may be found upon setting $s = 1$ in Eq. (13-277) (see Probs. 13-49 and 13-50).

A problem associated with the *convenience* of using Eq. (13-273) can be illustrated in the following example.

Example 13-12 A melt for which $K = 10^4$ dyne$\cdot$s$^{1/2}$/cm^2 and $n = \frac{1}{2}$ is injection molded through a runner of diameter 0.3 cm and length 7.5 cm at a rate of 20 cm^3/s. If the melt enters the runner at 250°C and the runner surface is at 250°C, estimate the average melt temperature leaving the runner. Take the thermal properties to be those of polypropylene (Fig. 13-3).

From Fig. 13-3 we find

$$k = 2.6 \times 10^{-4} \text{ cal/s} \cdot \text{cm} \cdot °\text{C}$$

$$\alpha_T = 5.25 \times 10^{-4} \text{ cm}^2/\text{s}$$

Since K is in mechanical units we convert k and find

$$k = 2.6 \times 10^{-4}(4.18 \times 10^7) = 1.1 \times 10^4 \text{ g} \cdot \text{cm/s}^3 \cdot °\text{C}$$

The Brinkman number is needed since

$$\langle T \rangle - T_0 = T_0 \text{ Br} \langle \tilde{T} \rangle$$

We find [Eq. (13-264)]

$$T_0 \text{ Br} = \frac{KR^2}{k}\left(\frac{1 + 3n}{n}\frac{U}{R}\right)^{1+n} = 1.9 \times 10^4 \text{ }°\text{C}$$

An accurate solution would have to come from Eq. (13-273) [and the corresponding expression for $\langle\theta\rangle_{\text{iso}}$ (Prob. 13-44)]. It is interesting to see what problems this raises. We begin by noting that

$$\frac{\xi}{G} = \frac{2(1 + 3n)}{1 + n}\zeta \tag{13-278}$$

so that $\xi/G = 1.2 \times 10^{-3}$. We can see, then, that the exponential terms in Eq. (13-273) are nearly unity at this small value of ξ/G, which means that the series converges very slowly.

From Fig. 13-56 we note that if the tube were long enough that the fully developed temperature field could be attained, the temperature rise would be predicted to be

$$T - T_0 = 0.034T_0 \text{ Br} = 646°C$$

This is a ridiculously high temperature rise, and the theory, which neglects the effect of temperature on viscosity, would be invalid.

But the tube is very short (thermally) since

$$\zeta = \frac{\alpha_T(1 + n)z}{R^2 U(1 + 3n)} = 3.7 \times 10^{-4}$$

From Fig. 13-56 it is apparent that $\langle \tilde{T} \rangle$ would be very small, probably less than 10^{-3}. Thus we might anticipate that the temperature rise would be less than

$$T - T_0 = 10^{-3}(1.9 \times 10^4) = 19°C$$

However, this estimate of $\langle \tilde{T} \rangle$ from the figure is very crude (it could be twice as large or 10 times smaller, since the figure is virtually unreadable in the "small-ζ corner.")

For very small ζ we may examine the limiting behavior of the series solution [Eq. (13-273)] and find, approximately, that the *average* temperature is given by

$$\langle \tilde{T} \rangle = 0.4\zeta \qquad \zeta < 0.05 \tag{13-279}$$

$$n = \tfrac{1}{2}$$

This allows us to estimate, for this specific example, that

$$\langle \tilde{T} \rangle \approx 1.5 \times 10^{-4}$$

or
$$T - T_0 = 2.8°C$$

It is important to take note of the enormous error that would have been made if the fully developed solution had been used (given above as 646°C).

For this example, the adiabatic assumption, using Eq. (13-257), gives $T - T_0 = 4.5°C$. As expected, the actual value is less than this.

A simple approximate solution for $\langle \tilde{T} \rangle$ If we examine Fig. 13-55 we make an observation that suggests an approximate solution in the entrance region. The observation is that as ζ gets small the maximum temperature moves toward the tube wall, and the profile becomes flat, except in the region of the maximum. The cup-mixing temperature is an average in which the temperature profile is weighted by the velocity profile, as in Eq. (13-266). For *very small* ζ we might suppose that the maximum temperature contributes relatively little to $\langle \tilde{T} \rangle$, since the velocity falls to zero as the tube wall is approached.

Let us develop a model based on the following idea: that in the central region of the tube the energy balance is primarily between the generation of heat and its axial convection. Except near the tube wall, where the radial temperature gradients are large, the conduction of heat is neglected. Then Eq. (13-270) becomes

$$(1 - s^a)\frac{\partial \tilde{T}}{\partial \zeta} = s^a \tag{13-280}$$

$$\tilde{T} = 0 \qquad \text{at } \zeta = 0$$

If we multiply both sides by $s \, ds$ and integrate over $0 \le s \le 1$, we find [noting Eq. (13-266)]

$$\frac{d\langle \tilde{T} \rangle}{d\zeta} = \frac{2}{\alpha}$$

or

$$\langle \tilde{T} \rangle = \frac{2}{\alpha}\zeta \qquad \left(\alpha = 1 + \frac{1}{n} \right) \tag{13-281}$$

We may evaluate this solution using the exact solution (Prob. 13-45) for $\langle \tilde{T} \rangle$. For the cases $n = 1$ and $\frac{1}{2}$ we find

$$\langle \tilde{T} \rangle = \begin{cases} 0.6\zeta & n = 1, \ \zeta < 0.05 \\ 0.4\zeta & n = \frac{1}{2}, \ \zeta < 0.05 \end{cases} \tag{13-282} \tag{13-283}$$

Equation (13-281) gives

$$\langle \tilde{T} \rangle = \begin{cases} \zeta & n = 1 \\ 0.67\zeta & n = \frac{1}{2} \end{cases} \tag{13-284} \tag{13-285}$$

Considering the simplicity of its derivation, we must regard Eq. (13-281) as a useful model. Thus, for arbitrary values of n, for which the coefficients of Table 13-5 have not been evaluated, we may use Eq. (13-281) as an approximate solution for $\langle \tilde{T} \rangle$.

Combined viscous heating and convective heating In previous models we have considered the effect of viscous heating separately from the effect of heating (or cooling) caused by the presence of a tube-wall temperature that might be different than the inlet temperature of the fluid. The "combined" problem could be formulated in the following way. The energy equation is

$$\rho C_p u_z \frac{\partial T}{\partial z} = k \frac{1}{r}\frac{\partial}{\partial r}\left(r\frac{\partial T}{\partial r} \right) + K\left(-\frac{\partial u_z}{\partial r} \right)^{1+n} \tag{13-286}$$

but *now* the boundary conditions are

$$T = \begin{cases} T_0 & \text{at } z = 0 \\ T_w & \text{at } r = R \end{cases}$$

$$\frac{\partial T}{\partial r} = 0 \qquad \text{at } r = 0$$

We may easily show that the solution of this problem is a linear combination of solutions already presented.

Let us define a dimensionless temperature as

$$\tilde{T}' = \frac{T - T_0}{T_w - T_0} \tag{13-287}$$

and a Brinkman number as

$$\text{Br}' = \frac{KR^2}{k(T_w - T_0)} \left(\frac{1 + 3n}{n} \frac{U}{R} \right)^{1+n} \tag{13-288}$$

The independent variables ζ and s are defined as before [Eq. (13-269) and $s = r/R$]. As a result we find (assuming a fully developed power law velocity profile)

$$(1 - s^\alpha) \frac{\partial \tilde{T}'}{\partial \zeta} = \frac{1}{s} \frac{\partial}{\partial s} \left(s \frac{\partial \tilde{T}'}{\partial s} \right) + \text{Br}' \, s^\alpha \tag{13-289}$$

The boundary conditions are

$$\tilde{T}' = \begin{cases} 0 & \text{at } \zeta = 0 \\ 1 & \text{at } s = 1 \end{cases}$$

$$\frac{\partial \tilde{T}'}{\partial s} = 0 \qquad \text{at } s = 0$$

For the case of negligible viscous heat generation in flow through a tube at constant wall temperature we have to solve

$$(1 - s^\alpha) \frac{\partial \tilde{T}_1}{\partial \zeta} = \frac{1}{s} \frac{\partial}{\partial s} \left(s \frac{\partial \tilde{T}_1}{\partial s} \right) \tag{13-290}$$

$$\tilde{T}_1 = \begin{cases} 0 & \text{at } \zeta = 0 \\ 1 & \text{at } s = 1 \end{cases}$$

$$\frac{\partial \tilde{T}_1}{\partial s} = 0 \qquad \text{at } s = 0$$

This differs from Eq. (13-218) in the definition of $\tilde{T}$ used there, as a result of which the boundary conditions are different. However, Fig. 13-47 may be used, since

$$\frac{T_1 - T_0}{T_w - T_0} = \tilde{T}_1 = 1 - \tilde{T} \tag{13-291}$$

The viscous heating problem with the isothermal-wall boundary condition has been solved above, but we reformulate it here in slightly different format:

$$(1 - s^\alpha)\frac{\partial \tilde{T}_2}{\partial \zeta} = \frac{1}{s}\frac{\partial}{\partial s}\left(s\frac{\partial \tilde{T}_2}{\partial s}\right) + \text{Br}' \, s^\alpha \tag{13-292}$$

$$\tilde{T}_2 = \begin{cases} 0 \text{ at } \zeta = 0 \\ 0 \text{ at } \zeta = 1 \end{cases}$$

$$\frac{\partial \tilde{T}_2}{\partial s} = 0 \qquad \text{at } s = 0$$

using

$$\tilde{T}_2 = \frac{T_2 - T_0}{T_w - T_0} \tag{13-293}$$

It is not difficult to see that $\tilde{T}_2/\text{Br}'$ is the solution of Eq. (13-270), for which Fig. 13-56 provides a graphical solution for $\langle \tilde{T}_2\rangle/\text{Br}'$. (For small ζ the analytical solutions discussed above must be used.)

Now, the relevant observation is that if we consider the sum $\tilde{T}_1 + \tilde{T}_2 = \tilde{T}_3$ we may easily verify that $\tilde{T}_3$ satisfies Eq. (13-289) *and* its boundary conditions. Hence

$$\tilde{T}' = \tilde{T}_3 = \tilde{T}_1 + \tilde{T}_2 \tag{13-294}$$

Thus we may evaluate $\langle \tilde{T}'\rangle$ from

$$\langle \tilde{T}'\rangle = \langle \tilde{T}_1\rangle + \text{Br}'\frac{\langle \tilde{T}_2\rangle}{\text{Br}'} \tag{13-295}$$

using the previously obtained solutions for $\langle \tilde{T}_1\rangle$ and $\langle \tilde{T}_2\rangle/\text{Br}'$.

Example 13-13 We repeat Example 13-12 except that the runner surface is taken to be 50°C.

For the viscous heating contribution we have found $\langle \tilde{T}\rangle = 1.5 \times 10^{-4}$, which is equivalent to (in the present notation)

$$\frac{\langle \tilde{T}_2\rangle}{\text{Br}'} = 1.5 \times 10^{-4}$$

In this example we find the Brinkman number defined by Eq. (13-288) is $\text{Br}' = -96$. Note that the Br' may be positive or negative, depending on the sign of $T_w - T_0$. For the contribution due to the cold wall we may use Fig. 13-47. Since ζ is so small we may approximate Eq. (13-221) by

$$\langle \tilde{T}\rangle = 1 - 14\zeta \qquad \zeta < 10^{-3} \tag{13-296}$$

Hence

$$\langle \tilde{T}_1\rangle = 1 - \langle \tilde{T}\rangle = 14\zeta$$

and we find $\langle \tilde{T}_1 \rangle = 5.2 \times 10^{-3}$. Putting these solutions together, we find

$$\langle \tilde{T}' \rangle = \langle \tilde{T}_1 \rangle + \mathrm{Br}' \frac{\langle \tilde{T}_2 \rangle}{\mathrm{Br}'}$$

$$= 5.2 \times 10^{-3} - 96(1.5)10^{-4}$$

$$= 5.2 \times 10^{-3} - 1.4 \times 10^{-2}$$

$$= -9.2 \times 10^{-3}$$

Thus

$$T - T_0 = (T_w - T_0)\langle \tilde{T}' \rangle = -200(-9.2)10^{-3} = 1.8°C$$

As expected, the cold tube wall offsets the effect of viscous heating, thereby reducing the mean temperature rise.

13-9 FREEZING AND MELTING OF POLYMERS

In many problems of interest the polymeric material undergoes a phase change. In injection molding, for example, the melt solidifies in the cavity after the filling stage. Filament solidification occurs during melt spinning. Often a simple model which ignores the effect of the heat of solidification can be used to estimate the behavior of such systems. We have already offered some examples of this in previous sections.

A more difficult problem is that of freezing *during* the filling process in injection molding. A solid layer grows in toward the center of runners and cavity, and the problem normally becomes one in at least two space variables as well as in time. Mathematically, then, such problems are quite complex, and simple models are not readily achieved. Several references to convective transport problems in the presence of solidification are given in the Bibliography.

An even more difficult problem is that which arises in the melting of granular polymeric solids. The development of a model for this type of process is central to the development of a model for the plasticating extruder. We will again have to be content with some references for further study, and the Bibliography contains several.

PROBLEMS

13-1 Show that for a simple shear flow the viscous dissipation G_v is proportional to II_Δ for purely viscous fluids.

13-2 For simple elongational flow of a purely viscous fluid, find G_v. Is G_v related to II_Δ in the same way as found in Prob. 13-1?

13-3 Formulate the equations and boundary conditions for the problem of solidification of a stationary melt confined to a cylindrical tube whose wall is at a constant temperature below the freezing temperature of the polymer. Do not attempt to solve the problem, but state how you would find the thickness of the frozen layer as a function of time, i.e., how the equations would be manipulated.

13-4 Give a derivation of Eq. (13-51).

13-5 Derive Eq. (13-122).

13-6 Derive Eq. (13-130).

13-7 With reference to Example 13-3, argue *intuitively* whether $T(z)$ (Fig. 13-18) will fall faster if U is increased. Then do the analysis and compare the conclusion with your intuition.

13-8 A theoretical paper by Rotte and Beek presents solutions for heat transfer coefficients to moving continuous surfaces, such as cylinders (fiber) or flat sheets (film). For their case 5 (flat plate of infinite heat capacity in a fluid of finite Prandtl number) they give

$$\frac{\sqrt{\pi}\, Nu_x}{\sqrt{Pe_x}} = f_5(Pr) \qquad \text{where } Pe_x = \frac{Ux}{\alpha_T}$$

From their fig. 2, at a Prandtl number of 0.72 (for air), one finds $f_5(0.72) = 0.745$.

Compare this prediction with that given from the use of Fig. 13-17, based on the theory of Shih and Middleman. Take note of differences in notation.

13-9 Derive Eq. (13-145) and its boundary conditions.

13-10 Derive Eq. (13-151).

13-11 Wanger gives the data, shown in Fig. 13-57, for melt spinning of polypropylene fibers. In both cases the initial melt temperature is 260°C, and the ambient air temperature is 25°C. Assess the ability of the simple models presented in Sec. 13-6 to predict temperature along the filament.

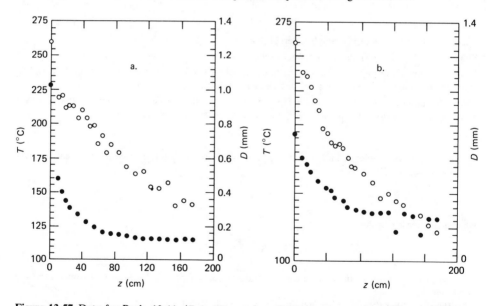

Figure 13-57 Data for Prob. 13-11. (*From Wanger.*) ○ Surface temperature; ● filament diameter. (*a*) $\pi R^2 \rho U = 2$ g/min; $U_L = 200$ m/min. (*b*) $\pi R^2 \rho U = 4$ g/min; $U_L = 100$ m/min.

13-12 The data shown in Fig. 13-58 were obtained by Wilhelm and replotted by Morrison. Force Eq. (13-152) to fit the data, and estimate thereby the value of the heat transfer coefficient. Is this heat transfer coefficient predictable on the basis of the model presented in Sec. 13-6?

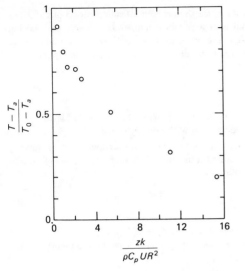

Figure 13-58 Data for Prob. 13-12.

The data correspond to the following conditions:

$$\pi R^2 \rho U = 2 \text{ g/min} \qquad T_0 = 290°\text{C}$$

$$U_0 = 800 \text{ m/min} \qquad T_a = 30°\text{C}$$

$$D_0 = 0.75 \text{ mm}$$

Use thermal properties of polyethylene terephthalate.

13-13 Verify Eq. (13-165) and the comment regarding n.

13-14 Continue Example 13-5 and plot the thickness of the solid polymer as a function of time. Define and give the value of the time at which the runner is frozen.

13-15 The boundary condition on Eq. (13-155) is written in a general notation so that a single equation may be written to stand for *both* heat and mass transport phenomena:

$$-\tilde{\nabla} X = \text{Sh} \ (X - X_a)$$

This creates some problems, and a source of confusion, in the notation used for the heat or mass transfer coefficient. For heat transfer

$$\text{Sh} = \text{Nu} = \frac{hL}{k}$$

whereas for mass transfer

$$\text{Sh} = \frac{k_x L}{\mathcal{D}}$$

(*a*) Give the units of the convective heat transfer coefficient h and the convective mass transfer coefficient k_x.

(*b*) If the Sherwood number is written in a generalized form as

$$\text{Sh} = \frac{h'L}{\alpha}$$

what are the units of the generalized convective coefficient h'?

(*c*) For the case of heat transfer, how is h' related to h?

13-16 Show that the single boundary condition

$$-\tilde{\nabla}X = \text{Sh}(X - X_a)$$

includes the special cases of

    (a) The adiabatic or inpenetrable boundary.

    (b) The boundary maintained at the ambient condition $X = X_a$.

    Give the values of Sh for cases (a) and (b).

13-17 Polystyrene is injection molded in a center-gated disk mold. The runner is $\frac{1}{8}$-in in diameter; the disk is $\frac{1}{4}$-in thick and 6 in in diameter. Assume that at some instant of time the melt is 170°C and the mold and runner surfaces are suddenly brought to a temperature of 50°C. Find the time required for the runner to freeze. Do the same for the disk.

13-18 Derive Eq. (13-165) by starting with Eqs. (13-158a) to (13-158c) (see Table 13-3) and taking the limit as Sh → 0. In doing so, establish a criterion for use of Eq. (13-165) in terms of Sh.

13-19 A strand of molten polyethylene is extruded into a water bath where it solidifies. In designing the bath it is necessary to have an estimate of a sufficient contact time for the process. Assume the polymer enters the bath at 170°C and that the bath is at 30°C. Estimate a contact time for strands of two diameters: $\frac{1}{8}$ in and $\frac{1}{16}$ in.

13-20 Begin with Eq. (13-186) as a transformation of variables, and show that an ordinary differential equation results from Eq. (13-183). Solve the equation to derive Eq. (13-185).

13-21 Derive Eq. (13-189).

13-22 Derive Eq. (13-194), and give the general result for a power law fluid.

13-23 Go back to Example 13-7 and carry through the calculations needed to determine the required length of the heat exchanger.

13-24 Suppose, in Prob. 13-23, the stream was split in two, and 750 lb/h of fluid was put through each of two parallel heat exchangers. What would be the required length? Is the short-time solution more or less applicable to this case than to that of the original example?

13-25 Repeat Prob. 13-23, but assume that the scrapers are not operating, so that no periodic mixing occurs. In what way do the scrapers enhance the efficiency of the system?

13-26 Beginning with the general transport equation, derive Eqs. (13-213) and (13-218).

13-27 Derive Eq. (13-217), and in doing so give the definition of Nu that appears in it.

13-28 Why does the model that leads to Eq. (13-236) fail to give a meaningful result in the limit of $n \to 0$?

13-29 For heat transfer in tube flow we have given the Graetz solution [Eq. (13-221)], useful for "large" z, and the Leveque solution, useful for "small" z.

    (a) Give the criteria for accurate and convenient use of the two solutions.

    (b) Molten Nylon 66 at 550°F is injected through a $\frac{1}{16}$-in-diameter runner of length 4 in. The flow rate is 4 in³/s. The runner surface is held at 125°F. Would the Graetz or Leveque model be more appropriate to this flow?

    (c) Estimate the temperature change of the melt leaving the runner.

13-30 For newtonian flow in a pipe with uniformly heated surface, plot the local Nusselt number, as well as $\overline{\text{Nu}}$ [Eq. (13-241)] and Nu_a [Eq. (13-247)] as a function of distance from the entrance to the pipe.

13-31 Develop a criterion for adiabatic flow in a pipe in terms of a ratio of the rate of heat transferred across the wall to the rate of flow of heat past any cross section. Use the Graetz solution given in Eq. (13-221).

13-32 Polypropylene is injection molded into a 32-oz cavity at a fill time of 1 s. The sprue-runner system may be considered to be a capillary of length 2 in and diameter $\frac{1}{8}$ in. Can the flow be considered adiabatic prior to the cavity?

13-33 Bassett and Welty give the data shown in Fig. 13-59 for heat transfer to polymer solutions flowing through a uniformly heated (constant flux) pipe. For the solutions used, n ranged from about 0.33 to 0.67. Compare the data to the simple model given by the Leveque solution (13-256).

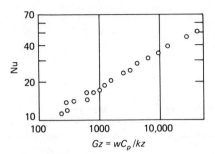

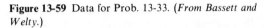

$Gz = wC_p/kz$

Figure 13-59 Data for Prob. 13-33. (*From Bassett and Welty.*)

13-34 Rearrange Eq. (13-235) so that it may be compared to Eq. (13-256).

13-35 Find Nu_a for the case of pipe flow with prescribed flux [Eq. (13-256)] and compare it to Eq. (13-251).

13-36 Show that in the case of flow through a conduit with constant heat flux, the cup-mixing average temperature in the fluid changes linearly with distance down the axis.

13-37 Carry out a rough first design of a heat exchanger for molten polyethylene. Use Fig. 13-46 for the viscometric data. Design under the following constraints:

We wish to lower the temperature of the melt from 400°F to 300°F.
The melt will be pumped through an exchanger consisting of parallel pipes connected to a common header at the inlet and outlet.
The pressure drop across the exchanger must be less than 300 psi.
The exchanger pipes are immersed in a very efficient heat exchange fluid maintained at 100°F.
The total flow rate is to be 1000 lb/h of polymer.

Specify the number of parallel pipes, their diameter, and their length.

Someone suggests that, instead of using a refrigeration unit to keep the cooling fluid at 100°F, ice be added periodically to the cooling fluid. At what rate must ice be supplied? (It is unfair to ask a sophomore chemical engineering student for help on this part.)

13-38 Give a careful derivation of Eq. (13-260) so that all assumptions can be stated.

13-39 Derive Eq. (13-262) and give the solutions for isothermal wall: Eq. (13-265) for $\tilde{T}$ and Eq. (13-267) for $\langle \tilde{T} \rangle$.

13-40 Give the solution of Eq. (13-262) for the case of adiabatic wall, for which the boundary condition at $s = 1$ is $\partial \tilde{T}/\partial s = 0$. Explain the result, or lack thereof. (This is another example of "throwing out the baby with the bathwater.")

13-41 Consider the case of fully developed temperature profile $[T \neq T(z)]$ with K a function of temperature according to Eq. (13-259). Make Eqs. (13-258) and (13-259) (and the appropriate boundary conditions for an isothermal wall) dimensionless and show that with an appropriate choice of dimensionless temperature the parameter $\mathscr{B}$ [Eq. (13-268)] enters the problem in a natural way.

13-42 Polyethylene is injection molded through a runner of length 6 in and diameter $\frac{1}{8}$ in at a rate of 5 oz/s. Take the temperature of the melt at the entrance to the runner to be 400°F, and assume the runner wall is at the same temperature. Estimate the extent of viscous heating through the runner. Use Fig. 13-46 for rheological data. Compare your estimate with the adiabatic temperature rise [Eq. (13-257)].

13-43 Give the relationship of θ [defined by Eq. (13-274)] to $\tilde{T}$ [Eq. (13-263)]. How are Br and G related?

13-44 Using the isothermal velocity profile (13-261), give the analytical form for the cup-mixing value of $\theta_{(iso)}$, using Eq. (13-273).

13-45 Using the result of Prob. 13-44 and the coefficients in Table 13-5, calculate $\langle \theta_{iso} \rangle$ as a function of ξ for $n = 1$ and $\frac{1}{2}$. Convert to $\langle \tilde{T} \rangle$ versus ζ, and compare with Fig. 13-56.

13-46 Using the isothermal velocity profile (13-261), give the analytical form for the cup-mixing value of $\theta_{(adi)}$, using Eq. (13-277).

13-47 Show that $\langle \theta_{(adi)} \rangle$, as obtained in Prob. 13-46, reduces to the expected value for the limit of $z \to \infty$.

13-48 Should $\langle \theta_{(adi)} \rangle$ differ from unity at finite values of z? Verify your answer by examining the result of Prob. 13-46, and comment on the result.

13-49 Prove that the maximum value of $\theta_{(adi)}$ [Eq. (13-277)] occurs at the tube wall.

13-50 For adiabatic flow in a tube, to what degree does the maximum temperature rise due to viscous heating, the wall value, exceed the value predicted from Eq. (13-257)?

13-51 Consider the possibility of a Leveque-type of solution to Eq. (13-270). What boundary condition is used for the core region (far from the tube wall)?

13-52 Repeat part *c* of Prob. 13-29, and include the effect of viscous heating.

BIBLIOGRAPHY

An excellent general reference for this chapter is

Bird, R. B., W. E. Stewart, and E. N. Lightfoot: "Transport Phenomena," John Wiley & Sons, Inc., New York, 1960.

13-2 Constitutive equations for diffusion

Shoulberg, R. H.: The Thermal Diffusivity of Polymer Melts, *J. Appl. Polym. Sci.*, **7**: 1597 (1963).
Crank, J., and G. S. Park (eds.): "Diffusion in Polymers," Academic Press, Inc., New York, 1968.
Boss, B. D., E. O. Stejskal, and J. D. Ferry: Self-Diffusion in High Molecular Weight Polyisobutylene-Benzene Mixtures, *J. Phys. Chem.*, **71**: 1501 (1967).
Duda, J. L., and J. S. Vrentas: Diffusion in Atactic Polystyrene above the Glass Transition Point, *J. Polym. Sci.*, *A-2*, **6**: 675 (1968).
Paul, D. R., V. Mavichak, and D. R. Kemp: Diffusion in Concentrated Polystyrene Solutions, *J. Appl. Polym. Sci.*, **15**: 1553 (1971).
Secor, R. M.: The Effect of Concentration on Diffusion Coefficient in Polymer Solutions, *AIChE J.*, **11**: 452 (1965).
Vanderkoui, W. N., M. W. Long, and R. A. Mock: The Concentration-dependent Diffusion of Styrene in Ethyl Cellulose, *J. Polym. Sci.*, **56**: 57 (1962).

A series of papers by Gainer provides a review of some principles of diffusion theory from a mechanistic view and leads to reasonably successful predictive methods.

Gainer, J. L., and A. B. Metzner: Diffusion in Liquids—Theoretical Analysis and Experimental Verification, *AIChE–Inst. Chem. Eng. Symp. Ser. No. 6*, p. 74 (1965).
Li, S. U., and J. L. Gainer: Diffusion in Polymer Solutions, *Ind. Eng. Chem. Fund.*, **7**: 433 (1968).
Navari, R. M., J. L. Gainer, and K. R. Hall: A Predictive Theory for Diffusion in Polymer and Protein Solutions, *AIChE J.*, **17**: 1028 (1971).

13-5 Dimensional analysis

The tubular reactor problem is treated numerically in

Wallis, J. P. A., R. A. Ritter, and H. Andre: Continuous Production of Polystyrene in a Tubular Reactor: Part II, *AIChE J.*, **21**: 691 (1975).

See also

Cintron-Cordero, R., R. A. Mostello, and J. A. Biesenberger: Reactor Dynamics and Molecular Weight Distributions: Some Aspects of Continuous Polymerization in Tubular Reactors, *Can. J. Ch. E.*, **46:** 434 (1968).

13-6 Convective transport coefficients

The analysis in the subsection "Transport Coefficients for a Moving Film" follows that given in chap. 19.3 of the Bird et al. reference. The effect of a high mass flux, due to evaporation of solvent from an extruded film, is analyzed in

Shih, N. C., and S. Middleman: Post Extrusion Heat and Solvent Transfer from Polymeric Film, *Polym. Eng. Sci.*, **10:** 4 (1970).

Theoretical treatments of the moving cylinder include

Sakiadis, B. C.: Boundary-Layer Behavior on Continuous Solid Surfaces, *AIChE J.*, **7:** 467 (1961).
Vasudevan, G., and S. Middleman: Momentum, Heat, and Mass Transfer to a Continuous Cylindrical Surface in Axial Motion, *AIChE J.*, **16:** 614 (1970).

This paper has a basic error in it, which was pointed out in

Fox, V. G., and F. Hagin: Similarity Transformations for Continuous Cylindrical Surfaces in Axial Motion, *AIChE J.*, **17:** 1014 (1971).
Rotte, J. W., and W. J. Beek: Some Models for the Calculation of Heat Transfer Coefficients to a Moving Continuous Cylinder, *Chem. Eng. Sci.*, **24:** 705 (1969).
Bourne, D. E., and D. G. Elliston: Heat Transfer through the Axially Symmetric Boundary Layer on a Moving Circular Fibre, *Int. J. Heat Mass Transfer*, **13:** 583 (1970).

Data on cooling of filaments may be found in

——— and H. Dixon: The Cooling of Fibres in the Formation Process, *Int. J. Heat Mass Transfer*, **14:** 1323 (1971).
Hill, J. W., and J. A. Cuculo: An Experimental Study of Threadline Dynamics, *J. Appl. Polym. Sci.*, **18:** 2569 (1974).
Lin, L. C. T., and J. Hauenstein: Cooling and Attenuation of a Threadline in Melt Spinning of Poly(ethylene Terephthalate), *J. Appl. Polym. Sci.*, **18:** 3509 (1974).
Wanger, W. H., Jr.: "Cooling of a Polymer Filament during Melt-Spinning," Ph.D. thesis, University of Denver, 1969.

The latter also gives numerical solutions of the energy equation for the fiber, as does

Morrison, M. E.: Numerical Evaluation of Temperature Profiles and Interface Position in Filaments Undergoing Solidification, *AIChE J.*, **16:** 57 (1970).

A simple theory and a significant amount of data are in

Wilhelm, G.: Die Abkühlung eines aus der Schmelze gesponnenen polymeren Fadens im Spinnschacht, *Kolloid Z.Z. Polym.*, **208:** 97 (1966).

Data on cooling of nonpolymeric liquid jets are available in

Kaplan, R. D., and L. H. Shendalman: Heat Transfer to a Cylindrical Laminar Liquid Jet Ejecting into a Gas, *Int. J. Heat Mass Transfer*, **16:** 1231 (1973).

13-7 Simple transport models

The reader with no prior background in transport phenomena or heat or mass transfer, may find it necessary to do some peripheral reading in the Bird et al. references, especially chaps. 11 and 19.

An example of some practical aspects of design of film-devolatilization equipment is in

Widmer, F.: Behavior of Viscous Polymers during Solvent Stripping or Reaction in an Agitated Thin Film, chap. 4 in *Adv. Chem. Ser. 128*, 51 (1973).

A variety of problems associated with heat transfer in confined flows is considered in

Vlachopoulos, J., and C. K. J. Keung: Heat Transfer to a Power-Law Fluid Flowing between Parallel Plates, *AIChE J.*, **18**: 1272 (1972).

Lyche, B. C., and R. B. Bird: The Graetz-Nusselt Problem for a Power-Law Non-Newtonian Fluid, *Chem. Eng. Sci.*, **6**: 35 (1956).

Griskey, R. G., and I. A. Wiehe: Heat Transfer to Molten Flowing Polymers, *AIChE J.*, **12**: 308 (1966).

Bassett, C. E., and J. R. Welty: Non-Newtonian Heat Transfer in the Thermal Entrance Region of Uniformly Heated, Horizontal Pipes, *AIChE J.*, **21**: 699 (1975).

Christiansen, E. B., G. E. Jensen, and F.-S. Tao: Laminar Flow Heat Transfer, *AIChE J.*, **12**: 196 (1966).

Mitsuishi, N., and O. Miyatake: Heat Transfer with Non-Newtonian Laminar Flow in a Tube Having a Constant Wall Heat Flux, *Intern. Chem. Eng.*, **9**: 352 (1969).

Matsuhisa, S., and R. B. Bird: Analytical and Numerical Solutions for Laminar Flow of the Non-Newtonian Ellis Fluid, *AIChE J.*, **11**: 588 (1965).

Mahalingam, R., L. O. Tilton, and J. M. Coulson: Heat Transfer in Laminar Flow of Non-Newtonian Fluids, *Chem. Eng. Sci.*, **30**: 921 (1975).

———, S. F. Chan, and J. M. Coulson: Laminar Pseudoplastic Flow Heat Transfer with Prescribed Wall Heat Flux, *Chem. Eng. J.*, **9**: 161 (1975).

Kannan, R., and M. R. Rao: Heat Transfer to Pseudoplastic Solutions in Horizontal Circular Tubes, *Indian J. Technol.*, **11**: 193 (1973).

13-8 Heat generation in nonadiabatic flows

Sukanek, P. C.: Poiseuille Flow of a Power-Law Fluid with Viscous Heating, *Chem. Eng. Sci.*, **26**: 1775 (1971).

Bird, R. B.: Viscous Heat Effects in Extrusion of Molten Plastics, *Soc. Plast. Eng. J.*, **11**: 35 (1955).

Galili, N., R. Takserman-Krozer, and Z. Rigbi: Heat and Pressure Effects in Viscous Flow through a Pipe, *Rheol. Acta*, **14**: 550, 816 (1975).

Cox, H. W., and C. W. Macosko: Viscous Dissipation in Die Flows, *AIChE J.*, **20**: 785 (1974).

13-9 Freezing and melting of polymers

A review paper with bibliography is

Muehlbauer, J. C., and J. E. Sunderland: Heat Conduction with Freezing or Melting, *Appl. Mech. Rev.*, **18**: 951 (1965).

Two papers on freezing in tube flow are

Zerkle, R. D., and J. W. Sunderland: The Effect of Liquid Solidification in a Tube upon Laminar-Flow Heat Transfer and Pressure Drop, *J. Heat Transfer*, **90**: 183 (1968).

Ozisik, M. N., and J. C. Mulligan: Transient Freezing of Liquids in Forced Flow Inside Circular Tubes, *J. Heat Transfer*, **91**: 385 (1969).

Models of melting of granular polymers are discussed in

Vermeulen, J. R., P. Gerson, and W. J. Beek: The Melting of a Bed of Polymer Granules on a Hot Moving Surface, *Chem. Eng. Sci.*, **26:** 1445 (1971).

———, P. G. Scargo, and W. J. Beek: The Melting of a Crystalline Polymer in a Screw Extruder, *ibid*, 1457.

Pearson, J. R. A.: On the Melting of Solids near a Hot Moving Interface, with Particular Reference to Beds of Granular Polymers, *Int. J. Heat Mass Transfer*, **19:** 405 (1976).

See also the references to plasticating extrusion in the Bibliography to Chap. 6.

Specific processing applications are discussed in

Gutfinger, C., E. Broyer, and Z. Tadmor: Melt Solidification in Polymer Processing, *Polym. Eng. Sci.*, **15:** 515 (1975).

FOURTEEN

ELASTIC PHENOMENA

If not now, when?

Hillel

In previous chapters we found that modeling of many processing flows was adequately carried out using purely viscous constitutive equations. This is equivalent to the notion that viscous effects, particularly shear viscosity, dominate the response of polymeric fluids in typical processes. Thus one did not need to introduce elasticity in developing models for these flows.

One does not always get away with this degree of simplification, for elastic phenomena do indeed occur in most processes and can sometimes play the *dominant* role in determining process operation. In this chapter we consider some aspects of polymer processing in which elasticity plays a major role.

We have already considered some examples where elastic effects are significant. These include

- Coating flows (Chap. 8)
- Fiber spinning (Chap. 9)
- Stirred-tank mixing (Chap. 12)

In this chapter we confine our attention to two elastic phenomena which affect the extrudate from a die.

14-1 DIE SWELL

It is well known that if a viscoelastic fluid is extruded from a die into air without subsequent drawing, the cross-sectional area of the extrudate will exceed that of the die exit, under most conditions of any interest. This phenomenon is usually called *die swell*. In the most common case, that of the circular die, diameter ratios (extrudate to die) in the range of 2 to 3 are often observed.

There seems to be general agreement that die swell is an elastic stress relaxation phenomenon. However, no single theory of die swell seems to be generally accepted, each theory being based on some assumption regarding the effect of stress relaxation on the dynamics of the extruded jet. A complication lies in the appearance, in several die swell theories, of the *recoverable shear* S_R, which we have defined as [see Eq. (3-156)]

$$S_R = \frac{\tau_{11} - \tau_{22}}{2\tau_{12}} \tag{14-1}$$

This ratio of the primary normal stress difference to the shear stress, the stresses being calculated at the same shear rate in a steady simple shear flow, is not always available to accompany die swell data. Consequently the comparison of a die swell theory to experimental data often involves an uncertainty in the value of S_R itself. Finally, we note that melt extrusion is usually into a cooler ambient medium, and it is possible that cooling of the extrudate can retard die swell. Unless the extrudate is collected and subsequently annealed† to complete the relaxation process, incorrect values of the true die swell are obtained. We comment further on this particular point below.

For the sake of providing some perspective we should note that newtonian fluids can exhibit die swell, with a maximum value observed to be about $D_j/D_0 \equiv \chi = 1.12$. Figure 14-1 shows experimental data which indicate that χ may take on values between 1.12 and 0.87, depending on the magnitude of the Reynolds number.

---

† Annealing is achieved by returning the sample to its original melt temperature for some time.

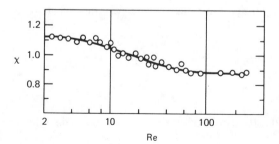

Figure 14-1 Data on die swell for several newtonian fluids extruded from a long capillary into air. (*From Middleman and Gavis.*)

The lower value of $\chi = 0.87$ can be predicted by theory, using a very simple combination of mass and momentum balances. The mass balance takes the form

$$\int_0^{R_0} \rho u_z(r) 2\pi r \, dr = \rho U_j \frac{\pi D_j^2}{4} \tag{14-2}$$

We are simply equating the mass flow out of the capillary to the mass flow downstream, at a point where the velocity field has flattened out to the constant value U_j. If the velocity profile in the capillary is taken to be parabolic, so that

$$u_z(r) = 2U \left[1 - \left(\frac{r}{R_0} \right)^2 \right] \tag{14-3}$$

then we find, assuming constant density,

$$U_j D_j^2 = U D_0^2 \tag{14-4}$$

[In fact, Eq. (14-4) holds for any velocity field so long as U is understood to be the average velocity.]

The momentum balance assumes that no external forces of any kind act on the jet between the capillary exit and some downstream position. Among the potential forces that could act, but are assumed not to do so to any significant extent, are surface tension, gravity, and ambient air resistance. The only momentum terms, then, arise from the flow itself, and conservation of momentum equates the two rates of flow of momentum in the form

$$\rho U_j^2 \frac{\pi D_j^2}{4} = \int_0^{R_0} \rho u_z^2(r) 2\pi r \, dr \tag{14-5}$$

If Eq. (14-3) is again used we find

$$U_j^2 D_j^2 = \tfrac{4}{3} U^2 D_0^2 \tag{14-6}$$

Using Eq. (14-4) to eliminate the velocities we obtain the observed result, namely,

$$\chi \equiv \frac{D_j}{D_0} = \frac{\sqrt{3}}{2} = 0.87 \tag{14-7}$$

If the fluid is assumed to follow the power law one can carry through an identical analysis and find

$$\chi \equiv \frac{D_j}{D_0} = \left(\frac{2n + 1}{3n + 1} \right)^{1/2} \tag{14-8}$$

This approach to the "die swell" problem (the quotes reflecting the fact that the "swelling" is really a contraction) is one which emphasizes the role of *convection*, there being no other mechanism for momentum transfer allowed in this model. For large Reynolds numbers (but still small enough that the flow is laminar) inertial, i.e. convective, effects dominate, and the theory is in complete agreement with experience.

At the other extreme, in the limit of very small Reynolds numbers, viscous effects become significant and must be accounted for in the momentum balance. Physically, the role of viscous effects is to cause the velocity field at, and near, the capillary exit to deviate from the fully developed profile. Tanner has solved the dynamic and continuity equations for a newtonian fluid in the neighborhood of the capillary exit by a numerical technique, and, in the limit of low Reynolds numbers, predicts a die swell of $\chi = 1.13$, in good agreement with data on newtonian fluids. Thus the newtonian flow problem is well understood, and the theory is in good agreement with data at both low and high Reynolds numbers.

Polymer *solutions* are observed to show a die swell which can exceed the upper newtonian limit of $\chi = 1.13$. This result is presumed to arise from the normal stresses generated by the shear flow within the capillary. The analysis can proceed along the lines of the momentum balance method, and the result, as given by Metzner et al., may be put in the form

$$S_R \equiv \left(\frac{\tau_{11} - \tau_{22}}{2\tau_{12}}\right)_{L, R_0} = \frac{1}{f}\left[\frac{1+n}{n}\frac{1+3n}{1+2n} - \frac{1}{n\chi^2}\left(1 + n - \frac{d \log \chi}{d \log \dot{\gamma}_N}\right)\right] \quad (14\text{-}9)$$

where f is the friction factor for the (assumed) fully developed flow within the capillary,

$$f = \frac{\tau_{12}}{\frac{1}{2}\rho U^2} \quad (14\text{-}10)$$

τ_{12} is the shear stress at the capillary wall, and $\dot{\gamma}_N$ is the *nominal* wall shear rate,

$$\dot{\gamma}_N = \frac{8U}{D_0} \quad (14\text{-}11)$$

In the derivation of Eq. (14-9) it is assumed that the velocity profile *right at the capillary exit* is that for fully developed power law flow. It is also assumed that the secondary normal stress difference $(\tau_{22} - \tau_{33})$ is small in magnitude relative to the primary normal stress difference. Finally, an assumption is made that the pressure within the fluid, on the axis at $r = 0$, right at the capillary exit, is equal to the external (ambient) pressure just outside the capillary.

Experiments with polymer *solutions* give fairly good agreement between normal stresses calculated from die swell, using Eq. (14-9), and normal stresses measured with conventional rheological instrumentation. A problem in making such a comparison lies in the fact that capillary flows of solutions which produce a *jet*, rather than a series of drips, are at shear rates of the order of 10^2 to 10^3 s^{-1} or higher, while standard instrumentation such as the cone-and-plate or torsional flow devices do not give reliable normal stress data in solutions at such high shear rates. Hence a "comparison" such as suggested above usually involves extrapolation over an interval of shear rate of more than one decade. Still, it would appear that order-of-magnitude agreement is achieved, indicating that the theory leading to Eq. (14-9) is basically correct, *when applied to solutions*.

Difficulty arises when one tries to use Eq. (14-9) for polymer melts. Graessley et al. show, for a polystyrene melt, that Eq. (14-9) underestimates the normal stresses by factors of as much as 10^7. This is such an intolerable error that one must conclude that the momentum balance method is not applicable to melts. Of course, the momentum balance must hold in principle; the assumptions made in deriving the specific form used [Eq. (14-9)] must be incorrect.

It seems likely that the incorrect assumption is with respect to conditions just within the capillary, near the exit. As in the case of the viscous newtonian problem, for which Tanner's theoretical analysis shows that the behavior of the free jet can perturb the flow and stress fields within the capillary near the exit, we expect also that the *elastic* fluid jet will perturb the dynamics of the exit region. On the grounds that the exit effects will complicate any momentum analysis to the point of gross inaccuracy, a completely different approach to the die swell problem has been presented by several investigators of this phenomenon.

The basic idea is to think of die swell as a problem in elastic recovery. The flow field within the capillary gives rise to stresses which strain the fluid with respect to some reference state. On leaving the capillary the fluid is supposed to "recover" and return to its reference state. A cylindrical element of fluid of diameter D_0 and length L_0 swells to a diameter D_j and a shorter length L_j. The several models of die swell which have been proposed all attempt to calculate the stress required to take the swollen cylinder of diameter D_j, as if it were an elastic solid, and "stretch" it axially so that it has the smaller capillary diameter D_0. This stress must then be related in some way to the elastic stresses developed within the fluid just prior to its ejection from the capillary.

We will illustrate here the theory due to Tanner, which may be written in the form

$$\chi = (1 + \tfrac{1}{2}S_R^2)^{1/6} \qquad (14\text{-}12)$$

We make no attempt to use Eq. (14-12) for values of χ below $\chi = 1.2$, since this approaches the region where purely viscous effects might be expected to give rise to a die swell of nearly this magnitude. In this regard we note that Tanner suggests replacing χ by $(\chi - 0.1)$, in using Eq. (14-12), as a means of compensating for the viscous contribution to die swell.

An important point to notice is the sensitivity of χ to S_R. Small changes in S_R cause even smaller changes in χ while, inversely, small changes in χ give rise to relatively large changes in S_R. For large values of χ a good approximation is $\chi \approx S_R^{1/3}$. Thus we have

$$\frac{d \log \chi}{d \log S_R} \approx \frac{1}{3} \qquad (14\text{-}13)$$

from which we see that the relative change in χ is one-third that in S_R. Actually, for moderate values of χ, say, $1.2 < \chi < 1.6$, the relative error is even smaller.

A consequence of this observation is that if S_R values are available, subject to some error, the resulting error in the predicted value of χ will be relatively small. Thus one might expect to be able to make at least rough predictions of χ if the basic rheological information is available. The inverse problem is more difficult to deal with: While in principle one could calculate S_R from measured values of χ,

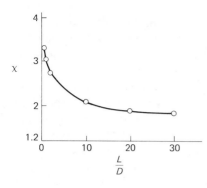

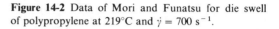

Figure 14-2 Data of Mori and Funatsu for die swell of polypropylene at 219°C and $\dot{\gamma} = 700$ s^{-1}.

the error would be relatively large, and the uncertainty as to whether the theory is realistic gives rise to a great uncertainty in the correct value of S_R.

Thus we must emphasize that the relationship of die swell to basic rheological properties is imperfectly known. With this in mind, however, let us examine some of the available data on this phenomenon. Figure 14-2 shows data of Mori and Funatsu for a polypropylene melt. The important point to note is the dependence of χ on capillary length. For a very long capillary the normal stress at the exit is in equilibrium with the fully developed shear flow within the capillary. For very short tubes, however, the large normal stresses that accompany the *acceleration* of the fluid through the entrance region of the capillary do not have sufficient residence time to relax, and the normal stress at the exit is considerably higher than that which would be expected simply on the basis of the steady shear flow. This particular set of data suggests that an L/D of 20 to 30 is required to give the equilibrium die swell, at shear rates of nearly 1000 s^{-1}. Equation (14-12) refers to equilibrium conditions since it is implied that S_R has been determined under *steady* shearing flow.

A second problem that complicates the attempt to relate χ to S_R lies in the methods used to obtain the measured values of χ. Typically the extrudate leaves the capillary as a molten polymer and is ejected into air at ambient temperatures far below its melting point. It is possible that the extrudate solidifies faster than the elastic stresses can relax, thus "freezing" the stresses into the solidified strand. If this occurs then stress relaxation is incomplete, and the extrudate does not achieve its equilibrium degree of expansion.

Data illustrating this problem are presented by White and Roman, and Fig. 14-3 shows some of their results obtained for a high-density polyethylene. The "isothermal" data refer to the case where the ambient medium was a silicone oil at the melt temperature. Photographs of the extrudate were taken 5 to 10 min after extrusion. It could be presumed that these data correspond to complete relaxation of stress. The "frozen" data represent the other extreme, where the extrusion was into cold air and the diameter of the frozen extrudate was measured with a micrometer. The "annealed" data correspond to taking the frozen extrudates and annealing them in a silicone bath at a temperature above the melting

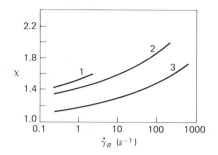

Figure 14-3 Effect of method of measurement on die swell. From the data of White and Roman on a high-density polyethylene extruded at 180°C. 1—isothermal, 2—annealed, 3—frozen. $\dot{\gamma}_R$ is the wall shear rate.

temperature for about 15 min. It is clear that there can be significant variations in χ due to the method of measurement. This point must be kept in mind when comparing data from several sources to a theory, such as Tanner's above.

Several such comparative studies have been made in an attempt to determine the " best " theory available for relating χ to S_R. Tanner's theory often comes out well in such tests, doing a better job than other available theories. Rather than illustrate some of the apparent successes of any theory, let us examine other data of White and Roman shown in Fig. 14-4.

We see that Tanner's theory does *not* do a good job of correlating the data. More to the point is the observation that the observed values of χ clearly depend on something other than S_R. We will not belabor this topic further here, important as it is. Several references are cited which present experimental data for a variety of materials. We must conclude that at the present time there is no adequate theory with which successful a priori predictions of die swell can be made from rheological data. This is notwithstanding the fact that *some* theories can be used to predict *some* data *some* of the time.

Instead, let us examine some observations, independent of theory, that allow useful generalizations to be drawn regarding die swell and from which some useful correlations may be found. Figure 14-5*a* shows data on a commercial polystyrene at three temperatures. As expected, the melt becomes more elastic at lower temperature, and χ is observed to increase (at fixed shear rate) to reflect that fact.

One of the most useful observations regarding die swell is that, *for a given polymer sample*, χ appears to be a unique function of shear *stress*. Figure 14-5*b*

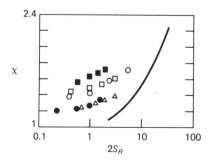

Figure 14-4 Data of White and Roman on die swell of various melts extruded isothermally into a silicone oil bath. ■, □ High-density polyethylenes; ● polystyrene; ○ low-density polyethylenes; △ polypropylene.

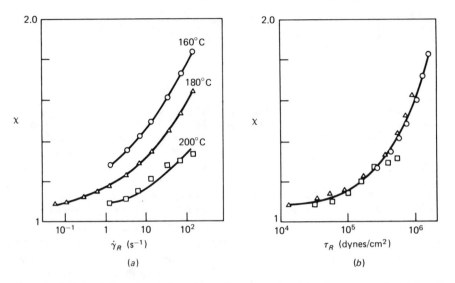

(a)

(b)

Figure 14-5 Data of Graessley et al. on die swell of a commercial polystyrene melt. (a) As a function of shear rate, at three melt temperatures. (b) Same data, replotted as a function of shear stress.

shows the data of Fig. 14-5a replotted to illustrate this point. Thus, if data are available on χ as a function of flow rate, at some temperature and for a given capillary of sufficient length that the equilibrium value of χ is attained, then one may *predict* the expected value of χ at another temperature, flow rate, and/or capillary diameter, so long as viscosity–shear rate–temperature data are available *for that polymer sample.*

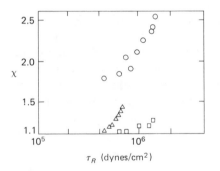

Figure 14-6 Data of Vlachopoulos et al. on die swell of various polystyrene melts.

	$10^{-5}M_w$	$\dfrac{M_z M_{z+1}}{M_w^2}$	T, °C
○	2.12	5.6	170
□	4.98	1.25	210
△	18	1.25	230

The observed correlation with τ_R does not hold if different molecular weight samples of the same polymer are studied. Figure 14-6 illustrates two aspects of this point. The two lowest sets of data are from narrow molecular weight distribution polystyrenes of different average molecular weight. The higher molecular weight material is the more elastic. The more striking result is seen for the lowest molecular weight material, which is a physical blend of narrow distribution samples. This blend shows markedly higher elasticity, even though it is a blend of materials of molecular weight no higher than the $M_w = 498,000$ material which gives the lowest χ in the figure. Can these results be rationalized?

To clarify this point we must turn to some results of Graessley regarding the role of molecular weight in affecting elastic properties of polymers. First, we note that most commercial polymers have a *distribution* of molecular weights, the breadth of the distribution depending in part on the design and operation of the polymerization reactor. As with any distributed quantity, we must characterize "the molecular weight" in terms of some average, and in terms of moments of the distribution. If we define $f(M)$ to be a normalized molecular weight distribution function,† so that

$$1 = \int_0^\infty f(M)\, dM \tag{14-14}$$

then we find that the *number average* and *weight average* molecular weights are given by

$$M_n = \int_0^\infty Mf(M)\, dM \tag{14-15}$$

$$M_w = \int_0^\infty \frac{M^2 f(M)\, dM}{M_n} \tag{14-16}$$

One also defines the *z average* and *z + 1 average* molecular weights as

$$M_z = \int_0^\infty \frac{M^3 f(M)\, dM}{M_n\, M_w} \tag{14-17}$$

$$M_{z+1} = \int_0^\infty \frac{M^4 f(M)\, dM}{M_n\, M_w\, M_z} \tag{14-18}$$

It should be clear that the higher averages are very sensitive to the presence of small fractions of high molecular weight material in the distribution.

One may define the so-called *steady-state shear compliance* J_0 as

$$S_R = J_0 \tau_{12} \tag{14-19}$$

---

† $f(M)\, dM$ is the *number* fraction of polymer chains with molecular weight in the range M to $M + dM$.

A molecular theory due to Rouse relates J_0 to the polymer properties through

$$J_0 \approx J_R = \frac{2}{5} \frac{M_w}{\rho RT} \frac{M_z M_{z+1}}{M_w^2} \tag{14-20}$$

where J_R is the *Rouse compliance*. Extensive experimental studies of Graessley show that J_0 is better calculated as

$$J_0 = \frac{2.2 J_R}{1 + 2.1 \times 10^{-5} \rho M_w} \qquad \text{polystyrene} \tag{14-21}$$

[Equation (14-21) holds for *solutions* of polystyrene if ρ is replaced by concentration c, g/ml.]

From Eq. (14-21) we can see that if M_w exceeds several hundred thousand (as in the data being considered above) then the additive factor of 1 in the denominator may be neglected, and we find

$$J_0 \approx \frac{2}{5} \times 10^5 \frac{1}{\rho^2 RT} \frac{M_z M_{z+1}}{M_w^2} \tag{14-22}$$

This remarkable result indicates that above some level of molecular weight, J_0, and hence S_R, becomes independent of M and depends only on the *distribution* of molecular weight. As the caption of Fig. 14-6 shows, the factor $M_z M_{z+1}/M_w^2$ is considerably higher for the low molecular weight sample, and the data on die swell are seen to be in agreement with these ideas.

This argument, of course, relates to die swell only through Eq. (14-19) which connects J_0 to S_R. Since we have already seen some data which show χ is not always a function *only* of S_R (Fig. 14-4) we must be careful in relying too heavily on a chain of logic which goes from Eq. (14-22) [or (14-21)] through Eq. (14-19) back to any theory of the form $\chi = \chi(S_R)$.

In fact, this point arises, perhaps, in Fig. 14-6, for the two sets of data on narrow distribution samples. Equation (14-22) holds approximately for both sets of data, and if $\chi = \chi(S_R)$ then we should not observe the differences that are seen, since the distributions of molecular weight are the same. This may reflect the idea that χ depends on factors other than S_R (consistent with observations on other materials, as in Fig. 14-4). In part, this may also reflect inaccuracy in the molecular weight distribution data. Small variations in the high molecular weight end of the distribution could significantly affect the higher moments of the distribution.

We may summarize some of the difficulties associated with developing a rational basis for understanding the die swell phenomenon.

1. Die swell appears to be an elastic phenomenon. However, data do not clearly indicate what elastic parameter(s) is (are) relevant. Recoverable shear S_R is not sufficient to completely define χ.
2. Experimental measurements of χ are subject to uncertainties due to incomplete relaxation of stresses. It is difficult to make quantitative comparisons among the results of several workers.

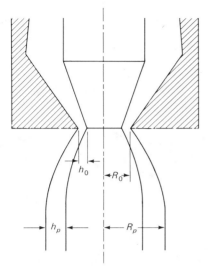

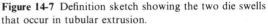

Figure 14-7 Definition sketch showing the two die swells that occur in tubular extrusion.

3. Die swell appears to be a strong function of molecular weight distribution. While the dependence of elastic parameters such as J_0 [Eqs. (14-20) and (14-21)] on molecular weight distribution is fairly well understood, the relationship of χ to J_0 (through S_R, for example) is not unequivocal.

And all this uncertainty attends the simplest possible die swell problem: steady extrusion from a long circular capillary or die.

If one is extruding odd-shaped profiles, and if die swell occurs, it becomes an exceedingly difficult task to design a die shape that, after stress relaxation of the extrudate, produces the desired profile shape. Some brief comments and results relevant to this point may be found in a paper by Han.

A particularly difficult problem arises in consideration of tubular extrusion. Figure 14-7 shows the geometry of the die and extrudate in a typical system used in blow molding. Two independent die swells can be defined. The *diameter swell* is given by the ratio R_p/R_0. In addition, the so-called *weight swell* is given by the thickness ratio h_p/h_0. Again, too little is known theoretically, and too few experiments have been performed, to provide any basis for prediction of the magnitude of these two die swells under a given set of operating conditions. The last three references in the Bibliography under Sec. 14-1 consider some aspects of this problem. We leave the discussion of die swell hanging, then, with the statement that we are not yet near a rational and comprehensive understanding of this important phenomenon.

14-2 MELT FRACTURE

In the previous section we have seen that if a polymeric fluid is extruded from a capillary into air, die swell will usually occur and cause the extrudate to have a larger diameter than that of the capillary. While no single theory appears to

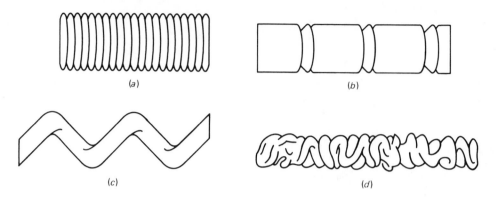

(a)

(b)

(c)

(d)

Figure 14-8 Examples of extrudate distortions.

describe the phenomenon adequately, the notion that die swell is an elastic phenomenon (at least for those situations in which $\chi > 1.1$) seems to be accepted. Another extrudate phenomenon of considerable importance, also believed to be an elastic "event" of some kind, is *melt fracture* or, more generally, *extrudate distortion.*

If an extrudate is observed as its throughput from a circular die is continuously increased, there is often a critical flow rate at which the extrudate surface is no longer smooth. Figure 14-8 shows sketches of several types of extrudates that have been observed with various polymers, or under different conditions on the same polymer. The first three distortions are fairly regular, and are often referred to as *ripple*, *bamboo*, and *screw* or *helix*, respectively. Case *d* represents a severe random surface roughness, for which the term *melt fracture* is most appropriate.

While it is somewhat of a misnomer, we will retain the tradition of using the term melt fracture to refer to the sudden onset of gross extrudate distortion. One sometimes observes a fine-scale surface roughness to appear at outputs considerably below that for the onset of melt fracture. This is probably a phenomenon distinct from melt fracture, and it is often called *matte* or *sharkskin*.

If the melt is extruded at constant output by a piston moving at constant speed, extrudate distortion is often accompanied by a fluctuation in pressure in the reservoir. If the extrudate has a varying cross-sectional area (as in cases *a* and *b* of Fig. 14-8) and is extruded at constant volumetric flow rate, then it must be true that the *linear* speed of extrusion is varying. Vinogradov et al. present evidence of a "stick-slip" phenomenon in the extrusion of rubbery polymers and associate this with melt fracture. The term *spurt* has also been introduced to reflect the idea that the average velocity may vary in a nearly discontinuous way under some conditions of melt fracture.

Melt fracture often appears to involve distinct mechanisms in different materials. Figure 14-9 shows flow rate–pressure drop data (in the form $\dot{\gamma}_N$ versus τ_R) for (a) a high-density polyethylene and (b) a low-density polyethylene. The arrow indicates, in each figure, the onset of melt fracture. For the low-density polyethy-

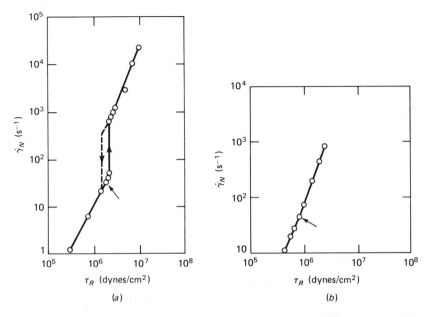

Figure 14-9 Output versus pressure drop for two polyethylenes. (*a*) High-density PE, 150°C, $L/D = 16$. (*b*) Low-density PE, 150°C, $L/D = 16$. The nominal shear rate is $\dot{\gamma}_N = 8U/D$.

lene (a branched polymer) the $\dot{\gamma}_N - \tau_R$ data are continuous (although there is a suggestion of a change in slope). The high-density material (a linear polymer), on the other hand, shows a discontinuity in output at a critical stress, and a hysteresis loop. Clearly there must be something different about these two flows.

Figure 14-10 shows the flow patterns typical of either linear or branched polymers. With branched polymers fluid enters the die through a conical region bounded by an annular region of *recirculating* fluid in the corner. This pattern exists whether the extrudate is smooth or not. However, when the flow rate gets high enough to produce melt fracture of the extrudate, it is observed that fluid

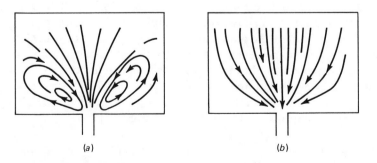

Figure 14-10 Flow patterns in the region just upstream of a capillary (*a*) typical of low-density PE and (*b*) typical of high-density PE.

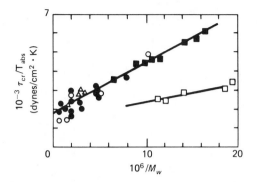

Figure 14-11 Critical stress dependence on molecular weight.

		M_w	M_w/M_n
△	Polypropylene	287,000–660,000	
■	High-density PE	60,000–150,000	4–7
●	Broad-distribution polystyrene	200,000–1,350,000	2.4–19
○	Narrow-distribution polystyrene	97,000–1,800,000	1.1–1.2
□	Low-density PE	50,000–90,000	3–6

from the vortices is periodically drawn into the die, and the streamlines waver from side to side.

Another interesting and important difference between linear and branched polymers is the dependence of the severity of melt fracture on die length. In branched polymers the severity decreases as the die is lengthened; in linear polymers the opposite occurs. These differences between linear and branched materials are subject to some exceptions: linear polypropylene, for example, does not show the flow discontinuity of Fig. 14-9a when exhibiting melt fracture. The Tordella, White, and Petrie and Denn references give more complete reviews and extensive bibliographies on the various melt fracture phenomena. We turn instead to some examples of specific results that are of interest.

Vlachopoulis and Alam present data on a series of polymers of well-defined molecular weight distribution. The critical shear stress at which melt fracture occurred was measured. Figure 14-11 shows their data, plotted in the form τ_{cr}/T_{abs} versus $1/M_w$. Two distinct lines fit the data, corresponding to linear or branched polymers. If these results have some general validity then it should be possible to use Fig. 14-11 to estimate critical conditions for the onset of melt fracture. At least for the polymers studied, one should hope to be able to use Fig. 14-11 for predictions with reasonable success, in view of the fairly wide range of molecular weights of the samples used. The equations for the two lines of Fig. 14-11 are

$$\frac{\tau_{cr}}{T_{abs}} = 1717 + \frac{2.7 \times 10^8}{M_w} \qquad \text{linear polymers} \qquad (14\text{-}23)$$

$$\frac{\tau_{cr}}{T_{abs}} = 1317 + \frac{10^8}{M_w} \qquad \text{branched PE} \qquad (14\text{-}24)$$

By measuring die swell at the onset of melt fracture and using Tanner's theory [Eq. (14-12)], Vlachopoulis and Alam calculate a critical value of recoverable shear for melt fracture and give this as

$$S_R = \frac{2.65}{M_z M_{z+1}/M_w^2} \qquad \text{for polystyrene} \qquad (14\text{-}25)$$

The result is given only for polystyrene because that is the only polymer for which the requisite measurements of M_z and M_{z+1} were available.

Since Eq. (14-25) begs the validity of Eq. (14-12), which we have already stated to be of limited value, one should avoid using Eq. (14-25) for prediction of critical conditions of melt fracture. Equation (14-23) is more likely to have general validity and requires only viscosity–shear rate information for its use. The result [Eq. (14-25)] is of interest principally because some stability theories (see Chap. 15) which attempt to "explain" melt fracture predict that capillary flow becomes unsteady at values of S_R in the neighborhood of 2.6. This result is in agreement with Eq. (14-25) for a monodisperse polymer. However, the stability theory does not really distinguish between monodisperse and polydisperse materials. Hence the result may be fortuitous.

A careful reading of a recent, comprehensive, and critical review of the melt fracture phenomenon, such as that of Petrie and Denn, shows that melt fracture is still poorly understood. Results are often contradictory, and generalizations of a quantitative and predictive nature are rare. Petrie and Denn suggest that two different phenomena occur, both of which are of the nature of elastic instabilities. In linear polymers the instability probably occurs in the shear flow of the die. In branched polymers the converging entry flow is probably unstable and leads to unsteady flow and melt fracture. A good understanding of the nature of these instabilities does not yet exist.

Example 14-1: Prediction of extrudate behavior from steady-shear data In both sections of this chapter we have emphasized the lack of a strong theoretical base for prediction, or even correlation, of die swell and melt fracture. In this example we consider some rheological data available for a low-density polyethylene, show how to manipulate the data in order to predict die swell and melt fracture, and then compare the predictions to experience.

Figure 14-12 shows viscosity and normal stress data for a commercial low-density polyethylene. The viscosity data were obtained at 150°C using several instruments over a very wide range of shear rate. Normal stress data were obtained at low shear rates and at a temperature of 130°C. The corresponding shear stress data at that temperature are plotted with the normal stress data.

We may begin by calculating the recoverable shear at 130°C, and Fig. 14-13 shows the results. We note that the S_R values are given at shear rates in the range 10^{-2} to 1 s^{-1}. This range is considerably below that of interest for extrusion through a die. We will have to extrapolate S_R to higher values of $\dot{\gamma}$.

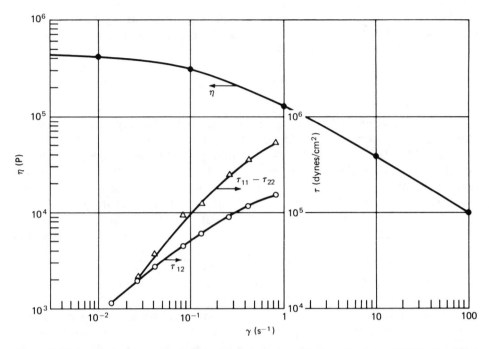

Figure 14-12 Steady shear data for a low-density polyethylene. ● Viscosity data at 150°C; ○ τ_{12} and △ τ_{11}-τ_{22} at 130°C.

There does not seem to be a better choice than to extrapolate the data on Fig. 14-13 in a linear fashion. It is necessary to extrapolate to an order-of-magnitude higher shear stress to get into the range of interest for die flows. The extrapolated values of S_R are certainly suspect.

Once S_R is available, die swell can be predicted using Eq. (14-12). Figure

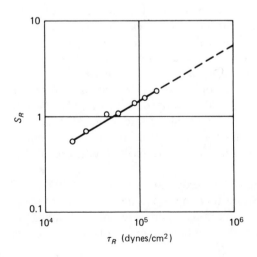

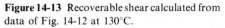

Figure 14-13 Recoverable shear calculated from data of Fig. 14-12 at 130°C.

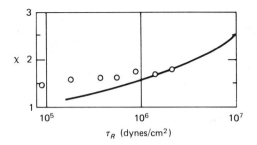

Figure 14-14 Die swell data $\bigcirc$ and prediction from Eq. (14-12).

14-14 shows the predicted and observed values. The agreement is not bad, but it is no great triumph if we consider the fact that we could predict almost *any* set of data within ± 25 percent by "predicting" $\chi = 1.6$.

Molecular weight measurements on this material give $M_n = 2 \times 10^4$, and M_w is "approximately" 10^6. If we use this value in Eq. (14-24) we may predict the critical shear stress for the onset of melt fracture. We see that the result is insensitive to the uncertainty in M_w, and we find, at 150°C, $\tau_R = 5.9 \times 10^5$ dynes/cm². Using the viscosity data of Fig. 14-12 we may calculate a critical shear *rate* of 30 s⁻¹. The observed value, from a capillary of $L/D = 25$, at 150°C, was 29 s⁻¹. This is considerably better agreement than one could have hoped for and must be regarded as a stroke of luck.

All the above comments are based on available observations of melt fracture from circular dies. Little in the way of quantitative data is available regarding melt fracture of odd-shaped extrusions, although the phenomenon clearly occurs in such cases. Data on melt fracture of polystyrene extruded from tubes of noncircular cross section are available and suggest that the critical shear stress at fracture is independent of shape (see Ramsteiner).

PROBLEMS

14-1 Carry through the derivation of Eq. (14-7) for the case that the densities within the capillary and outside it are different.

14-2 Evaluate the importance of the forces due to surface tension and frictional drag of air on the jet, for the simple case of a newtonian fluid. Take the following conditions:

$$\chi = 0.87 \qquad \rho = 1 \text{ g/cm}^3$$

$$U_j = 400 \text{ cm/s} \qquad \sigma = 65 \text{ dynes/cm}$$

$$D_0 = 0.1 \text{ cm} \qquad \text{Ambient fluid is air at 25°C and 1 atm}$$

14-3 Derive Eq. (14-8).

14-4 The data shown in Fig. 14-15 were obtained under the following conditions:

$$n = 1/3 \qquad K = 10 \text{ dyne} \cdot \text{s}^{1/3}/\text{cm}^2 \qquad D_0 = 0.086 \text{ cm}$$

Calculate and plot S_R versus $\dot{\gamma}_R$ by using Eq. (14-9). Do the same, using Eq. (14-12).

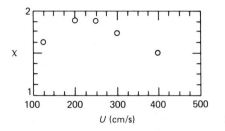

U (cm/s)   **Figure 14-15** Data for Prob. 14-4.

14-5 In discussing the data of Fig. 14-5a, it was asserted that "the melt becomes more elastic at lower temperatures." If we take S_R as the measure of elasticity and assume the validity of Eqs. (14-19) and (14-20), we may evaluate the effect of T on S_R at fixed $\dot{\gamma}_R$.

Argue that the dominant effect is actually through the temperature dependence of viscosity rather than through the temperature dependence of the compliance.

14-6 A process has been designed for a polystyrene having a fairly narrow molecular weight distribution, such that

$$f_1(M) = \begin{cases} 10^{-5} & 2 \times 10^5 \le M \le 3 \times 10^5 \\ 0 & \text{all other } M \end{cases}$$

The supplier of the polymer sends through a batch of material containing some high molecular weight polymer, such that

$$f_2(M) = \begin{cases} 0.9 \times 10^{-5} & 2 \times 10^5 \le M \le 3 \times 10^5 \\ 10^{-5} & 10^6 \le M \le 1.01 \times 10^6 \\ 0 & \text{all other } M \end{cases}$$

For each batch of material calculate (a) M_w; (b) M_w/M_n; (c) $M_z M_{z+1}/M_w^2$.

The process involves continuous extrusion of the melt through a cylindrical die. Estimate the change in S_R experienced by the new polymer, assuming that the process is controlled at fixed pressure at the die. Is the result different if the process is controlled at fixed output?

14-7 Using data from Fig. 14-5 predict die swell as a function of flow rate from a die of length 2 in and diameter 0.1 in, at $T = 155°C$, for Q in the range 10^{-4} to 10^{-2} in³/s. Note that $\dot{\gamma}_R$ of Fig. 14-5a is $[(3n + 1)/4n](32Q/\pi D^3)$.

14-8 Estimate the maximum output for a distortion-free extrudate of polystyrene under the following conditions:

$$T = 200°C \qquad D = 0.1 \text{ cm}$$

$$n = 1/3 \qquad L = 1.5 \text{ cm}$$

$$M_w = 580,000 \qquad K = 10^5 \text{ dyne} \cdot \text{s}^{1/3}/\text{cm}^2$$

Repeat the calculation at $T = 215°C$, using $K = K_0 \exp \dfrac{\Delta E}{R} (1/T - 1/T_0)$ with $\Delta E = 10$ kcal/g mol.

14-9 Consider a polymer whose (discrete) molecular weight distribution is characterized in terms of the *weight* fraction w_i of chains of molecular weight M_i. Then the various averages are defined by

$$M_n = \frac{\sum w_i}{\sum w_i/M_i}$$

$$M_w = \frac{\sum w_i M_i}{\sum w_i}$$

$$M_z = \frac{\sum w_i M_i^2}{\sum w_i M_i}$$

$$M_{z+1} = \frac{\sum w_i M_i^3}{\sum w_i M_i^2}$$

and $\sum w_i = 1$.

Suppose 1 percent by weight of polymer whose molecular weight is 10^6 is added to a narrow-distribution polymer for which $M_w = 5 \times 10^5$. Find the change in $M_z M_{z+1}/M_w^2$. If the material is polystyrene, find the change in S_R. Would this have a significant effect on die swell? On the critical stress for melt fracture?

14-10 The White-Metzner viscoelastic fluid [Eq. (3-149)] leads to the prediction that (assuming $G = $ constant)

$$\Psi_{12} = \frac{2\eta^2}{G} \qquad \text{[Eq. (3-152)]}$$

Show that this, in turn, leads to the prediction that

$$S_R = \frac{\tau_{12}}{G}$$

Do the data of Fig. 14-13 agree with this?

14-11 Suppose no normal stress data were available for use in Example 14-1. Show, first, that Eq. (3-152) is consistent with

$$\frac{\Psi_{12}}{\Psi_{12}^0} = \left(\frac{\eta}{\eta_0}\right)^2$$

and $\Psi_{12}^0 = 2\lambda_0 \eta_0$ where λ_0 is the relaxation time at low shear rate. A method sometimes used for estimating λ_0 from viscosity data involves extrapolating the power law region of $\eta(\dot{\gamma})$ toward low $\dot{\gamma}$. The reciprocal of the shear rate $\dot{\gamma}_0$ at which this extrapolated line crosses the line $\eta = \eta_0$ is often used as the relaxation time of the fluid. Use this method, with the data of Fig. 14-12, and plot S_R versus τ_R and χ versus τ_R. Compare to the data of Figs. 14-13 and 14-14.

BIBLIOGRAPHY

14-1 Die swell

Middleman, S., and J. Gavis: Expansion and Contraction of Capillary Jets of Newtonian Liquids, *Phys. Fluids*, **4:** 355 (1961).

Nickell, R. E., R. I. Tanner, and B. Caswell: The Solution of Viscous Incompressible Jet and Free-Surface Flows Using Finite Element Methods, *J. Fluid Mech.*, **65:** 189 (1974).

Metzner, A. B., W. T. Houghton, R. A. Sailor, and J. L. White: A Method for the Measurement of Normal Stresses in Simple Shearing Flow, *Trans. Soc. Rheol.*, **5:** 133 (1961).

Graessley, W. W., S. D. Glasscock, and R. L. Crawley: Die Swell in Molten Polymers, *Trans. Soc. Rheol.*, **14:** 519 (1970).

Tanner, R. I.: A Theory of Die Swell, *J. Polym. Sci.*, A-2, **8:** 2067 (1970).

Mori, Y., and K. Funatsu: On Die Swell in Molten Polymer, *J. Appl. Polym. Sci. (Appl. Polym. Symp.)*, **20:** 209 (1973).

An interesting study of the effect of take-up force on die swell is

White, J. L., and J. F. Roman: Extrudate Swell During the Melt Spinning of Fibers—Influence of Rheological Properties and Take-up Force, *J. Appl. Polym. Sci.*, **20:** 1005 (1976).

Vlachopoulos, J., M. Morie, and S. Lidorikis: An Evaluation of Expressions Predicting Die Swell, *Trans. Soc. Rheol.*, **16**: 669 (1972).

Mendelson, R. A., F. L. Finger, and E. B. Bagley: Die Swell and Recoverable Shear Strain in Polyethylene Extrusion, *J. Polym. Sci. (Pt. C)*, **35**: 177 (1971).

Abdel-Khalik, S. I., O. Hassager, and R. B. Bird: Prediction of Melt Elasticity from Viscosity Data, *Polym. Eng. Sci.*, **14**: 859 (1974).

Han, C. D.: Rheology of Shaped Fiber Formation, *J. Appl. Polym. Sci.*, **15**: 1091 (1971).

Worth, R. A., and J. Parnaby: The Design of Dies for Polymer Processing Machinery, *Trans. Inst. Chem. Eng.*, **52**: 368 (1974).

Pritchatt, R. J., J. Parnaby, and R. A. Worth: Design Considerations in the Development of Extrudate Wall-Thickness Control in Blow Moulding, *Plast. Polym.*, **43**: 1 (1975).

An interesting set of papers relevant to the topic of the role of elastic phenomena in process development is

Symposium on What We Don't Know in Blow Molding, *Trans. Soc. Rheol.*, **19**: 337–371 (1975).

14-2 Melt fracture

General surveys of this topic, with extensive bibliographies and interesting photographs of the die flow and the emerging extrudates, can be found in

Tordella, J. P.: Unstable Flow in Molten Polymers, in F. R. Eirich (ed.), "Rheology," vol. V, Academic Press, Inc., New York, 1969.

White, J. L.: Critique on Flow Patterns in Polymer Fluids at the Entrance of a Die and Instabilities Leading to Extrudate Distortion, *J. Appl. Polym. Sci. (Appl. Polym. Symp.)*, **20**: 155 (1973).

Petrie, C. J. S., and M. M. Denn: Instabilities in Polymer Processing, *AIChE J.*, **22**: 209 (1976).

An interesting set of experiments on melt fracture, and a discussion which ascribes the phenomenon to a "high elastic state," is given in

Vinogradov, G. V., et al.: Viscoelastic Properties and Flow of Narrow Distribution Polybutadienes and Polyisoprenes, *J. Polym. Sci.*, A-2, **10**: 1061 (1972).

———: Critical Regimes of Shear in Linear Polymers, *Polym. Eng. Sci.*, **12**: 323 (1972).

The best available data on well-defined melts is

Vlachopoulos, J., and M. Alam: Critical Stress and Recoverable Shear for Polymer Melt Fracture, *Polym. Eng. Sci.*, **12**: 184 (1972).

The data from which Example 14-1 is drawn may be found in

Meissner, J.: Basic Parameters, Melt Rheology, Processing and End-Use Properties of Three Similar Low Density Polyethylene Samples, *Pure Appl. Chem.*, **42**: 553 (1975).

Studies of elastic phenomena in extrusion from noncircular dies may be found in

Ramsteiner, F.: Fliessverhalten von Kunststoffschmelzen durch Düsen mit kreisformigen, quadratischen, rechteckigen oder dreieckigen Querschnitt, *Kunststoffe*, **61**: 943 (1971); see also *ibid*, **62**: 766 (1972).

FIFTEEN

STABILITY OF FLOWS

Everything subject to time is liable to change.

Albo

In many of the models which have been presented throughout this text the assumption is made that the system is in a steady state, by which we mean that the dependent variables are independent of time. The assumption is based on the appealing intuitive idea that steady causes produce steady effects. Examples of the failure of this premise are well known, however. The most common such observation is that of turbulent pipe flow: Under a steady driving pressure the velocity and pressure distributions within the fluid are observed to be randomly fluctuating functions of time if some critical value of a Reynolds number is exceeded.

Unstable flows occur in polymer processes on occasion, and it is important that the parameters that control stability be identified and that the critical values of these parameters be known. In this chapter we describe some processing instabilities and outline the methods of stability analysis. The mathematical treatment is more complex, in some respects, than that of most of the problems considered elsewhere in the book, and some new ideas must be introduced. An outstanding review of instabilities in polymer processing, with special attention to fiber spinning and melt fracture, is presented by Petrie and Denn. It should be read along with this chapter.

We begin with a simple definition of stability. Let us suppose that there is a steady-state solution $(\mathbf{u}^s, p^s)$ of the dynamic equations and the continuity equation, subject to some set of boundary conditions, for a fluid satisfying a particular constitutive equation. Suppose that at some time t_0 the flow is disturbed so that new values of **u** and p, say, **u'** and p', are observed. In a free boundary flow the disturbance may also be to the position of the boundary. The flow is said to be stable with respect to the imposed disturbances if

$$\lim_{t - t_0 \to \infty} \mathbf{u}' = \mathbf{u}^s$$

$$\lim_{t - t_0 \to \infty} p' = p^s \tag{15-1}$$

It is important to keep in mind at the outset that the phrase "with respect to the imposed disturbances" is significant. A flow may be stable to one *type* of disturbance but unstable to another, or the stability may depend on the magnitude of the imposed disturbance.

We turn, then, to a series of examples which will introduce the methods of stability analysis, in the context of some flows of interest in polymer process analysis.

15-1 STABILITY OF A VISCOUS CAPILLARY JET

Figure 15-1 shows a jet of fluid issuing from a capillary into air. It is observed that the jet surface, initially cylindrical, develops a varicose swelling which grows in amplitude until the jet is disrupted into a series of droplets. This instability arises from the surface tension of the jet interface. A stability analysis may be carried out which relates the growth rate of a disturbance to the fluid physical properties.

We begin by considering a cylindrical jet of constant radius R, translating along its axis at constant velocity U. We assume that exit effects associated with the ejection of the fluid from the capillary (a change in radius of the jet and a flattening of the velocity profile) are completed within a few diameters of the exit, and we confine our attention to the region of constant R and uniform velocity field $\mathbf{u} = (U, 0, 0)$.

It is convenient to carry out the analysis of this flow in a coordinate system translating at the velocity U. In that coordinate system the steady velocity field vanishes identically. There is, in addition, a uniform pressure σ/R due to the curvature of the free surface. We now assume that the flow is disturbed in some manner, as a consequence of which the velocity field (in the translating system) is (u_z, u_r) and the pressure is p. Our two major assumptions are:

1. The disturbances are of small amplitude.
2. The disturbances are (somehow) axisymmetric, i.e., independent of θ.

Figure 15-1 Capillary jet, showing the growth of unstable varicose disturbances.

The disturbed flow field will either damp out toward the undisturbed flow (stability) or else the disturbances will grow until the jet is disrupted (instability). In any event, the dynamics of the flow are governed by the continuity equation and the dynamic equations. We introduce a third assumption by selecting a specific constitutive equation, that of a newtonian fluid, so that the dynamic equations are (for axisymmetric flow)

$$\frac{\partial u_z}{\partial t} = -\frac{1}{\rho}\frac{\partial p}{\partial z} + v\left[\frac{1}{r}\frac{\partial}{\partial r}\left(r\frac{\partial u_z}{\partial r}\right) + \frac{\partial^2 u_z}{\partial z^2}\right] \tag{15-2}$$

$$\frac{\partial u_r}{\partial t} = -\frac{1}{\rho}\frac{\partial p}{\partial r} + v\left[\frac{\partial}{\partial r}\left(\frac{1}{r}\frac{\partial}{\partial r}ru_r\right) + \frac{\partial^2 u_r}{\partial z^2}\right] \tag{15-3}$$

where v is the kinematic viscosity μ/ρ, and the continuity equation is

$$\frac{1}{r}\frac{\partial}{\partial r}(ru_r) + \frac{\partial u_z}{\partial z} = 0 \tag{15-4}$$

Note that the inertial terms have been dropped in view of assumption 1 above. We now seek a general solution of these equations.

There is some convenience to be gained if we first obtain the solution in the absence of viscous effects. Thus we are solving the two dynamic equations with $v = 0$, along with the continuity equation. The dynamic equations become

$$\frac{\partial u_z^\circ}{\partial t} = -\frac{1}{\rho}\frac{\partial p^\circ}{\partial z} \tag{15-5}$$

$$\frac{\partial u_r^\circ}{\partial t} = -\frac{1}{\rho}\frac{\partial p^\circ}{\partial r} \tag{15-6}$$

We use a superscript $\circ$ to denote the inviscid solution.

We may define a *velocity potential* φ by

$$u_r^\circ = \frac{\partial \varphi}{\partial r} \qquad u_z^\circ = \frac{\partial \varphi}{\partial z} \tag{15-7}$$

in consequence of which the continuity equation gives

$$\frac{1}{r}\frac{\partial}{\partial r}\left(r\frac{\partial \varphi}{\partial r}\right) + \frac{\partial^2 \varphi}{\partial z^2} = 0 \tag{15-8}$$

If φ is introduced into either dynamic equation the pressure is found to be

$$p^\circ = -\rho\frac{\partial \varphi}{\partial t} + \frac{\sigma}{R} \tag{15-9}$$

Consistent with observations we seek solutions which are spatially periodic and which either grow or decay in time. A useful trial function is

$$\varphi = \Phi(r)e^{ikz + \alpha t} \tag{15-10}$$

Assuming for the moment that such a function is indeed a solution, we see that k is a *wave number* of the disturbance and α is a *growth rate*. If $\alpha < 0$ then the disturbances vanish, the system is stable, and α is a *damping rate*. The main goal of a stability analysis is to determine the combination of parameters, if any, that yield positive values of α so that conditions of instability may be defined.

If φ is substituted into Eq. (15-8) the result is

$$\frac{1}{r}\frac{d}{dr}\left(r\frac{d\Phi}{dr}\right) - k^2\Phi = 0 \tag{15-11}$$

the solution of which is

$$\Phi = C_1 I_0(kr) \tag{15-12}$$

where I_0 is the "hyperbolic" Bessel function, defined so that $I_0(x) = J_0(ix)$. In solving the second-order Eq. (15-11) we have imposed the boundary condition that Φ be finite along the axis, $r = 0$.

With Eq. (15-12) the pressure and velocities may be written as

$$p^\circ = -C_1 \rho\alpha I_0(kr)E + \frac{\sigma}{R} \tag{15-13}$$

$$u_r^\circ = C_1 kI_1(kr)E \tag{15-14}$$

$$u_z^\circ = C_1 ikI_0(kr)E \tag{15-15}$$

where $E \equiv e^{ikz + \alpha t}$.

The surface disturbance has been assumed to be symmetric about the axis. We write this in the form

$$R^\circ = R + \zeta^\circ(z, t) \tag{15-16}$$

and take ζ° to be given by the kinematic condition

$$\frac{\partial\zeta^\circ}{\partial t} = u_r^\circ \qquad \text{at } r = R \tag{15-17}$$

This gives ζ° as

$$\zeta^\circ = C_1 \frac{k}{\alpha} I_1(kR)E \tag{15-18}$$

We may eliminate the constant C_1 by imposing a boundary condition on pressure, namely, that [see Eq. (3-75)]

$$p^\circ = p_\sigma = \sigma\left(\frac{1}{R_1} + \frac{1}{R_2}\right) \qquad \text{at } r = R \tag{15-19}$$

This gives p_σ as

$$p_\sigma = \frac{\sigma}{R} - \frac{\sigma}{R^2}\left(\zeta^\circ + R^2 \frac{\partial^2\zeta^\circ}{\partial z^2}\right) \tag{15-20}$$

and, using Eq. (15-18), we find

$$\frac{\sigma}{R^2}(1 - k^2 R^2)\frac{k}{\alpha}I_1(kR) = \rho\alpha I_0(kR) \tag{15-21}$$

We may solve for α^2 to obtain

$$\alpha^2 = \frac{\sigma k}{\rho R^2}(1 - k^2 R^2)\frac{I_1(kR)}{I_0(kR)} \tag{15-22}$$

It is apparent that α is a function of the wave number of the imposed disturbance. So long as $kR < 1$ we find a positive α. (The ratio I_1/I_0 is always positive.) Hence this flow is unstable to any disturbance for which $kR < 1$, so long as the assumptions of the analysis—small disturbances, axisymmetry—are valid. For $kR > 1$ we find α to be imaginary, corresponding to *periodic* disturbances which do not damp out (capillary waves).

The function $\alpha(k)$ has a maximum in the range $[0 < kR < 1]$, given by

$$\alpha_{max} = 0.34\left(\frac{\sigma}{\rho R^3}\right)^{1/2} \tag{15-23}$$

which occurs for a wave number

$$(kR)_{max} = 0.69 \tag{15-24}$$

Thus, while *all* disturbances satisfying $kR < 1$ will disrupt the jet, one disturbance grows more rapidly than the others. The presumption (borne out by observation) is that a jet subject to a *spectrum* of disturbances of various wave numbers will ultimately be disrupted by the fastest-growing disturbance.

If ζ_0 is the initial magnitude of the disturbance of wave number k_{max}, then at some time t the disturbance will have grown to magnitude

$$\frac{\zeta^\circ}{\zeta_0} = e^{\alpha_{max}t} \tag{15-25}$$

(We have shifted now to the laboratory coordinate system. The initial disturbance was at $z = 0$, and the time variable is now understood to be just z/U.) We may take the time of disruption of the jet to be defined as the point where $\zeta^\circ = R$. Solving for the time from Eq. (15-25), we find

$$t^* = \frac{L^*}{U} = \frac{1}{\alpha_{max}}\ln\frac{R}{\zeta_0} \tag{15-26}$$

L^* is the distance from the exit to the point at which the jet breaks into droplets. The value of ζ_0 is unknown, but experiments suggest that for an apparently stationary capillary, subject to no obvious macroscopic disturbance, a suitable value is such that

$$\ln\frac{R}{\zeta_0} = 13 \quad \text{or} \quad \zeta_0 = 2.2 \times 10^{-6}R \tag{15-27}$$

If we wish to estimate how far downstream a jet will travel before a disturbance of magnitude $\zeta°/R = f$ will appear, we find

$$L_f^* = \frac{U}{\alpha_{max}} \ln \frac{fR}{\zeta_0}$$

or
$$\frac{L_f^*}{D} = We^{1/2}(\ln f + 13) \tag{15-28}$$

where $We = \rho U^2 D/\sigma$ is the Weber number, and D is the jet diameter. If we take $f = 10^{-2}$ we find $L^*/D = 8.4\,We^{1/2}$. Because of the exponential character of the growth the distance to reach $f = 10^{-2}$ is 65 percent of the distance required to disrupt the jet completely. This result says that the jet is disrupted not far beyond the point of first measurable appearance of the disturbance.

The analysis outlined here is valid for an inviscid fluid, and so is of little interest except that it facilitates the solution for the stability of a newtonian jet. We return to Eqs. (15-2) and (15-3) now and anticipate that the velocity components may be written as

$$u_r = u_r° + u_r^1 \tag{15-29}$$

$$u_z = u_z° + u_z^1 \tag{15-30}$$

and the pressure as $p = p° + p^1$, where $\mathbf{u}°$, $p°$ is the inviscid solution just given.

We solve for the "viscous" part of the velocity by introducing a *stream function* ψ, defined such that

$$u_r^1 = -\frac{1}{r}\frac{\partial\psi}{\partial z} \qquad u_z^1 = \frac{1}{r}\frac{\partial\psi}{\partial r} \tag{15-31}$$

It is easy to verify that, with this definition of ψ, the continuity equation is satisfied identically.

If Eqs. (15-29) and (15-30) [along with (15-31)] are substituted into Eqs. (15-2) and (15-3), and the resulting equations are stared at and thought about, one discovers that $p^1 = 0$ (the pressure is uninfluenced by viscosity). The stream function is found to obey the following equation:

$$\frac{\partial^2\psi}{\partial r^2} - \frac{1}{r}\frac{\partial\psi}{\partial r} + \frac{\partial^2\psi}{\partial z^2} = \frac{1}{\nu}\frac{\partial\psi}{\partial t} \tag{15-32}$$

If we again anticipate a functionality for ψ as

$$\psi = \Psi(r)e^{ikz + \alpha t} \tag{15-33}$$

we find $\Psi(r)$ as the solution to

$$\frac{d^2\Psi}{dr^2} - \frac{1}{r}\frac{d\Psi}{dr} - \left(k^2 + \frac{\alpha}{\nu}\right)\Psi = 0 \tag{15-34}$$

This is seen to be a Bessel equation whose solution is

$$\Psi = C_2 r I_1(k'r) \tag{15-35}$$

where

$$k'^2 = k^2 + \frac{\alpha}{\nu} \qquad (15\text{-}36)$$

(We have imposed the boundary condition requiring that Ψ be finite along the jet axis, $r = 0$.)

If we go back to Eqs. (15-13) through (15-15) for the inviscid solution, we may write the newtonian solution as

$$u_r = -ik[iC_1 I_0(kr) + C_2 I_1(k'r)]E \qquad (15\text{-}37)$$

$$u_z = k\left\{iC_1 I_0(kr) + C_2\left[\frac{I_1(k'r)}{kr} + \frac{k'}{k} I_1'(k'r)\right]\right\}E \qquad (15\text{-}38)$$

$$p = -C_1 \rho \alpha I_0(kr)E \qquad (15\text{-}39)$$

where $E = e^{ikz + \alpha t}$, as before.

In the viscous case a new boundary condition must be introduced that does not apply to an inviscid fluid: The shear stress in the liquid must vanish at the jet boundary (we neglect the viscosity of the surrounding air). Thus we have

$$\tau_{rz} = \mu\left(\frac{\partial u_z}{\partial r} + \frac{\partial u_r}{\partial z}\right)\bigg|_{r=R} = 0 \qquad (15\text{-}40)$$

This gives us a relationship between C_1 and C_2:

$$C_1 = C_2 \frac{I_1(k'R)(k'^2 + k^2)}{2ik^2 I_1(kR)} \qquad (15\text{-}41)$$

To eliminate C_1 we again impose a boundary condition on the radial normal stress, which now takes the form [see Eq. (15-19)]

$$-\tau_{rr} + p^\circ = p_\sigma \qquad \text{at } r = R \qquad (15\text{-}42)$$

where $\tau_{rr} = 2\mu\, \partial u_r/\partial r$. p_σ is calculated as in the inviscid case.

When the algebraic dust has settled the result is a transcendental equation for α of the form

$$\alpha^2 + \frac{2\nu k^2}{I_0(kR)}\left[I_1'(kR) - \frac{2kk'}{k^2 + k'^2}\frac{I_1(kR)}{I_1(k'R)}I_1'(k'R)\right]\alpha$$

$$= \frac{\sigma k}{\rho R^2}(1 - k^2 R^2)\frac{I_1(kR)}{I_0(kR)}\frac{k'^2 - k^2}{k'^2 + k^2} \qquad (15\text{-}43)$$

Since k' depends upon α the quadratic format of this equation is deceptive. Since k appears in the argument of the Bessel functions, as does α (through k'), it is necessary to find $\alpha_{\max}$ numerically.

A useful approximation is possible, however, which takes account of the fact that kR is small compared to unity and is smaller the larger the viscosity. For small kR the Bessel functions may be approximated by $I_0(kR) = 1$, $I_1(kR) = kR/2$, and $I'_1(kR) = \frac{1}{2}$. Since, for large viscosity, k' is comparable to k, similar approximations hold for the Bessel functions of argument $k'R$. As a consequence, a quadratic equation for α is obtained of the form

$$\alpha^2 + 3k^2 v\alpha = \frac{\sigma k^2}{2\rho R}(1 - k^2 R^2) \tag{15-44}$$

One may easily verify that

$$\alpha_{max} = 0.34\left(\frac{\sigma}{\rho R^3}\right)^{1/2}\left[1 + \tfrac{3}{2}v\left(\frac{2\rho}{R\sigma}\right)^{1/2}\right]^{-1} \tag{15-45}$$

and the breakup length is given by

$$\frac{L^*}{D} = \left(\ln \frac{R}{\zeta_0}\right) We^{1/2}(1 + 3Z) \tag{15-46}$$

where $Z = We^{1/2}/Re = \mu/(\rho D\sigma)^{1/2}$ is called the *Ohnesorge* number.

The analysis outlined here is a good example of a successful stability theory. Data obtained with low-speed jets of newtonian fluids ejected into air are in good agreement with the result given in Eq. (15-46). Closer examination of the data suggests that $\ln (R/\zeta_0)$ is a function of Z, and Sterling and Sleicher suggest an explanation of this observation, associated with the effect of the relaxation of the parabolic profile upon ejection of the jet.

Since the presentation of this analysis is for the purpose of illustrating the main ideas of a hydrodynamic stability analysis, we turn to consideration of some of the assumptions made in simplifying the model. The neglect of the inertial terms in the dynamic equations is valid so long as the amplitude of the disturbance velocities is small. This gives the so-called *linearized stability analysis*. It is much more difficult to carry out the nonlinear analysis, but this has been done most successfully by Lafrance. One interesting result is that the nonlinear analysis shows that the $\alpha(k)$ relationship is not greatly altered. Of greater interest, however, is the ability of the nonlinear theory to predict a phenomenon which does not appear in the linear stability analysis but which does in fact appear in the laboratory.

Figure 15-2 shows the appearance of "satellite" drops, small drops that occur between the large "primary" drops of the disrupted jet. The nonlinear analysis of Lafrance not only predicts the *existence* of these drops, but it is also quite successful in predicting their size. This is an important point, for it is often the case that the linearized stability analysis fails to account, even qualitatively, for certain observed features of an unstable flow. Hence one must view the results of a linear analysis with some caution.

Figure 15-2 Satellite drops in an unstable capillary jet. (*From Golden et al.*)

A second restrictive feature of the analysis outlined above is the assumption of axisymmetric disturbances. If that assumption is to be relaxed then it is necessary to include a term in Eq. (15-10) which depends upon the angular coordinate θ. The usual form is taken to be

$$\varphi = \Phi(r)e^{ikz + in\theta + \alpha t} \tag{15-47}$$

where n is an integer. The case $n = 0$, already considered, corresponds to axisymmetry (varicose disturbances). For $n = 1$ the disturbances correspond to a jet whose cross section remains of constant radius but whose *axis* is now sinusoidal instead of straight. This is the so-called *sinuous disturbance*. For $n = 2$ the axis is straight and the cross section is elliptical, the major axis "rotating" in the cross-sectional plane with a wave number k. The sinuous disturbance can be imposed in the laboratory by oscillating the capillary tip normal to the axis. The elliptical disturbance can be imposed by ejecting the jet from a capillary whose cross section is elliptical.

If Eq. (15-47) is used the analysis proceeds as before, and one may obtain an equation for the growth rate α in the form

$$\alpha^2 + 3k^2v\alpha = \frac{\sigma k^2}{2\rho R}(1 - n^2 - k^2R^2) \tag{15-48}$$

which is only slightly different from Eq. (15-44). Inspection shows, however, that for $n > 1$ there is no real positive root of the quadratic, for any wave number disturbance. This means that the jet is stable to sinuous, or elliptic, or more complex disturbances. Conversely we may say that the jet is unstable only to varicose disturbances ($n = 0$), although we must keep in mind that the statement is established from Eq. (15-48), which holds for *small* disturbances only. Experiments are reported which support this result.

A comment is in order regarding the form assumed for the disturbed variables, Eq. (15-47). The dynamic equations are linear (by the assumption of small disturbances) and first order in time. This guarantees that any solution will have the form $e^{\alpha t}$. The term e^{ikz} implies that the disturbance is spatially periodic at the time $t = 0$ of imposition of the disturbance. (Keep in mind that a moving coordinate system is in use here.) There is, of course, no reason why the fluid *must* be subjected to such a uniform disturbance. However, we could express a more complicated initial perturbation as an infinite series of Fourier components, of which $\Phi_j(r)e^{ik_jz}$ would be one term, of amplitude $\Phi_j(r)$ and wave number k_j. Since the equations are linear (again, in the small disturbance case) the net result would be that *every* Fourier component would satisfy Eq. (15-48). Hence the value of $k_{j(\max)}$ could be regarded as that wave number, out of the entire spectrum of Fourier components, which grows most rapidly. Again the assumption is implied that the jet ultimately breaks down under the most rapidly growing mode.

Experimental evidence suggests that if the system forming the jet (fluid reservoir, capillary, etc.) is isolated from macroscopic laboratory disturbances (e.g., pump vibrations, or fluctuations in reservoir pressure) the jet is observed to break down according to theory. We note that Eq. (15-27) suggests that the initial disturbances are of exceedingly small amplitude.

It is possible to impose macroscopic varicose disturbances of any wave number on the jet as it leaves the capillary. Such disturbances will be so large in amplitude that they will lead to breakdown of the jet before the infinitesimal disturbance of wave number k_{max} can become macroscopic. If these macroscopic disturbances are, nevertheless, small compared to the jet radius, the breakdown occurs according to the linearized theory. It is possible, then, to measure the amplitude $\bar{\zeta}$ of the varicose disturbance as a function of distance from its point of imposition and to plot $\ln \bar{\zeta}$ versus t. The slope gives the growth rate α. Lafrance shows such experimental data, which adequately confirm the linearized stability analysis.

In summary, then, we have illustrated the general features of a stability analysis. The primary (undisturbed) flow must be known, after which it is assumed that some perturbation or disturbance is added to the primary flow. With the assumption of small disturbances a linearized stability analysis follows, in which a growth rate of a disturbance is determined as a function of material and operating parameters and as a function of the wave number of the imposed disturbance. A "most unstable" disturbance may be determined from the maximum (if one exists) in the $\alpha(k)$ solution. The linearization may remove certain features from the problem, as in the failure to predict "satellite" drops, noted above. Indeed, a linear stability analysis may lead falsely to the conclusion that a dynamic system is stable to all disturbances, although that is not the case in the example chosen.

If we return to Eq. (15-10) for a moment we may recall that the question of stability is determined by the growth rate α, since the disturbances, at least in the linear stability analysis, behave as $e^{\alpha t}$. If $\alpha < 0$ the disturbances decay and the system is stable; if $\alpha > 0$ the opposite case is found. Actually, α may be a complex number, as Eq. (15-22) shows for the case $kR > 1$. In that *specific* case (the inviscid jet) α is purely imaginary for $kR > 1$, corresponding to sustained oscillations of the disturbance with neither damping nor growth. For the newtonian jet, α [from Eq. (15-44)] is a complex number with the real part negative so long as $kR < 1$, leading to damped oscillations of the disturbance.

In general, α depends upon system parameters. Thus there is usually some set of conditions which determines whether the real part of α is positive (instability) or negative (stability). There is also, usually, a set of conditions under which the real part of α vanishes. This is referred to as *neutral stability*, and the functional relationship among the parameters defines the *neutral stability curve*. The word *curve* may refer to something other than a two-dimensional curve in a plane if more than two parameters determine neutral stability.

The capillary jet problem is very simple with regard to neutral stability. In the newtonian case, for example, Eq. (15-44) may be written as

$$\alpha \left(\rho \frac{R^3}{\sigma} \right)^{1/2} = \mathscr{F}(kR, Z) \qquad (15\text{-}49)$$

where Z is the Ohnesorge number defined previously. Thus the condition of neutral stability is given by the solution to

$$\mathscr{F}(kR, Z) = 0 \tag{15-50}$$

The neutral stability curve, in this case, is a two-dimensional curve in the (kR, Z) space. However, since $\alpha = 0$ for $kR \equiv 1$ for *all* Z, the "curve" is simply a straight line, $kR = 1$, and there is no value to plotting the neutral stability curve. In some subsequent analyses we will have occasion to refer to the neutral stability curve, and we will find that its dependence on parameters is more complex than in the case of the capillary jet.

15-2 STABILITY OF WIRE COATING

In Chap. 5 the process of coating a uniform annular layer of fluid onto a solid cylindrical core, as in wire coating, was described. In that section attention was confined to the die flow, which determines the thickness of coating deposited on the moving core. In this section we consider the stability of that coating. The physical problem and its mathematical analysis are similar to the problem of the liquid jet treated in Sec. 15-1. The algebraic detail is considerably more tedious.

The dynamic equations and the continuity equation are identical to those for the jet problem [Eqs. (15-2) to (15-4)]. Two new boundary conditions appear, expressing the no-slip condition at the solid core:

$$u_z = u_r = 0 \qquad \text{at } r = R_i \tag{15-51}$$

At the outer (free) surface the boundary conditions are with respect to the stresses, as in the case of the jet:

$$\tau_{rz} = 0 \tag{15-52}$$

$$-p + \tau_{rr} = -\sigma\left(\frac{1}{R_1} + \frac{1}{R_2}\right) \qquad \text{at } r = R_0 \tag{15-53}$$

For the newtonian fluid the general solution, in terms of the stream function, may be written as

$$\psi = (\varphi_1 + \varphi_2)e^{\alpha t + ikz} \tag{15-54}$$

where

$$\varphi_1 = A_1 r I_1(kr) + B_1 r K_1(kr) \tag{15-55}$$

$$\varphi_2 = A_2 r I_1(k'r) + B_2 r K_1(k'r) \tag{15-56}$$

$$k'^2 = k^2 + \frac{\alpha}{\nu} \tag{15-57}$$

I_1 and K_1 are the modified (hyperbolic) Bessel functions of the first and second kind.

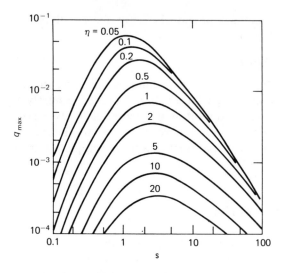

Figure 15-3 Maximum growth rate (non-dimensionalized as q) as a function of η and s for an annular layer of newtonian liquid on a cylinder. (*From Roe.*)

If this solution is substituted into the four boundary conditions, one may obtain a transcendental equation for $\alpha(k)$ of the form [see Eq. (15-44)]

$$\alpha^2 + \frac{2v}{R_0^2}(2 - Q_1)k^2R_0^2\alpha = \frac{\sigma}{\rho R_0^3}Q_1 k^2 R_0^2(1 - k^2R_0^2) - \frac{4v^2}{R_0^4}k^4R_0^4\left(1 - \frac{Q_1}{Q_2}\right)$$

(15-58)

where Q_1 and Q_2 are complicated functions of kR_0 and $k'R_0$ and of the geometrical parameter R_0/R_i. As in the case of the jet, the quadratic format in α is deceptive since Q_2 depends upon $k'(\alpha)$.

Numerical differentiation of Eq. (15-58) allows evaluation of α_{max} as a function of fluid and geometric parameters. Such calculations were carried out by Roe and are shown in Fig. 15-3, in terms of the following dimensionless variables:

$$q = \alpha\left(\frac{R_i^3\rho}{\sigma}\right)^{1/2}$$

(15-59)

$$s = \frac{R_0}{R_i} - 1$$

(15-60)

$$\eta = \frac{\mu}{(\rho\sigma R_i)^{1/2}} = \sqrt{2}\,Z$$

(15-61)

As expected, the growth rate is reduced by viscous effects. The more interesting prediction is with respect to the maximum in q_{max}, for a given fluid, at a value of s in the range of about 1 to 3. Very thin or very thick coatings are relatively stable in comparison to coatings for which s is of the order of unity.

As in the case of the jet, the growth of a varicose disturbance may be written as

$$\frac{\zeta}{\zeta_0} = e^{\alpha_{max}\,z/U}$$

(15-62)

If we are interested in the point at which the disturbance is some fraction of the coating radius, say, $\zeta = f R_0$, then

$$L_f^* = \left(\ln f \, \frac{R_0}{\zeta_0} \right) \frac{U}{\alpha_{max}} \tag{15-63}$$

The magnitude of ζ_0 is unknown, and no experiments are available with which its value may be determined. While it is tempting to suggest that the disturbances to the coating might be quite similar to those imposed onto the jet (for which $\ln R/\zeta_0 \approx 13$), there is no basis for this assumption. Indeed, the presence of the wire, subject to fluctuations in speed, probably controls the magnitude of the initial disturbances.

Let us put together some earlier modeling of the wire-coating process (Chap. 5) with the results of this stability analysis and examine the implications that can be extracted from the models.

Example 15-1: Stability of the coating on a wire We consider a wire-coating system operating under conditions similar to those described in Sec. 5-3:

$$R_i = 0.07 \text{ cm} \qquad \rho = 1 \text{ g/cm}^3$$

$$U = 10 \text{ ft/s} \qquad \sigma = 50 \text{ dynes/cm}$$

$$\mu = 1 \text{ P}$$

Take the coating thickness to be 0.021 cm, so that $s = 0.3$.

Suppose the wire speed U is subject to fluctuations of the form $U = U_0(1 + A \sin \omega t)$, where we take $A = 0.045$ and $\omega = 2 \times 10^3 \text{ s}^{-1}$.

It can be established, as earlier, that such an amplitude would lead to variations in coating thickness of about 1 percent of the wire radius: $\zeta_0 = 0.01 R_i$. A frequency ω will produce a spatial periodicity of the coating on the wire of wave number

$$kz = \omega t = \frac{\omega z}{U} \tag{15-64}$$

or $k = \omega/U = 2 \times 10^3/3 \times 10^2 = 6.7 \text{ cm}^{-1}$. This gives a value of kR_0 of

$$kR_0 = k(s + 1)R_i = 6.7(1.3)(0.07) = 0.63$$

Since $kR_0 < 1$ the coating is unstable to this disturbance. Further, this value of kR_0 gives nearly the maximum growth rate q_{max}. For $s = 0.3$ and $\eta = 1/(50 \times 0.07)^{1/2} = 0.53$ we find

$$q_{max} = 1.5 \times 10^{-3} = \alpha_{max} \left(\frac{R_i^3 \rho}{\sigma} \right)^{1/2}$$

or

$$\alpha_{max} = 1.5 \times 10^{-3} \left[\frac{50}{(0.07)^3} \right]^{1/2} = 0.57 \text{ s}^{-1}$$

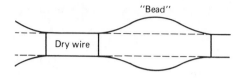

Figure 15-4 Beading of an annular coating on a wire.

From Eq. (15-63), then, and choosing $f = 0.05$ as an acceptable degree of varicosity, we find

$$L_f^* = \left[\ln \frac{0.05(0.091)}{0.01(0.07)} \right] \frac{300}{0.57} = 985 \text{ cm}$$

This is quite a long distance and suggests that one need not worry about this form of instability for the conditions of the problem. Indeed, for nearly any reasonable set of conditions under which a moderately viscous fluid is coated onto a wire or fiber, varicose disturbances would not be expected to grow to a significant degree.

Of what value, then, is the stability theory? In the first place, there may be processes in which low-viscosity lubricants or finishes are applied to a fiber prior to subsequent processing. If the calculations of the example above are reworked for a coating of viscosity 1 cP, the length L_f^* is found to be about 10 cm. If a condition of "beading" is defined (as in Fig. 15-4) to occur when $f = s$, the above calculations give $L_s^* = 20$ cm. With low-viscosity coatings it is quite possible that nonuniform wetting occurs as a result of this type of instability.

A second application of the stability analysis is with respect to the notion of "leveling." One expects, on the basis of intuition, that a wave imposed on a free surface will level out—that distortions on a coated surface will somehow smooth out by themselves. The stability analysis indicates the possible failure of this intuitive notion. According to Eq. (15-62), a disturbance will "level," i.e., damp out, only if $\alpha < 0$.

15-3 MELT FRACTURE

In Chap. 14 we discussed observations on melt fracture which, we should now see, might properly be interpreted as an instability of the primary flow. It would appear that melt fracture is an "elastic instability," associated in some way with the development and relaxation of normal stresses in the flow. The observations are not unequivocal on this point, since one can in fact correlate most data crudely with the assertion that melt fracture occurs at a critical *shear* stress (not *normal* stress) of 10^6 dynes/cm². Since there are so many complicating features of melt fracture which are subject to variation from one study to the next (e.g., inlet angle and channel length of the die, molecular weight distribution, chain branching in the polymer, the general lack of *complete* rheological characterization of the fluids studied), it is really quite difficult to make simple statements regarding the conditions for onset of this complex phenomenon. Hence one might turn to stability theory to see if any results can be generated that could suggest further experimental studies or, perhaps, different methods of correlation of existing data. An analysis of the stability of Poiseuille flow of a viscoelastic fluid has been carried out by Denn and his coworkers, and we present their results here. The review paper of Petrie and Denn gives a more critical discussion of this analysis and considers several alternate mechanisms of melt fracture as well.

We consider Poiseuille flow down the axis of a long circular pipe or between parallel plates, due to a pressure ΔP imposed on the plane $z = 0$. Both the steady state and perturbed flows must satisfy the dynamic and continuity equations:

$$\rho \frac{D\mathbf{u}}{Dt} = -\nabla \cdot \boldsymbol{\tau} \tag{15-65}$$

$$\nabla \cdot \mathbf{u} = 0 \tag{15-66}$$

A constitutive equation is required, and we select one which is a compromise between simplicity and minimal agreement with reality. A nonlinear Maxwell model is chosen, of the form

$$\boldsymbol{\tau} + \lambda \frac{\delta \boldsymbol{\tau}}{\delta t} = \eta_0 \, \boldsymbol{\Delta} \tag{15-67}$$

The nonlinearity enters through the *convected time derivative*, defined by [see Eq. (3-126) and the discussion of rate equations]

$$\frac{\delta \tau_{ij}}{\delta t} = \frac{\partial \tau_{ij}}{\partial t} + u_k \frac{\partial \tau_{ij}}{\partial x_k} - \tau_{kj} \frac{\partial u_i}{\partial x_j} - \tau_{ik} \frac{\partial u_j}{\partial x_k} \tag{15-68}$$

in cartesian components. The steady-state flow is a simple shear flow for which the velocity and stress fields are given by

$$\text{Slit } (-H \le y \le H) \qquad \text{Pipe } (0 \le r \le R)$$

$$\mathbf{u}^s = [u_z(y), 0, 0] \qquad \mathbf{u}^s = [u_z(r), 0, 0]$$

$$u_z^s(y) = \tfrac{3}{2} U \left[1 - \left(\frac{y}{H} \right)^2 \right] \qquad u_z^s(r) = 2U \left[1 - \left(\frac{r}{R} \right)^2 \right]$$

$$p^s = \Delta P + \frac{dp}{dz} z$$

$$\boldsymbol{\tau}^s = \eta_0 \dot{\gamma} \begin{pmatrix} 2\lambda\dot{\gamma} & 1 & 0 \\ 1 & 0 & 0 \\ 0 & 0 & 0 \end{pmatrix}$$

$$\dot{\gamma} = \frac{du_z}{dy} = -\frac{3Uy}{H^2} \qquad \dot{\gamma} = \frac{du_z}{dr} = -\frac{4Ur}{R^2}$$

We note, for this flow, that this particular Maxwell model predicts

$$\tau_{11} - \tau_{22} = 2\lambda\eta_0 \dot{\gamma}^2$$

$$\tau_{22} - \tau_{33} = 0$$

$$\tau_{12} = \eta_0 \dot{\gamma}$$

Thus we have a fluid which is newtonian in shear but which shows a finite primary normal stress difference and a zero secondary normal stress difference. These results are at least qualitatively correct for viscoelastic fluids at small deformation rates.

The stability analysis proceeds in the usual way. Each dependent variable is written as a sum of a steady-state term plus a perturbation term:

$$u = u^s + u' \tag{15-69}$$

The variables may then be substituted in this form into Eqs. (15-65) through (15-67), and these equations are then linearized with respect to the variables u'. It is assumed that the disturbances are two-dimensional, so that (in the slit case, e.g.) $u'_x = 0$ and $\partial/\partial x = 0$. The disturbance variables are all written in the form

$$u' = U(x_2)e^{ik(x_1 - ct)} \tag{15-70}$$

where x_1 = primary flow direction (z)

x_2 = direction of primary shear (y in the slit; r in the pipe)

k = a wave number of a disturbance

kc = a growth rate

c = a complex number: $c = c_R + ic_I$

If the stress components are written as in Eq. (15-70) the constitutive equations become algebraic equations in the stress components. One may then solve for the stresses and substitute them directly into the dynamic equations. This eliminates the stresses in favor of the velocities. A stream function is then introduced (for slit flow, for example, by defining $\partial\psi/\partial z = u'_y$ and $\partial\psi/\partial y = -u'_z$), and when pressure is eliminated from the dynamic equations (by differentiating the z equation with respect to y and the y equation with respect to z), the result is a fourth-order homogeneous ordinary differential equation for the stream-function amplitude Ψ, defined by

$$\psi = \Psi(y)e^{ik(z - ct)} \tag{15-71}$$

The details of this procedure are presented in the work of Rothenberger. In contrast to the simpler problem of jet stability treated in Sec. 15-1, the differential equation for Ψ does not yield to an analytical solution, and as a consequence an approximate numerical method must be employed.

The solution is most conveniently formulated by first nondimensionalizing Eqs. (15-65) through (15-67). If this is done as illustrated in Chap. 4 one finds a

Reynolds number and a Weissenberg number appearing. The Weissenberg number may be defined as

$$\text{Ws} = \begin{cases} \left| \dfrac{U\lambda}{R} \right| & \text{(pipe)} \\[2ex] \dfrac{U\lambda}{H} & \text{(slit)} \end{cases}$$

If we examine the steady-state solutions given above we can show easily that the Weissenberg number is proportional, in this problem, to the recoverable shear at the wall:

$$S_R = \frac{\tau_{11} - \tau_{22}}{2|\tau_{12}|} \bigg|_{\substack{y=H \\ \text{or} \\ r=R}} = \lambda |\dot{\gamma}| = \begin{cases} \left| \dfrac{4U\lambda}{R} \right| & \text{(pipe)} \\[2ex] \dfrac{3U\lambda}{H} & \text{(slit)} \end{cases}$$

Since the primary flow normally occurs at Reynolds numbers very small compared to unity, while the recoverable shear is of the order of 1 to 10, it is possible to neglect the Reynolds number terms in the equation for Ψ and consider the solution to be a function of S_R only (in addition to being a function of the independent variable y or r). The solution procedure leads to a constraining relationship between the recoverable shear S_R and the parameters k and c. By fixing k, and setting the imaginary part of c, c_I, equal to zero, a value of $S_R(k)$ may be found which satisfies the equation. The family of values $S_R(k)$ defines the neutral stability curve. Figure 15-5 shows neutral stability curves for the pipe and slit problems. Because of the approximate nature of the numerical methods employed in solving this problem, the k scale is rather imprecisely located. The critical value of S_R, especially in the slit case, is also believed to be inaccurate.

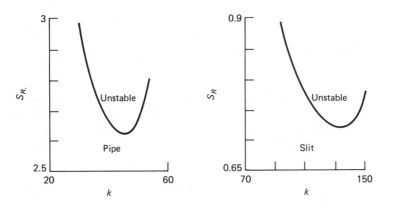

Figure 15-5 Neutral stability curves for low Reynolds number flow of a Maxwell fluid defined by Eq. (15-67). (*From Rothenberger et al.*)

Still, one can draw some useful conclusions from the stability analysis. First, the flow of a viscoelastic fluid is indeed predicted to be unstable to two-dimensional disturbances and to exhibit its instability at a critical value of recoverable shear. The predicted value of S_R of 2.6 for pipe flow is of the observed order of magnitude noted in Chap. 14. The slit flow is predicted to be unstable at a much lower value of S_R, approximately $S_R = 0.7$. Insufficient data are available (as from annular dies or flat-film dies) to evaluate this feature of the theory. Considering the relative simplicity of the constitutive equation used here, the agreement with experience must be seen to be quite reasonable.

15-4 SPINNING INSTABILITY (DRAW RESONANCE)

The melt fracture phenomenon described in the previous section often limits the output rate of a fiber-spinning system, since successful spinning usually requires a smooth extrudate. Another type of spinning instability is possible, which arises not from the die flow but from the free boundary flow in the drawing region between the spinneret and the take-up point. The observation is that there is a critical draw ratio beyond which stable operation is impossible. When that draw ratio is exceeded, a periodic variation in fiber diameter is observed, which is referred to as *draw resonance.*

The draw resonance phenomenon appears to be a viscous instability in the sense that it is observed in nonelastic melts, and even the stability theory for a newtonian fluid predicts the existence of a critical draw ratio. Unlike the melt fracture instability it would appear, in both observation and theory, that elasticity has a stabilizing influence on draw resonance. A second stabilizing factor, and one which makes comparison between theory and experiment difficult, arises from the nonisothermal nature of most real spinning systems. Cooling of the threadline enhances the stability of the flow, and an adequate predictive theory must account for heat transfer to the surroundings. As suggested in Sec. 13-6, the solution of the heat transfer problem itself is subject to uncertainties, and so any nonisothermal stability theory will include parameters which are not normally known with much precision.

With these comments and reservations in mind, let us examine the stability analysis of isothermal newtonian spinning. As in any stability analysis of fluids we begin with the relevant momentum and mass balances. In Chap. 9 we have already seen that, to a good approximation, the momentum balance may be reduced to

$$\frac{\partial}{\partial z}(R^2 T_{zz}) = 0 \tag{15-72}$$

[This is equivalent to Eq. (9-11) with the inertial terms neglected.]

The unsteady-state mass balance is easily derived in the form

$$\frac{\partial R^2}{\partial t} + \frac{\partial}{\partial z}(R^2 U) = 0 \tag{15-73}$$

For the newtonian fluid Eq. (9-18) gives

$$T_{zz} = 3\mu \frac{\partial u_z}{\partial z} = 3\mu \frac{\partial U}{\partial z} \tag{15-74}$$

Since U is twice differentiated with respect to z a typical set of boundary conditions specifies U at the ends of the spinning path:

$$U = \begin{cases} U_0 & \text{at } z = 0 \tag{15-75} \\ U_L & \text{at } z = L \tag{15-76} \end{cases}$$

The single required boundary condition on R is usually taken as an initial condition:

$$R = R_0 \qquad \text{at } z = 0 \tag{15-77}$$

Because die swell may occur, the value of R_0 appropriate to the analysis may not necessarily be the die radius. We will assume that in a spinning system the die swell ratio is not great in comparison to the draw ratio, and that any die swell phenomena, including relaxation of the stresses generated prior to extrusion, occur over a distance which is short with respect to L. Under these assumptions it is reasonable to use the die radius for R_0. In any event, conservation of mass relates R_0 and U_0 through the volumetric flow rate:

$$\pi R_0^2 U_0 = Q \tag{17-78}$$

The results of Sec. 9-1 may be used to give the steady-state solutions

$$a^s = \left(\frac{R}{R_0}\right)^2 = D_R^{-\bar{z}} = e^{-\bar{z}\ln D_R} \tag{15-79}$$

$$u^s = \frac{U}{U_0} = D_R^{\bar{z}} \tag{15-80}$$

$$T^s = \frac{T_{zz}}{\mu U_L / L} = \frac{3 \ln D_R}{D_R} D_R^{\bar{z}} \tag{15-81}$$

where we have introduced $\bar{z} = z/L$, and a dimensionless area a, velocity $\mathbf{u}$, and tensile stress T.

The stability of Eqs. (15-72) through (15-74), subject to boundary conditions given in Eqs. (15-75) through (15-77), has been studied by Pearson and coworkers, who also examined the effects of nonnewtonian viscosity as well as heat transfer, and by Denn and coworkers, who also studied the viscoelastic fluid case and examined stability to *finite* amplitude disturbances. For the newtonian isothermal case it is most convenient to write the momentum and mass balances [after substituting Eq. (15-74) into Eq. (15-72)] in the dimensionless form

$$\frac{\partial}{\partial \tilde{z}}\left(a\frac{\partial u}{\partial \tilde{z}}\right) = 0 \tag{15-82}$$

$$\frac{\partial a}{\partial \tilde{t}} + \frac{\partial}{\partial \tilde{z}}(au) = 0 \tag{15-83}$$

where $\tilde{t} = tU_0/L$.

Perturbation variables are now introduced in the form

$$a = a^s + \varphi \tag{15-84}$$

$$u = u^s + \psi \tag{15-85}$$

On substitution of these definitions of φ and ψ into Eqs. (15-82) and (15-83), and taking account of Eqs. (15-79) and (15-80), the perturbation variables are found to satisfy the following *linearized* differential equations:

$$\psi_{\tilde{z}\tilde{z}} - (\ln D_R)\psi_{\tilde{z}} + (\ln D_R)D_R^{2\tilde{z}}\varphi_{\tilde{z}} + (\ln D_R)^2 D_R^{2\tilde{z}}\varphi = 0 \tag{15-86}$$

$$\varphi_{\tilde{t}} + D_R^{\tilde{z}}\varphi_{\tilde{z}} + (\ln D_R)D_R^{\tilde{z}}\varphi + \frac{\psi_{\tilde{z}}}{D_R^{\tilde{z}}} - \frac{\ln D_R}{D_R^{\tilde{z}}}\psi = 0 \tag{15-87}$$

Subscripts $\tilde{z}$ or $\tilde{t}$ on φ and ψ refer to partial differentiation (for example, $\psi_{\tilde{z}\tilde{z}} \equiv \partial^2\psi/\partial\tilde{z}^2$).

It is assumed that the perturbed area and velocity may be written in the form

$$\varphi = \Phi(\tilde{z})e^{\alpha\tilde{t}} \tag{15-88}$$

$$\psi = \Psi(\tilde{z})e^{\alpha\tilde{t}} \tag{15-89}$$

where α is a dimensionless growth (or decay) rate. When Eqs. (15-88) and (15-89) are substituted into Eqs. (15-86) and (15-87) the form is unchanged, except that φ and ψ are replaced by Φ and Ψ, and the term $\varphi_{\tilde{t}}$ in Eq. (15-87) becomes $\alpha\Phi$. The boundary conditions take the form

$$\Phi(0) = \Psi(0) = 0 \tag{15-90}$$

$$\Psi(1) = 0 \tag{15-91}$$

The system of linear homogeneous equations (15-86) and (15-87) (in terms of Φ and Ψ) and the homogeneous boundary conditions give us an *eigenvalue* problem, in which the growth rate parameter α really represents an infinite set of complex eigenvalues. The only parameter that appears in the problem is the draw ratio, upon which each set of the eigenvalues must therefore depend.

The method of solution is discussed in several references. The result of interest is that at a critical value $D_R = 20.2$ the real part of the first eigenvalue changes sign and becomes positive (instability) for $D_R > 20.2$. The second and higher eigenvalues are all negative at this draw ratio. We conclude that the flow is unstable to draw ratios in excess of $D_R = 20.2$.

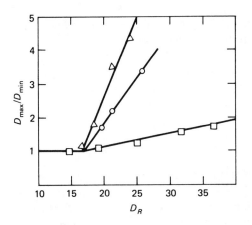

Figure 15-6 Draw resonance data of Donnelly and Weinberger.

	Q, mm³/s	L, mm
△	64.1	20
○	44.4	20
□	30.8	40

Experimental data on isothermal newtonian spinning are rather hard to obtain. The major problem is finding a fluid of high enough viscosity that the surface-tension-driven instability of Sec. 15-1 is suppressed without having to use a polymer fluid which might be nonnewtonian. One set of data are available, due to Donnelly and Weinberger, using a silicone oil (Dow Corning 200) which seems to show newtonian shear behavior, with a viscosity of 1000 P at room temperature. Because the oil is polymeric (polydimethylsiloxane) it is possible that some viscoelastic phenomena occur at high deformation rates, but the oil seems likely to be newtonian under the conditions of the experiment described.

Experiments were conducted under isothermal conditions, the "fiber" being extruded vertically through a short capillary of diameter 0.28 cm and a length-diameter ratio of 2. A take-up roll drew the fluid filament, and the controlled variables were flow rate Q, spinning path L, and draw ratio D_R (through the take-up speed). The principal measurements were of the filament diameter variation along the filament axis.

Figure 15-6 shows the data in terms of the ratio of maximum to minimum filament diameter at a fixed point along the spinning path as a function of D_R. Stable operation is characterized by no variation in diameter ratio at any fixed axial position, or max diam. to min diam. = 1. The data suggest that values of this ratio in excess of unity occur at a critical draw ratio of approximately $D_R = 17$.

Donnelly and Weinberger observe a die swell of about 14 percent in their experiments. If the die swell diameter is used for calculation of D_R instead of the capillary diameter, the values of D_R are increased by about 28 percent, and the critical value of D_R is estimated to be approximately $D_R = 22$. These two values of D_R (either 17 or 22) bracket the predicted value of 20.2, and although the amount of experimental data is small, and only the single fluid was studied, the results encourage the belief that the essence of the spinning stability analysis is correct.

Data are available, however, which suggest that under some conditions it is not possible to spin a stable filament at draw ratios anywhere near the critical value of 20. One often observes draw resonance at values of D_R in the range of 3 to

5. One should examine, then, the results of relaxing some of the assumptions of the simplest theory. In particular, let us examine the effect of

- Cooling of the fiber
- Nonnewtonian viscous behavior
- Viscoelasticity
- Finite-amplitude disturbances

Nonisothermal Spinning

For a Newtonian melt spun under nonisothermal conditions we must add an energy equation to Eqs. (15-72) through (15-74), which are still valid. If we let θ be the melt temperature relative to the ambient temperature and assume that for a thin fiber θ is independent of radial position, the energy balance becomes

$$\rho C_p\left(\frac{\partial \theta}{\partial t} + U\frac{\partial \theta}{\partial z}\right) = -\frac{2h}{R}\theta \tag{15-92}$$

where h is an appropriate convective heat transfer coefficient. Using the same nondimensionalization as before, and using θ_0 (initial melt temperature minus the ambient temperature) to normalize θ, we find

$$\frac{\partial \tilde{\theta}}{\partial \tilde{t}} + u\frac{\partial \tilde{\theta}}{\partial \tilde{z}} = -H\tilde{\theta}a^{-1/2} \tag{15-93}$$

where H is given by

$$H = 2\pi^{1/2}\frac{L\alpha_T}{Q}\frac{hR_0}{k} \tag{15-94}$$

We note that the heat transfer coefficient might be a function of axial position $\tilde{z}$ and of the velocity u. Hence H might not be a constant.

In the nonisothermal analysis it is necessary to account for the temperature dependence of viscosity, for example, using

$$\mu = \mu_0 e^{-b(T-T_0)} = \mu_0 e^{-b(\theta-\theta_0)} = \mu_0 e^{-b\theta_0(\theta-1)} \tag{15-95}$$

This model makes μ_0 the viscosity of the melt at $z = 0$ and introduces a new parameter $b\theta_0$ into the analysis. The momentum balance, Eq. (15-72), takes the form [see Eq. (15-82)]

$$\frac{\partial}{\partial \tilde{z}}\left[ae^{-b\theta_0(\theta-1)}\frac{\partial u}{\partial \tilde{z}}\right] = 0 \tag{15-96}$$

while the mass balance, Eq. (15-73), remains unchanged. With Eq. (15-93) these three equations provide the starting point of the nonisothermal stability analysis of Pearson and his coworkers.

We will not pursue the details further here but note instead that the main result is that the critical draw ratio can exceed the predicted isothermal value of 20.2 when the parameter

$$\mathcal{H} \equiv b\theta_0 He^{-H} \tag{15-97}$$

becomes comparable to unity, say, $\mathcal{H} > 0.5$. The specific results presented by Pearson are based on the assumption that the heat transfer coefficient h is proportional to $U^{2/3}$. Other assumptions will, of course, modify the results quantitatively, but the basic qualitative idea would be expected to hold.

Example 15-2: Drawing of polyester fiber It is interesting, at this point, to estimate the value of $\mathcal{H}$ for some set of spinning data, to see if $\mathcal{H} > 0.5$ is within the realm of possible operating conditions. We use the data of Example 13-4 for polyester fiber. The observed draw ratio was about 60. The following parameters are needed:

$$\frac{k}{\alpha_T} = \rho C_p = 1.3(0.4) \text{ cal/cm}^3 \cdot {}^\circ\text{C}$$

(from Kase's work on polyethylene terepthalate)

$$Q = 5 \times 10^{-3} \text{ cm}^3/\text{s}$$

$$L = 15 \text{ cm (the actual drawing length, from Fig. 13-21)}$$

$$\text{Nu} = \frac{2hR_0}{k_a} = 0.2 \text{ (taking } \eta_0 \approx 10^{-3}, \text{ using Fig. 13-20).}$$

Note that Nu is based on the air conductivity k_a. In Eq. (15-94) we need hR_0, so

$$hR_0 = \tfrac{1}{2} \text{ Nu } k_a = 0.1(6.5 \times 10^{-5}) = 6.5 \times 10^{-6} \text{ cal/cm} \cdot \text{s} \cdot {}^\circ\text{C}$$

The value of θ_0 in this case is $290 - 22 = 268^\circ\text{C}$. For b we take $b = 0.02^\circ\text{C}^{-1}$. From Eqs. (15-94) and (15-97) we find $\mathcal{H} \approx 0.6$. An $\mathcal{H}$ value comparable to unity could be easily attained. Pearson's theoretical work shows the possibility of draw ratios of the order of 100 at such values of $\mathcal{H}$.

Effect of Viscoelasticity on Spinning Stability

Fisher and Denn have analyzed the spinning stability of a fluid described by a viscoelastic model of the form

$$\tau + \lambda \frac{\delta\tau}{\delta t} = \mu \, \Delta \tag{15-98}$$

Superficially, this is the same nonlinear Maxwell model discussed in Sec. 15-3 on melt fracture. However, the viscosity coefficient μ is taken, in this case, to be a *function* of the deformation rate through a power law of the form

$$\mu = K(\tfrac{1}{2}\text{II}_\Delta)^{n-1} \tag{15-99}$$

The relaxation time is also taken to be a function of the deformation rate by using the definition

$$\lambda = \frac{\mu}{G} \tag{15-100}$$

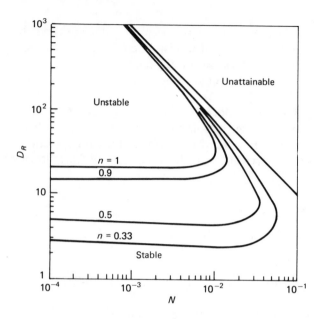

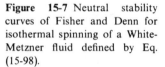

Figure 15-7 Neutral stability curves of Fisher and Denn for isothermal spinning of a White-Metzner fluid defined by Eq. (15-98).

where G is a constant elastic modulus. When μ and λ are taken as material *functions* instead of constants, this constitutive equation is referred to as the *White-Metzner model*.

The stability analysis proceeds as in the simpler newtonian case, with some additional algebraic and computational complexity. For isothermal spinning the linearized stability equations include three parameters: the draw ratio D_R, the constant n, and a viscoelastic parameter defined by

$$N = 3^{(n-1)/2n}\left(\frac{K}{G}\right)^{1/n}\frac{U_0}{L} \tag{15-101}$$

[This same parameter appeared as α in Eq. (9-45).] Figure 15-7 shows the neutral stability curves calculated by Fisher and Denn. There are three features of interest. First, we find that there is an upper limit to the attainable stretch rate under steady spinning conditions, defined by

$$D_R^* = 1 + \frac{1}{N} \tag{15-102}$$

[This is equivalent to Eq. (9-53), the constitutive equation there being the same as Eq. (15-98), with $n = 1$ in Eq. (15-99).]

The next feature of interest is the effect of a deformation rate–dependent viscosity. For $n < 1$ the fluid is expected to show an instability at a smaller draw ratio than for $n = 1$. Thus the newtonian fluid is the most stable to spinning instability at low values of N, according to this theory, and the observation that draw resonance occurs at D_R as low as $3 < D_R < 5$ might simply reflect a nonnewtonian shear effect.

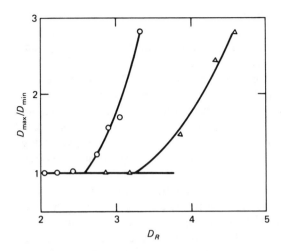

Figure 15-8 Draw resonance data on two polymer melts spun under isothermal conditions. $T = 218°C$, $L = 16.5$ cm, $D_0 = 0.159$ cm, and $U_0 = 5$ cm/s. △ Polypropylene (Hercules 6523); ○ polystyrene (Monsanto HF 77). (*From Cruz-Saenz et al.*)

Finally, and perhaps of greatest interest, each neutral stability curve "doubles back" and leaves a stable operating region just under the limiting D_R^* boundary. This suggests that one could "spin through" the draw resonance region and find an upper, high-draw-ratio region of stable operation. Alternatively, for N high enough, the critical D_R can be increased by increasing N. Thus, for a fluid for which $n = 0.5$, for example, the theory predicts that if $N = 0.02$ one will find instability at $D_R = 5$. However, if N is doubled, e.g., by halving the spinning length L [see Eq. (15-101)], one gets past the "nose" of the stability envelope and stable operation is possible up to D_R^*. It is known that one can stabilize a fiber-spinning operation by reducing the length of the spinning path, although it would be speculative to take this point as a confirmation of the theory.

Data on isothermal spinning of polymer melts are provided by Weinberger and coworkers (see Cruz-Saenz et al.) and are shown in Fig. 15-8. The most significant point is the small draw ratio at which draw resonance is observed. Since the observed values of the power law index are approximately $n = 0.5$ for both melts, the results are in crude agreement with the predictions of Fig. 15-7 for small N. The only rheological data presented (Fig. 15-9) show the two melts to have practically identical shear viscosity behavior (see Prob. 15-7).

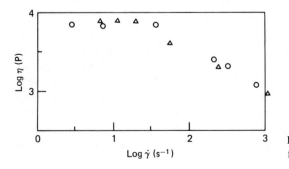

Figure 15-9 Shear viscosity data at 218°C for the melts of Fig. 15-8 (same symbols).

Analysis of Finite-amplitude Disturbances

A linearized stability analysis gives information regarding the response of a system to *small* disturbances. When such an analysis predicts stability, the prediction is relevant only to the case studied, i.e., to small disturbances. The system may be unstable to *large* disturbances. Although an analysis may give *some* results under the assumption of linearity, there is usually a limitation to how much *meaningful* information may be extracted from a linear stability analysis. Thus, as noted in Sec. 15-1, the linearized analysis of jet stability does a good job of predicting growth rates of small disturbances, but it fails to predict the appearance of "satellite drops" in the disrupted jet.

If a system *is* unstable, the instability usually "appears," in the literal sense of being measurably important, only after the disturbance has grown to a macroscopic size. By *macroscopic* we mean that the disturbance variable is comparable to the steady-state value of that variable. At this point the linear stability analysis is not applicable, and subsequent features of the growth of the macroscopic disturbance are usually not accommodated by the theory.

Thus, the prediction of a critical draw ratio in spinning, based on the linearized equations, may be irrelevant to the actual behavior of the system. The linear theory, by its nature, does not predict the appearance of sustained periodic variations in filament diameter, so it is not clear that the instability predicted by the linearized model is the same as the instability that occurs. Such information must come from a nonlinear analysis. Fisher and Denn have carried out such an analysis for isothermal newtonian spinning.

A major result of the finite-amplitude analysis is that, for $D_R < 20.2$, the system is stable to finite-amplitude perturbations. Thus the linearized analysis is adequate to determine the critical draw ratio for unstable spinning. The second significant result is that, for $D_R > 20.2$, disturbances grow and approach a sustained finite-amplitude oscillation whose period and magnitude are in the range of experimental observations of draw resonance. It would appear, then, that the stability analysis of spinning is indeed relevant to the problem of draw resonance.

Observations of draw resonance and comparison to a finite-amplitude analysis are provided in an interesting paper by Kase. Polyethylene terepthalate (PET) was melt spun and quenched in a water bath. Figure 15-10 shows the spinning conditions and the measured variation in filament thickness. (The thickness was not directly measured, but the weight of filament in a 10-cm length was measured and used as a measure of nonuniformity. Since the wavelength of oscillation was about 70 cm, the 10-cm weight provides a reasonable measure.)

We note that the ratio of maximum to minimum filament radius corresponding to these data is approximately 2. Considerably larger ratios are often observed, but we chose to present Kase's data because they were obtained under well-controlled and well-documented operating conditions. Over the air gap of 2 cm the temperature of the filament falls slightly, perhaps 20°C, and then upon entering the water the fiber is rapidly quenched and solidified. It is assumed that no

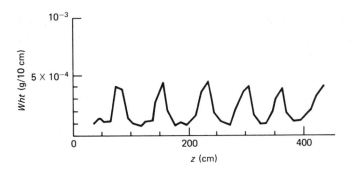

Figure 15-10 Draw resonance data of Kase. PET filament spun under following conditions: $D_0 = 0.03$ cm; $L = 2$ cm; $U_L = 500$ cm/s; $D_R = 76$; $T_0 = 284°C$; $T_a = 30°C$.

drawing occurs within the water bath. Hence the experimental results correspond to nonisothermal spinning, although the extent of cooling does not permit stable operation at the imposed draw ratio of 76.

15-5 STABILITY OF FILM CASTING

Polymeric film may be produced by a "casting" process, as suggested in Fig. 15-11. In a typical operation the film is extruded toward a cold roll, the "chill" roll, which quenches the melt and prepares it for further processing or windup. Film is normally drawn to reduce its thickness to the desired value and, possibly, to impart orientation to the material so as to alter its mechanical properties.

Despite the "necking in" suggested in the front view of the process, the flow may be considered two-dimensional to a good approximation, especially for a very wide film. The *steady* flow analysis is nearly trivial, so we have chosen to introduce it here with the stability analysis itself.

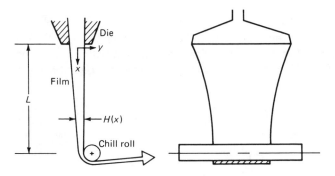

Figure 15-11 Side and front views of a film-casting process.

We assume that the velocity profile is flat across the film thickness, so that u is a function only of x. We will consider the simplest case: isothermal newtonian flow. For two-dimensional flow the continuity equation is

$$\frac{\partial u}{\partial x} + \frac{\partial v}{\partial y} = 0 \qquad (15\text{-}103)$$

We let $u = U(x)$, and on integrating the continuity equation we find

$$v = -\frac{dU}{dx}y \qquad (15\text{-}104)$$

At $y = \frac{1}{2}H$ this gives the maximum y-directed velocity

$$v_H = -\frac{1}{2}\frac{dU}{dx}H \qquad (15\text{-}105)$$

The volumetric flow rate (per unit width) is

$$Q = 2\int_0^{H/2} u \, dy = UH \qquad (15\text{-}106)$$

and at steady state we take Q to be constant. It follows then that

$$\frac{dQ}{dx} = 0 = H\frac{dU}{dx} + U\frac{dH}{dx} \qquad (15\text{-}107)$$

or

$$\frac{dH}{dx} = -\frac{H}{U}\frac{dU}{dx} \qquad (15\text{-}108)$$

The dynamic equations for this flow reduce to

$$\frac{\partial}{\partial y}\tau_{xy} + \frac{\partial}{\partial x}T_{xx} = 0 \qquad (15\text{-}109)$$

if inertial terms are neglected. As in the analysis of fiber spinning in Sec. 9-1, the shear stress term τ_{xy} arises from the geometry. In fact, Eqs. (9-1) through (9-9) are valid for both the cylindrical fiber and the two-dimensional film with r replaced by y. Upon integrating Eq. (15-109) with respect to y (from 0 to $\frac{1}{2}H$), one obtains the integrated force balance in the forms

$$\frac{d}{dx}HT_{xx} = 0 \qquad (15\text{-}110)$$

or

$$HT_{xx} = f \qquad (15\text{-}111)$$

where f is the axial drawing force per unit width of film.

For the newtonian fluid the three normal stresses are

$$T_{xx} = -p + 2\mu \frac{\partial u}{\partial x} \tag{15-112}$$

$$T_{yy} = -p + 2\mu \frac{\partial v}{\partial y} = -p - 2\mu \frac{\partial u}{\partial x} \tag{15-113}$$

$$T_{zz} = -p + 2\mu \frac{\partial w}{\partial z} = -p \tag{15-114}$$

Equation (15-113) follows from application of Eq. (15-103), while Eq. (15-114) follows from the assumption that the flow is only two-dimensional.

The stress difference

$$T_{xx} - T_{yy} = 4\mu \frac{dU}{dx} \tag{15-115}$$

is seen to be independent of y. However, at $y = \frac{1}{2}H(x)$ we may assume that no stresses normal to the free surface act (surface tension effects are assumed negligible), and, to a good approximation, this implies that $T_{yy} = 0$ at $y = \frac{1}{2}H$ and, indeed, for all y. This leads, then, to

$$p = -2\mu \frac{dU}{dx} \tag{15-116}$$

and

$$T_{xx} = 4\mu \frac{dU}{dx} \tag{15-117}$$

Upon combining Eqs. (15-106), (15-108), and (15-117) with Eq. (15-111) we find

$$\frac{1}{H}\frac{dH}{dx} = -\frac{1}{U}\frac{dU}{dx} = -\frac{T_{xx}}{4\mu U} = \frac{f}{4\mu HU} = \frac{f}{4\mu Q} \tag{15-118}$$

Integration yields

$$\frac{H}{H_0} = e^{-fx/4\mu Q} \tag{15-119}$$

and

$$\frac{U}{U_0} = e^{fx/4\mu Q} \tag{15-120}$$

From Eq. (15-104) we find

$$v = -\frac{U_0 f}{4\mu Q} y e^{fx/4\mu Q} \tag{15-121}$$

This completes the steady-state analysis. It is convenient to introduce dimensionless variables, so we define

$$\tilde{x} = \frac{x}{L} \qquad \tilde{u} = \frac{U}{U_0}$$

$$\tilde{y} = \frac{2y}{H_0} \qquad \tilde{v} = \frac{v}{U_0}$$

$$\tilde{H} = \frac{H}{H_0} \qquad \beta = \frac{fL}{4\mu Q}$$

The steady-state solutions then become

$$\tilde{u}^s = e^{\beta \tilde{x}} \tag{15-122}$$

$$\tilde{H}^s = e^{-\beta \tilde{x}} \tag{15-123}$$

$$\tilde{v}^s = y e^{\beta \tilde{x}} \tag{15-124}$$

In this format it is easy to see that the draw ratio, defined as

$$D_R = \frac{1}{\tilde{H}(1)} \tag{15-125}$$

is simply related to β through

$$\ln D_R = \beta \tag{15-126}$$

The film-casting process is the two-dimensional counterpart of fiber spinning, and like fiber spinning the physical process exhibits instabilities. An analysis of the stability of the film-casting process has been carried out by Yeow. We illustrate here a simpler case of his more general analysis. We consider two-dimensional perturbations only, of the form

$$\tilde{u} = \tilde{u}^s[1 + u^*(\tilde{x})e^{i\Omega \tilde{t}}] \tag{15-127}$$

$$\tilde{H} = \tilde{H}^s[1 + H^*(\tilde{x})e^{i\Omega \tilde{t}}] \tag{15-128}$$

$$\tilde{v} = \tilde{v}^s[1 + v^*(\tilde{x})e^{i\Omega \tilde{t}}] \tag{15-129}$$

We define a dimensionless time as $\tilde{t} = tU_0/L$. The dimensionless growth rate parameter is Ω, and instability corresponds to negative values of the imaginary part of Ω.

The unsteady-state mass balance takes the form [see Eq. (15-73)]

$$\frac{\partial H}{\partial t} + \frac{\partial}{\partial x}(HU) = 0 \tag{15-130}$$

When Eqs. (15-127) and (15-128) are substituted into this mass balance the result, after some algebra including linearization, becomes

$$H^{*\prime} + u^{*\prime} + i\Omega e^{-\beta \tilde{x}} H^* = 0 \tag{15-131}$$

(The prime denotes $d/d\tilde{x}$.)

If inertial terms are neglected Eq. (15-110) still constrains the perturbed flow, and Eq. (15-115) is still valid, if only two-dimensional disturbances are assumed. The axial stress is then

$$T_{xx} = \frac{4\mu U_0}{L} \frac{d}{d\tilde{x}} \tilde{u}^s(1 + u^* e^{i\Omega i}) = \frac{4\mu U_0}{L} e^{\beta \tilde{x}}[\beta + (\beta u^* + u^{*\prime})e^{i\Omega i}] \quad (15\text{-}132)$$

Using Eq. (15-128) for H and Eq. (15-132) for T_{xx}, one can carry out the indicated differentiations of Eq. (15-110), and after linearization the result is

$$u^{*\prime\prime} + \beta u^{*\prime} + \beta H^{*\prime} = 0 \quad (15\text{-}133)$$

Equations (15-131) and (15-133) constitute a pair of linear homogeneous equations in the two perturbation variables H^* and u^*. Appropriate boundary conditions are

$$\begin{aligned} u^* = 0 \quad\quad H^* = 0 \quad\quad &\text{at } \tilde{x} = 0 \\ u^* = 0 \quad\quad &\text{at } \tilde{x} = 1 \end{aligned} \quad (15\text{-}134)$$

which correspond to fixing the velocity at the two ends of the process and holding film thickness constant at $\tilde{x} = 0$ only.

Equations (15-131) and (15-133) constitute an eigenvalue problem with the draw ratio (through β) being the parameter that determines the values of Ω for which the equations have nontrivial solutions satisfying the boundary conditions. Stability exists so long as the imaginary part of Ω is positive. Yeow has shown that the neutral stability condition (the point where the imaginary part of Ω passes from positive to negative) occurs at a value $\beta = 3$. Thus the critical draw ratio is [from Eq. (15-126)] $D_R = 20.2$.

This critical draw ratio for film casting is identical to that for newtonian isothermal spinning. This follows from the fact that in the case of a two-dimensional disturbance† the two problems are mathematically identical with respect to the perturbation variables. Yeow states that this is not necessarily the case for the nonnewtonian fluid. One might expect qualitative similarity, however, between spinning of fiber and casting of film, with respect to the effects of nonnewtonian viscosity, viscoelasticity, and cooling on stability.

Several sets of experimental data are available which show the existence of draw resonance in film casting and which shed light on several features of the phenomenon. Figure 15-12 shows film thickness variations in a polypropylene melt cast through air onto a chill roll. It is apparent that draw resonance occurs, and the ratio of maximum to minimum film thickness is seen to be about 1.8.

Bergonzoni and DiCresce show even more severe oscillations in polypropylene " ribbon " (narrow-width film) which was drawn through air and quenched in

---

† The dimensionality of the disturbance refers to the fact that both u and v are perturbed, not to the fact that the perturbation variables are functions only of x. For a three-dimensional disturbance a perturbed velocity w is assumed to exist, and all perturbed quantities include a z variation of the form e^{iaz}.

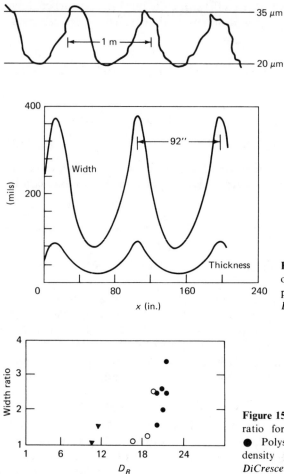

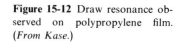

Figure 15-12 Draw resonance observed on polypropylene film. (*From Kase.*)

Figure 15-13 Effect of draw resonance on thickness and width of a narrow polypropylene ribbon. (*From Bergonzoni and DiCresce.*)

Figure 15-14 Width ratio as a function of draw ratio for various melts extruded as ribbon. ● Polystyrene; ○ polypropylene; ▼ high-density polyethylene. (*From Bergonzoni and DiCresce.*)

water. Figure 15-13 indicates ratios of about 4.5 for the maximum to minimum film thickness. Bergonzoni's data also indicate that for a variety of polymeric melts a critical draw ratio of about 20 is evident, with some variation that could probably be ascribed to nonnewtonian or viscoelastic phenomena. Figure 15-14 illustrates this point.

Kase has carried out a theoretical analysis through numerical solution of the perturbation equations, and subsequently through solution of the full nonlinear equations, for newtonian isothermal drawing. He shows generally good agreement between theory and observation. One useful result is a prediction for the spatial periodicity of the oscillation on the final film or fiber, which Kase gives as

$$\frac{L_p}{L} = \frac{D_R}{\ln D_R} f(D_R) \qquad (15\text{-}135)$$

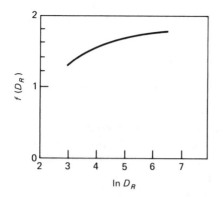

Figure 15-15 The function $f(D_R)$ in Eq. (15-135).

L_p is the distance between successive maxima or minima along the quenched film (or fiber), and $f(D_R)$ is a function given in Fig. 15-15. Agreement between Eq. (15-135) and observed values of L_p is quite good. Theoretical predictions of amplitude ratios are much less successful. Kase's work overestimates the ratio of maximum to minimum thickness by an unacceptable amount.

15-6 OTHER STABILITY PROBLEMS

We mention here briefly some other stability problems of importance in the polymer-processing industries, for which earlier chapters considered only the analysis of the steady state. Bibliographical references provide an opportunity for further study.

Coating flows Figure 15-16 shows a sketch of a common occurrence in roll-coating systems: the development of waves oriented in the direction of motion of the sheet. This instability is associated with interfacial tension and occurs in both

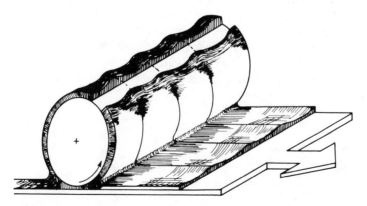

Figure 15-16 Instability observed in roll coating.

Figure 15-17 Two-roll system studied by Pitts and Greiller.

newtonian and viscoelastic fluids. Some aspects of this problem have been treated both theoretically and experimentally by Pitts and Greiller and by Pearson (see Chap. 8 Bibliography), and by Mill and South. In none of these papers was the specific case of the sheet-roll system treated, but the broad similarity of this type of coating system to the geometrics considered in those papers makes their results relevant to our discussion here. In a two-roll system, operating as shown in Fig. 15-17, Pitts and Greiller observe unstable coatings on the rolls when

$$\frac{\mu U R}{\sigma H_0} > 62 \tag{15-136}$$

Mill and South, in later experiments, correlate observations on this instability with the criterion

$$\frac{\mu U}{\sigma} \left(\frac{R}{H_0}\right)^{3/4} > 10.3 \tag{15-137}$$

Pearson studied coating with a "wedge spreader," as depicted in Fig. 15-18. He found an instability of the type considered here when

$$\frac{\mu U}{2\sigma\alpha} > 0.3 \tag{15-138}$$

where $\alpha = \tan\theta$.

Pearson carries out a moderately successful stability analysis of his problem, but the most significant remarks are his qualitative comments at the end of his paper. He points out that diverging flows (as in the spreader or roll configurations shown in Figs. 15-17 and 15-18) are subject to instability, whereas converging flow (as in blade coating) should be stable to interfacial instability. Pearson also notes that stability should occur in a two-roll system if the rolls move in opposite directions. This is consistent with the common observation that reverse-roll coating, in which the sheet and roll move counter to each other, is a much more stable process than the case of concurrent motion.

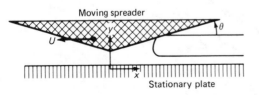

Moving spreader

Stationary plate

Figure 15-18 The "wedge spreader."

Tubular film blowing As discussed in Chap. 10, tubular blown film is drawn, and hence one might anticipate the possibility of draw resonance. This is, in fact, observed, and Han and Park show photographs of this phenomenon and have carried out some preliminary studies.

PROBLEMS

15-1 Derive Eq. (15-34), and prove that $p' = 0$.

15-2 By considering the boundary conditions for the problem of newtonian jet stability, derive Eq. (15-43).

15-3 Derive Eq. (15-44) and its solution. Then verify Eq. (15-45).

15-4 A polyester monofilament, of diameter 5 mils, has a "finish" coated on it by a withdrawal-coating operation. The filament speed is 100 cm/s, and the "finish" solution has a viscosity of 3 cP and a surface tension of 30 dynes/cm. Is the coating still uniform at a distance of 3 m above the bath?

15-5 Give the derivation of Eq. (15-73).

15-6 Using Fig. 15-7, and taking $n = \frac{1}{2}$, $K = 10^4$ dyne·s$^{1/2}$/cm^2, and $G = 2 \times 10^4$ dynes/cm^2, plot the limiting value of D_R as a function of spinning length L. Take Q to be fixed at 10^{-2} cm^3/s, and $R_0 = 0.02$ cm. Suppose a slight modification in the polymer leaves K and n unchanged but causes G to increase by a factor of 2. Plot the limiting value of D_R versus L and compare to the first case.

15-7 A crude estimate of a relaxation time for the melts shown in Fig. 15-9 is the reciprocal of the shear rate at which nonnewtonian behavior is first observed. Estimate N [Eq. (15-101)] for each melt, and compare the predicted critical D_R with the data of Fig. 15-8.

15-8 For the data of Kase (Fig. 15-10) calculate the temperature profile along the spinning path.

15-9 The data of Fig. 15-14 were obtained at a draw ratio of 20. Predict the spatial period of the draw resonance and compare to the observed value of 92 in. L was 12 in. Assume isothermal conditions.

15-10 Evaluate the isothermal assumption in Prob. 15-9 by calculating the temperature in the ribbon as a function of distance from the die. Use the following data:

$$T_0 = 454°F \qquad D_R = 20$$
$$U_L = 100 \text{ cm/s} \qquad \rho Q = 4.2 \text{ g/s}$$
$$\rho = 0.83 \text{ g/cm}^3 \qquad C_P = 0.7 \text{ cal/g·°C}$$
$$T_a = 80°F$$

BIBLIOGRAPHY

An excellent review article with an extensive bibliography is

Petrie, C. J. S., and M. M. Denn: Instabilities in Polymer Processing, *AIChE J.*, **22**: 209 (1976).

15-1 Stability of a viscous capillary jet

This is a classical stability problem, and one of the best examples of correspondence of stability analysis with observation.

McCarthy, M. J., and N. A. Molloy: Review of Stability of Liquid Jets and the Influence of Nozzle Design, *Chem. Eng. J.*, **7**: 1 (1974).
Grant, R. P., and S. Middleman: Newtonian Jet Stability, *AIChE J.*, **12**: 669 (1966).

Sterling, A. M., and C. A. Sleicher: The Instability of Capillary Jets, *J. Fluid. Mech.*, **68:** 477 (1975).

Goldin, M., J. Yerushalmi, R. Pfeffer, and R. Shinnar: Breakup of a Laminar Capillary Jet of a Viscoelastic Fluid, *J. Fluid Mech.*, **38:** 689 (1969).

Kroesser, F. W., and S. Middleman: Viscoelastic Jet Stability, *AIChE J.*, **15:** 383 (1969).

Goedde, E. F., and M. C. Yuen: Experiments on Liquid Jet Instability, *J. Fluid Mech.*, **40:** 495 (1970).

Lafrance, P.: Non-Linear Breakup of a Laminar Liquid Jet, *Phys. Fluids*, **18:** 428 (1975).

Wang, D. P.: Finite Amplitude Effect on the Stability of a Jet of Circular Cross-Section, *J. Fluid Mech.*, **34:** 299 (1968).

15-2 Stability of wire coating

Roe, R.-J.: Wetting of Fine Wires and Fibers by a Liquid Film, *J. Colloid Interface Sci.*, **50:** 70 (1975).

Dumbleton, J. H., and J. J. Hermans: Capillary Instability of a Thin Annular Layer of Liquid Around a Solid Cylinder, *Ind. Eng. Chem., Fundamen.*, **9:** 466 (1970).

15-3 Melt fracture

Rothenberger, R., D. H. McCoy, and M. M. Denn: Flow Instability in Polymer Melt Extrusion, *Trans. Soc. Rheol.*, **17:** 259 (1973).

15-4 Spinning instability (draw resonance)

The principal theoretical studies are

Pearson, J. R. A., and Y. T. Shah: Stability Analysis of the Fiber Spinning Process, *Trans. Soc. Rheol.*, **16:** 519 (1972).

Fisher, R. J., and M. M. Denn: A Theory of Isothermal Melt Spinning and Draw Resonance, *AIChE J.*, **22:** 236 (1976).

—— and ——: Finite-Amplitude Stability and Draw Resonance in Isothermal Melt Spinning, *Chem. Eng. Sci.*, **30:** 1129 (1975).

Kase, S.: Studies on Melt Spinning. IV. On the Stability of Melt Spinning, *J. Appl. Polym. Sci.*, **18:** 3279 (1974).

Ishihara, H., and S. Kase: Studies on Melt Spinning. V. Draw Resonance as a Limit Cycle, *J. Appl. Polym. Sci.*, **19:** 557 (1975).

Data are presented in the above paper, as well as in

Donnelly, G. J., and C. B. Weinberger: Stability of Isothermal Fiber Spinning of a Newtonian Fluid, *Ind. Eng. Chem., Fundamen.*, **14:** 334 (1975).

Cruz-Saenz, G. F., G. J. Donnelley, and C. B. Weinberger: Onset of Draw Resonance During Isothermal Melt Spinning: A Comparison between Measurements and Predictions, *AIChE. J.*, **22:** 441 (1976).

15-5 Stability of film casting

Yeow, Y. L.: On the Stability of Extending Films: A Model for the Film Casting Process, *J. Fluid. Mech.*, **66:** 613 (1974).

Bergonzoni, A., and A. J. DiCresce: The Phenomenon of Draw Resonance in Polymeric Melts, *Polym. Eng. Sci.*, **6:** 45 (1966).

15-6 Other stability problems

Mill, C. C., and G. R. South: Formation of Ribs on Rotating Rollers, *J. Fluid Mech.*, **28:** 523 (1967).

Han, C. D., and J. Y. Park: Studies on Blown Film Extrusion. III. Bubble Instability, *J. Appl. Polym. Sci.*, **19:** 3291 (1975).

INDEX

INDEX

Page numbers in *italic* refer to bibliographic references; page numbers in **bold-face** refer to illustrations.